T0375212

Geometrical Theory of Dynamical Systems and Fluid Flows

ADVANCED SERIES IN NONLINEAR DYNAMICS*

Editor-in-Chief: R. S. MacKay *(Univ. Warwick)*

Published

Vol. 5 Combinatorial Dynamics & Entropy in Dimension One
L. Alseda, J. Llibre & M. Misiurewicz

Vol. 6 Renormalization in Area-Preserving Maps
R. S. MacKay

Vol. 7 Structure & Dynamics of Nonlinear Waves in Fluids
eds. A. Mielke & K. Kirchgässner

Vol. 8 New Trends for Hamiltonian Systems & Celestial Mechanics
eds. J. Llibre & E. Lacomba

Vol. 9 Transport, Chaos and Plasma Physics 2
S. Benkadda, F. Doveil & Y. Elskens

Vol. 10 Renormalization and Geometry in One-Dimensional and Complex Dynamics
Y.-P. Jiang

Vol. 11 Rayleigh–Bénard Convection
A. V. Getling

Vol. 12 Localization and Solitary Waves in Solid Mechanics
A. R. Champneys, G. W. Hunt & J. M. T. Thompson

Vol. 13 Time Reversibility, Computer Simulation, and Chaos
W. G. Hoover

Vol. 14 Topics in Nonlinear Time Series Analysis – With Implications for EEG Analysis
A. Galka

Vol. 15 Methods in Equivariant Bifurcations and Dynamical Systems
P. Chossat & R. Lauterbach

Vol. 16 Positive Transfer Operators and Decay of Correlations
V. Baladi

Vol. 17 Smooth Dynamical Systems
M. C. Irwin

Vol. 18 Symplectic Twist Maps
C. Gole

Vol. 19 Integrability and Nonintegrability of Dynamical Systems
A. Goriely

Vol. 20 The Mathematical Theory of Permanent Progressive Water-Waves
H. Okamoto & M. Shoji

Vol. 21 Spatio-Temporal Chaos & Vacuum Fluctuations of Quantized Fields
C. Beck

Vol. 22 Energy Localisation and Transfer
eds. T. Dauxois, A. Litvak-Hinenzon, R. MacKay & A. Spanoudaki

*For the complete list of titles in this series, please write to the Publisher.

ADVANCED SERIES IN
NONLINEAR DYNAMICS
VOLUME 23

Geometrical Theory of Dynamical Systems and Fluid Flows

Tsutomu Kambe
Institute of Dynamical Systems, Tokyo, Japan

World Scientific
NEW JERSEY • LONDON • SINGAPORE • BEIJING • SHANGHAI • HONG KONG • TAIPEI • CHENNAI

Published by

World Scientific Publishing Co. Pte. Ltd.
5 Toh Tuck Link, Singapore 596224
USA office: 27 Warren Street, Suite 401–402, Hackensack, NJ 07601
UK office: 57 Shelton Street, Covent Garden, London WC2H 9HE

British Library Cataloguing-in-Publication Data
A catalogue record for this book is available from the British Library.

GEOMETRICAL THEORY OF DYNAMICAL SYSTEMS AND FLUID FLOWS

ISBN 981-238-806-0

Printed in Singapore by World Scientific Printers (S) Pte Ltd

Preface

This is an introductory textbook on the geometrical theory of dynamical systems, fluid flows, and certain integrable systems. The subjects are interdisciplinary and extend from mathematics, mechanics and physics to mechanical engineering. The approach is very fundamental and would be traced back to the times of Poincaré, Weyl and Birkhoff in the first half of the 20th century. The theory gives geometrical and frame-independent characterizations of various dynamical systems and can be applied to chaotic systems as well from the geometrical point of view. For integrable systems, similar but different geometrical theory is presented.

Underlying concepts of the present subject are based on the *differential geometry* and the *theory of Lie groups* in mathematical aspect and based on the *gauge theory* in physical aspect. Usually, those subjects are not easy to access, nor familiar to most students in physics and engineering. A great deal of effort has been directed to make the description elementary, clear and concise, so that beginners have easy access to the subject. This textbook is intended for upper level undergraduates and postgraduates in physics and engineering sciences, and also for research scientists interested in the subject.

Various dynamical systems often have common geometrical structures that can be formulated on the basis of Riemannian geometry and Lie group theory. Such a dynamical system always has a symmetry, namely it is invariant under a group of transformations, and furthermore it is necessary that the group manifold is endowed with a Riemannian metric. In this book, pertinent mathematical concepts are illustrated and applied to physical problems of several dynamical systems and integrable systems.

The present text consists of four parts: I. *Mathematical Bases*, II. *Dynamical Systems*, III. *Flows of Ideal Fluids*, and IV. *Geometry of*

Integrable Systems. Part I is composed of three chapters where basic mathematical concepts and tools are described. In Part II, three dynamical systems are presented in order to illustrate the fundamental idea on the basis of the mathematical framework of Part I. Although those systems are well-known in mechanics and physics, new approach and formulation will be provided from a geometrical point of view. Part III includes two new theoretical formulations of flows of ideal fluids: one is a variational formulation on the basis of the gauge principle and the other is a geometrical formulation based on a group of diffeomorphisms and associated Riemannian geometry. Part IV aims at presenting a different geometrical formulation for integrable systems. Its historical origin is as old as the Riemannian geometry and traced back to the times of Bäcklund, Bianchi and Lie, although modern theory of geometry of integrable systems is still being developed.

More details of each Part are as follows. In Part I, before considering particular dynamical systems, mathematical concepts are presented and reviewed concisely. In the first chapter, basic mathematical notions are illustrated about flows, diffeomorphisms and the theory of Lie groups. In the second chapter, the geometry of surface in Euclidian space $\mathbb{R}^3$ is summarized with special emphasis on the Gaussian cuvature which is one of the central objects in this treatise. This chapter presents many elementary concepts which are developed subsequently. In the third chapter, theory of Riemannian differential geometry is summarized concisely and basic concepts are presented: the first and second fundamental forms, commutator, affine connection, geodesic equation, Jacobi field, and Riemannian curvature tensors.

The three dynamical systems of Part II are fairly simple but fundamental systems known in mechanics. They were chosen to illustrate how the geometrical theory can be applied to dynamical systems. The first system in Chapter 4 is a free rotation of a rigid body (Euler's top). This is a well-known problem in physics and one of the simplest nonlinear integrable systems of finite degrees of freedom. Chapter 5 illustrates derivation of the KdV equation as a geodesic equation on a group (actually an extended group) of diffeomorphisms, which gives us a geometrical characterization of the KdV system. The third example in Chapter 6 is a geometrical analysis of chaos of a Hamiltonian system, which is a self-gravitating system of a finite number of point masses.

Part III is devoted to Fluid Mechanics which is considered to be a central part of the present book. In Chapter 7, a new gauge-theoretical formulation is presented, together with a consistent variational formulation in terms of variation of material particles. As a result, Euler's equation of motion is derived for an isentropic compressible flow. This formulation implies that the vorticity is a gauge field. Chapter 8 is a Riemannian-geometrical formulation of the hydrodynamics of an incompressible ideal fluid. This gives us not only geometrical characterization of fluid flows but also interpretation of the origin of Riemannian curvatures of flows. Chapter 9 is a geometrical formulation of motions of a vortex filament.

It is well known that some soliton equations admit a geometric interpretation. In Part IV, Chapter 10 reviews a classical theory of the sine–Gordon equation and the Bäcklund transformation which is an oldest example of geometry of a pseudo-spherical surface in $\mathbb{R}^3$ with the Gaussian curvature of a constant negative value. Chapter 11 presents a geometric and group-theoretic theory for integrable systems such as sine–Gordon equation, nonlinear Schrödinger equation, nonlinear sigma model and so on. Final section presents a new finding [CFG00] that all integrable systems described by the $su(2)$ algebra are mapped to a spherical surface.

Highlights of this treatise would be: (i) Geometrical formulation of dynamical systems; (ii) Geometric description of ideal-fluid flows and an interpretation of the origin of Riemannian curvatures of fluid flows; (iii) Various geometrical characterizations of dynamical fields; (iv) Gauge-theoretic description of ideal fluid flows; and (v) Modern geometric and group-theoretic formulation of integrable systems.

It is remarkable that the present geometrical formulations are successful for all the problems considered here and give insight into common background of the diverse physical systems. Furthermore, the geometrical formulation opens a new approach to various dynamical systems.

Parts I–III of the present monograph were originally prepared as *lecture notes* during the author's stay at the Isaac Newton Institute in the programme "Geometry and Topology of Fluid Flow" (2000). After that, the manuscript had been revised extensively and published as a *Review* article in the journal, *Fluid Dynamics Research.* In addition, the present book includes Part IV, which describes geometrical theory of *Integrable Systems.* Thus, this covers an extensive area of dynamical systems and reformulates those systems on the basis of *geometrical concepts.*

Tsutomu Kambe
Former Professor (Physics)*
University of Tokyo

December 2003

*Visiting Professor, Nankai Institute of Mathematics (Tienjin, China)

Contents

II. Dynamical Systems

IV. Geometry of Integrable Systems

Part I

Mathematical Bases

Chapter 1

Manifolds, Flows, Lie Groups and Lie Algebras

In geometrical theory of dynamical systems, fundamental notions and tools are manifolds, diffeomorphisms, flows, exterior algebras and Lie algebras.

1.1. Dynamical Systems

In mechanics, we deal with physical systems whose state at a time t is specified by the values of n real variables,

$$x^1, x^2, \ldots, x^n,$$

and furthermore the system is such that its time evolution is *completely determined* by the values of the n variables. In other words, the rate of change of these variables, i.e. $\mathrm{d}x^1/\mathrm{d}t, \ldots, \mathrm{d}x^n/\mathrm{d}t$, depends on the values of the variables themselves, so that the equations of motion can be expressed by means of n differential equations of the first order,

$$\frac{\mathrm{d}x^i}{\mathrm{d}t} = X^i(x^1, x^2, \ldots, x^n), \quad (i = 1, 2, \ldots, n). \tag{1.1}$$

A system of time evolution of variables, such as $(x^1(t), \ldots, x^n(t))$ described by (1.1), is termed a *dynamical system* [Birk27]. A simplest example would be the rectlinear motion of a point mass m located at x under a restoring force $-kx$ of a spring:

$$\mathrm{d}x/\mathrm{d}t = y, \quad \mathrm{d}y/\mathrm{d}t = -kx,$$

where k is a spring constant. A system of N point masses under self-interaction governed by Newton's equations of motion is another example.

However, the notion of the dynamical system is more general, and not restricted to such Newtonian dynamical system.

The space where the n-tuple of real numbers $(x^1, \ldots, x^n)$ reside is called a n-dimensional manifold M^n which will be detailed in the following sections. The space is also called the *configuration space* of the system, while the *physical state* of the system is determined by the $2n$ variables: the coordinates $(x^1, \ldots, x^n)$ and the velocities $(\dot{x}^1, \ldots, \dot{x}^n)$ where $\dot{x}^i = \mathrm{d}x^i/\mathrm{d}t$. Such a system is said to have n *degrees of freedom*. It is of fundamental importance how the differential equations are determined from basic principles, and in fact this is the subject of the present monograph.

Study of dynamical systems may be said to have started with the work of Henri Poincaré at the turn of the 19th to 20th century. Existence of very complicated orbits was disclosed in the problem of interacting three celestial bodies. After Poincaré, Birkhoff studied an exceedingly complex structure of orbits arising when an integrable system is perturbed [Birk27; Ott93]. Later, the basic question of how prevalent integrability is, was given a mathematical answer by Kolmogorov (1954), Arnold (1963) and Moser (1973), which is now called the KAM theorem and regarded as a fundamental theorem of chaos in Hamiltonian systems (e.g. [Ott93]).

The present approach to the dynamical systems is based on a geometrical point of view.[1] The geometrical frameworks concerned here were founded earlier in the 19th century by Gauss, Riemann, Jacobi and others. However in the 20th century, stimulated by the success of the theory of general relativity, the gauge theory (a geometrical theory) has been developed in theoretical physics. It has now become possible to formulate a geometrical theory of dynamical systems, mainly due to the work of Arnold [Arn66].

1.2. Manifolds and Diffeomorphisms

A fundamental object in the theory of dynamical systems is a manifold. A *manifold* M^n is an n-dimensional space that is locally an n-dimensional euclidean space $\mathbb{R}^n$ in the sense described just below, but is not necessarily $\mathbb{R}^n$ itself.[2] A unit n-sphere S^n in $(n+1)$-dimensional euclidean space $\mathbb{R}^{n+1}$ is a typical example of the n-dimensional manifold M^n. Consider

---

[1] In this context, the following textbooks may be useful: [Fra97; AK98; AM78].

[2] The euclidean space $\mathbb{R}^n$ is endowed with a global coordinate system $(x^1, \ldots, x^n)$ and is basically an important manifold. Henceforth the lower case e is used as "euclidean" because of its frequent occurrence.

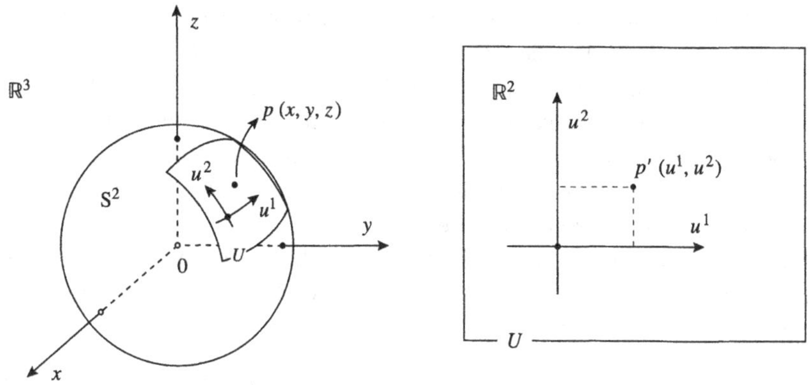

Fig. 1.1. Two-sphere S^2 and local coordinates.

a unit two-sphere $\mathbf{S}^2$ which is a two-dimensional object imbedded in three-dimensional space $\mathbf{R}^3$ (Fig. 1.1). Denoting a point in $\mathbb{R}^3$ by $p = (x, y, z)$, the two-sphere S^2 is defined by all points p satisfying $\|p\|^2 = x^2 + y^2 + z^2 = 1$, where $\|\cdot\|$ is the euclidean norm. The two-sphere S^2 is not a part of the euclidean space $\mathbb{R}^2$. However, an observer on S^2 would see that the immediate neighborhood is described by two coordinates and cannot be distinguished from a small domain of $\mathbb{R}^2$. A point p' in a patch U (an open subset of S^2) is represented by (u^1, u^2).

In general, an n-dimensional manifold M^n is a topological space (Appendix A.1), which is covered with a collection of open subsets $U_1, U_2, \ldots$ such that each point of M^n lies in at least one of them (Fig. 1.2). Using a map F_U, called a *homeomorphism* (Appendix A.2), each open

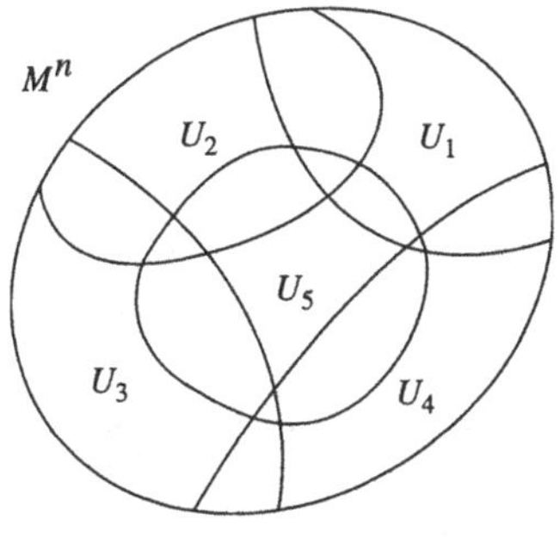

Fig. 1.2. Atlas.

subset U is in one-to-one correspondence with an open subset $F_U(U)$ of $\mathbb{R}^n$. Each pair (U, F_U), called a *chart*, defines a *coordinate patch* on M. To each point p ($\in U \subset M$), we may assign the n coordinates of the point $F_U(p)$ in $\mathbb{R}^n$. For this reason, we call F_U a *coordinate map* with the jth component written as x_U^j. This is often described in the following way. On the patch U, a point p is represented by a *local coordinate*, $p = (x_p^1, \ldots, x_p^n)$. The whole system of charts is called an *atlas*.

The unit circle in the plane $\mathbb{R}^2$ is a manifold of one-sphere S^1. The S^1 has a local coordinate $\theta \in [0,1]$ (with the ends 0 and 1 identified) $\subset \mathbb{R}^1$. Consider a map by a complex function $f(\theta)$,

$$f(\theta) = e^{i2\pi\theta}, \quad f : \theta \in [0,1] \subset \mathbb{R}^1 \to p(x,y) \in S^1 \subset \mathbb{R}^2 \tag{1.2}$$

where $e^{i2\pi\theta} = x + iy$ ($i = \sqrt{-1}, x^2 + y^2 = 1$). The map is one-to-one and onto if we identify the endpoints by $f(0) = f(1) \to (1,0) \in \mathbb{R}^2$ (Fig. 1.3). Choosing a patch (open subset) $U \subset S^1$, a homeomorphism map F_U is given by $f^{-1}(U)$.

It is readily seen that the unit circle $\mathbf{S}^1$ (a connected space[3]) is *covered* by the real axis $\mathbf{R}^1$ (another connected space) an infinite number of times by the map $f : \mathbb{R}^1 \to S^1$. Corresponding to an open subset $U \subset S^1$, the preimage $f^{-1}(U)$ consists of infinite number of disjoint open subsets $\{U_\alpha\}$ of $\mathbb{R}^1$, each U_α being diffeomorphic with U under $f : U_\alpha \to U$. It is said that the $\mathbb{R}^1$ is an *infinite-fold cover* of S^1.

Suppose that a patch U with its local coordinates $p = x = (x^1, \ldots, x^n)$ overlap with another patch V with local coordinates $p = y = (y^1, \ldots, y^n)$.

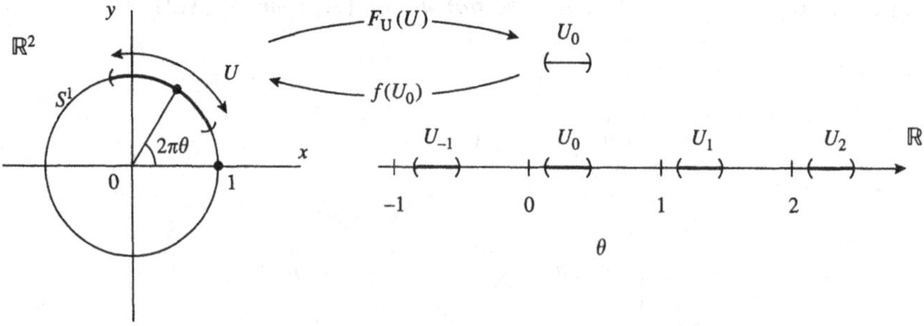

Fig. 1.3. Manifold S^1.

[3]A manifold M is said to be (*path-*)*connected* if any two points in M can be joined by a (piecewise smooth) curve belonging to M.

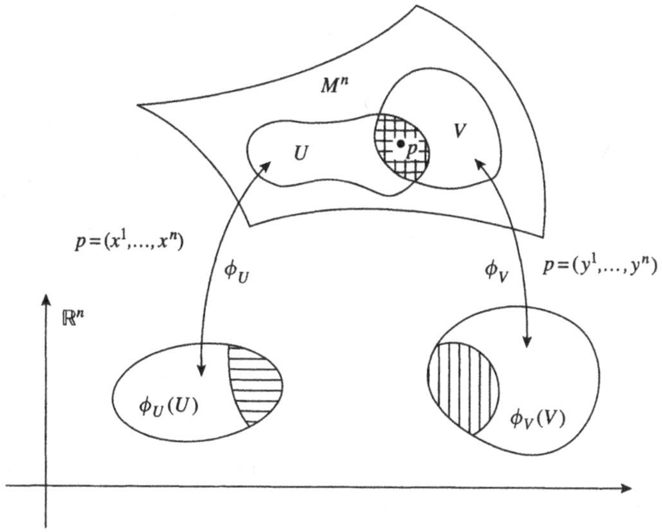

Fig. 1.4. Coordinate maps.

Then, a point p lying in the overlapping domain can be represented by both systems of x and y (Fig. 1.4). In particular, y^i is expressed in terms of x as

$$y^i = y^i(x^1, \ldots, x^n), \quad (i = 1, \ldots, n). \tag{1.3}$$

We require that these functions are smooth and differentiable, and that the Jacobian determinant

$$|J| = \frac{\partial(y)}{\partial(x)} = \frac{\partial(y^1, \ldots, y^n)}{\partial(x^1, \ldots, x^n)}. \tag{1.4}$$

does not vanish at any point $p \in U \cap V$ [Fla63, Ch. V].

Let $F : M^n \to W^r$ be a smooth map from a manifold M^n to another W^r. In local coordinates $x = (x^1, \ldots, x^n)$ in the neighborhood of the point $p \in M^n$ and $z = (z^1, \ldots, z^r)$ in the neighborhood of $F(p)$ on W^r, the map F is described by r functions $F^i(x), (i = 1, \ldots, r)$ of n variables, abbreviated to $z = F(x)$ or $z = z(x)$, where F^i are differentiable functions of x^j $(j = 1, \ldots, n)$.

When $n = r$, we say that the map F is a *diffeomorphism*, provided F is differentiable (thus continuous), one-to-one, onto, and in addition F^{-1} is differentiable (Fig. 1.5). Such an F is a *differentiable homeomorphism.* If the inverse F^{-1} does exist and the Jacobian determinant does not vanish, then the inverse function theorem would assure us that the

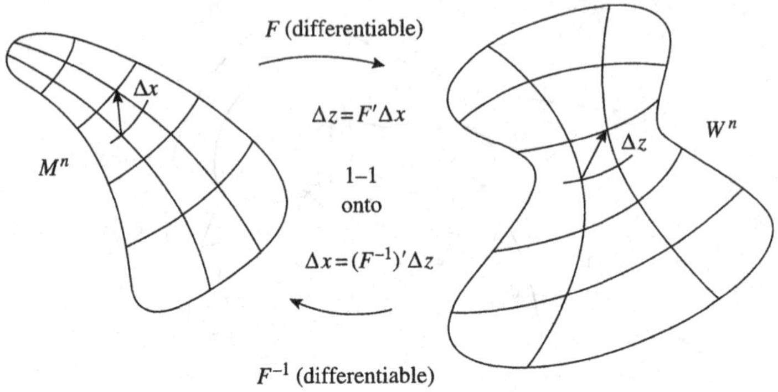

Fig. 1.5. Diffeomorphism.

inverse is differentiable. In the next section, the fluid flow is described to be a smooth sequence of diffeomorphisms of particle configuration (of infinite dimension).

1.3. Flows and Vector Fields

The vector field we are going to consider is not an object residing in a flat euclidean space and is different from a field of simple n-tuple of real numbers.

1.3.1. *A steady flow and its velocity field*

Given a steady flow[4] of a fluid in $\mathbb{R}^3$, one can construct a one-parameter family of maps: $\phi_t : \mathbb{R}^3 \to \mathbb{R}^3$, where ϕ_t takes a fluid particle located at p when $t = 0$ to the position $\phi_t(p)$ of the same particle at a later time $t > 0$ (Fig. 1.6). The family of maps are the so-called *Lagrangian* representation of motion of fluid particles. In terms of local coordinates, the jth coordinate of the particle is written as $x^j \circ \phi_t(p) = x_t^j(p)$, where "$x^j \circ$" denotes a *projection map* to take the jth component.

Associated with any such flow, we have a velocity at p,

$$v(p) := \frac{\mathrm{d}}{\mathrm{d}t}\phi_t(p)\bigg|_{t=0}.$$

In terms of the coordinates, we have $v^j(p) = (\mathrm{d}x_t^j(p)/\mathrm{d}t)|_{t=0}$. Taking a smooth function $f(x) = f(x^1, x^2, x^3)$, i.e. $f : \mathbb{R}^3 \to \mathbb{R}$ and differentiating

[4] *Steady* velocity field does not depend on time t by definition.

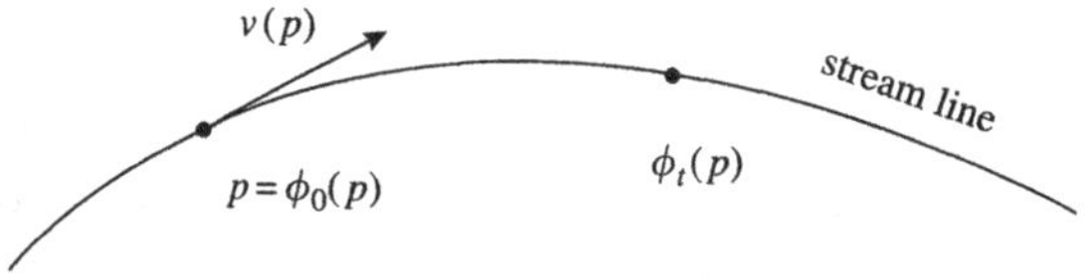

Fig. 1.6. Map ϕ_t.

$f(\phi_t(p))$ with respect to t, we have[5]

$$\frac{\mathrm{d}}{\mathrm{d}t} f(\phi_t(p))\bigg|_{t=0} = \sum_j \frac{\mathrm{d}x_t^j(p)}{\mathrm{d}t} \frac{\partial f}{\partial x^j} = \sum_j v^j(p) \frac{\partial}{\partial x^j} f \tag{1.5}$$

$$=: X(p)f, \quad X(p) := \sum_j v^j(p) \frac{\partial}{\partial x^j}. \tag{1.6}$$

This is also written in the following way by bearing in mind that f is a map, $f : \mathbb{R}^3 \to \mathbb{R}$:

$$Xf = \frac{\mathrm{d}}{\mathrm{d}t} f(\phi_t) := \frac{\mathrm{d}}{\mathrm{d}t} f \circ \phi_t. \tag{1.7}$$

The differential operator X is written also as v by the reason described in the next subsection.

Conversely, to each vector field $v(x) = (v^j)$ in $\mathbb{R}^3$, one may associate a flow $\{\phi_t\}$ having v as its velocity field. The map $\phi_t(p)$ with t as an *integration parameter* can be found by solving the system of ordinary differential equations,

$$\frac{\mathrm{d}x^j}{\mathrm{d}t} = v^j(x^1(t), x^2(t), x^3(t))$$

with the initial condition, $x(0) = p$. Thus one finds an integral curve (called a *stream line*) in a neighborhood of $t = 0$, which is a one-parameter family of maps $\phi_t(p)$ for any $p \in \mathbb{R}^3$, called a flow *generated* by the vector field v, where $v = \dot{\phi}_t$ (Fig. 1.7). The map ϕ_t is a diffeomorphism, because $\phi_t(p)$ is differentiable, one-to-one, onto and ϕ_t^{-1} is differentiable, with respect to every point $p \in \mathbb{R}^3$.[6] This is assured in flows of a fluid by its physical property that each fluid particle is a physical entity which keeps its identity

[5]We use the symbol := to define the left side by the right side, and =: to define the right side by the left side.

[6]The flow $\{\phi_t\}$ is considered to be diffeomorphisms of Sobolev class H^s in Chapter 8 ($s > n/2 + 1$ in $\mathbb{R}^n$, Appendix F).

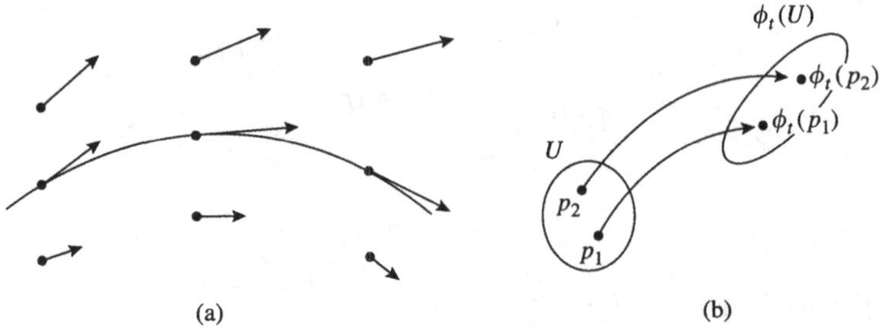

Fig. 1.7. (a) An integral curve and (b) a flow ϕ_t.

during the motion, as long as two particles do not come to occupying an identical point simultaneously.[7]

Remark. Continuous distribution of fluid particles in a three-dimensional euclidean space has infinite degrees of freedom. Therefore, the velocity field of all the particles as a whole is regarded to be of infinite dimensions. In this context, the set of diffeomorphisms ϕ_t forms an infinite dimensional manifold $D^{(\infty)}$ and a point $\eta = \phi_t \in D^{(\infty)}$ represents a configuration (as a whole) of all particles composing the fluid at a given time t.

1.3.2. *Tangent vector and differential operator*

The vector fields we are going to consider on M^n are not an object residing in a flat euclidean space. We need a sophisticated means to represent vectors which are different from a simple n-tuple of real numbers. In general, on a manifold M^n, one can define a vector v tangent to the parameterized curve ϕ_t at any point x on the curve. We motivate the definition of vector as follows.

A *flow* $\phi_t(p) = (x_t^j(p))$ on an n-dimensional manifold M^n is described by the system of ordinary differential equations,

$$\frac{\mathrm{d}x_t^j}{\mathrm{d}t} = v^j(x_t^1, \ldots, x_t^n), \quad (j = 1, \ldots, n), \tag{1.8}$$

with the initial condition, $\phi_0 = p$. The one-to-one correspondence between the tangent vector $v = (v^j)$ to M^n at x and the first order differential

[7]The present formulation is relevant to the time before a spontaneous formation of singularity (if any).

operator $\sum_j v^j(x)\partial/\partial x^j$, mediated by (1.8) and the n-dimensional version of (1.6), implies the following representation,

$$v(x) := \sum_j v^j(x)\frac{\partial}{\partial x^j}, \tag{1.9}$$

which defines the *vector field* $v(x)$ as a differential operator $v^j(x)\partial/\partial x^j$.

In fact, with a local coordinate patch (U, x_U) in the neighborhood of a point p, a curve will be described by n differentiable functions $(x_U^1(t), \ldots, x_U^n(t))$. The tangent vector at p is described by $v_U = (\dot{x}_U^1(0), \ldots, \dot{x}_U^n(0))$ where $\dot{x}(0) = \mathrm{d}x/\mathrm{d}t|_{t=0}$. If p also lies in the coordinate patch (V, x_V), then the same tangent vector is described by another n-tuple $v_V = (\dot{x}_V^1(0), \ldots, \dot{x}_V^n(0))$. In terms of the transformation function (1.3) on the overlapping domain which is now represented by $x_V^i = x_V^i(x_U^j)$, the two sets of tangent vectors are related by the chain rule,

$$v_V^i = \left.\frac{\mathrm{d}x_V^i}{\mathrm{d}t}\right|_{t=0} = \sum_j \left(\frac{\partial x_V^i}{\partial x_U^j}\right)\left.\frac{\mathrm{d}x_U^j}{\mathrm{d}t}\right|_{t=0} = \sum_j \left(\frac{\partial x_V^i}{\partial x_U^j}\right) v_U^j. \tag{1.10}$$

This suggests a transformation law of a tangent vector. Owing to this transformation, the definition (1.9) of a vector v is frame-independent, i.e. independent of local coordinate basis. In fact, by the transformation $x_V^i = x_V^i(x_U^j)$, we obtain

$$\sum_j v_U^j \frac{\partial}{\partial x_U^j} = \sum_j v_U^j(x)\sum_i \left(\frac{\partial x_V^i}{\partial x_U^j}\right)\frac{\partial}{\partial x_V^i} = \sum_i v_V^i \frac{\partial}{\partial x_V^i}. \tag{1.11}$$

It is not difficult to see that the properties of the linear vector space are satisfied by the representation (1.9).[8] Usually, in the differential geometry, no distinction is made between a vector and its associated differential operator. The vector $v(x)$ thus defined at a point $x \in M^n$ is called a *tangent* vector.

1.3.3. *Tangent space*

Each one of the n operators $\partial/\partial x^\alpha$ ($\alpha = 1, \ldots, n$) defines a vector. The αth vector $\partial/\partial x^\alpha$ ($v^\alpha = 1$ and $v^i = 0$ for $i \neq \alpha$) is the tangent vector to the αth coordinate curve parameterized by x^α. This curve is described by

[8]It is evident from (1.9) that the sum of two vectors at a point is again a vector at that point, and that the product of a vector by a real number is a vector.

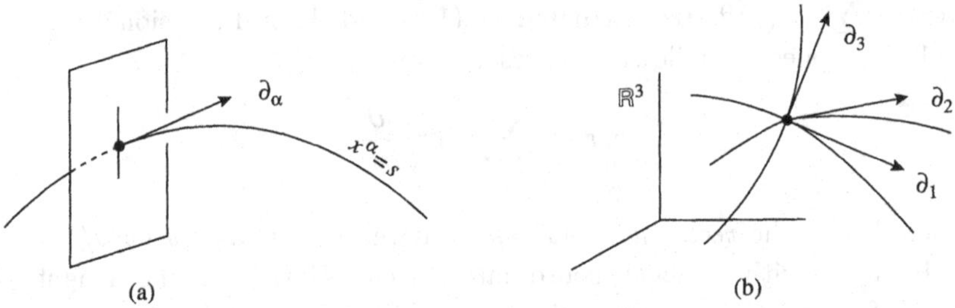

Fig. 1.8. (a) αth coordinate curve, (b) coordinate basis in $\mathbb{R}^3$.

$x^\alpha = s$ and $x^i = \text{const}$ for $i \neq \alpha$. Then the tangent vector $\partial/\partial x^\alpha$ for the αth curve has components $dx^\alpha/ds = 1$ and $dx^i/ds = 0$ for $i \neq \alpha$ (Fig. 1.8(a)). The n vectors $\partial/\partial x^1, \ldots, \partial/\partial x^n$ form a basis of a vector space, and this base is called a *coordinate* basis (Fig. 1.8(b)). The basis vector $\partial/\partial x^\alpha$ is simply written as ∂_α. A tangent vector X is written in general as[9]

$$X = X^j \partial_j, \quad \text{or} \quad X_x = X^j(x)\partial_j.$$

If $\boldsymbol{r} = (r^1, \ldots, r^N)$ is a position vector in the euclidean space $\mathbb{R}^N$ and M^n is a submanifold of $\mathbb{R}^N$: $M^n \subset \mathbb{R}^N$ $(n \leq N)$, the vector $\partial/\partial x^\alpha$ is understood as $\partial_\alpha \equiv \partial/\partial x^\alpha = \partial \boldsymbol{r}/\partial x^\alpha = (\partial/\partial x^\alpha)(r^1, \ldots, r^N)$, where $r^i = r^i(x^1, \ldots, x^n)$.[10]

The *tangent space* is defined by a vector space consisting of all tangent vectors to M^n at x and is written as $\mathbf{T}_x\mathbf{M}^n$.[11] When the coefficients X^j are smooth functions $X^j(x)$ for $x \in M^n$, the $X(x)$ is called a *vector field.*

1.3.4. *Time-dependent (unsteady) velocity field*

In most dynamical systems, a parameter t called the *time* plays a special role, and the tangent vector $v = (v^j)$ is called the *velocity.* A velocity field is said to be *time-dependent*, or *unsteady*, when v^j depends on t (an integration parameter) as well as space coordinates (Fig. 1.9). In the unsteady problem, an additional coordinate x^0 is introduced, and the n equations of (1.8) for

[9] The summation convention is used hereafter, i.e. the summation with respect to j is understood for the pair of double indices like j without the summation symbol $\sum$.
[10] The parameters $(x^1, \ldots, x^n)$ form a *curvilinear* coordinate system.
[11] It is useful in later sections to keep in mind that the tangent space T_xM^n is the usual n-dimensional *affine* subspace of $\mathbb{R}^N$.

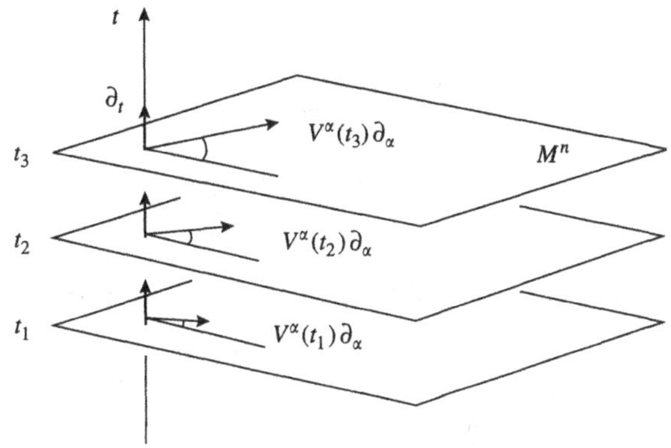

Fig. 1.9. Time-dependent velocity field.

$v^j \in \mathbb{R}^n$ are replaced by the following $(n+1)$ equations,

$$\frac{\mathrm{d}x^j}{\mathrm{d}t} = v^j(x^0(t), x^1(t), \dots, x^n(t)), \quad \text{with } v^0 = 1, \tag{1.12}$$

for $j = 0, 1, \dots, n$. It is readily seen that the newly added equation reduces to $x^0 = t$. Correspondingly, the tangent vector in the time-dependent case is written as, using the *tilde* symbol,

$$\tilde{v} := \tilde{v}^i \partial_i = v^0 \partial_0 + v^\alpha \partial_\alpha = \partial_t + v^\alpha \partial_\alpha, \tag{1.13}$$

where the index α denotes the spatial components $1, \dots, n$.[12]

1.4. Dynamical Trajectory

A fundamental space of the theory of dynamical systems is a fiber bundle. How is the phase space of Hamiltonian associated with it?

1.4.1. *Fiber bundle (tangent bundle)*

In mechanics, a Lagrangian function L of a dynamical system of n degrees of freedom is usually defined in terms of generalized coordinates $q = (q^1, \dots, q^n)$ and generalized velocities $\dot{q} = (\dot{q}^1, \dots, \dot{q}^n)$ such as $L(q, \dot{q})$, while a Hamiltonian function is usually represented as $H(q, p)$, where

[12] *Nonvelocity* tangent vector such as the Jacobi vector $\tilde{J}$ is written simply as $\tilde{J} = J^\alpha \partial_\alpha$ (see the footnote of §8.3.4).

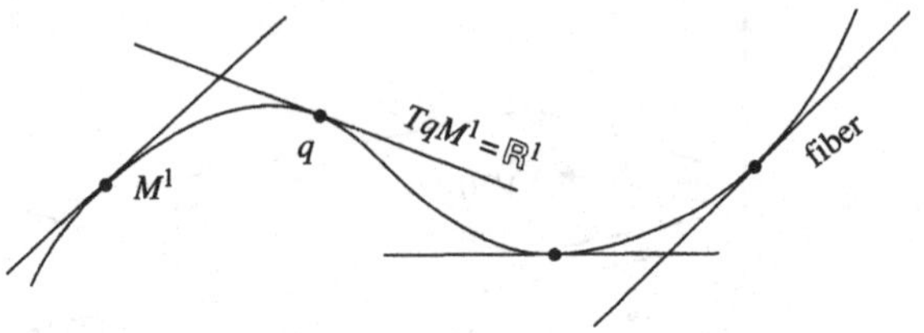

Fig. 1.10. A tangent bundle TM^1 for M^1 (a curve).

$p = (p_1, \dots, p_n)$ are generalized momenta. Is there any significant difference between the pairs of independent variables?

Suppose that $q = (q^1, \dots, q^n)$ is a point in an n-dimensional manifold U^n, which is a coordinate patch of a manifold M^n and a portion of $\mathbb{R}^n$, and that $\dot{q} = (\dot{q}^1, \dots, \dot{q}^n)$ is a tangent vector to M^n at q. The pair $(q, \dot{q})$ is an element of a tangent bundle, $\mathbf{TM}^n$. Namely, a *tangent bundle* TM^n is defined as the collection of all tangent vectors at all points of M^n, called a *base manifold.*[13] A schematic diagram of a tangent bundle TM is drawn in Fig. 1.10 (see also Fig. 1.23 for a tangent bundle TS^1).

Associated with any bundle space TM, a projection map $\pi : TM \to M$ is defined by $\boldsymbol{\pi}(Q) = q$, where $Q \in TM, q \in M$. On the other hand, the inverse map $\pi^{-1}(q)$ represents all vectors v tangent to $M = M^n$ (base manifold) at q, i.e. a vector space $\mathbf{T}_q\mathbf{M} = \mathbb{R}^n$. It is called the *fiber* over q. In this regard, the tangent bundle is also called a *fiber bundle*, or a *vector bundle*[14] (Fig. 1.11). Since $\pi^{-1}(U^n)$ is topologically $U^n \otimes \mathbb{R}^n$, the tangent bundle is locally a product. However, this is not so in general (see [Fla63, Ch. 2; Sch80, Ch. 2; NS83, Ch. 7]).

1.4.2. *Lagrangian and Hamiltonian*

The space of generalized coordinates $q = (q^1, \dots, q^n)$ is called the *configuration space* in mechanics (also called a base space), whereas the space $(q, \dot{q})$ is called the tangent bundle, a *mathematical* term. The Lagrangian $L(q, \dot{q})$ is a function on the tangent bundle to M^n, namely $L : TM^n \to \mathbb{R}$.

[13] If a point of TM^n is represented globally as $(q, \dot{q})$, i.e. a global product bundle $q \otimes \dot{q}$, the tangent bundle is called a *trivial* bundle. Note that the first n coordinates $(q^1, \dots, q^n)$ take their values in a portion $U^n \in \mathbb{R}^n$, whereas the second set $(\dot{q}^1, \dots, \dot{q}^n)$ take any value in $\mathbb{R}^n$. Thus, the patch is of the form, $U^n \otimes \mathbb{R}^n$.

[14] A fiber is not necessarily a simple vector. It takes, for example, even an element of Lie algebra. See Chapter 9.

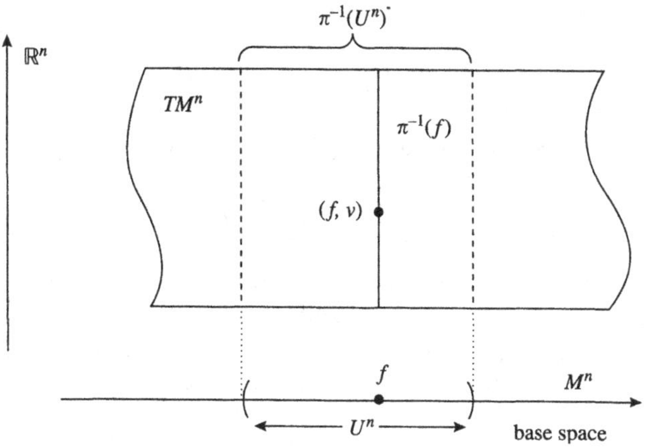

Fig. 1.11. A fiber bundle TM^n.

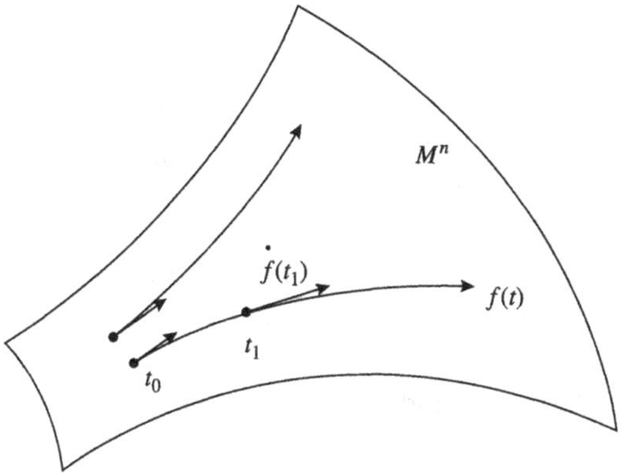

Fig. 1.12. Dynamical trajectories.

If we consider a specific trajectory $q(t)$ in the configuration space with t as the time parameter, then we have $\dot{q} = \mathrm{d}q/\mathrm{d}t$. Thus, the pair $(q, \dot{q})$ has a certain *geometrical* significance (Fig. 1.12).

Dynamical trajectory of the point $q(t)$ is determined by the following Lagrange's equation of motion (see §7.2.3):

$$\frac{\mathrm{d}}{\mathrm{d}t}\left(\frac{\partial L}{\partial \dot{q}^i}\right) - \frac{\partial L}{\partial q} = 0. \tag{1.14}$$

The Hamiltonian function $H(q,p)$ is defined by

$$H(q,p) = \sum_i p_i \dot{q}^i - L(q,\dot{q}), \tag{1.15}$$

where p_i is an ith component of the generalized momentum defined by

$$p_i(q,\dot{q}) := \frac{\partial}{\partial \dot{q}^i} L(q,\dot{q}). \tag{1.16}$$

Change of variables from $(q,\dot{q})$ for the Lagrangian $L(q,\dot{q})$ to (q,p) for the Hamiltonian $H(q,p)$ has a certain significance more than a mere change of coordinates. Consider a coordinate transformation from q_U to q_V by $q_V = q_V(q_U)$. Correspondingly, the change of velocity, $\dot{q}_U \to \dot{q}_V$, is represented by

$$\dot{q}_V^i = \sum_k \frac{\partial q_V^i}{\partial q_U^k} \dot{q}_U^k. \tag{1.17}$$

On the other hand, the generalized momentum is transformed as follows,

$$\begin{aligned} (p_V)_i &= \frac{\partial}{\partial \dot{q}_V^i} L(q_U, \dot{q}_U) = \sum_k \frac{\partial \dot{q}_U^k}{\partial \dot{q}_V^i} \frac{\partial L}{\partial \dot{q}_U^k} \\ &= \sum_k \frac{\partial \dot{q}_U^k}{\partial \dot{q}_V^i} (p_U)_k = \sum_k \frac{\partial q_U^k}{\partial q_V^i} (p_U)_k, \end{aligned} \tag{1.18}$$

since $q_U = q_U(q_V)$ and therefore $\partial q_U^k / \partial \dot{q}_V^i = 0$ in the second equality, and (1.17) is used to obtain the last equality since $\partial \dot{q}_U^k / \partial \dot{q}_V^i = \partial q_U^i / \partial q_V^k$. Thus, it is found that the transformation matrix for p is the inverse of that of $\dot{q}$.

The expression (1.17) represents the transformation law of vectors and characterizes the tangent bundle, while the expression (1.18) characterizes the transformation law of covectors (see §1.5.2). The two transformation laws imply that the product $\sum_i p_i \dot{q}^i$ would be a scalar, an invariant under a coordinate transformation, since $\sum_i (p_V)_i \dot{q}_V^i = \sum_i (p_U)_i \dot{q}_U^i$ can be shown. A covector and a vector are associated with each other by means of a metric tensor (see §1.4.2).

1.4.3. *Legendre transformation*

Mathematically, the change $\dot{q} \to p$ is interpreted as a *Legendre transformation* (e.g. [Arn78, §14]). Consider a function $l(x)$ of a single variable x, where $l''(x) > 0$, i.e. $l(x)$ is *convex*. Let p be a given real number and define the function $h(x,p) = px - l(x)$. The function $h(x,p)$ has a maximum with respect to x at a point $x_*(p)$. The point x_* is determined uniquely by the

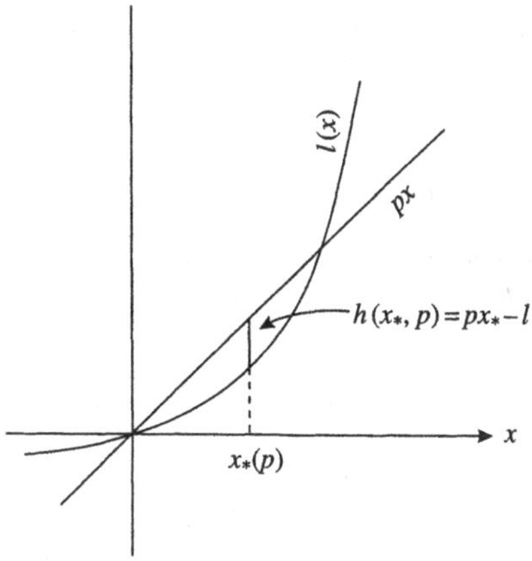

Fig. 1.13. Legendre transformation.

condition, $\partial h/\partial x = p - l'(x_*) = 0$, since $l'(x)$ is a monotonically increasing function by the convexity (Fig. 1.13). Thus, $p = l'(x_*)$. If we write $x = \dot{q}$, the relation $p = l'(x)$ is equivalent to (1.16) as far as the variable $\dot{q}$ is concerned.

By the Legendre transformation, the Lagrangian $L(q, \dot{q})$ on a vector space is transformed to the Hamiltonian $H(q,p)$ on the dual space, defined by (1.15) and (1.16). In mechanics, the space (q,p) is called the *phase space*. The equations of motion in the phase space are derived as follows:

$$\mathrm{d}H(q,p) = \sum_i \left(\frac{\partial H}{\partial q^i}\,\mathrm{d}q^i + \frac{\partial H}{\partial p_i}\,\mathrm{d}p_i \right). \tag{1.19}$$

On the other hand, taking the differential of the right-hand side of (1.15) and using (1.16), we obtain

$$\begin{aligned} \mathrm{d}\left(\sum_i p_i \dot{q}^i - L(q,\dot{q})\right) &= \sum_i \left(p_i \mathrm{d}\dot{q}^i + \dot{q}^i \mathrm{d}p_i - \frac{\partial L}{\partial q^i}\,\mathrm{d}q^i - \frac{\partial L}{\partial \dot{q}^i}\,\mathrm{d}\dot{q}^i \right) \\ &= \sum_i \left(-\frac{\partial L}{\partial q^i}\,\mathrm{d}q^i + \dot{q}^i \mathrm{d}p_i \right). \end{aligned} \tag{1.20}$$

Equating the right sides of the above two equations, we obtain the following Hamilton's equations of motion,

$$\frac{\mathrm{d}q^i}{\mathrm{d}t} = \frac{\partial H}{\partial p_i}, \quad \frac{\mathrm{d}p_i}{\mathrm{d}t} = -\frac{\partial H}{\partial q^i}, \tag{1.21}$$

since $\mathrm{d}p_i/\mathrm{d}t = \partial L/\partial q^i$ by using (1.16) and Lagrange's equation of motion (1.14).

1.5. Differential and Inner Product

A basic tool of a dynamical system is a metric. How are vectors and covectors related to it?

1.5.1. *Covector (1-form)*

Differential $\mathrm{d}f$ of a function f on M^n is *defined* by $\mathrm{d}f[v] := vf$ and is regarded as a linear functional $T_xM^n \to \mathbb{R}$ for any vector $v \in E = T_xM^n$. In local coordinates, we have $v = v^j\partial_j$. Using (1.9), we obtain

$$\mathrm{d}f[v] = \mathrm{d}f[v^j\partial_j] = vf = \sum_j v^j(x)\frac{\partial f}{\partial x^j}. \tag{1.22}$$

This is a basis-independent definition (see (1.11)). The differential $\mathrm{d}f[v^j\partial_j]$ is linear with respect to the scalar coefficient v^j. In particular, if f is the coordinate function x^i, we obtain

$$\mathrm{d}x^i[v] = \mathrm{d}x^i[v^j\partial_j] = v^j\,\mathrm{d}x^i\left[\frac{\partial}{\partial x^j}\right] = v^j\frac{\partial x^i}{\partial x^j} = v^j\delta^i_j = v^i \tag{1.23}$$

by replacing f with x^i. Namely the operator $\mathrm{d}x^i$ reads off the ith component of any vector v (Fig. 1.14). It is seen that[15]

$$\mathrm{d}x^i[\partial_j] = \delta^i_j.$$

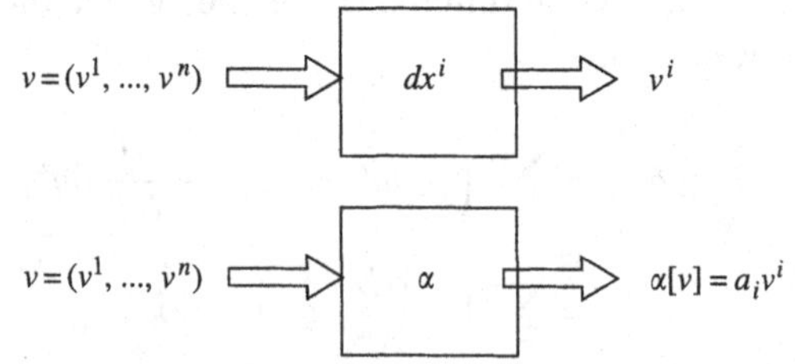

Fig. 1.14. 1-forms: $\mathrm{d}x^i$ and $\alpha = a_i\mathrm{d}x^i$.

[15]The symbols δ_{ij}, δ^{ij} and δ^i_j are identity tensors of rank 2, i.e. second order *covariant*, second order *contravariant* and *mixed* (first order covariant and first order contravariant) unit tensor, respectively (see §1.10).

Thus, the n functionals $\mathrm{d}x^i$ $(i = 1, \ldots, n)$ yield the dual bases corresponding to the coordinate bases $(\partial_1, \ldots, \partial_n)$ of a vector space T_xM^n, in the sense described below. The dual bases $(\mathrm{d}x^1, \ldots, \mathrm{d}x^n)$ form a dual space $(T_xM^n)^*$. The most general linear functional, $\alpha : T_xM^n \to \mathbb{R}$, is expressed in coordinates as

$$\alpha := a_1 \mathrm{d}x^1 + \cdots + a_n \mathrm{d}x^n. \tag{1.24}$$

The α is called a *covector*, or a *covariant vector*, or a differential *one-form* (1-form),[16] and is an element of the *cotangent* space $E^* = (T_xM^n)^*$. Corresponding to the covariant vector α, the vector v is also called a *contravariant vector*. Given a contravariant vector $v = v^j \partial_j$, the 1-form $\alpha \in E^*$ takes the value,

$$\alpha[v] = \sum_i a_i \, \mathrm{d}x^i [v^j \partial_j] = a_i v^i. \tag{1.25}$$

Correspondingly, a contravariant vector $v \in E$ can be considered as a linear functional on the covariant vectors with the definition of the same value as (1.25)[17]:

$$v[\alpha] \equiv \alpha[v] = a_i v^i. \tag{1.26}$$

When the coefficients a_i are smooth functions $a_i(x)$, the α is a 1-form *field* and an element of the *cotangent bundle* $(TM^n)^*$.

Appendix B describes exterior forms, products and differentials in some detail. A function $f(x)$ on $x \in M^n$ is a zero-form. Differential of a function $f(x)$ is a typical example of the covector (1-form):

$$\mathrm{d}f = \frac{\partial f}{\partial x^i} \mathrm{d}x^i = \partial_i f \, \mathrm{d}x^i, \quad \partial_i f = \frac{\partial f}{\partial x^i}, \tag{1.27}$$

where $\mathrm{d}x^i$ is a basis covector and $\partial f / \partial x^i$ is its component. This form holds in any manifold. In the next subsection, a vector grad f is defined as one corresponding to the covector $\mathrm{d}f$.

---

[16] The differential one-form is called also *Pfaff form*, and the equation $a_1 \mathrm{d}x^1 + \cdots + a_n \mathrm{d}x^n = 0$ is called *Pfaffian equation* on M^n.

[17] In Eqs. (1.25) and (1.26), the Einstein's *summation convention* is used, i.e. a summation is implied over a pair of double indices (i in the above cases) appearing in a lower (covariant) and an upper (contravariant) index in a single term, and is used henceforth.

1.5.2. *Inner (scalar) product*

Let the vector space T_xM^n be endowed with an inner (scalar) product $\langle \cdot, \cdot \rangle$. For each pair of vectors $X, Y \in T_xM^n$, the inner product $\langle X, Y \rangle$ is a real number, and it is bilinear and symmetric with respect to X and Y. Furthermore, the $\langle X, Y \rangle$ is *nondegenerate* in the sense that

$$\langle X, Y \rangle = 0 \quad \text{for } {}^\forall Y \in T_xM^n, \quad \text{only if } X = 0.$$

Writing $X = X^i \partial_i$ and $Y = Y^j \partial_j$, the inner product is given by

$$\langle X, Y \rangle := g_{ij} X^i Y^j, \tag{1.28}$$

where

$$g_{ij} := \langle \partial_i, \partial_j \rangle = g_{ji} \tag{1.29}$$

is the *metric tensor*. If it happens that the tensor is the unit matrix,

$$g_{ij} = \delta_{ij}, \quad \text{i.e.} \quad g = (\delta_{ij}) = I, \tag{1.30}$$

we say that the metric tensor is the *euclidean metric*, where δ_{ij} is the Kronecker's delta: $\delta_{ij} = 1$ (if $i = j$), 0 (if $i \neq j$).

By definition, the inner product $\langle A, X \rangle$ is linear with respect to X when the vector A is fixed. Then the following α-operation on X, $\alpha[X] = \langle A, X \rangle$, is a linear functional: $\alpha = \langle A, \cdot \rangle$. In other words, to each vector $A = A^j \partial_j$, one may associate a covector α. By definition, $\alpha[X] = g_{ij} A^j X^i = (g_{ij} A^j) X^i$. On the other hand, for a covector of the form (1.24), one has $\alpha[X] = a_i \, \mathrm{d}x^i[X] = a_i X^i$, in terms of the basis $\mathrm{d}x^i$. Thus one obtains

$$a_i = g_{ij} A^j = g_{ji} A^j =: A_i, \tag{1.31}$$

which defines a covector A_i, and the component a_i is given by $g_{ij} A^j$ and written as A_i using the same letter A. The covector $\alpha = A_i \mathrm{d}x^i = (g_{ij} A^j) \mathrm{d}x^i$ is called the covariant version of the vector $A = A^j \partial_j$. In tensor analysis, Eq. (1.31) is understood as indicating that the upper index j is lowered by means of the metric tensor g_{ij}. In other words, *a covector A_i is obtained by lowering the upper index of a vector A^j by means of g_{ij}*. In summary, the inner product is represented as

$$\langle X, Y \rangle = g_{ij} X^i Y^j = X^i Y_i = X_j Y^j. \tag{1.32}$$

On the other hand, a vector A^j is obtained by raising the lower index of the covector A_i as

$$A^j = g^{ji} A_i, \tag{1.33}$$

which is equivalent to solving Eq. (1.31) to obtain A^j. This is verified by the property that the metric tensor $g = (g_{ij})$ is assumed nondegenerate, therefore the inverse matrix g^{-1} must exist and is symmetric. The inverse is written as $g^{-1} =: (g^{ji})$ in Eq. (1.33) using the same letter g. As an example, we obtain the expression of the vector grad f as

$$(\text{grad } f)^j = g^{ji} \frac{\partial f}{\partial x^i}. \tag{1.34}$$

1.6. Mapping of Vectors and Covectors

Dynamical development is a smooth sequence of maps from one state to another with respect to a parameter "time". Here we consider general rules of mappings (transformations).

1.6.1. *Push-forward transformation*

Let $\phi : M^n \to V^r$ be a smooth map. In addition, let us define the differential of the map ϕ by $\phi_* : T_x M^n \to T_y V^r$. In local coordinates, the map ϕ is represented by a function $F(x)$ as $y = \phi(x) = F(x)$, where $x \in M^n$ and $y \in V^r$. Let $p(t)$ be a curve on M^n with $p(0) = p$ and $\dot{p}(0) = X$ (a *tangent* vector), where $X \in T_p M^n$. The differential map ϕ_* at p is defined by

$$Y = \phi_* X (= F_* X) := \frac{\mathrm{d}}{\mathrm{d}t}(F(p(t))|_{t=0}. \tag{1.35}$$

This is called a *push-forward* transformation (Fig. 1.15) of the velocity vector X to the vector Y (the velocity vector of the image curve at $F(p)$).

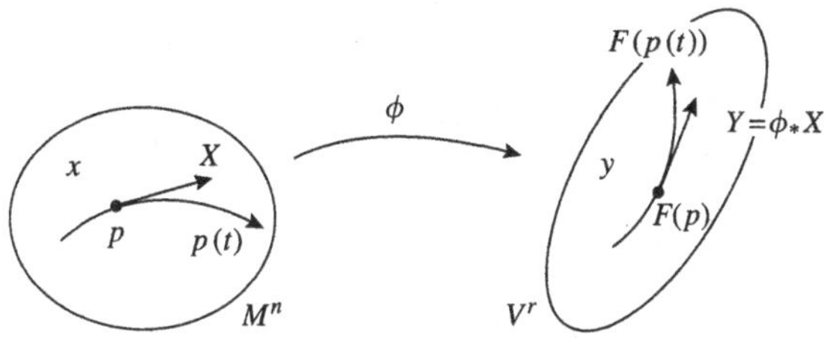

Fig. 1.15. Push-forward transformation ϕ by a function $y = F(x)$.

(a) Let us consider the case $n = r$. Suppose that the transformation is given by $x \mapsto y = (F^k(x))$ within the same reference frame ∂_k, and that the tangent vector $X = X^j \partial_j$ is mapped to $Y = Y^k \partial_k$. Then the components are transformed as (see §4.2.1)

$$Y^k = (\phi_* X)^k = \left(\frac{\partial F^k}{\partial x^j}\right) X^j. \tag{1.36}$$

(b) Next, consider a transformation between two basis vectors for $n = r$ again. The transformation ϕ_* applies to the basis vectors $\partial/\partial x^j$, and we have

$$Y = \phi_* X = \phi_* \left[X^j \frac{\partial}{\partial x^j}\right] = X^j \phi_* \left[\frac{\partial}{\partial x^j}\right] = X^j \frac{\partial y^k}{\partial x^j} \frac{\partial}{\partial y^k} = Y^k \frac{\partial}{\partial y^k}. \tag{1.37}$$

The components of Y are given by

$$Y^k = \frac{\partial y^k}{\partial x^j} X^j = J^k_j X^j, \quad J^k_j := \frac{\partial y^k}{\partial x^j}. \tag{1.38}$$

This is also written as $Y = JX$, where $J = (J^k_j)$.

In particular, setting $X^i = 1$ (for an integer i) and others as zero in (1.37), it is found that the bases $(\partial/\partial x^i)$ are transformed as

$$\phi_* \left[\frac{\partial}{\partial x^i}\right] = \frac{\partial y^k}{\partial x^i} \frac{\partial}{\partial y^k}. \tag{1.39}$$

If we write this in the form,

$$\frac{\partial}{\partial y^k} = B^i_k \frac{\partial}{\partial x^i}, \tag{1.40}$$

the matrix $B^i_k = \partial x^i / \partial y^k$ is the inverse of J since

$$BJ = \frac{\partial x^i}{\partial y^k} \frac{\partial y^k}{\partial x^j} = \frac{\partial x^i}{\partial x^j} = \delta^i_j = I. \tag{1.41}$$

A physical example of the transformations (a) and (b) is seen in §4.2.1 for rotations of a rigid body. Equation (1.39) is also written as

$$\phi_* \left[\frac{\partial}{\partial x^i}\right] f = \frac{\partial y^k}{\partial x^i} \frac{\partial f}{\partial y^k} = \frac{\partial}{\partial x^i} f(\phi(x)) \equiv \frac{\partial}{\partial x^i} f \circ \phi(x). \tag{1.42}$$

Writing as $X = X^j(x) \partial/\partial x^j$,

$$\phi_* X[f] = X[f \circ \phi]. \tag{1.43}$$

(c) A manifold M^n is called a *submanifold* of a manifold V^r (where $n < r$) provided that there is a one-to-one smooth mapping $\phi : M^n \to V^r$ in which the matrix J has (maximal) rank n at each point. We refer to ϕ as an *imbedding* or an *injection*. This appears often when $V = \mathbb{R}^r$ so that we consider submanifolds of an euclidean space $\mathbb{R}^r$.

1.6.2. *Pull-back transformation*

Corresponding to the *push-forward* ϕ_*, one can define the *pull-back* ϕ^*, which is the linear transformation taking a covector at y back to a covector at x, i.e. $\phi^* : (T_y V)^* \to (T_x M)^*$. Suppose that a vector X at $x \in M$ is transformed to $Y = \phi_*(X)$ at $y = \phi(x) \in V$, then the pull-back ϕ^* of a covector α (one-form) is defined, using the push-forward $\phi_*(X)$, by

$$(\phi^*\alpha)[X] := \alpha[\phi_*(X)], \tag{1.44}$$

for any one-form $\alpha = A_i \mathrm{d}y^i$. This defines an invariance of the pull-back transformation. Namely, the value of the covector $\alpha = A_i \mathrm{d}y^i$ at the vector $Y = \phi_* X$ (in V) is equal to the value of the pull-back covector $\phi^*\alpha$ at the original vector X (in M).

Note that, owing to $\mathrm{d}x^i[\partial_j] = \delta^i_j$, one has

$$\alpha\left[\frac{\partial}{\partial y^k}\right] = A_i \mathrm{d}y^i \left[\frac{\partial}{\partial y^k}\right] = A_k. \tag{1.45}$$

Writing

$$\phi^*\alpha = a_i\, \mathrm{d}x^i, \tag{1.46}$$

one obtains $a_i = \phi^*\alpha[\partial/\partial x^i]$, and furthermore one can derive the following transformation of the components of covectors by using (1.39) and (1.45):

$$\begin{aligned} a_i = \phi^*\alpha\left[\frac{\partial}{\partial x^i}\right] &= \alpha\left[\phi_* \frac{\partial}{\partial x^i}\right] = \alpha\left[\frac{\partial y^k}{\partial x^i}\frac{\partial}{\partial y^k}\right] \\ &= \frac{\partial y^k}{\partial x^i}\alpha\left[\frac{\partial}{\partial y^k}\right] = A_k \frac{\partial y^k}{\partial x^i}. \end{aligned} \tag{1.47}$$

Thus, using J^k_i of (1.38), we have the transformation law,

$$a_i = A_k J^k_i. \tag{1.48}$$

Substituting the expression $A_k \mathrm{d}y^k$ for α in (1.46) and using (1.47), we have

$$\phi^*(A_k \mathrm{d}y^k) = A_k \frac{\partial y^k}{\partial x^j} \mathrm{d}x^j. \tag{1.49}$$

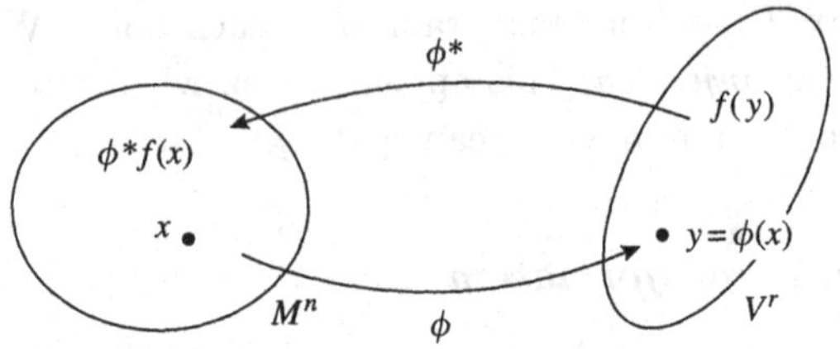

Fig. 1.16. Pull-back of a function $f(y)$ to $(\phi^* f)(x)$.

Setting A_i (only) $= 1$ (the other components being zero) as before (for an integer k), it is found that the bases $(\mathrm{d}y^i)$ are transformed as

$$\phi^*[\mathrm{d}y^i] = \frac{\partial y^i}{\partial x^j}\mathrm{d}x^j. \tag{1.50}$$

The *pull-back of a function* $f(y)$ (Fig. 1.16) is given by

$$(\phi^* f)(x) = f(\phi(x)),$$

where a scalar function $f(y)$ is a zero-form. If one sets $A_i = \partial f/\partial y^i$ in (1.44), Eq. (1.44) expresses invariance of the differential:

$$\begin{aligned}\phi^*(\mathrm{d}f)_y = \phi^*\left[\left(\frac{\partial f}{\partial y^i}\right)\mathrm{d}y^i\right] &= \left(\frac{\partial f}{\partial y^i}\right)\left(\frac{\partial y^i}{\partial x^j}\right)\mathrm{d}x^j \\ &= \left(\frac{\partial f}{\partial x^j}\right)\mathrm{d}x^j = (\mathrm{d}f)_x.\end{aligned}$$

Based on this invariance, the general pull-back formula is defined for the integral of a form (covector) α over a curve σ as

$$\int_{\phi(\sigma)} \alpha = \int_\sigma \phi^*\alpha, \tag{1.51}$$

where $\phi : \sigma \subset M \to \phi(\sigma) \subset V$. Namely, the integral of a form α over the image $\phi(\sigma)$ is the integral of the pull-back $\phi^*\alpha$ over the original σ. See the next section 1.6.3 for $M = V$, and Appendix B.7 for an integral of a general form α.

1.6.3. *Coordinate transformation*

Change of coordinate frame can be regarded as a mapping $y = y(x) : x \in U^n \to y \in V^n$, where (U^n, x) and (V^n, y) are two identical coordinate patches. A same vector is denoted by $X = (X^j)$ in U^n and by $Y = (Y^k)$

in V^n. Transformation of the components of the same vector $X = Y$ is described by Eq. (1.38) (using W in place of J):

$$Y^k = W_j^k X^j, \quad W_j^k = \frac{\partial y^k}{\partial x^j}, \tag{1.52}$$

which is equivalent to (1.17). Correspondingly, transformation of bases is described by (1.39):

$$\frac{\partial}{\partial x^j} = W_j^k \frac{\partial}{\partial y^k}, \quad \text{or} \quad \frac{\partial}{\partial y^k} = (W^{-1})_k^j \frac{\partial}{\partial x^j}, \tag{1.53}$$

where $W = (W_j^k)$. It is easy to see the identity: $Y^k \partial/\partial y^k = X^j \partial/\partial x^j$.

The transformation (1.18) corresponds to Eq. (1.48) which describes the transformation of components of a covector. Solving (1.48) for A_k, we obtain

$$A_k = a_i \, (W^{-1})_k^i. \tag{1.54}$$

Thus, we find the invariance of inner product:

$$A_k Y^k = a_i (W^{-1})_k^i W_j^k X^j = a_i \delta_j^i X^j = a_i X^i. \tag{1.55}$$

1.7. Lie Group and Invariant Vector Fields

Dynamical evolution of a physical system is described by a trajectory over a manifold, which is often represented by a space of Lie group, a symmetry group of the system. This and the following section are a concise account of some aspects of the theory of Lie group and Lie algebra related to the present subject.

We consider various Lie groups G associated with various physical systems below. In abstract terms, a group **G** of smooth transformations (maps) of a manifold M into itself is called a group, provided that (i) with two maps $g, h \in G$, the product $gh = g \circ h$ belongs to G: $G \times G \to G$, (ii) for every $g \in G$, there is an inverse map $g^{-1} \in G$. From (i) and (ii), it follows that the group contains an *identity* map id, which is often called *unity* denoted by e. Thus, $gg^{-1} = g^{-1}g = e$.

A *Lie group* is a group which is a differentiable manifold, for which the operations (i) and (ii) are differentiable. Some lists of typical Lie groups are given in Appendix C. A Lie group always has two families of diffeomorphisms, the left and right translations. Namely, with a fixed element $h \in G$,

$$L_h(g) = hg \quad (\text{or } R_h(g) = gh), \quad \text{for any } g \in G,$$

where $\mathbf{L}_h$ (or $\mathbf{R}_h$) denotes the *left-* (or *right-*) translation of the group onto itself, respectively. Note that $L_g(h) = R_h(g) = gh$. The operation inverse to L_h (or R_h) is simply $L_{h^{-1}}$ (or $R_{h^{-1}}$), respectively.

Suppose that g_t is a curve on G described in terms of a parameter t. The left translation of g_t by $g_{\Delta t}$ for an infinitesimal Δt is given by $g_{\Delta t} \circ g_t$. Hence the t-derivative is expressed as

$$\dot{g}_t = \lim_{\Delta t \to 0} \frac{g_{t+\Delta t} - g_t}{\Delta t} = X \circ g_t, \quad X = \lim_{\Delta t \to 0} \frac{g_{\Delta t} - id}{\Delta t}, \tag{1.56}$$

where id denotes the identity map. Thus, the *left*-translation leads to the *right*-invariant vector field [AzIz95] in the sense defined just below. Similarly, the *right*-translation leads to the *left*-invariant vector field. The $\dot{g}_t$ is said to be a tangent vector at a point g_t.

A vector field X^{L}, or X^{R} on G is *left-invariant*, or *right-invariant*, if it is invariant under all left-translations, or right-translations respectively, namely for all $g, h \in G$, if

$$(\mathbf{L}_h)_* X_g^{\mathrm{L}} = X_{hg}^{\mathrm{L}}, \quad \text{or} \quad (\mathbf{R}_h)_* X_g^{\mathrm{R}} = X_{gh}^{\mathrm{R}}, \tag{1.57}$$

respectively. Given a tangent vector X to G at e, one may left-translate or right-translate X to every point $g \in G$ as

$$X_g^{\mathrm{L}} = (L_g)_* X = g \circ X = gX, \tag{1.58}$$

$$X_g^{\mathrm{R}} = (R_g)_* X = X \circ g = Xg, \tag{1.59}$$

respectively. It is readily seen from (1.59) that $(R_h)_* X_g^{\mathrm{R}} = X_{gh}^{\mathrm{R}}$, hence the transformation (1.59) gives a right-invariant field generated by X. Similarly, the transformation (1.58) gives a left-invariant field.

Consider a curve $\xi_t : t \in \mathbb{R} \to G$ with the tangent $\dot{\xi}_0 = X$ at $t = 0$. The left-invariant field is given by $X_s^L = (\mathrm{d}/\mathrm{d}t)(g_s \circ \xi_t)$ for $g_s \in G$ (s: a parameter), whereas the right-invariant vector field is represented by $X_s^R = (\mathrm{d}/\mathrm{d}t)(\xi_t \circ g_s)$. Examples of such invariant fields are given by (3.86) and (3.87) in §3.7.3.

The left-translation $(L_g)_* X$ is understood as a transformation of a vector X located at x under the push-forward to gX at $g(x)$ (Fig. 1.17(a)). On the other hand, the right-translation $(R_g)_* X = X \circ g\ (x)$ is understood as follows: first let the map g act on the point x and then the vector X is taken at the point $g(x)$ (Fig. 1.17(b)). This is something like a change of variables when g is an element of a transformation group.

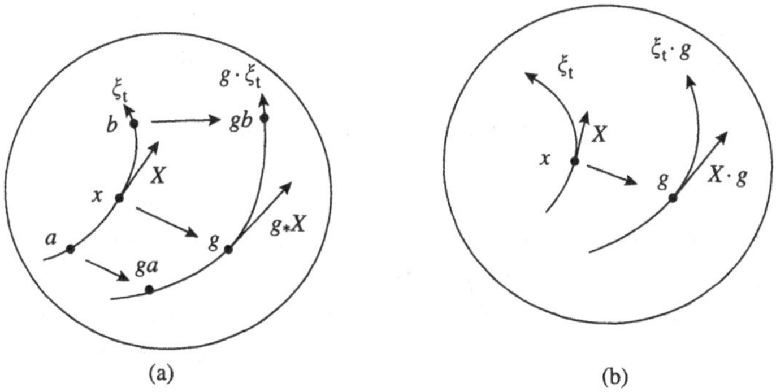

Fig. 1.17. (a) Left- and (b) Right-translation of a vector field X.

For two right-invariant tangent vectors X_g^{R} and Y_g^{R}, the metric (1.28) is called *right-invariant* if

$$\langle X_g^{\mathrm{R}}, Y_g^{\mathrm{R}}\rangle = \langle X_e, Y_e\rangle.$$

Similarly, the metric is *left-invariant* if $\langle X_g^{\mathrm{L}}, Y_g^{\mathrm{L}}\rangle = \langle X_e, Y_e\rangle$ for left-invariant vectors, $X_g^{\mathrm{L}}, Y_g^{\mathrm{L}}$. Examples of the left-invariant field are given in Chapters 4 and 9.

1.8. Lie Algebra and Lie Derivative

1.8.1. *Lie algebra, adjoint operator and Lie bracket*

Every pair of vector fields defines a new vector field called the Lie bracket $[\cdot,\cdot]$. More precisely, the tangent space $\mathbf{T}_e\mathbf{G}$ at the identity e of a Lie group G is called the *Lie algebra* g of the group G. The *Lie algebra* g $(= T_eG)$ is equipped with the bracket operation $[\cdot,\cdot]$ of bilinear skew-symmetric pairing, $[\cdot,\cdot] : \mathsf{g}\times\mathsf{g} \to \mathsf{g}$, defined below. The bracket satisfies the *Jacobi identity*,

$$[[X,Y],Z] + [[Y,Z],X] + [[Z,X],Y] = 0, \tag{1.60}$$

for any triplet of $X, Y, Z \in \mathsf{g}$.

Any element of the Lie algebra $X \in \mathsf{g}$ defines a one-parameter subgroup (Appendix C.2, Eq. (C.4)):

$$\xi_t = \exp[tX] = e + tX + \frac{1}{2!}t^2X^2 + O(t^3), \quad X \in \mathsf{g} \tag{1.61}$$

where ξ_t is a curve $t \to G$ with the tangent $\dot{\xi}_0 = X \in \mathsf{g}$ at $t = 0$. In this sense, the element X is called an (infinitesimal) *generator* of the subgroup.

A Lie group G acts as a group of linear transformations on its own Lie algebra g. Namely for $^\forall g \in G$, there is an operator Ad_g, such that

$$Ad_g Y := (L_g)_* \circ (R_{g^{-1}})_* Y = gYg^{-1}, \tag{1.62}$$

for $^\forall Y \in$ g.[18] The operator Ad_g transforms $Y \in$ g into $Ad_g Y \in$ g linearly (Fig. 1.18). The set of all such Ad_g, i.e. $Ad(G)$, is called the *adjoint* representation of G, an *adjoint group*. Setting g with the inverse $\xi_t^{-1} := (\xi_t)^{-1}$, the adjoint transformation $Ad_{\xi_t^{-1}} Y$ is a function of t. Its derivative with respect to t is a linear transformation from Y to $ad_X Y$ defined by

$$ad_X Y = \frac{\mathrm{d}}{\mathrm{d}t} \xi_t^{-1} Y \xi_t \bigg|_{t=0} := [X, Y]. \tag{1.63}$$

This defines the Lie bracket $[X, Y]$.[19] Its explicit expression depends on each group or each dynamical system considered. It can be shown that the bracket $[X, Y]$ thus introduced satisfies all the properties required for the Lie bracket in each example considered below. The bracket operation is usually called the *commutator*. The ad_X is a linear transformation, g $\to$ g,

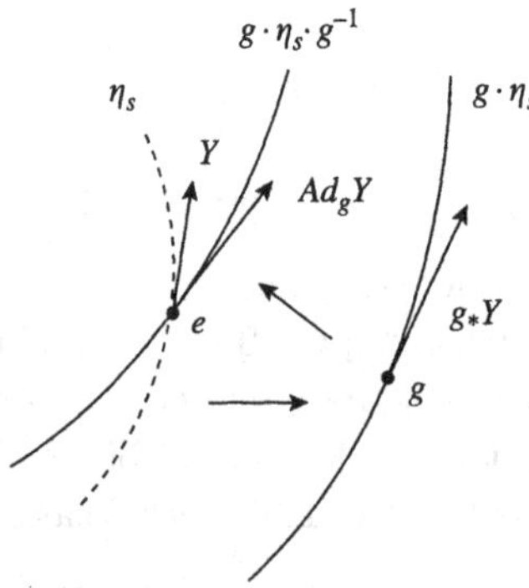

Fig. 1.18. Adjoint transformation $Ad_g Y$, where $Y = \mathrm{d}\eta_s/\mathrm{d}s|_{s=0}$ and $Ad_g Y = (\mathrm{d}/\mathrm{d}s)(g \circ \eta_s \circ g^{-1})|_{s=0}$.

[18] g is the Lie algebra and g is an element of the group G. The operator gYg^{-1} may be better written as the push-forward notation, $g_* Y g^{-1}$.

[19] Most textbooks in mathematics adopt this definition. Arnold [Arn66; Arn78] uses the definition of its opposite sign which is convenient for physical systems related to rotation group (see (1.64)) in Chapters 4 and 9, characterized with the left-invariant metric. In fact, the difference between the left-invariant and right-invariant field, (1.58) and (1.59), results in different signs of the Lie bracket of right- and left-invariant fields (see [AzIz95]).

by the representation, $Y \to ad_X Y = [X, Y]$. The operator ad_X stands for the image of an element X under the linear *ad*-action.

1.8.2. *An example of the rotation group SO(3)*

Consider the rotation group $G = \mathbf{SO(3)}$. Any element $A \in SO(3)$ is represented by a 3×3 orthogonal matrix ($AA^T = I$) of det $A = 1$ (Appendix C), where A^T denotes transpose of A, i.e. $(A^T)^i_k = A^k_i$. Let $K = (\partial_x, \partial_y, \partial_z)$ be a cartesian right-handed frame. By the element A, the coordinate frame K is transformed to another frame $K' = (\partial_{x'}, \partial_{y'}, \partial_{z'}) = AK$, and a point $X = (x, y, z)$ in K is transformed to $X' = (x', y', z') = WX$ by the rules in §1.6.3, where $W = (A^{-1})^T$. Then, we have

$$(X')^T K' = (WX)^T AK = X^T W^T AK = X^T A^{-1} AK = XK.$$

Consider successive transformations $A' = A_2 A_1$, i.e. A_1 followed by A_2. Then we have $(X')^T = X^T (A_2 A_1)^{-1} = X^T A_1^{-1} A_2^{-1}$, that is, the components X evolve by the right-translation, resulting in the left-invariant vector field (§1.6).[20]

Let $\xi(t)$ be a curve (one-parameter subgroup) issuing from $e = \xi(0)$ with a tangent vector $\mathbf{a} = \dot{\xi}(0)$ on $SO(3)$. Then one has $\xi(t) = \exp[t\mathbf{a}] = e + t\mathbf{a} + O(t^2)$ for an infinitesimal parameter (time) t, where $\mathbf{a}$ is an element of the algebra $\mathbf{g}$ (usually written as $\mathbf{so}(3)$) and a skew-symmetric matrix due to the orthogonality of $\xi(t)$ (Fig. 1.19).

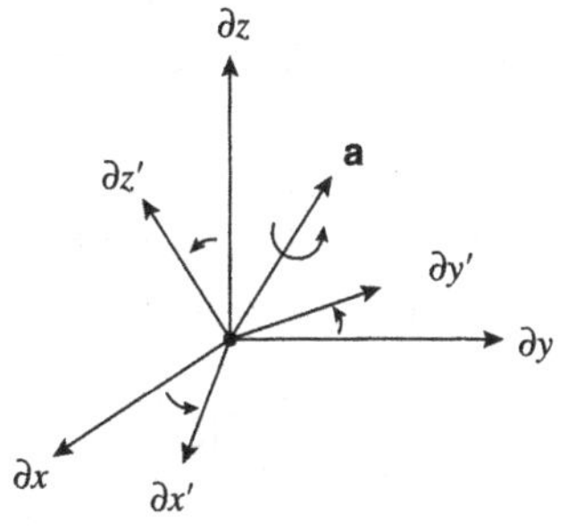

Fig. 1.19. $K = (\partial_x, \partial_y, \partial_z), K' = \xi(t)K$ with $e = \xi(0), \mathbf{a} = \dot{\xi}(0)$.

[20]The length of vector is also invariant by this transformation, i.e. *isometry*, since $\langle X', X' \rangle = \langle X, (W^T W) X \rangle = \langle X, X \rangle$, where $W^T W = (A^{-1})^T A^{-1} = (AA^T)^{-1} = I$.

Then, for $\mathbf{a}, \forall\mathbf{b} \in \mathbf{so}(3)$, the operation $ad_{\mathbf{a}} : \mathrm{g} \to \mathrm{g}$ is represented by

$$ad_{\mathbf{a}}\mathbf{b} = [\mathbf{a}, \mathbf{b}] = -(\mathbf{ab} - \mathbf{ba}), \tag{1.64}$$

where the minus sign in front of $(\mathbf{ab} - \mathbf{ba})$ is due to the definition (1.63). This is verified as follows. Since $\xi(t)^{-1} = \exp[-t\mathbf{a}] = (e - t\mathbf{a} + \cdots)$, we have

$$\begin{aligned} \xi(t)^{-1}\mathbf{b}\xi(t) &= (e - t\mathbf{a} + \cdots)\mathbf{b}(e + t\mathbf{a} + \cdots) \\ &= \mathbf{b} - t(\mathbf{ab} - \mathbf{ba}) + O(t^2). \end{aligned} \tag{1.65}$$

Its differentiation with respect to t results in Eq. (1.64).

In Chapter 4, we consider time trajectories over the rotation group $SO(3)$ such as $\xi(t)$ with time t. In such a case, it is convenient to define the bracket $[\mathbf{a}, \mathbf{b}]^{(\mathrm{L})}$ for the left-invariant field defined as

$$[\mathbf{a}, \mathbf{b}]^{(\mathrm{L})} := -[\mathbf{a}, \mathbf{b}] = \mathbf{ab} - \mathbf{ba} = \mathbf{c}. \tag{1.66}$$

In Appendix C.3, it is shown that, for $\mathbf{a}, \mathbf{b} \in \mathbf{so}(3)$, $\mathbf{c}$ is also skew-symmetric, and that the matrix equation $\mathbf{ab} - \mathbf{ba} = \mathbf{c}$ is equivalent to the cross-product equation (C.14),

$$\hat{\mathbf{c}} = \hat{\mathbf{a}} \times \hat{\mathbf{b}}, \tag{1.67}$$

where $\hat{\mathbf{a}}, \hat{\mathbf{b}}$ and $\hat{\mathbf{c}}$ are three-component (axial) vectors associated with the skew-symmetric matrices $\mathbf{a}$, $\mathbf{b}$ and $\mathbf{c}$, respectively.

1.8.3. *Lie derivative and Lagrange derivative*

(a) **Derivative of a scalar function $f(x)$**

Suppose that a vector field $X = X^i\partial_i$ is given on a manifold M^n. As described in §1.2, with every such vector field, one can associate a *flow*, or one-parameter group of *diffeomorphisms* $\xi_t : M^n \to M^n$, for which $\xi_0 = id$[21] and $(\mathrm{d}/\mathrm{d}t)\xi_t x|_{t=0} = X(x)$. A first order differential operator $\mathcal{L}_X$ on a scalar function $f(x)$ on M (a function of coordinates x only) is defined as

$$\mathcal{L}_X f(x) := \frac{\mathrm{d}}{\mathrm{d}t}(\xi_t)^* f(x)\bigg|_{t=0} = \frac{\mathrm{d}}{\mathrm{d}t} f(\xi_t x)\bigg|_{t=0} = X^i \frac{\partial}{\partial x^i} f(x), \tag{1.68}$$

(see (1.44) and below, and (1.5)). This defines the derivative $\mathcal{L}_X f$ of a function f (a zero-form) by the time derivative of its pull-back $\xi_t^* f$ at $t = 0$, where the point $\xi_t x$ moves forward in accordance with the flow of

[21] The *id* is used here in order to emphasize that this is an *identity map*.

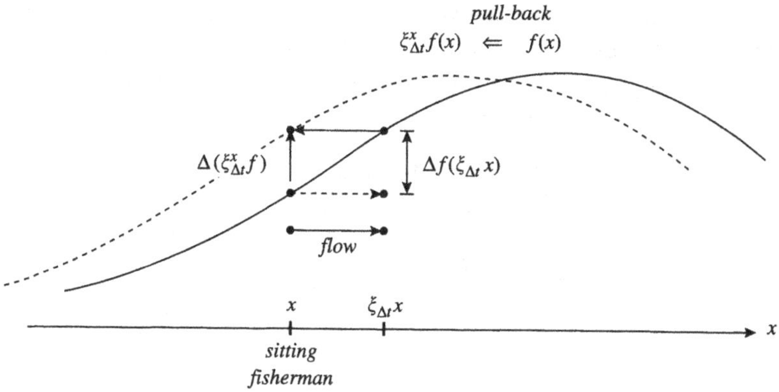

Fig. 1.20. Fisherman's derivative.

velocity X. Relatively observing, the pull-back $\xi_t^* f$ is evaluated at x, and its time derivative is defined by the Lie derivative. This is sometimes called as a *derivative of a fisherman* [AK98] sitting at a fixed place x (Fig. 1.20).[22]

In *fluid dynamics* however, the same derivative is called the *Lagrange* derivative, which refers to the third and fourth expressions,

$$\frac{\mathrm{D}f}{\mathrm{D}t} := \frac{\mathrm{d}}{\mathrm{d}t} f(\xi_t x) = X^i \frac{\partial}{\partial x^i} f(x).$$

Therefore we obtain that $(\mathrm{D}f/\mathrm{D}t)f = \mathcal{L}_X f$, which is valid for scalar functions. But this does not hold for vectors, as shown in the next subsection.

In the unsteady problem, the right-hand side is written as $(\partial_t + X^i \partial_i) f(x,t)$. The Lagrange derivative is understood as denoting the time derivative, with respect to the fluid particle $\xi_t x$ moving with the flow, of the function $f(x,t)$.

(b) **Derivatives of a vector field $Y(x)$**

Now, suppose that we are given a second vector field $Y(x) = Y^i \partial_i$, and consider its time derivative along the X-flow generated by $X(x)$. To that end, let us denote the second Y-flow generated by $Y(x)$ as η_s with $\eta_0 = e = id$. The first flow ξ_t transports the vector $Y(x)$ in front of a fisherman sitting at a point x. After an infinitesimal time t, the fluid particle at x

[22] The Lie derivative $\mathcal{L}_X$ also acts on any form field α in the same way, $(\mathrm{d}/\mathrm{d}t)(\xi_t)^* \alpha$ as (1.78). On the contrary, to a vector field Y, the Lie derivative is defined by (1.69) in terms of the push-forward $(\xi_t)_*$. This definition is different from that of Arnold [1978, 1966] by the sign, but consistent with the present definition of (1.63).

will arrive at $\xi_t x$. We take the vector Y at this point $\xi_t x$ and translate it backwards to the original point x by the inverse map of the push-forward, that is $(\xi_t)^{-1}Y(\xi_t x)$, in precise $(\xi_t)_*^{-1}Y(\xi_t x)$. Its time derivative is the *Lie derivative of a vector* Y, given by

$$\begin{aligned} \mathcal{L}_X Y := \lim_{t\to 0} \frac{\xi_{t*}^{-1}Y(\xi_t x) - Y(x)}{t} &= \lim_{t\to 0} \xi_{t*}^{-1} \left. \frac{Y\xi_t - \xi_{t*}Y}{t} \right|_x \\ &= \lim_{t\to 0} \frac{1}{t}(Y\xi_t - \xi_{t*}Y). \end{aligned} \tag{1.69}$$

The first expression is nothing but that of $ad_X Y(x)$ according to (1.63). Thus we have

$$\mathcal{L}_X Y = \left. \frac{\mathrm{d}}{\mathrm{d}t} \xi_{t*}^{-1} Y \xi_t \right|_{t=0} = ad_X Y = [X, Y], \tag{1.70}$$

where $[X, Y]$ is the *Lie bracket* (see (1.63)).

The last expression of (1.69) suggests another useful expression of $[X, Y]$, which is given by

$$\begin{aligned} \mathcal{L}_X Y = [X, Y] := \lim_{t\to 0, s\to 0} \frac{1}{st}(\eta_s \xi_t - \xi_t \eta_s) \\ = \frac{\partial}{\partial t}\frac{\partial}{\partial s} \left. (\eta_s \xi_t - \xi_t \eta_s) \right|_{t=0, s=0}. \end{aligned} \tag{1.71}$$

According to Appendix C, the two flows ξ_t and η_s generated by X and Y can be written in the form [AK98, §2]:

$$\xi_t : x \mapsto x + tX(x) + O(t^2), \quad t \to 0, \tag{1.72}$$

$$\eta_s : x \mapsto x + sY(x) + O(s^2), \quad s \to 0. \tag{1.73}$$

Recalling that $\eta_s \xi_t(x)$ for diffeomorphisms is given by $\eta_s(\xi_t(x))$, i.e. the composition rule, we have

$$\begin{aligned} \eta_s \xi_t &= e + tX + O(t^2) + sY(\xi_t) + O(s^2) \\ &= e + tX + sY + stX^j \partial_j Y + O(t^2, s^2), \end{aligned} \tag{1.74}$$

where $\xi_t = e + tX + O(t^2)$. The expression of $\xi_t \eta_s$ is obtained by exchanging the pairs (t, X) and (s, Y). Thus finally, we have

$$\eta_s \xi_t - \xi_t \eta_s = st \left(X^j \frac{\partial Y}{\partial x^j} - Y^j \frac{\partial X}{\partial x^j} \right) + O(st^2, s^2 t). \tag{1.75}$$

The first term may be written as $st[X,Y]$ according to (1.71). Thus the non-commutativity of two diffeomorphisms ξ_t and η_s is proportional to $[X,Y]$, where

$$[X,Y] = \{X,Y\} := \{X,Y\}^k\partial_k, \tag{1.76}$$

$$\{X,Y\}^k := X^j\frac{\partial Y^k}{\partial x^j} - Y^j\frac{\partial X^k}{\partial x^j} \tag{1.77}$$

and $\{X,Y\}$ is the *Poisson bracket.* The degree of non-commutativity of ξ_t and η_s is interpreted graphically in Fig. 1.21. According to the definition (1.70), by using the expression $L_X := X^i\partial_i$, the Lie derivative of the vector field Y with respect to X is given by

$$\mathcal{L}_X Y = [X,Y] = L_X L_Y - L_Y L_X = [L_X, L_Y] = L_{\{X,Y\}}. \tag{1.78}$$

If they commute, i.e. $\xi_t\eta_s = \eta_s\xi_t$, then obviously we have $[X,Y]=0$. This suggests that the coordinate bases commute since the coordinate curves are defined to intersect. In fact, for $X = \partial_\alpha, Y = \partial_\beta$, we obtain from (1.78)

$$[\partial_\alpha, \partial_\beta] = \partial_\alpha\partial_\beta - \partial_\beta\partial_\alpha = 0. \tag{1.79}$$

In general, we have

$$\xi_t\eta_s - \eta_s\xi_t = [X,Y]st + O(st^2, s^2t).$$

If $[X,Y]=0$, then we obtain $\xi_t\eta_s - \eta_s\xi_t = O(st^2, s^2t)$.

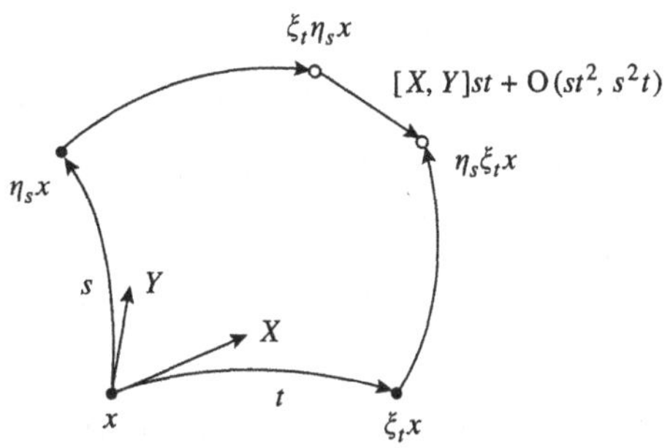

Fig. 1.21. Graphic interpretation of $[X,Y]$ for infinitesimal s and t.

In unsteady problem of *fluid dynamics*, the Lagrange derivative of the vector $Y = Y^k(x,t)\partial_k$ is defined by

$$\frac{\mathrm{D}}{\mathrm{D}t}Y = \frac{\mathrm{D}}{\mathrm{D}t}Y^k(\xi_t x)\partial_k := \frac{\partial Y^k}{\partial t}\partial_k + X^j\frac{\partial Y^k}{\partial x^j}\partial_k. \tag{1.80}$$

This derivative makes sense in the gauge-theoretical formulation described in §7.5 and denotes the derivative following a fluid particle moving with the velocity $X^j\partial_j$, whereas the Lie derivative characterizes a frozen field (see the remark just below).

Remark. A vector field Y defined along the integral curve ξ_t generated by the tangent field X is said to be *invariant* if $Y(\xi_t x) = (\xi_t)_* Y(x)$. Substituting this in the previous expression of (1.69), it is readily seen that $\mathcal{L}_X Y = 0$, or rewriting it,

$$\mathcal{L}_X Y = \left(X^j\frac{\partial Y^i}{\partial x^j} - Y^j\frac{\partial X^i}{\partial x^j}\right)\partial_i = 0. \tag{1.81}$$

In unsteady problem, $X^j\partial_j Y^i$ is also written as $\mathrm{D}Y^i/\mathrm{D}t$ given by the right-hand side of (1.80). Then, using the operator $\mathrm{D}/\mathrm{D}t = \partial_t + X^j\partial_j$, the above equation (1.81) becomes

$$\frac{\mathrm{D}}{\mathrm{D}t}Y = (Y^j\partial_j)X. \tag{1.82}$$

In *fluid dynamics*, the equation $\mathcal{L}_X Y = 0$ is called the equation of *frozen field* (Fig. 1.22).[23] If we set $\phi = \xi_t$ in (1.37) together with $X = Y(x)$ and $Y = Y(\xi_t x)$, then the equation $Y(\xi_t x) = (\xi_t)_* Y(x)$ represents the

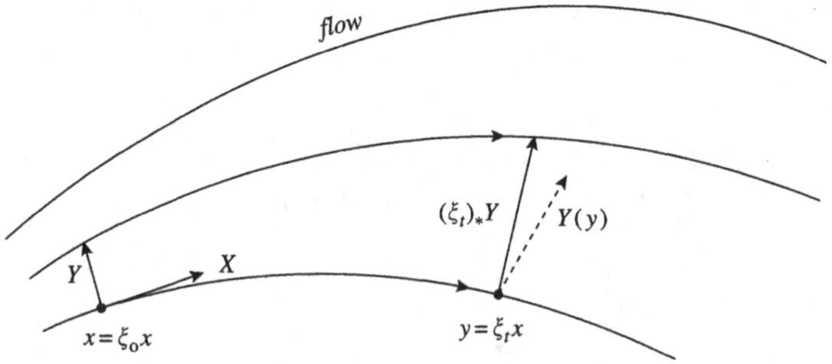

Fig. 1.22. Frozen field $(\xi_t)_* Y$ (push-forward of Y) coincides with $Y(y)$ at $y = \xi_t x$.

[23]The Jacobi field Y ($= J$ below) satisfies this equation (see §8.4).

push-forward transformation. Therefore, writing $\xi_t x = y_t(x)$, the solution of (1.82) is given by Eq. (1.38),

$$Y^\alpha(t) = Y^j(0)\frac{\partial y_t^\alpha}{\partial x^j}, \tag{1.83}$$

which is called the *Cauchy's solution* [Cau1816] in the fluid dynamics.

1.9. Diffeomorphisms of a Circle S^1

A smooth sequence of diffeomorphisms is a mathematical concept of a flow and the unit circle S^1 is one of the simplest base manifolds for physical fields.

Diffeomorphism of the manifold S^1 (a unit circle in $\mathbb{R}^2$, see Fig. 1.3) is represented by a map $g : x \in \mathbb{R}^1 \to g(x) \in \mathbb{R}^1$ (where $g \in C^\infty$) with every point of x or $g(x)$ is identified with $x+1$ or $g(x)+1$ respectively.[24] Collection of all such maps constitutes a group $\mathcal{D}(S^1)$ of diffeomorphisms with the composition law:

$$h = g \circ f, \quad \text{i.e.} \quad h(x) = g(f(x)) \in \mathcal{D}(S^1),$$

for $f, g \in \mathcal{D}(S^1)$. The diffeomorphism is a map of infinite degrees of freedom (i.e. having *pointwise* degrees of freedom). In Chapter 5, the diffeomorphism is assumed to be orientation-preserving in the sense that $g'(x) > 0$, where the prime denotes $\partial/\partial x$.

Consider a flow $\xi_t(x)$ which is a smooth sequence of diffeomorphisms with the time parameter as t (see (1.72)). Its tangent field at ξ_t is defined by

$$\dot{\xi}_t(x) := \frac{\mathrm{d}}{\mathrm{d}t}\xi_t(x)\bigg|_t = \lim_{\tau\to 0}\frac{\xi_\tau(x) - id}{\tau} \circ \xi_t(x) = u(x) \circ \xi_t(x),$$

in a right-invariant form. The tangent field $X(x)$ at the identity (id) is given by $u(x) = \mathrm{d}\xi_t(x)/\mathrm{d}t|_{t=0}$.

Alternatively, with the language of differentiable manifolds, the tangent field $X(x)$ is represented as

$$X(x) = u(x)\partial_x \in TS^1, \tag{1.84}$$

where TS^1 is a tangent bundle (§1.3.1) over the manifold S^1. The tangent bundle TS^1 allows a global product structure $S^1 \times \mathbb{R}^1$ as shown in Fig. 1.23(a). The figure (b) is obtained from (a) by cutting it along one

[24]By the map $\phi(x) = e^{i2\pi x}$, there is a perioditity $\phi(x+1) = \phi(x)$ for $x \in \mathbb{R}^1$.

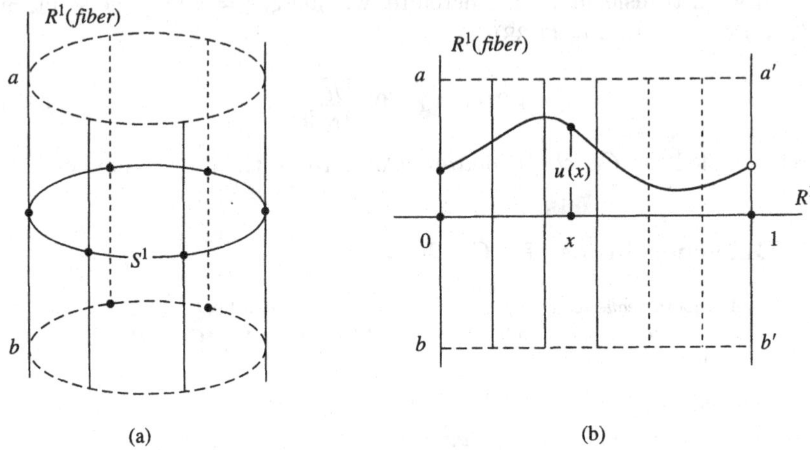

Fig. 1.23. Tangent bundle $S^1 \times \mathbb{R}^1$ with the circle S^1 and the fiber $\mathbb{R}^1$.

fiber ab and developing it flat, where $a'b'$ is identified with ab. The solid curve in the figure describes a particular vector field $u(x)$ on S^1, which is called a *cross-section* of the tangent bundle TS^1.

If $a'b'$ is identified with ba by twisting the strip, then a Möbius band is formed. The resulting fiber bundle is not trivial, i.e. not a product space (see, e.g. [Sch80]). The Möbius band is a *double-fold cover*, i.e. two-sheeted cover of the circle S^1 (see Fig. 1.4).

For two diffeomorphisms ξ_t and η_t corresponding to the vector fields X and Y respectively, the *Lie bracket* (commutator) is given by (1.76) and (1.77):

$$[X, Y] = (uv' - vu')\partial_x, \tag{1.85}$$

where $X = u(x)\partial_x, Y = v(x)\partial_x \in TS^1$. This is sometimes called *Witt algebra* [AzIz95].

1.10. Transformation of Tensors and Invariance

1.10.1. *Transformations of vectors and metric tensors*

We considered the transformations of vectors and covectors in §1.4 (see also (1.10)) together with the invariance of the value of covectors based on (1.44).

Here, we consider a transformation and an invariant property of tensors.[25] Let M be an n-manifold with a Riemannian metric and covered with a family of local (curvilinear) coordinate systems, $\{U : x^1, \dots, x^n\}$, $\{V : y^1, \dots, y^n\}, \dots$, where $U, V, \dots$ are open sets (called *patches*) with coordinates $x, y, \dots$. A point $p \in U \cap V$, lying in two overlapping patches U and V, has two sets of coordinates $x_{(p)}$ and $y_{(p)}$ which are related differentiably by the functions $y^k(x)$:

$$y^k_{(p)} = y^k(x^1_{(p)}, \dots, x^n_{(p)}), \quad k = 1, \dots, n.$$

In the corresponding tangent spaces, the vectors are represented as $X = X^i \partial/\partial x^i \in T_pU$ and $Y = Y^k \partial/\partial y^k \in T_pV$. The coordinate bases are transformed according to

$$\frac{\partial}{\partial x^i} = \frac{\partial y^k}{\partial x^i}\frac{\partial}{\partial y^k} = W^k_i \frac{\partial}{\partial y^k}, \quad \text{where} \quad W^k_i = \frac{\partial y^k}{\partial x^i}, \tag{1.86}$$

by the chain rule (in an analogous way to (1.39)), W^k_i being the transformation matrix. Suppose that the components of the vectors are related by

$$Y^k = W^k_i X^i, \quad (\text{written as } Y = WX), \tag{1.87}$$

as is the case of the push-forward transformation (1.38), then the vectors are invariant in the sense:

$$X = X^i \frac{\partial}{\partial x^i} = X^i W^k_i \frac{\partial}{\partial y^k} = Y^k \frac{\partial}{\partial y^k} = Y.$$

Equation (1.87) is the rule of transformation of *vector components*. In physical problems, the logic is reversed. The vector, like the velocity vector of a particle, should be the same (may be written as $X = Y$) in both coordinate frames. Then the components must be transformed according to the rule (1.87).

The metric tensor is defined by (1.29). According to the basis transformation (1.86), we obtain

$$\begin{aligned} g_{ij}(x) &= \left\langle \frac{\partial}{\partial x^i}, \frac{\partial}{\partial x^j} \right\rangle = \left\langle W^k_i \frac{\partial}{\partial y^k}, W^l_j \frac{\partial}{\partial y^l} \right\rangle \\ &= W^k_i W^l_j \left\langle \frac{\partial}{\partial y^k}, \frac{\partial}{\partial y^l} \right\rangle = W^k_i W^l_j g_{kl}(y). \end{aligned} \tag{1.88}$$

[25] See §3.11 for differentiation of tensors.

This is the transformation rule of the tensor g_{ij}. Using (1.87) for the transformation of two pairs (X, Y) and (ξ, η) of tangent vectors, where $X, \xi \in T_pU$ and $Y, \eta \in T_pV$, we have the invariance of the inner product with the transformation (1.88):

$$\begin{aligned} G(\xi, X) = \langle \xi, X\rangle(x) = g_{ij}(x)\xi^i X^j &= W_i^k W_j^l g_{kl}(y)\xi^i X^j \\ &= g_{kl}(y)\eta^k Y^l = \langle \eta, Y\rangle(y), \end{aligned} \tag{1.89}$$

where $Y^k = W_i^k X^i$ and $\eta^l = W_j^l \xi^j$.

1.10.2. *Covariant tensors*

The inner product $G(\xi, X)$ in the previous section is an example of covariant tensor of rank 2. In general, a *covariant tensor* of rank n is defined by

$$Q^{(n)} : E_1 \times E_2 \times \cdots \times E_n \to \mathbb{R},$$

a multilinear real-valued function of n-tuple vectors, written as $Q^{(n)}(\boldsymbol{v}_1, \ldots, \boldsymbol{v}_n)$ which is linear in each entry $\boldsymbol{v}_i$ $(i = 1, \ldots, n)$, where E_k is the tangent vector space for the kth entry.

A covector $\alpha = a_i \mathrm{d}x^i$ on a vector $\boldsymbol{v} = v^j \partial_j = v^j \partial/\partial x^j$ is an example of $Q^{(1)}$, a covariant tensor of rank 1. In fact, we have $\alpha(\boldsymbol{v}) = a_i v^j \mathrm{d}x^i(\partial_j) = a_i v^i$. An example of $Q^{(2)}$ is $G(A, X) = g_{ij} A^i X^j$. Both are shown to be invariant with the coordinate transformation (see §1.6.3 for $\alpha(\boldsymbol{v})$).

In general, the values of $Q^{(n)}$ must be independent of the basis with respect to which components of the vectors are expressed. In components, we have

$$\begin{aligned} \bar{Q}^{(n)}(x) := Q^{(n)}(\boldsymbol{v}_1(x), \ldots, \boldsymbol{v}_n(x)) &= Q^{(n)}(v_1^{k_1}\partial_{k_1}, \ldots, v_n^{k_n}\partial_{k_n}) \\ &= Q^{(n)}_{k_1,\ldots,k_n}(x) v_1^{k_1} \cdots v_n^{k_n} \end{aligned}$$

at $x \in U$, where

$$Q^{(n)}_{k_1,\ldots,k_n}(x) = Q^{(n)}(\partial/\partial x^{k_1}, \ldots, \partial/\partial x^{k_n}).$$

Considering that the bases are transformed according to (1.86) and $Q^{(n)}$ is multilinear, we have the transformation rule,

$$Q^{(n)}_{k_1,\ldots,k_n}(x) = W_{k_1}^{l_1} \cdots W_{k_n}^{l_n} Q^{(n)}_{l_1,\ldots,l_n}(y). \tag{1.90}$$

Owing to the transformation (1.87), it is obvious that we have the invariance, $\bar{Q}^{(n)}(x) = \bar{Q}^{(n)}(y)$.

From two covectors $\alpha = a_i \mathrm{d}x^i$ and $\beta = b_j \mathrm{d}x^j$, one can form a covariant tensor of rank 2 by the *tensor product* $\otimes$ as follows: $\alpha \otimes \beta : E \times E \to \mathbb{R}$, defined by

$$\begin{aligned} \alpha \otimes \beta(\boldsymbol{v}, \boldsymbol{w}) &:= \alpha(\boldsymbol{v})\beta(\boldsymbol{w}) = a_i \mathrm{d}x^i \otimes b_j \mathrm{d}x^j(\boldsymbol{v}, \boldsymbol{w}) \\ &= Q_{kl} v^k w^l, \qquad (1.91) \\ Q_{kl} = Q^{(2)}&(\partial_k, \partial_l) = a_i b_j \mathrm{d}x^i \otimes \mathrm{d}x^j(\partial_k, \partial_l) \end{aligned}$$

where $\boldsymbol{v} = v^k \partial_k, \boldsymbol{w} = w^l \partial_l \in E$.

1.10.3. *Mixed tensors*

A mixed tensor of rank 2 is defined by

$$M^i_j(x) = M^{(2)}\left(\mathrm{d}x^i, \frac{\partial}{\partial x^j}\right),$$

which is a first order covariant and first order contravariant tensor. According to (1.50), a 1-form base $\mathrm{d}x^i$ is transformed as

$$\mathrm{d}x^i = \frac{\partial x^i}{\partial y^j}\,\mathrm{d}y^j = \hat{W}^i_j\,\mathrm{d}y^j, \quad \text{where} \quad \hat{W}^i_j := \partial x^i/\partial y^j. \qquad (1.92)$$

Thus, using (1.86) and (1.92), we obtain the transformation rule of the mixed tensor M:

$$M^i_j(x) = W^\beta_j \hat{W}^i_\alpha M^\alpha_\beta(y). \qquad (1.93)$$

The transformation matrix $\hat{W} = \partial x/\partial y$ is the inverse of $W = \partial y/\partial x$, i.e. $\hat{W} = W^{-1}$, since one can verify $W\hat{W} = I$, i.e.

$$(W\hat{W})^k_j = \frac{\partial y^k}{\partial x^\beta}\frac{\partial x^\beta}{\partial y^j} = \frac{\partial y^k}{\partial y^j} = \delta^k_j.$$

Let us consider such mixed tensors through several examples.

(i) *Transformation*: A mixed tensor $M^{(2)}$ of rank 2 arises from the matrix of a linear transformation $W = (W^k_i)$. In the coordinate patch V, taking a covariant vector $\alpha = A_i \mathrm{d}y^i \in E^*$ (cotangent space, §1.5.1) and a contravariant vector $Y = Y^k \partial_k \in E$, the mixed tensor $M^{(2)} : E^* \times E \to R$ is defined by $M^{(2)}(\alpha, Y) \equiv \alpha[Y] = A_i Y^i$. Next, consider the transformation ϕ and its matrix $W(= \phi_*) : E(U) \to E(V)$, i.e. $Y = WX$ defined by (1.86) and

(1.87). The corresponding pull-back is given by $\phi^*\alpha = a_i \mathrm{d}x^i$ (see (1.46)), and the component a_i is expressed by (1.47):

$$a_i = A_j \frac{\partial y^j}{\partial x^i} = A_j W_i^j. \tag{1.94}$$

Thus, we have the invariance of the value of the mixed tensor as follows (using $Y^j = W_i^j X^i$):

$$M_W^{(2)}(\alpha, Y) := \alpha[Y] = A_j W_i^j X^i = a_i X^i = \phi^*\alpha[X]. \tag{1.95}$$

According to (1.93), the transformation of the tensor $W = (W_\beta^\alpha)$ is $(W_j^\beta \hat{W}_\alpha^i)\ W_\beta^\alpha = W_j^i$. Namely, the transformation of W is an identity: $W \to W$.

(ii) *Vector-valued one-form*: Next example of the mixed tensor is the tensor product, $M^{(2)} = \boldsymbol{v} \otimes \alpha : E^* \times E \to \mathbb{R}$, of a vector and a covector, defined by

$$\begin{aligned} M_V^{(2)}(\beta, \boldsymbol{w}) &:= \boldsymbol{v} \otimes \alpha(\beta, \boldsymbol{w}) = v^j \partial_j \otimes a_i \mathrm{d}x^i(\beta, \boldsymbol{w}) \\ &= \partial_j(\beta) v^j a_i \mathrm{d}x^i(\boldsymbol{w}) = b_j M_i^j w^i, \end{aligned} \tag{1.96}$$

where $\boldsymbol{w} = w^i \partial_i, \beta = b_i \mathrm{d}x^i$ and $M_i^j = v^j a_i$. The value of the tensor $\boldsymbol{v} \otimes \alpha$ on a vector $X = X^i \partial_i$ takes the value of a vector (rather than a scalar) as follows:

$$M_V^{(2)}(X) = \boldsymbol{v} \otimes \alpha(X) = \boldsymbol{v} \otimes a_i \mathrm{d}x^i(X) = \boldsymbol{v} a_i X^i. \tag{1.97}$$

In this sense, $M_V^{(2)} = \boldsymbol{v} \otimes \alpha$ is interpreted also as a *vector-valued 1-form*.

In particular, the following I is the identity mixed-tensor:

$$I := \partial_i \otimes \mathrm{d}x^i. \tag{1.98}$$

In fact, we have

$$I(X) = I(X^\alpha \partial_\alpha) = \partial_i \otimes \mathrm{d}x^i(X^\alpha \partial_\alpha) = X^\alpha \delta_\alpha^i \partial_i = X.$$

(iii) *Covariant derivative*: The third example is the covariant differentiation ∇, which is an essential building block in the differential geometry and also in *Physics*, and investigated in the subsequent chapters as well.

Consider a vector $X = X^k \partial_k$ and the transformation $X = \hat{W} Y$ with $\hat{W}_i^k = \partial x^k / \partial y^i$. It may appear that, just like the tensor $W_i^k = \partial y^k / \partial x^i$, the derivative $\partial X^k/\partial x^j$ is also a mixed tensor. But this is not the case. In fact, we have

$$\frac{\partial X^k}{\partial x^j} = \frac{\partial}{\partial x^j} Y^\alpha \hat{W}^k_\alpha = \frac{\partial}{\partial x^j}\left(Y^\alpha \frac{\partial x^k}{\partial y^\alpha}\right)$$
$$= \frac{\partial Y^\alpha}{\partial y^\beta}\frac{\partial y^\beta}{\partial x^j}\frac{\partial x^k}{\partial y^\alpha} + Y^\alpha \frac{\partial^2 x^k}{\partial y^\alpha \partial y^\beta}\frac{\partial y^\beta}{\partial x^j} = W^\beta_j \hat{W}^k_\alpha \frac{\partial Y^\alpha}{\partial y^\beta} + W^\beta_j \frac{\partial^2 x^k}{\partial y^\alpha \partial y^\beta} Y^\alpha. \tag{1.99}$$

Only the first term follows the transformation rule (1.93), while the second does not. In order to overcome this difficulty, the differential geometry introduces the following *linear* operator ∇_{∂_j} on the product of a scalar X^k and a vector ∂_k, defined by

$$\nabla_{\partial_j}(X^k \partial_k) = (\nabla_{\partial_j} X^k)\partial_k + X^\alpha(\nabla_{\partial_j}\partial_\alpha)$$
$$:= \frac{\partial X^k}{\partial x^j}\partial_k + X^\alpha \Gamma^k_{j\alpha}\partial_k, \tag{1.100}$$

where Γ^k_{ij} is the *Christoffel symbol*, which can be represented in terms of the derivatives of the metric tensors $g_{\alpha\beta}$ (see §2.4 and 3.3.2). From (1.100), the following mixed tensor is defined:

$$X^k_{;j} := \frac{\partial X^k}{\partial x^j} + \Gamma^k_{j\alpha} X^\alpha. \tag{1.101}$$

In fact, it can be verified that this tensor is transformed like a mixed tensor according to $X^k_{;j}(x) = X^\alpha_{;\beta}(y) W^\beta_j \hat{W}^k_\alpha$ (see e.g. [Eis47]).

In order to write it in the form of a vector-valued 1-form just as (1.97), it is useful to define

$$\nabla X = \partial_k \otimes (\nabla X^k), \tag{1.102}$$
$$\nabla X^k = \mathrm{d}X^k + \Gamma^k_{i\alpha} X^\alpha \mathrm{d}x^i = \frac{\partial X^k}{\partial x^i}\mathrm{d}x^i + \Gamma^k_{i\alpha} X^\alpha \mathrm{d}x^i = X^k_{;i}\mathrm{d}x^i.$$

Then the value of ∇X on a vector $v = v^j \partial_j$ is found to be a vector, which is given by

$$\nabla X(v) = \partial_k \otimes \nabla X^k(v^j \partial_j) = v^j \nabla X^k(\partial_j)\partial_k$$
$$= \left(v^j \frac{\partial X^k}{\partial x^j} + \Gamma^k_{j\alpha} X^\alpha v^j\right)\partial_k = v^j X^k_{;j}\partial_k. \tag{1.103}$$

The operator ∇X^k is called a *connection 1-form* (see §3.5), and $\nabla X(v)$ is called the covariant derivative of X with respect to the vector v.

(iv) *Riemann tensors*: In the differential geometry, a fourth order tensor called the Riemann's curvature tensor plays a central role. This is defined as $R^l_{ijk} = \partial_j \Gamma^l_{ik} - \partial_k \Gamma^l_{jl} + \Gamma^l_{jm}\Gamma^m_{ki} - \Gamma^l_{km}\Gamma^m_{ji}$ (see §2.4 and 3.9.2). It can be verified (e.g. [Eis47]) that this tensor is transformed according to

$$R^l_{ijk}(x) = R^\delta_{\alpha\beta\gamma}(y)\hat{W}^l_\delta W^\alpha_i W^\beta_j W^\gamma_k, \tag{1.104}$$

showing that R^l_{ijk} is a mixed tensor of rank 4, the third order covariant and the first order contravariant tensor.

(v) *General mixed tensor*: In general, a *mixed tensor* of rank n is defined by

$$M^{(n)} : E_1^* \times \cdots \times E_q^* \times E_1 \times \cdots \times E_p \to \mathbb{R}.$$

This is a p times covariant and q times contravariant tensor $(p + q = n)$ and a multilinear real-valued function of p-tuple vectors and q-tuple covectors, written as $M^{(n)}(\alpha_1, \ldots, \alpha_q, \boldsymbol{v}_1, \ldots, \boldsymbol{v}_p)$, which is linear in each entry α_i $(i = 1, \ldots, q)$ and $\boldsymbol{v}_i$ $(i = 1, \ldots, p)$. The values of $M^{(n)}$ is independent of the basis by which the components of the vectors are expressed. In components, we have

$$\hat{M}^{(n)}(x) = M^{(n)}(\alpha_1, \ldots, \alpha_q, \boldsymbol{v}_1, \ldots, \boldsymbol{v}_p) = a_{1k_1} \cdots a_{qk_q} M^{k_1 \cdots k_q}_{l_1 \cdots l_p} v_1^{l_1} \cdots v_p^{l_p},$$

where

$$M^{k_1 \cdots k_q}_{l_1 \cdots l_p} = M^{(n)}(\mathrm{d}x^{k_1}, \ldots, \mathrm{d}x^{k_q}, \partial_{l_1}, \ldots, \partial_{l_p}).$$

1.10.4. *Contravariant tensors*

In the second example (ii) of the mixed tensor, we obtained the expression, $M^{(2)} = b_j M^j_i w^i$. According to the rule (1.31) of §1.3, the lower-index component b_j is related to the upper-index component B^k (the vector counterpart of b_j) by means of the metric tensor g_{jk} as $b_j = g_{jk}B^k$. Similarly, according to (1.33), the upper-index component w^i is related with its covector counter part $W = W_l \mathrm{d}x^l$ as $w^i = g^{il}W_l$ by means of the inverse of the metric tensor $g^{il} = (g^{-1})^{il}$. Hence, we have

$$M^{(2)} = b_j M^j_i w^i = (g_{jk}M^j_i)B^k w^i = (g^{il}M^j_i)b_j W_l. \tag{1.105}$$

Thus it is found that a covariant tensor M_{ki} of rank 2 is obtained by lowering the upper index of the mixed tensor of rank 2:

$$M_{ki} = g_{jk}M^j_i.$$

Similarly, a contravariant tensor M^{jl} of rank 2 is obtained by raising the lower index:

$$M^{jl} = g^{il} M_i^j.$$

In this way, we have found the equivalence:

$$M^{jl} b_j W_l = M_i^j b_j w^i = M_{ki} B^k w^i.$$

In tensor analysis, one can use the same letter M for the derived tensors by lowering or raising the indices by means of the metric tensor.

In general, a *contravariant tensor* of rank n is defined by

$$P^{(n)} : E_1^* \times E_2^* \times \cdots \times E_n^* \to \mathbb{R},$$

a multilinear real-valued function of n-tuple covectors, written as $P^{(n)}(\alpha_1, \dots, \alpha_n)$ which is linear in each entry α_i $(i = 1, \dots, n)$. The values of $P^{(n)}$ is independent of the basis. In components, we have

$$\bar{P}^{(n)} = P(\alpha_1, \dots, \alpha_n) = a_{1\ k_1} \cdots a_{k\ k_n} P^{k_1, \dots, k_n},$$

where

$$P^{k_1, \dots, k_n} = P(\mathrm{d}x^{k_1}, \dots, \mathrm{d}x^{k_n}).$$

From the two vectors $\boldsymbol{v} = v^i \partial_i$ and $\boldsymbol{w} = w^j \partial_j$, one can form a contravariant tensor of rank 2 by the *tensor product*: $\boldsymbol{v} \otimes \boldsymbol{w}$, defined by

$$\boldsymbol{v} \otimes \boldsymbol{w}\ (\mathrm{d}x^i, \mathrm{d}x^j) := \mathrm{d}x^i(\boldsymbol{v})\, \mathrm{d}x^j(\boldsymbol{w}) = v^i w^j. \tag{1.106}$$

Chapter 2

Geometry of Surfaces in $\mathbb{R}^3$

Most concepts in differential geometry spring from the geometry of surfaces in $\mathbb{R}^3$.

2.1. First Fundamental Form

Consider a two-dimensional surface Σ^2 parameterized with u^1 and u^2 in the three-dimensional euclidean space $\mathbb{R}^3$. A point p on $\Sigma^2 \in \mathbb{R}^3$ is denoted by a three-component vector $\boldsymbol{x}(u^1, u^2)$, which is represented as $\boldsymbol{x} = x\mathbf{i}+y\mathbf{j}+z\mathbf{k} = x^1\mathbf{i}+x^2\mathbf{j}+x^3\mathbf{k}$ with respect to a *cartesian* frame with the orthonormal right-handed basis $(\mathbf{i}, \mathbf{j}, \mathbf{k})$ (Fig. 2.1). The vector $\boldsymbol{x}$ is also written as

$$\boldsymbol{x}(u^1, u^2) = (x^1(u^1, u^2), x^2(u^1, u^2), x^3(u^1, u^2)), \tag{2.1}$$

where $(u^1, u^2) \in U \subset \mathbb{R}^2$. A surface defined by $(x, y, f(x, y))$ is considered in Appendix D.3.

A curve C lying on Σ^2, denoted by $\boldsymbol{x}(t)$ with the parameter t, is the image of the curve C_U: $u^\alpha(t)$ on U. The tangent vector $\boldsymbol{T}$ to the curve C is given by

$$\boldsymbol{T} = \frac{\mathrm{d}\boldsymbol{x}}{\mathrm{d}t} = \frac{\mathrm{d}u^\alpha}{\mathrm{d}t}\boldsymbol{x}_\alpha = \dot{u}^\alpha \boldsymbol{x}_\alpha = \dot{u}^\alpha \partial_\alpha,$$

where

$$\boldsymbol{x}_\alpha = (x^i_\alpha) := \frac{\partial \boldsymbol{x}}{\partial u^\alpha} = \boldsymbol{x}_{u^\alpha} = \partial_\alpha.$$

The two tangent vectors $(\boldsymbol{x}_1, \boldsymbol{x}_2)$ form a basis for the tangent space $T_p\Sigma^2$ to Σ^2 (Fig. 2.2).

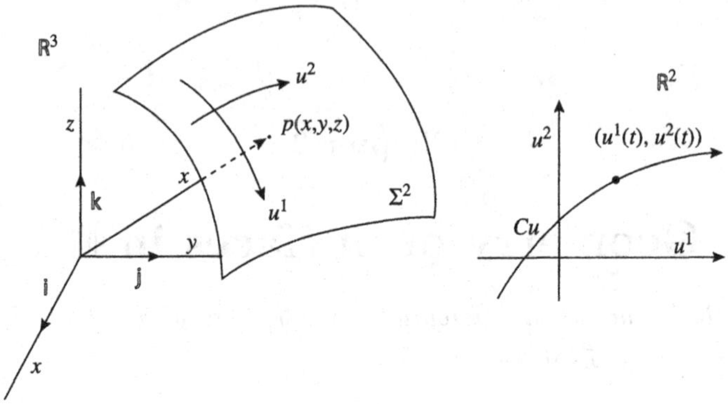

Fig. 2.1. Surface Σ^2 in $\mathbb{R}^3$.

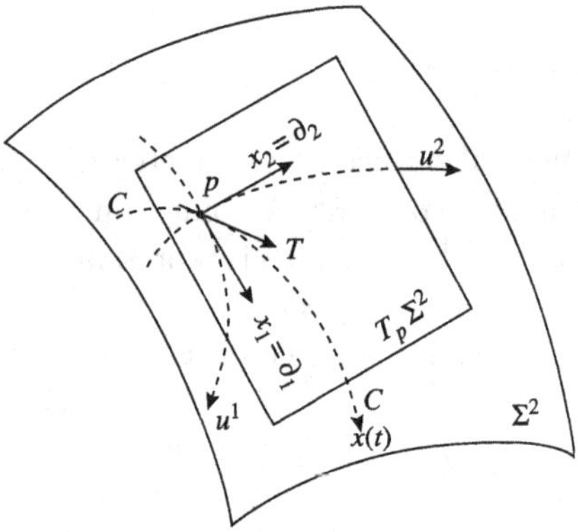

Fig. 2.2. Tangent space $T_p\Sigma^2$.

A scalar product in $\mathbb{R}^3$ is defined for a pair of general tangent vectors $\mathbf{A} = A^\alpha \boldsymbol{x}_\alpha = A^\alpha \partial_\alpha$ and $\mathbf{B} = B^\alpha \boldsymbol{x}_\alpha = B^\alpha \partial_\alpha$ as

$$\langle \mathbf{A}, \mathbf{B} \rangle = \langle A^\alpha \boldsymbol{x}_\alpha, B^\beta \boldsymbol{x}_\beta \rangle = g_{\alpha\beta} A^\alpha B^\beta, \tag{2.2}$$

where the *metric tensor* $g = \{g_{\alpha\beta}\}$ is defined by

$$g_{\alpha\beta}(\Sigma^2) = \langle \partial_\alpha, \partial_\beta \rangle = \langle \boldsymbol{x}_\alpha, \boldsymbol{x}_\beta \rangle := (\boldsymbol{x}_\alpha, \boldsymbol{x}_\beta)_{R^3} \tag{2.3}$$

$$(\boldsymbol{x}_\alpha, \boldsymbol{x}_\beta)_{R^3} := \sum_{i=1}^{3} \delta_{ij}\, x^i_\alpha\, x^j_\beta = \sum_{i=1}^{3} \frac{\partial x^i}{\partial u^\alpha} \frac{\partial x^i}{\partial u^\beta}, \tag{2.4}$$

where the symbol $(\cdot,\cdot)_{R^3}$ is defined by (2.4), the scalar product with the euclidean metric (§1.5.2) in $\mathbb{R}^3$. This is so that the euclidean metric *induces* the metric tensor $g_{\alpha\beta}$ on the surface Σ^2. The metric tensor is obviously symmetric by the definition (2.4): $g_{\alpha\beta} = g_{\beta\alpha}$.

Denoting a differential 1-form as $\mathrm{d}\boldsymbol{x} = \boldsymbol{x}_\alpha \mathrm{d}u^\alpha$, the *first fundamental form* is defined by

$$\begin{aligned} \mathrm{I} &:= \langle \mathrm{d}\boldsymbol{x}, \mathrm{d}\boldsymbol{x}\rangle = \langle \boldsymbol{x}_\alpha \mathrm{d}u^\alpha, \boldsymbol{x}_\beta \mathrm{d}u^\beta\rangle = g_{\alpha\beta}\mathrm{d}u^\alpha \mathrm{d}u^\beta \\ &= g_{11}\mathrm{d}u^1\mathrm{d}u^1 + 2g_{12}\mathrm{d}u^1\mathrm{d}u^2 + g_{22}\mathrm{d}u^2\mathrm{d}u^2. \end{aligned} \tag{2.5}$$

Note that $\mathrm{d}u^\alpha$ is a 1-form (§1.5.1), i.e. $\mathrm{d}u^\alpha(\mathbf{A}) = \mathrm{d}u^\alpha(A^\beta \boldsymbol{x}_\beta) = A^\alpha$. The first fundamental form $\mathrm{I} = \langle \mathrm{d}\boldsymbol{x}, \mathrm{d}\boldsymbol{x}\rangle$ is a quadratic form associated with the metric. This quadratic form I is interpreted, in a mathematical term, as a second-rank covariant tensor (§1.10.2) taking a real value on a pair of vectors **A** and **B**,

$$\begin{aligned} \mathrm{I}(\mathbf{A},\mathbf{B}) &= g_{\alpha\beta}\mathrm{d}u^\alpha \otimes \mathrm{d}u^\beta(\mathbf{A},\mathbf{B}) \\ &= g_{\alpha\beta}\mathrm{d}u^\alpha(\mathbf{A}) \otimes \mathrm{d}u^\beta(\mathbf{B}) = g_{\alpha\beta}A^\alpha B^\beta, \end{aligned} \tag{2.6}$$

by using the definition (1.91) of the tensor product $\otimes$. In this context, $\mathrm{d}\boldsymbol{x}$ is a 1-form yielding a vector, defined by a mixed tensor: $\mathrm{d}\boldsymbol{x} = \boldsymbol{x}_\alpha \otimes \mathrm{d}u^\alpha$, and we have

$$\mathrm{d}\boldsymbol{x}(\mathbf{A}) = \boldsymbol{x}_\alpha \otimes \mathrm{d}u^\alpha(\mathbf{A}) = \boldsymbol{x}_\alpha A^\alpha = \mathbf{A}. \tag{2.7}$$

Thus, *eating* a vector **A**, $\mathrm{d}\boldsymbol{x}$ *yields* the same vector **A**. In this language, $\mathrm{d}\boldsymbol{x}$ is not an *infinitesimal* increment vector, but a *vector-valued 1-form* (§1.10.3(ii)).

Example. *Torus* (I). Consider a surface of revolution called a *torus* Σ^{tor}, obtained by rotating a circle C of radius a about the z-axis, where the circle C is in the x–z plane with its center located at $(x,y,z) = (R,0,0)$ with $R > a$ (Fig. 2.3). The surface is represented as $\boldsymbol{x}^{\mathrm{tor}}(u,v) = (x(u,v), y(u,v), z(u,v))$, where

$$\boldsymbol{x}^{\mathrm{tor}} = (x,y,z) = ((R + a\cos u)\cos v, (R + a\cos u)\sin v, a\sin u). \tag{2.8}$$

The parameter u is an angle-parameter of the circle C (e.g. $u \in [0,2\pi]$) and v denotes the angle of rotation of C around the z-axis (e.g. $v \in [0,2\pi]$). In this sense, the surface Σ^{tor} is denoted as T^2 in a standard notation: $T^2([0,2\pi],[0,2\pi])$ with 2π being identified with 0 for both parameters.

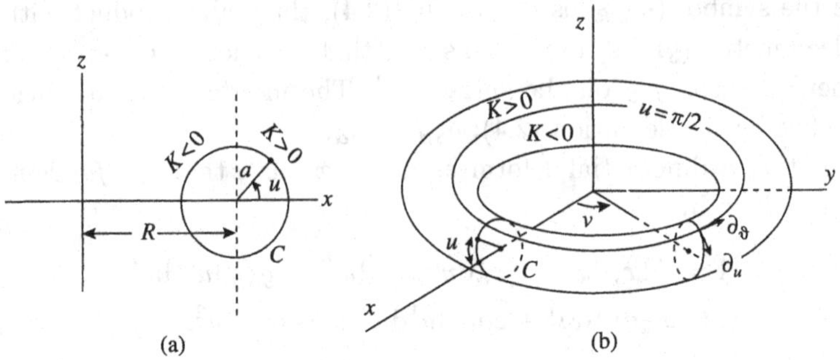

Fig. 2.3. Torus. $(\boldsymbol{x}_u, \boldsymbol{x}_v) = (\partial_u, \partial_v)$.

The basis vectors to $T_p\Sigma_{\text{tor}}$ are

$$\boldsymbol{x}_u = \partial_u \boldsymbol{x}^{\text{tor}} = (-a \sin u \cos v, -a \sin u \sin v, a \cos u), \tag{2.9}$$

$$\boldsymbol{x}_v = \partial_v \boldsymbol{x}^{\text{tor}} = (-(R + a\cos u)\sin v, (R + a \cos u)\cos v, 0). \tag{2.10}$$

The metric tensors $g_{\alpha\beta} = \langle \boldsymbol{x}_\alpha, \boldsymbol{x}_\beta \rangle$ are

$$g_{uu} = a^2, \quad g_{uv} = 0, \quad g_{vv} = (R + a\cos u)^2, \tag{2.11}$$

which are independent of v, and g_{uu} is a constant. The first fundamental form is

$$\mathrm{I} = a^2(\mathrm{d}u)^2 + (R + a\cos u)^2(\mathrm{d}v)^2. \tag{2.12}$$

□

A line element of the curve $\boldsymbol{x}(t)$ (on Σ^2) for an increment Δt is denoted as $\Delta\boldsymbol{x} = (\Delta u^\alpha)\boldsymbol{x}_\alpha$, and its length Δs is defined by

$$\Delta s = \|\Delta \boldsymbol{x}\| := \langle \Delta\boldsymbol{x}, \Delta\boldsymbol{x}\rangle^{1/2} \tag{2.13}$$

$$\langle \Delta\boldsymbol{x}, \Delta\boldsymbol{x}\rangle = g_{\alpha\beta}(\Delta u^\alpha)(\Delta u^\beta) \tag{2.14}$$

(Fig. 2.4). Thus the length of an infinitesimal element of u^1-coordinate curve $(\Delta_1 u^1, 0)$ is given by $(\Delta_1 s)^2 = g_{11}(\Delta_1 u^1)^2$. Similarly, the length of an infinitesimal element $(0, \Delta_2 u^2)$ is given by $(\Delta_2 s)^2 = g_{22}(\Delta_2 u^2)^2$. Consequently, we have $g_{11} > 0$ and $g_{22} > 0$.

The angle θ between two tangent vectors $\mathbf{A}$ and $\mathbf{B}$ is defined by

$$\cos\theta = \frac{\langle \mathbf{A}, \mathbf{B}\rangle}{\|\mathbf{A}\| \, \|\mathbf{B}\|}. \tag{2.15}$$

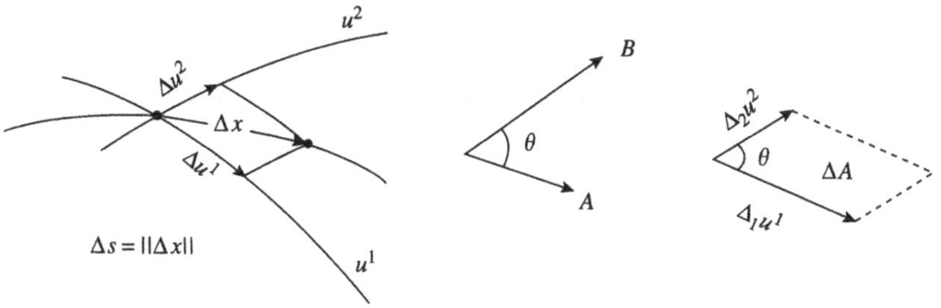

Fig. 2.4. Length, angle and area over Σ^2, $(\Delta s)^2 = g_{11}(\Delta u^1)^2 + 2g_{12}\Delta u^1 \Delta u^2 + g_{22}(\Delta u^2)^2$.

So the intersecting angle θ of two coordinate curves $u^1 = \text{const}$ and $u^2 = \text{const}$ is given by

$$\cos\theta = \frac{\langle \boldsymbol{x}_1, \boldsymbol{x}_2 \rangle \Delta_1 u^1 \Delta_2 u^2}{(\Delta_1 s)(\Delta_2 s)} = \frac{g_{12}}{\sqrt{g_{11}}\sqrt{g_{22}}}. \tag{2.16}$$

Hence, it is necessary for *orthogonality* of two coordinate curves on a surface to be $g_{12} = 0$ at each point. It is also the sufficient condition.

Consider a small *parallelogram* spanned by two infinitesimal line-elements $(\Delta_1 u^1, 0)$ and $(0, \Delta_2 u^2)$. Its area ΔA is given by

$$\Delta A = \sin\theta(\sqrt{g_{11}}\Delta_1 u^1)(\sqrt{g_{22}}\Delta_2 u^2) = \sqrt{g}\Delta_1 u^1 \Delta_2 u^2, \tag{2.17}$$

where $g = g_{11}g_{22} - (g_{12})^2 = \det(g_{\alpha\beta})$, where g can be shown to be positive.[1]

The metric properties of a surface, such as the length, angle and area, can be expressed completely by means of the first fundamental form of the surface. It is true that these quantities are embedded in the enveloping space $\mathbb{R}^3$, where the metric tensor is euclidean, i.e. $g_{ij}(R^3) = \delta_{ij}$ (Kronecker's delta) with $i, j = 1, 2, 3$. In this sense, the metric $g_{\alpha\beta}(\Sigma^2) = (\boldsymbol{x}_\alpha, \boldsymbol{x}_\beta)_{R^3}$ of the surface Σ^2 is *induced* by the *euclidean* metric of the space $\mathbb{R}^3$. Consider two surfaces, and suppose that there exists a coordinate system on each surface which gives an identical first fundamental form for the two surfaces, i.e. their metric tensors are identical, then it is said that the two surfaces have the same *intrinsic* geometry. Two such surfaces are said to be *isometric*. As far as the measurement of the above three quantities are concerned, there is no difference in the two surfaces, no matter how different the surfaces may appear when viewed from the enveloping space. Furthermore, we can

---

[1] $g = (A^{12})^2 + (A^{23})^2 + (A^{31})^2 > 0$, where $A^{ik} = x_1^i x_2^k - x_1^k x_2^i$. This is verified by using (2.4) and the identity, $\det(g_{\alpha\beta}) = \det(\sum_{i=1}^3 x_\alpha^i x_\beta^i) = (A^{12})^2 + (A^{23})^2 + (A^{31})^2$.

define an intrinsic curvature called the Gaussian curvature for the isometric surfaces, which is considered in §2.5. It will be seen later that the Gaussian curvature is one of the central geometrical objects in the analysis of time evolution of various dynamical systems.

2.2. Second Fundamental Form

The vector product of two tangent vectors $\boldsymbol{x}_1(=\boldsymbol{x}_{u^1})$ and $\boldsymbol{x}_2(=\boldsymbol{x}_{u^2})$ yields a vector directed to the normal to the surface Σ^2 with the magnitude $\|\boldsymbol{x}_1 \times \boldsymbol{x}_2\|$ (Appendix B.5). Thus, the unit normal to Σ^2 is defined as

$$\boldsymbol{N} = \frac{\boldsymbol{x}_1 \times \boldsymbol{x}_2}{\|\boldsymbol{x}_1 \times \boldsymbol{x}_2\|} \quad \left(= \frac{\boldsymbol{x}_{u^1} \times \boldsymbol{x}_{u^2}}{\|\boldsymbol{x}_{u^1} \times \boldsymbol{x}_{u^2}\|} \right), \tag{2.18}$$

at a point $\boldsymbol{x}(u^1, u^2)$. Since $\langle \boldsymbol{N}, \boldsymbol{N} \rangle = 1$, we obtain $\langle \boldsymbol{N}_\alpha, \boldsymbol{N} \rangle = 0$ by differentiating it with respect to u^α, where $\boldsymbol{N}_\alpha = \partial \boldsymbol{N}/\partial u^\alpha$. Hence, $\boldsymbol{N}_\alpha \perp \boldsymbol{N}$ for $\alpha = 1$ and 2, i.e. $\boldsymbol{N}_1$ and $\boldsymbol{N}_2$ are tangent to Σ^2. Thus, we have two pairs of tangent vectors, $(\boldsymbol{x}_1, \boldsymbol{x}_2)$ and $(\boldsymbol{N}_1, \boldsymbol{N}_2)$ (Fig. 2.5). Using the pair $(\boldsymbol{N}_1, \boldsymbol{N}_2)$, one can define a mixed tensor, $\mathrm{d}\boldsymbol{N} := \boldsymbol{N}_\beta \otimes \mathrm{d}u^\beta$ (see §1.10.3(ii)). The assignment $b(\boldsymbol{x}_\alpha) : \boldsymbol{x}_\alpha \mapsto -\mathrm{d}\boldsymbol{N}(\boldsymbol{x}_\alpha) = -\boldsymbol{N}_\alpha$ $(\alpha = 1, 2)$ defines a linear transformation between $(\boldsymbol{x}_1, \boldsymbol{x}_2)$ and $(\boldsymbol{N}_1, \boldsymbol{N}_2)$, where the equation

$$\boldsymbol{N}_\alpha = -b(\boldsymbol{x}_\alpha), \quad \text{or} \quad \boldsymbol{N}_\alpha = -b^\beta_\alpha \boldsymbol{x}_\beta = -b^1_\alpha \boldsymbol{x}_1 - b^2_\alpha \boldsymbol{x}_2 \tag{2.19}$$

$(\alpha = 1, 2)$ is called the *Weingarten equation*. The (b^β_α) is the transformation matrix from $(\boldsymbol{x}_1, \boldsymbol{x}_2)$ to $(\boldsymbol{N}_1, \boldsymbol{N}_2)$, whose elements are related to the

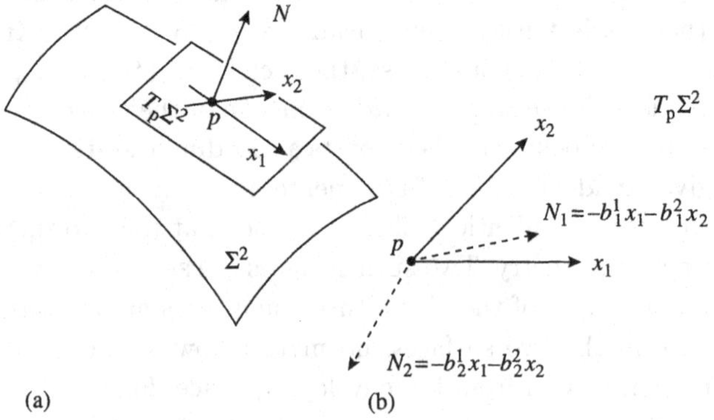

Fig. 2.5. (a) Unit normal $\boldsymbol{N}$, (b) Two pairs: $(\boldsymbol{x}_1, \boldsymbol{x}_2)$ and $(\boldsymbol{N}_1, \boldsymbol{N}_2)$.

coefficients of the second fundamental form defined below. For $^\forall \boldsymbol{A} \in T\Sigma^2$, this assigns another tangent vector $b(\boldsymbol{A}) = -\mathrm{d}\boldsymbol{N}(\boldsymbol{A}) = -A^\alpha \boldsymbol{N}_\alpha$.

With two tangents $\mathrm{d}\boldsymbol{x} = \boldsymbol{x}_\alpha \mathrm{d}u^\alpha$ and $\mathrm{d}\boldsymbol{N} = \boldsymbol{N}_\beta \mathrm{d}u^\beta$, the *second fundamental form* is defined by

$$\mathrm{II} = -\langle \mathrm{d}\boldsymbol{x}, \mathrm{d}\boldsymbol{N}\rangle = -\langle \boldsymbol{x}_\alpha \mathrm{d}u^\alpha, \boldsymbol{N}_\beta \mathrm{d}u^\beta\rangle = b_{\alpha\beta}\mathrm{d}u^\alpha \mathrm{d}u^\beta, \tag{2.20}$$

where

$$b_{\alpha\beta} = -\langle \boldsymbol{x}_\alpha, \boldsymbol{N}_\beta\rangle. \tag{2.21}$$

In the language of the tensor product, the II is a second-rank covariant tensor taking a real value on a pair of vectors, i.e.

$$\begin{aligned}\mathrm{II}(\boldsymbol{A},\boldsymbol{B}) &= -\langle \mathrm{d}\boldsymbol{x}, \mathrm{d}\boldsymbol{N}\rangle(\boldsymbol{A},\boldsymbol{B}) = \langle \boldsymbol{A}, b(\boldsymbol{B})\rangle \\ &= b_{\alpha\beta}\mathrm{d}u^\alpha \otimes \mathrm{d}u^\beta(\boldsymbol{A},\boldsymbol{B}) = b_{\alpha\beta}A^\alpha B^\beta.\end{aligned} \tag{2.22}$$

Owing to the orthogonality $\langle \boldsymbol{x}_\alpha, \boldsymbol{N}\rangle = 0$, we obtain

$$0 = \partial_{u^\beta}\langle \boldsymbol{x}_\alpha, \boldsymbol{N}\rangle = \langle \boldsymbol{x}_{\alpha\beta}, \boldsymbol{N}\rangle + \langle \boldsymbol{x}_\alpha, \boldsymbol{N}_\beta\rangle = \langle \boldsymbol{x}_{\alpha\beta}, \boldsymbol{N}\rangle - b_{\alpha\beta}.$$

Note that the equality $\boldsymbol{x}_{\alpha\beta} = \partial^2\boldsymbol{x}/\partial u^\alpha \partial u^\beta = \partial^2\boldsymbol{x}/\partial u^\beta \partial u^\alpha$ is considered to be a condition of integrability to obtain a smooth surface (see §2.7). This leads to the symmetry of the tensor $b_{\alpha\beta}$:

$$b_{\alpha\beta} = \langle \boldsymbol{x}_{\alpha\beta}, \boldsymbol{N}\rangle = b_{\beta\alpha}. \tag{2.23}$$

Thus, it is found that the second fundamental form is symmetric:

$$\mathrm{II}(\boldsymbol{A},\boldsymbol{B}) = \langle \boldsymbol{A}, b(\boldsymbol{B})\rangle = b_{\alpha\beta}A^\alpha B^\beta = \langle \boldsymbol{B}, b(\boldsymbol{A})\rangle = \mathrm{II}(\boldsymbol{B},\boldsymbol{A}). \tag{2.24}$$

Taking the scalar product of (2.19) with $\boldsymbol{x}_\gamma$, and using (2.4) and (2.21), we have the relation

$$\langle \boldsymbol{N}_\alpha, \boldsymbol{x}_\gamma\rangle = -\langle b_\alpha^\beta\, \boldsymbol{x}_\beta, \boldsymbol{x}_\gamma\rangle, \quad \text{or} \quad b_{\gamma\alpha} = b_\alpha^\beta\, g_{\beta\gamma}.$$

Solving this, we obtain $b_\alpha^\beta = g^{\beta\gamma}\, b_{\gamma\alpha}$, where $g^{\alpha\beta}$ denotes the components of the inverse g^{-1} of the metric tensor g, that is,

$$g^{\alpha\beta} = (g^{-1})^{\alpha\beta} \quad \text{and} \quad g^{\alpha\beta}g_{\beta\gamma} = g_{\gamma\beta}g^{\beta\alpha} = \delta_\gamma^\alpha, \tag{2.25}$$

where δ_γ^α is the Kronecker's delta. Thus,

$$g^{11} = \frac{g_{22}}{\det g}, \quad g^{12} = g^{21} = -\frac{g_{12}}{\det g}, \quad g^{22} = \frac{g_{11}}{\det g}. \tag{2.26}$$

Example. *Torus* (II). In §2.1, we considered a torus surface $\Sigma^{\rm tor}$ and its metric tensor. According to the formula (2.18), unit normal to the surface $\Sigma^{\rm tor}$ is obtained as

$$\boldsymbol{N} = (-\cos u \cos v, -\cos u \sin v, -\sin u), \tag{2.27}$$

where the expressions (2.9) and (2.10) are used for the two basis vectors $\boldsymbol{x}_u, \boldsymbol{x}_v \in T_p\Sigma_{\rm tor}$. The coefficients $b_{\alpha\beta}$ of the second fundamental form are obtained by using (2.23) and the derivatives of $\boldsymbol{x}_u$ and $\boldsymbol{x}_v$ as

$$b_{uu} = a, \quad b_{uv} = 0, \quad b_{vv} = (R + a\cos u)\cos u. \tag{2.28}$$

Using the metric tensor g of (2.11), we have $\det g = a^2(R + a\cos u)^2$. Hence, the inverse g^{-1} is given by

$$g^{uu} = \frac{1}{a^2}, \quad g^{uv} = g^{vu} = 0, \quad g^{vv} = \frac{1}{(R + a\cos u)^2}. \tag{2.29}$$

□

2.3. Gauss's Surface Equation and an Induced Connection

What is an induced *derivative for a curved surface in* $\mathbb{R}^3$.

Although the vectors $\boldsymbol{x}_1$ and $\boldsymbol{x}_2$ are tangent, the second derivatives $\boldsymbol{x}_{\alpha\beta}$ are not necessarily so, and we have the following expression,

$$\boldsymbol{x}_{\alpha\beta} = \partial_\beta \boldsymbol{x}_\alpha = b_{\alpha\beta}\,\boldsymbol{N} + \Gamma^\gamma_{\alpha\beta}\,\boldsymbol{x}_\gamma. \tag{2.30}$$

This represents four equations, called the *Gauss's surface equations* (Fig. 2.6). The first normal part is consistent with (2.23) and the second tangential part introduces the coefficient $\Gamma^\gamma_{\alpha\beta}$, the *Christoffel symbols*

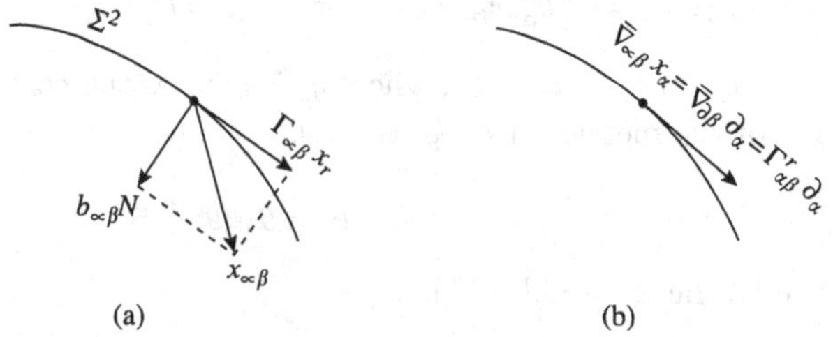

Fig. 2.6. (a) $\partial_\beta \boldsymbol{x}_\alpha$, (b) covariant derivative $\nabla_{\partial_\beta} \boldsymbol{x}_\alpha$.

defined by

$$\langle \boldsymbol{x}_\nu, \boldsymbol{x}_{\alpha\beta} \rangle = \langle \boldsymbol{x}_\nu, \boldsymbol{x}_\gamma \rangle \Gamma^\gamma_{\alpha\beta} = g_{\nu\gamma} \Gamma^\gamma_{\alpha\beta} =: \Gamma_{\alpha\beta,\nu}.$$

Evidently we have the symmetry, $\Gamma^\gamma_{\alpha\beta} = \Gamma^\gamma_{\beta\alpha}$ and $\Gamma_{\alpha\beta,\nu} = \Gamma_{\beta\alpha,\nu}$.

Using the expression (2.30), one can introduce an *induced covariant derivative* $\bar{\nabla}$ of a base vector $\boldsymbol{x}_\alpha$ with respect to a second base $\boldsymbol{x}_\beta$ defined as

$$\bar{\nabla}_{\boldsymbol{x}_\beta} \boldsymbol{x}_\alpha (= \bar{\nabla}_{\partial_\beta} \partial_\alpha) := \boldsymbol{x}_{\alpha\beta} - b_{\alpha\beta} \boldsymbol{N} = \Gamma^\gamma_{\alpha\beta} \boldsymbol{x}_\gamma, \tag{2.31}$$

where $\bar{\nabla}$ (the nabla with an *overbar*) denotes the tangential part of the derivative. The equation $\bar{\nabla}_{\boldsymbol{x}_\beta} \boldsymbol{x}_\alpha = \bar{\nabla}_{\partial_\beta} \partial_\alpha = \Gamma^\gamma_{\alpha\beta} \partial_\gamma$ is consistent with an equality in (1.100) for $X^k = 1$, that is $\nabla_{\partial_j} \partial_\alpha = \Gamma^k_{j\alpha} \partial_k$. This is the property of (an affine) connection considered in §1.10.3(iii).[2]

For two tangent vectors $X, Y \in T\Sigma^2$, one can introduce a vector-valued 1-form,

$$\bar{\nabla} X = \partial_k \otimes \bar{\nabla} X^k, \quad \bar{\nabla} X^k = \mathrm{d}X^k + \Gamma^k_{\beta\alpha} X^\alpha \mathrm{d}x^\beta,$$

where $\bar{\nabla} X^k$ is a (connection) 1-form. Then, we have the induced *covariant derivative* of X with respect to Y:

$$\bar{\nabla}_Y X := \partial_k \otimes \bar{\nabla} X^k(Y) = \bar{\nabla} X^k(Y) \partial_k \tag{2.32}$$

$$\bar{\nabla} X^k(Y) = \mathrm{d}X^k(Y) + \Gamma^k_{\beta\alpha} X^\alpha Y^\beta. \tag{2.33}$$

The covariant derivative $\bar{\nabla}_Y X$ is considered to be a map, $\bar{\nabla} : T\Sigma^2 \times T\Sigma^2 \to T\Sigma^2$, namely $\bar{\nabla}_Y X$ represents a tangent vector as evident by (2.32), and the operator $\bar{\nabla}$ is called an *induced connection*. Taking $X^\alpha = 1$ and $Y^\beta = 1$, Eq. (2.33) reduces to $\bar{\nabla}_{\partial_\beta} \partial_\alpha = \Gamma^k_{\beta\alpha} \partial_k$, which is equivalent to (2.31), the tangential part of (2.30).

For later application, let us rewrite the expression (2.30) (or (2.31)) by multiplying $X^\alpha Y^\beta$ on both sides. Then we obtain

$$X^\alpha Y^\beta \partial_\beta \boldsymbol{x}_\alpha = X^\alpha Y^\beta b_{\alpha\beta} \boldsymbol{N} + X^\alpha Y^\beta \bar{\nabla}_{\boldsymbol{x}_\beta} \boldsymbol{x}_\alpha. \tag{2.34}$$

The first term on the right-hand side is equal to the second fundamental form $\mathrm{II}(X, Y)\boldsymbol{N}$ by (2.24). Using (2.31)–(2.33), the second term is found to be equal to $\bar{\nabla}_Y X - \mathrm{d}X(Y)$, where $\mathrm{d}X(Y) = \mathrm{d}X^\alpha(Y) \boldsymbol{x}_\alpha = Y^\beta \partial_\beta X^\alpha \boldsymbol{x}_\alpha$.

---

[2]See §3.5 for general definition of affine connection or covariant derivative.

Now let us define the nabla ∇ in the ambient space (in the ordinary sense) by

$$\nabla_Y X = Y^\beta\,\partial_\beta(X^\alpha \boldsymbol{x}_\alpha) = \mathrm{d}X(Y) + X^\alpha Y^\beta \partial_\beta \boldsymbol{x}_\alpha.$$

Collecting these, it is found that Eq. (2.34) is written as

$$\nabla_Y X = \bar{\nabla}_Y X + \mathrm{II}(X,Y)\boldsymbol{N}, \tag{2.35}$$

$$\bar{\nabla}_Y X = \mathrm{d}X^\alpha(Y)\boldsymbol{x}_\alpha + X^\alpha\,\bar{\mathrm{d}}\boldsymbol{x}_\alpha(Y), \tag{2.36}$$

$$\bar{\mathrm{d}}\boldsymbol{x}_\alpha(Y) := Y^\beta\,\bar{\nabla}_{\boldsymbol{x}_\beta}\,\boldsymbol{x}_\alpha \text{ (tangential part)}.$$

The first is called the *Gauss's surface equation*, and the second gives another definition of the covariant derivative.

A vector field $X(x)$ is said to be *parallel* along the curve generated by $Y(x)$ if the covariant derivative $\bar{\nabla}_Y X$ vanishes:

$$\bar{\nabla}_Y X = 0. \tag{2.37}$$

See §3.5 and 3.8.1 for more details of the parallel translation.

Sometimes (very often in the cases considered below), it is possible to formulate the curved space (such as Σ^2) without taking into consideratiaon the enveloping flat space $\mathbb{R}^3$. In those cases, the nabla ∇ (without the overbar) is used to denote the covariant derivative.

2.4. Gauss–Mainardi–Codazzi Equation and Integrability

We consider some integrability condition of a surface.

Differentiating the metric tensor $g_{\nu\alpha} = \langle \boldsymbol{x}_\nu, \boldsymbol{x}_\alpha\rangle$ with respect to u^β, we obtain

$$\begin{aligned}\partial_\beta g_{\nu\alpha} = \partial_\beta\langle \boldsymbol{x}_\nu, \boldsymbol{x}_\alpha\rangle &= \langle \boldsymbol{x}_{\nu\beta}, \boldsymbol{x}_\alpha\rangle + \langle \boldsymbol{x}_\nu, \boldsymbol{x}_{\alpha\beta}\rangle \\ &= \Gamma_{\nu\beta,\alpha} + \Gamma_{\alpha\beta,\nu}.\end{aligned} \tag{2.38}$$

It is readily shown that $\partial_\beta g_{\nu\alpha} + \partial_\nu g_{\alpha\beta} - \partial_\alpha g_{\beta\nu} = 2\Gamma_{\beta\nu,\alpha} = 2g_{\alpha\lambda}\,\Gamma^\lambda_{\beta\nu}$. Hence, we find

$$\Gamma^\mu_{\beta\nu} = g^{\mu\alpha}\,\Gamma_{\beta\nu,\alpha}, \tag{2.39}$$

$$\Gamma_{\beta\nu,\alpha} = \frac{1}{2}(\partial_\beta g_{\nu\alpha} + \partial_\nu g_{\alpha\beta} - \partial_\alpha g_{\beta\nu}), \tag{2.40}$$

which are well-known representations of the Christoffel symbols in terms of the metric tensors.

Example. *Torus* (III). Using the metric tensor g of (2.11), we can calculate the Christoffel symbols $\Gamma_{\alpha\beta,u}$ and $\Gamma_{\alpha\beta,v}$ for the torus Σ^{tor}, as follows:

$$\Gamma_{\alpha\beta,u} = \begin{pmatrix} 0 & 0 \\ 0 & aQ\sin u \end{pmatrix}, \qquad \Gamma_{\alpha\beta,v} = \begin{pmatrix} 0 & -aQ\sin u \\ -aQ\sin u & 0 \end{pmatrix}, \tag{2.41}$$

where $Q = R + a\cos u$. Next, using the inverse g^{-1} of (2.29), we obtain $\Gamma^u_{\alpha\beta} = g^{uu}\,\Gamma_{\alpha\beta,u} + g^{uv}\,\Gamma_{\alpha\beta,v} = g^{uu}\,\Gamma_{\alpha\beta,u}$ since $g^{uv} = 0$. Similarly, we have $\Gamma^v_{\alpha\beta} = g^{vv}\,\Gamma_{\alpha\beta,v}$. Thus,

$$\Gamma^u_{\alpha\beta} = \frac{1}{a}\begin{pmatrix} 0 & 0 \\ 0 & Q\sin u \end{pmatrix}, \qquad \Gamma^v_{\alpha\beta} = -\frac{1}{Q}\begin{pmatrix} 0 & a\sin u \\ a\sin u & 0 \end{pmatrix}. \tag{2.42}$$

□

As a result, it is found that all the coefficients of the Gauss equation (2.30) are represented in terms of $g_{\alpha\beta}$ and $b_{\alpha\beta}$ (the first and second fundamental forms). The following equality of third order derivatives

$$\boldsymbol{x}_{\alpha\beta\gamma} = \partial_\gamma\partial_\beta\partial_\alpha\boldsymbol{x} = \boldsymbol{x}_{\alpha\gamma\beta}, \tag{2.43}$$

is considered to be a condition of *integrability* of a surface.[3]

When two sets of tensors $g_{\alpha\beta}$ and $b_{\alpha\beta}$ are given, one can derive another important equation of the surface. Using the Weingarten equation (2.19) and the Gauss equation (2.30), one can derive the following expression:

$$0 = \boldsymbol{x}_{\alpha\beta\gamma} - \boldsymbol{x}_{\alpha\gamma\beta} = (R^\nu_{\alpha\gamma\beta} - B^\nu_{\alpha\beta\gamma})\boldsymbol{x}_\nu + V_{\alpha\beta\gamma}\boldsymbol{N}, \tag{2.44}$$

where

$$R^\nu_{\alpha\gamma\beta} = \partial_\gamma\Gamma^\nu_{\beta\alpha} - \partial_\beta\Gamma^\nu_{\gamma\alpha} + \Gamma^\nu_{\gamma\mu}\Gamma^\mu_{\beta\alpha} - \Gamma^\nu_{\beta\mu}\Gamma^\mu_{\gamma\alpha}, \tag{2.45}$$

$$B^\nu_{\alpha\beta\gamma} = b^\nu_\gamma\, b_{\alpha\beta} - b^\nu_\beta\, b_{\alpha\gamma}, \tag{2.46}$$

$$V_{\alpha\beta\gamma} = \Gamma^\nu_{\alpha\beta}b_{\nu\gamma} + \partial_\gamma b_{\alpha\beta} - \Gamma^\nu_{\alpha\gamma}b_{\nu\beta} - \partial_\beta b_{\alpha\gamma}. \tag{2.47}$$

The tensors $R^\nu_{\alpha\gamma\beta}$ defined by (2.45) are called the *Riemann–Christoffel curvature tensors*. Because the left-hand side of Eq. (2.44) is zero, both of the tangential and normal components of (2.44), i.e. the coefficients of $\boldsymbol{x}_\nu$ and

[3]In §2.7, integrability conditions are given in different ways.

$\boldsymbol{N}$ respectively, must vanish. Thus we obtain the Gauss's equation,

$$R^{\nu}_{\alpha\gamma\beta} = b^{\nu}_{\gamma}\, b_{\alpha\beta} - b^{\nu}_{\beta}\, b_{\alpha\gamma}, \tag{2.48}$$

and the *Mainardi–Codazzi* equation,[4]

$$\partial_\gamma b_{\alpha\beta} - \Gamma^{\nu}_{\alpha\gamma} b_{\nu\beta} = \partial_\beta\, b_{\alpha\gamma} - \Gamma^{\nu}_{\alpha\beta} b_{\nu\gamma}. \tag{2.49}$$

Introducing the notation $b_{\alpha\beta;\gamma} := \partial_\gamma b_{\alpha\beta} - \Gamma^{\nu}_{\alpha\gamma} b_{\nu\beta}$ for the left side, Eq. (2.49) reduces to $b_{\alpha\beta;\gamma} = b_{\alpha\gamma;\beta}$. Since all the indices take only 1 and 2, the Mainardi–Codazzi equation consists of only two equations: $b_{11;2} = b_{12;1}$ and $b_{22;1} = b_{21;2}$.

Remark. When the two sets of tensors $g_{\alpha\beta}$ and $b_{\alpha\beta}$ are given as the first and second fundamental tensors (see §2.11 for the uniqueness theorem), respectively, Eqs. (2.48) and (2.49) must be satisfied for the existence of the integral surface (represented by Eq. (2.43)), and therefore they are considered as the *integrability conditions.* In the geometrical theory of soliton systems considered in Part IV, the Gauss equation and Mainardi–Codazzi equation yield a nonlinear partial differential equation, usually called the *soliton* equation.

2.5. Gaussian Curvature of a Surface

We consider various curvatures in order to characterize a surface and curves lying on the surface.

2.5.1. *Riemann tensors*

From the Riemann tensor $R^{\nu}_{\alpha\beta\gamma}$ (third order covariant and first order contravariant), one can define the fourth order covariant tensor by

$$R_{\delta\alpha\beta\gamma} = g_{\delta\nu} R^{\nu}_{\alpha\beta\gamma}.$$

In terms of the Christoffel symbols, we have

$$R_{\delta\alpha\beta\gamma} = \partial_\beta \Gamma_{\gamma\alpha,\delta} - \partial_\gamma \Gamma_{\beta\alpha,\delta} + \Gamma^{\mu}_{\alpha\beta}\Gamma_{\delta\gamma,\mu} - \Gamma^{\mu}_{\alpha\gamma}\Gamma_{\delta\beta,\mu}, \tag{2.50}$$

which can be shown by using the equation,

$$\partial_\beta \Gamma_{\gamma\alpha,\delta} = \partial_\beta (g_{\delta\nu}\Gamma^{\nu}_{\gamma\alpha}) = g_{\delta\nu}\partial_\beta \Gamma^{\nu}_{\gamma\alpha} + \Gamma^{\nu}_{\gamma\alpha}(\Gamma_{\nu\beta,\delta} + \Gamma_{\delta\beta,\nu}).$$

This equation can be verified by using (2.38) and (2.40).

[4]This is sometimes called as Codazzi equation [Cod1869]. However, the equivalent equation had been derived earlier by Mainardi [Mai1856]. (See [Eis47, Ch. IV, §39].)

Furthermore, using (2.40) again, one can derive another expression of $R_{\delta\alpha\beta\gamma}$: [Eis47]

$$R_{\delta\alpha\beta\gamma} = \frac{1}{2}(\partial_\beta\partial_\alpha g_{\delta\gamma} - \partial_\beta\partial_\delta g_{\alpha\gamma} - \partial_\gamma\partial_\alpha g_{\delta\beta} + \partial_\gamma\partial_\delta g_{\alpha\beta}) + g^{\mu\nu}(\Gamma_{\alpha\beta,\nu}\Gamma_{\delta\gamma,\mu} - \Gamma_{\alpha\gamma,\nu}\Gamma_{\delta\beta,\mu}). \tag{2.51}$$

The last expression shows a remarkable property that the tensor $R_{\delta\alpha\beta\gamma}$ is skew-symmetric not only with respect to the pair of indices β and γ, changing sign with the exchange of both indices, but also with respect to the pair of indices α and δ.

From (2.51), it can be shown that

$$R_{\delta\alpha\beta\gamma} + R_{\delta\beta\gamma\alpha} + R_{\delta\gamma\alpha\beta} = 0. \tag{2.52}$$

It is obvious in the expression (2.51) that the Riemann tensors are represented in terms of the metric tensors $g_{\alpha\beta}$ only, characterizing the first fundamental form, because the Christoffel symbols are also represented in terms of $g_{\alpha\beta}$ by (2.40).[5]

Considering that the indices $\alpha, \beta \cdots$ take two values 1 and 2, the tensor $R_{\delta\alpha\beta\gamma}$ has nominally sixteen components, among which the number of nonvanishing independent components is only *one*. This can be verified as follows. Because of the above skew-symmetries with respect to two pairs of indices (α, δ) and (β, γ), all the tensors with $\alpha = \delta$ or $\beta = \gamma$ of the form $R_{\alpha\alpha\beta\gamma}$ or $R_{\delta\alpha\beta\beta}$ vanish, and the remaining nonvanishing components are only four. However, we obviously have the following relations among the four by the skew-symmetries:

$$R_{1212} = -R_{1221} = R_{2121} = -R_{2112}.$$

Hence all the Riemann tensors are given, once R_{1212} is known.

When a surface is isometric with the plane, i.e. when their first fundamental forms are identical to each other, there necessarily exists a coordinate system for which $g_{11} = g_{22} = 1$ and $g_{12} = 0$. Then, we have $R_{1212} = 0$, since R_{1212} is represented in terms of the Christoffel symbols and their derivatives (see (2.50)) and the Christoffel symbols, in turn, are given by derivatives of the metric tensors (see (2.40)) which vanish identically. In general, a surface is isometric with the plane if and only if the Riemann tensor is a zero tensor.

[5]This remarkable property is called the Gauss's *Theorema Egregium*.

From a sheet of paper, we can form a cylinder or cone, but it is not possible to form a spherical surface without stretching, folding or cutting. The geometrical property which can be expressed entirely in terms of the first fundamental form is called the *intrinsic geometry* of the surface. As shown in the subsequent sections, the Gaussian curvature of a surface is an intrinsic quantity. The Gaussian curvature of a plane sheet, circular cylinder or a cone are all zero, while that of the sphere takes a positive value.

2.5.2. *Gaussian curvature*

The *Gaussian curvature* K of a surface is defined by

$$K := \det(b^\alpha_\beta) = R^{12}_{12}, \tag{2.53}$$

which will be given another meaning by (2.62) below in terms of the normal curvatures. Here the following equation, obtained from (2.48), is used:

$$R^{12}_{12} := g^{2\alpha} R^1_{\alpha 12} = g^{2\alpha}(b^1_1 b_{\alpha 2} - b^1_2 b_{\alpha 1}) = b^1_1 b^2_2 - b^1_2 b^2_1. \tag{2.54}$$

Using (2.48) again and the relation $b_{\alpha\beta} = g_{\alpha\gamma} b^\gamma_\beta$ defined in §2.2 together with $\det(b_{\alpha\beta}) = \det(g_{\alpha\gamma})\,\det(b^\gamma_\beta)$, one obtains another form of the Gauss equation (2.48),

$$\begin{aligned} R_{1212} &= g_{1\nu} R^\nu_{212} = g_{1\nu}(b^\nu_1 b_{22} - b^\nu_2 b_{21}) = g_{1\nu} g^{\nu\mu}(b_{1\mu} b_{22} - b_{2\mu} b_{21}) \\ &= b_{11} b_{22} - b_{21} b_{21} = \det(b_{\alpha\beta}) = \det(g_{\alpha\gamma})\,\det(b^\gamma_\beta), \end{aligned} \tag{2.55}$$

since $g_{1\nu} g^{\nu\mu} = \delta^\mu_1$. Thus it is found that

$$K = \det(b^\alpha_\beta) = \frac{\det(b_{\alpha\beta})}{\det(g_{\alpha\beta})} = \frac{R_{1212}}{g}, \tag{2.56}$$

where $g = \det(g_{\alpha\beta}) = g_{11} g_{22} - (g_{12})^2$.

Example. *Torus* (IV). Regarding the torus surface Σ^{tor}, we have calculated already the tensor coefficients of the first and second fundamental forms in §2.1 and 2.2. Using (2.11) and (2.28), the Gaussian curvature is found immediately as

$$K^{\mathrm{tor}} = \frac{\det(b_{\alpha\beta})}{\det(g_{\alpha\beta})} = \frac{a(R + a\cos u)\cos u}{a^2(R + a\cos u)^2} = \frac{\cos u}{a(R + a\cos u)}. \tag{2.57}$$

The curvature K^{tor} takes both positive and negative values according to $\cos u > 0$ or < 0 (Fig. 2.3). □

2.5.3. *Geodesic curvature and normal curvature*

In order to find another useful expression of K, consider a *space curve* C defined by $\boldsymbol{x}(s)$ with s the arc length parameter (see Appendix D.1). The unit tangent is given by $\boldsymbol{T} = \mathrm{d}\boldsymbol{x}/\mathrm{d}s = \boldsymbol{x}_\alpha \mathrm{d}u^\alpha/\mathrm{d}s = \boldsymbol{x}_\alpha \dot{u}^\alpha$ (where $\dot{u}^\alpha = \mathrm{d}u^\alpha/\mathrm{d}s$). The curvature κ of C at $\boldsymbol{x}$ is defined by

$$\begin{aligned}\kappa\boldsymbol{n} := \frac{\mathrm{d}\boldsymbol{T}}{\mathrm{d}s} &= \boldsymbol{x}_{\alpha\beta}\frac{\mathrm{d}u^\alpha}{\mathrm{d}s}\frac{\mathrm{d}u^\beta}{\mathrm{d}s} + \boldsymbol{x}_\alpha\frac{\mathrm{d}^2u^\alpha}{\mathrm{d}s^2} \\ &= (b_{\alpha\beta}\boldsymbol{N} + \Gamma^\gamma_{\alpha\beta}\boldsymbol{x}_\gamma)\dot{u}^\alpha\dot{u}^\beta + \boldsymbol{x}_\alpha\ddot{u}^\alpha = \kappa_N\boldsymbol{N} + \boldsymbol{\kappa}_g \qquad (2.58)\end{aligned}$$

(using (2.30)), where $\boldsymbol{n}$ is the unit principal normal to the space curve C (cf. Serret–Frenet formula), and the *geodesic curvature* defined by

$$\boldsymbol{\kappa}_g = (\Gamma^\gamma_{\alpha\beta}\,\dot{u}^\alpha\dot{u}^\beta + \ddot{u}^\gamma)\boldsymbol{x}_\gamma \qquad (2.59)$$

is the tangential component of $\mathrm{d}\boldsymbol{T}/\mathrm{d}s$ (Fig. 2.7). The normal $\boldsymbol{n}$ of the curve is not necessarily parallel to the normal $\boldsymbol{N}$ of the surface on which the curve C is lying, and the component of the vector $\kappa\boldsymbol{n}$ in the direction to the surface normal $\boldsymbol{N}$ is given by

$$\kappa_N = \langle\kappa\boldsymbol{n}, \boldsymbol{N}\rangle = \langle\boldsymbol{x}_{\alpha\beta}, \boldsymbol{N}\rangle\dot{u}^\alpha\dot{u}^\beta = \mathrm{II}(\boldsymbol{T}, \boldsymbol{T}), \qquad (2.60)$$

in view of the definition (2.23). Equivalently, we have

$$\kappa_N = \frac{\langle\boldsymbol{x}_{\alpha\beta}, \boldsymbol{N}\rangle\mathrm{d}u^\alpha\mathrm{d}u^\beta}{\mathrm{d}s^2} = \frac{b_{\alpha\beta}\,\mathrm{d}u^\alpha\mathrm{d}u^\beta}{g_{\alpha\beta}\,\mathrm{d}u^\alpha\mathrm{d}u^\beta} = \mathrm{II}(\boldsymbol{T}, \boldsymbol{T}).$$

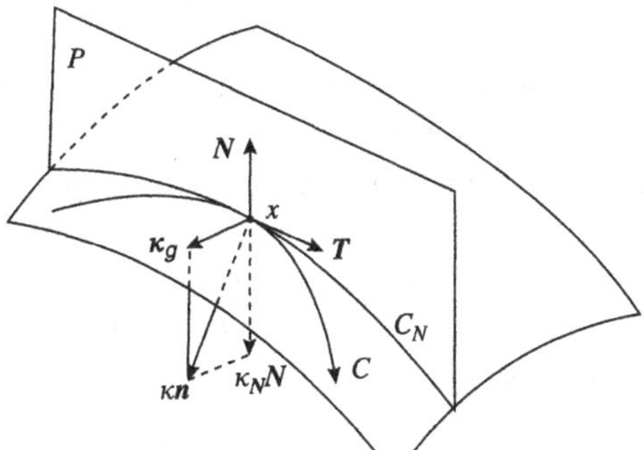

Fig. 2.7. Geodesic curvature $\boldsymbol{\kappa}_g$ and normal curvature $\kappa_N(< 0)$.

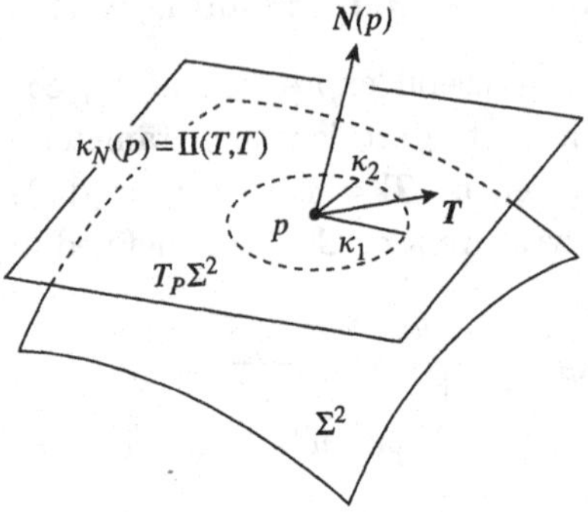

Fig. 2.8. $\kappa_N(p) = \mathrm{II}(\boldsymbol{T}, \boldsymbol{T})$ at $p \in \Sigma^2$.

The κ_N is called the *normal curvature* and understood as the second fundamental form on the unit tangent $\boldsymbol{T}$ (Fig. 2.8). In general, the curvature vector $\kappa \boldsymbol{n}$ has a tangential component $\boldsymbol{\kappa}_g$ (the geodesic curvature vector). The above derivation shows that the normal curvature is given by $\kappa_N = \mathrm{II}(\boldsymbol{T}, \boldsymbol{T})$ under the restricting condition, $\mathrm{I}(\boldsymbol{T}, \boldsymbol{T}) = \langle \boldsymbol{T}, \boldsymbol{T} \rangle = 1$.

2.5.4. *Principal curvatures*

The plane P spanned by the vectors $\boldsymbol{T}$ and $\boldsymbol{N}$ at a point $p \in \Sigma^2$ cuts the surface Σ^2 with a section called a *normal section* (a curve C_N, Fig. 2.7). The second fundamental form takes

$$\mathrm{II}(\boldsymbol{T}, \boldsymbol{T}) = \pm \kappa,$$

where the signs $\pm$ depend on whether the curve C is curving toward $\boldsymbol{N}(+)$ or not $(-)$. Next, rotating the tangent direction $\boldsymbol{T}$ with p fixed, the normal curvature $\kappa_N = \mathrm{II}(\boldsymbol{T}, \boldsymbol{T})$ changes, in general, with keeping $\langle \boldsymbol{T}, \boldsymbol{T} \rangle = 1$ fixed, and takes a maximum κ_1 in one direction, and a minimum κ_2 in another direction (Fig. 2.8). These values and directions are called the *principal* values and directions respectively, and are determined as follows.

Observe that the above extremum problem is equivalent to finding the extrema of the following function,

$$\lambda = \frac{b_{\alpha\beta} T^\alpha T^\beta}{g_{\alpha\beta} T^\alpha T^\beta}$$

without the restricting condition $\langle \boldsymbol{T}, \boldsymbol{T} \rangle = 1$ (multiplication of T^α with an arbitrary constant c does not change the value λ). The extremum condition given by $\partial\lambda/\partial T^\alpha = 0$ yields the following equation for the principal direction T^β:

$$(b_{\alpha\beta} - \lambda\, g_{\alpha\beta})\, T^\beta = 0, \quad (\alpha = 1, 2). \tag{2.61}$$

The condition of nontrivial solution of T^β, i.e. vanishing of the coefficient determinant, becomes the equation for the eigenvalue λ yielding a quadratic equation of λ (Appendix K), which has two roots κ_1 and κ_2 (principal curvatures). Product of the two roots is given by $\kappa_1\kappa_2 = \det(b_{\alpha\beta})/\det(g_{\alpha\beta})$ from the quadratic equation, which is nothing but the Gaussian curvature K of (2.56).[6] Thus it is found that

$$K = \frac{\det(b_{\alpha\beta})}{\det(g_{\alpha\beta})} = \kappa_1\kappa_2. \tag{2.62}$$

In Appendix D.3, for the surface defined by $(x, y, f(x, y))$, a formula of its Gaussian curvature K is given, together with the tensors of the second fundamental form

Example. *Torus* (V). For the torus Σ^{tor}, by using the first and second fundamental tensors, (2.11) and (2.28), Eq. (2.61) reduces to

$$(a - \lambda a^2)\, T^u = 0,$$
$$(R + a\cos u)(\cos u - \lambda(R + a\cos u)T^v = 0.$$

Thus we obtain the eigenvalues (principal values) κ_1 and κ_2 and associated eigen directions (T^u, T^v) given by

$$\kappa_1 = \frac{1}{a}, \quad (1, 0); \qquad \kappa_2 = \frac{\cos u}{R + a\cos u}, \quad (0, 1)$$

Naturally, the product $\kappa_1\kappa_2$ is equal to K^{tor} of (2.57). □

Figure 2.9 shows two surfaces of *positive* Gaussian curvature ($K > 0$). Figure 2.10 shows a surface of *negative* Gaussian curvature ($K < 0$) and an example of $\mathrm{II}(\boldsymbol{T}, \boldsymbol{T})$ taking both signs.

Although the definition (2.53) of the Gaussian curvature K is given by the coefficients b^α_β derived from the second fundamental form $b_{\alpha\beta}$, the curvature K is determined entirely when the metric tensor fields $g_{\alpha\beta}(u^1, u^2)$

---

[6] Sum of the two roots divided by 2, $(\kappa_1 + \kappa_2)/2 = H$, is called the *mean* curvature.

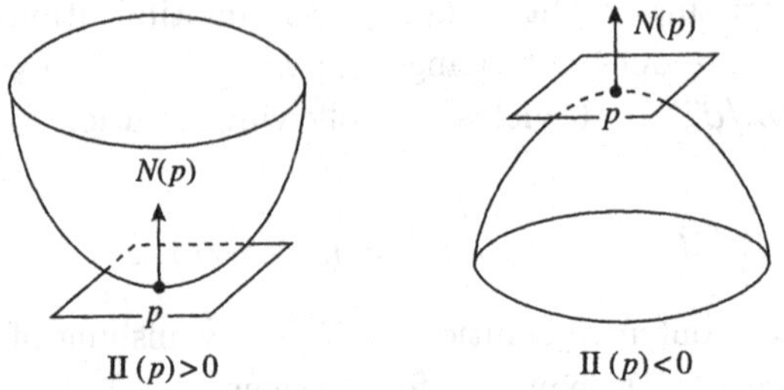

Fig. 2.9. Surfaces of positive Gaussian curvature ($K > 0$).

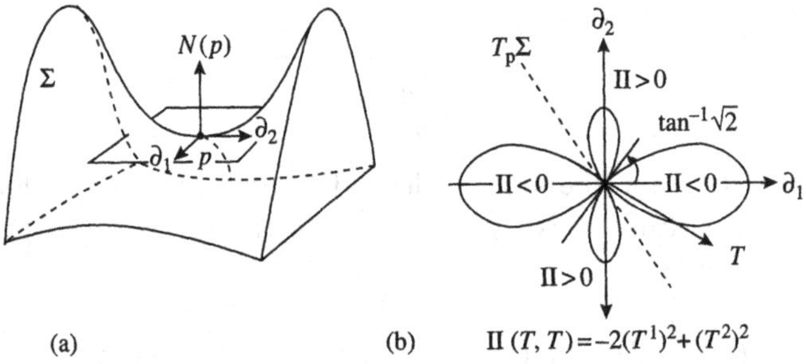

Fig. 2.10. A surface of negative Gaussian curvature ($K < 0$).

are given, i.e. K is intrinsic. This is understood by recalling the observation given in §2.5.1 (below (2.51)) that R_{1212} (or equivalently R^{12}_{12}) is determined by the metric tensors $g_{\alpha\beta}$. This remarkable property is called the Gauss's *Theorema Egregium*. This is essential in the Riemannian geometry.

According to (2.61), it can be shown that the two principal directions are orthogonal unless $b_{\alpha\beta} = c g_{\alpha\beta}$ with c a constant (see also [Eis47]). When $K = 0$ at every point, the surface is called *flat*, while if $H = 0$ at every point, the surface is called a *minimal* surface.

From the definition of the Gaussian curvature (2.53) and the expression of Riemannian tensor (2.51), the Gaussian curvature is given the following expression (Liouville–Beltrami formula):

$$K = \frac{1}{2\sqrt{g}}\left[\frac{\partial}{\partial u^1}\left(\frac{g_{12}}{g_{11}\sqrt{g}}\frac{\partial g_{11}}{\partial u^2} - \frac{1}{\sqrt{g}}\frac{\partial g_{22}}{\partial u^1}\right) + \frac{\partial}{\partial u^2}\left(\frac{2}{\sqrt{g}}\frac{\partial g_{12}}{\partial u^1} - \frac{1}{\sqrt{g}}\frac{\partial g_{11}}{\partial u^2} - \frac{g_{12}}{g_{11}\sqrt{g}}\frac{\partial g_{11}}{\partial u^1}\right)\right]. \tag{2.63}$$

2.6. Geodesic Equation

Geometrical object "geodesics" governs the dynamics of physical systems to be considered in later chapters.

Geodesic curves are characterized by the property of vanishing geodesic curvature $\kappa_g = 0$ at every point. Namely, the derivative of the unit tangent T along the geodesic curve has no component tangent to Σ^2 (Fig. 2.11), from (2.58). Hence, using the induced covariant derivative $\bar{\nabla}$ of (2.33), the geodesic curve is described by $\bar{\nabla}_T T = 0$. This implies another definition of the geodesics according to (2.37) of parallel translation, namely the tangent vector T on the geodesic curve is translated parallel along it. When the curve is parametrized by the arc length s, we have $T = \mathrm{d}\boldsymbol{x}/\mathrm{d}s = (\mathrm{d}u^\alpha/\mathrm{d}s)\boldsymbol{x}_\alpha = T^\alpha \boldsymbol{x}_\alpha$, where $T^\alpha = \mathrm{d}u^\alpha/\mathrm{d}s$. Then Eq. (2.33) reduces to

$$\bar{\nabla}_T T = \boldsymbol{x}_\gamma[\mathrm{d}T^\gamma(T) + \Gamma^\gamma_{\alpha\beta} T^\beta T^\alpha] = 0,$$

where $\mathrm{d}T^\gamma(T) = T^\alpha \partial T^\gamma/\partial u^\alpha = \mathrm{d}T^\gamma/\mathrm{d}s$. Thus, we obtain the *geodesic equation*:

$$\frac{\mathrm{d}T^\gamma}{\mathrm{d}s} + \Gamma^\gamma_{\alpha\beta} T^\alpha T^\beta = 0, \tag{2.64}$$

which is written in another form by using $T^\gamma = \mathrm{d}u^\gamma/\mathrm{d}s$ as

$$\frac{\mathrm{d}^2 u^\gamma}{\mathrm{d}s^2} + \Gamma^\gamma_{\alpha\beta}\frac{\mathrm{d}u^\alpha}{\mathrm{d}s}\frac{\mathrm{d}u^\beta}{\mathrm{d}s} = 0. \tag{2.65}$$

This is consisitent with the expression (2.59) for $\kappa_g = 0$.

The above definition of the geodesic curve is a generalization of a rectilinear line on a flat plane. Since such a rectilinear line is not curved,

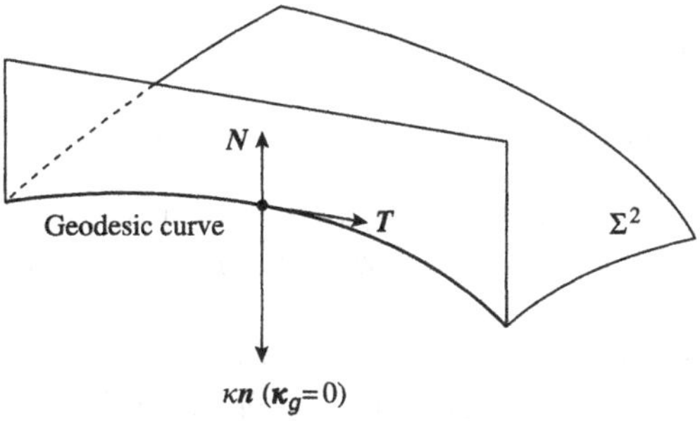

Fig. 2.11. Geodesic curve ($\kappa_g = 0$).

its geodesic curvature is zero and its tangent vector is translated parallel along it. Another important property of a rectilinear line on a plane is that it is a shortest line among those curves connecting two points on a plane. In §3.8, it will be verified in a general space that the arc-length along a curve is extremum if the property $\bar{\nabla}_T T = 0$ is satisfied.

Example. *Torus* (VI). For the surface of a torus Σ^{tor}, the Christoffel symbols $\Gamma^{\gamma}_{\alpha\beta}$ are already given in (2.42). The geodesic equation (2.64) for the tangent vector $T(s) = T^u(s)\boldsymbol{x}_u + T^v(s)\boldsymbol{x}_v$ can be immediately written down as follows:

$$\frac{\mathrm{d}T^u}{\mathrm{d}s} + \frac{Q(u)}{a}\sin u\, T^v T^v = 0, \tag{2.66}$$

$$\frac{\mathrm{d}T^v}{\mathrm{d}s} - \frac{2a}{Q(u)}\sin u\, T^u T^v = 0, \tag{2.67}$$

where $(T^u, T^v) = (u'(s), v'(s))$, $Q(u) = R + a\cos u$ and $(\mathrm{d}s)^2 = a^2(\mathrm{d}u)^2 + Q^2(\mathrm{d}v)^2$. □

2.7. Structure Equations in Differential Forms

Differential geometry in $\mathbb{R}^3$ is reformulated with differential forms.

2.7.1. *Smooth surfaces in $\mathbb{R}^3$ and integrability*

In this section, differential geometry of surfaces in $\mathbb{R}^3$ is reformulated by means of the differential forms. Suppose we have a smooth surface Σ^2 represented by (2.1), and choose a *right-handed* orthonormal moving frame $K_x : \{\boldsymbol{e}_1, \boldsymbol{e}_2, \boldsymbol{e}_3\}$ at each point $\boldsymbol{x}$ on the surface Σ^2 in such a way that $\boldsymbol{e}_3$ is always normal to Σ^2:

$$K_x : \{\boldsymbol{e}_1, \boldsymbol{e}_2, \boldsymbol{e}_3\} \quad \text{where} \quad (\boldsymbol{e}_i, \boldsymbol{e}_j) = \delta_{ij},$$

where $\boldsymbol{e}_1$ and $\boldsymbol{e}_2$ span the tangent plane at each point of Σ^2. This frame is analogous to the orthonormal $(\boldsymbol{t}, \boldsymbol{n}, \boldsymbol{b})$-frame for a space curve to obtain the Frenet–Serret equations (Appendix D.1).

The vector $\mathrm{d}\boldsymbol{x} = \boldsymbol{x}_1 \mathrm{d}u^1 + \boldsymbol{x}_2 \mathrm{d}u^2$ lies in the tangent plane $T_x\Sigma^2$ where $\boldsymbol{x}_\alpha = \partial\boldsymbol{x}/\partial u^\alpha \in T_x\Sigma^2$. Therefore, we can also represent $\mathrm{d}\boldsymbol{x}$ in terms of $\boldsymbol{e}_1$ and $\boldsymbol{e}_2$ as

$$\mathrm{d}\boldsymbol{x} = \sigma^1 \boldsymbol{e}_1 + \sigma^2 \boldsymbol{e}_2, \tag{2.68}$$

where σ^1 and σ^2 are 1-forms and $\sigma^3 = 0$ by definition. If we write $\boldsymbol{x}_k = c_k^1 \boldsymbol{e}_1 + c_k^2 \boldsymbol{e}_2$, we obtain

$$\sigma^1 = c_1^1 \mathrm{d}u^1 + c_2^1 \mathrm{d}u^2, \quad \sigma^2 = c_1^2 \mathrm{d}u^1 + c_2^2 \mathrm{d}u^2. \tag{2.69}$$

The *first fundamental form* is given by

$$\mathrm{I} = \langle \mathrm{d}\boldsymbol{x}, \mathrm{d}\boldsymbol{x} \rangle = \sigma^1 \sigma^1 + \sigma^2 \sigma^2. \tag{2.70}$$

Analogously to the expression (2.68) for $\mathrm{d}\boldsymbol{x}$, one may write differential forms for each of the basis vectors $\boldsymbol{e}_i$ as follows:

$$\mathrm{d}\boldsymbol{e}_i = \omega_i^1 \boldsymbol{e}_1 + \omega_i^2 \boldsymbol{e}_2 + \omega_i^3 \boldsymbol{e}_3, \quad (i = 1, 2, 3), \tag{2.71}$$

where the ω_i^k are 1-forms[7] and can be represented by a linear combination of $\mathrm{d}u^1$ and $\mathrm{d}u^2$ like (2.69). Since $(\mathrm{d}\boldsymbol{e}_i, \boldsymbol{e}_j) + (\boldsymbol{e}_i, \mathrm{d}\boldsymbol{e}_j) = 0$ due to the orthonomality $(\boldsymbol{e}_i, \boldsymbol{e}_j) = \delta_{ij}$, we have the anti-symmetry:

$$\omega_i^j = -\omega_j^i, \quad i, j = 1, 2, 3. \tag{2.72}$$

Hence, we have $\omega_1^1 = \omega_2^2 = \omega_3^3 = 0$.

The *second fundamental form* (see (2.20)) is

$$\begin{aligned} \mathrm{II} = -\langle \mathrm{d}\boldsymbol{x}, \mathrm{d}\boldsymbol{e}_3 \rangle &= -\langle (\sigma^1 \boldsymbol{e}_1 + \sigma^2 \boldsymbol{e}_2), (\boldsymbol{e}_1 \omega_3^1 + \boldsymbol{e}_2 \omega_3^2) \rangle \\ &= -\sigma^1 \omega_3^1 - \sigma^2 \omega_3^2. \end{aligned} \tag{2.73}$$

From the differential calculus (Appendix B.3), we have the relation,

$$\mathrm{d}(\sigma^k \boldsymbol{e}_k) = \mathrm{d}\sigma^k \boldsymbol{e}_k - \sigma^k \wedge \mathrm{d}\boldsymbol{e}_k.$$

If we take exterior differentiation of (2.68), the left-hand side vanishes identically,[8] and we have

$$\begin{aligned} 0 = \mathrm{d}(\mathrm{d}\boldsymbol{x}) = (\mathrm{d}\sigma^1 - \sigma^2 \wedge \omega_2^1)\boldsymbol{e}_1 + (\mathrm{d}\sigma^2 - \sigma^1 \wedge \omega_1^2)\boldsymbol{e}_2 \\ - (\sigma^1 \wedge \omega_1^3 + \sigma^2 \wedge \omega_2^3)\boldsymbol{e}_3. \end{aligned} \tag{2.74}$$

Thus, from the first two terms, one obtains the first integrability equations:

$$\mathrm{d}\sigma^1 = \sigma^2 \wedge \omega_2^1 = -\varpi \wedge \sigma^2, \qquad \mathrm{d}\sigma^2 = \sigma^1 \wedge \omega_1^2 = \varpi \wedge \sigma^1, \tag{2.75}$$

---

[7]According to the notation of §1.8.3(iii), $\mathrm{d}\boldsymbol{e}_i (= \nabla \boldsymbol{e}_i) = \boldsymbol{e}_k \otimes \omega_i^k$, where ω_i^k is called a connection 1-form.

[8]The property $\mathrm{d}(\mathrm{d}\boldsymbol{x}) = \mathrm{d}^2\boldsymbol{x} = 0$ is nothing more than the equality of mixed partial deivatives. It is the source of most *integrability conditions* in partial differential equations and differential geometry.

where $\varpi := \omega_2^1 = -\omega_1^2$. In addition, the third term gives the equation,

$$-\sigma^1 \wedge \omega_1^3 - \sigma^2 \wedge \omega_2^3 = \sigma^1 \wedge \omega^1 + \sigma^2 \wedge \omega^2 = 0, \tag{2.76}$$

where $\omega^1 := \omega_3^1 = -\omega_1^3, \omega^2 := \omega_3^2 = -\omega_2^3$. Both pairs of 1-forms, (σ^1, σ^2) and (ω^1, ω^2), can be represented by a linear combination of du^1 and du^2. Hence, (ω^1, ω^2) are expressed by a linear combination of σ^1 and σ^2 with constant matrix coefficients β_{ij} (say) as follows:

$$\omega^1 = -\beta_{11}\sigma^1 - \beta_{12}\sigma^2, \qquad \omega^2 = -\beta_{21}\sigma^1 - \beta_{22}\sigma^2. \tag{2.77}$$

Substituting this into (2.76), we have $(\beta_{21} - \beta_{12})\sigma^1 \wedge \sigma^2 = 0$. Hence, we obtain the symmetry, $\beta_{12} = \beta_{21}$, and the second fundamental form (2.73) is written as

$$\mathrm{II} = \beta_{11}\sigma^1\sigma^1 + 2\beta_{12}\sigma^1\sigma^2 + \beta_{22}\sigma^2\sigma^2. \tag{2.78}$$

Comparing the present first and second fundamental forms (2.70) and (2.78) (in the local orthonormal frame $\{\boldsymbol{e}_1, \boldsymbol{e}_2\}$) with the previous (2.5) and (2.20) respectively, it is found that the present metric tensor is $g_{ij} = \delta_{ij}$ and

$$b_{ij} = \beta_{ij}. \tag{2.79}$$

2.7.2. *Structure equations*

It is convenient to use matrix notations and write as follows:

$$\boldsymbol{e} = \begin{pmatrix} \boldsymbol{e}_1 \\ \boldsymbol{e}_2 \\ \boldsymbol{e}_3 \end{pmatrix}, \quad \boldsymbol{\sigma} = (\sigma^1, \sigma^2, \sigma^3), \quad \Omega = \begin{pmatrix} 0 & -\varpi & -\omega^1 \\ \varpi & 0 & -\omega^2 \\ \omega^1 & \omega^2 & 0 \end{pmatrix}, \tag{2.80}$$

where $\Omega = (\omega_j^i)$ and $\sigma^3 = 0$. Then, the first structure equation (2.68) is

$$d\boldsymbol{x} = \boldsymbol{\sigma}\boldsymbol{e}, \quad \text{with} \quad \sigma^3 = 0, \tag{2.81}$$

where $\boldsymbol{\sigma}$ is the basis of 1-forms dual to the frame $\boldsymbol{e}$. One may recall that $d\boldsymbol{x}$ is a 1-form having a vector value (*a vector-valued* 1-*form*). Note that $\sigma^1 \wedge \sigma^2$ represents the surface element of Σ.

Similarly, the second structure equation, (2.71) and (2.72), is

$$d\boldsymbol{e} = \Omega\, \boldsymbol{e}, \quad \Omega = -\Omega^T, \tag{2.82}$$

where the left-hand superscript T denotes *transpose* of the matrix. This is also a vector-valued 1-form. In components, we have

$$\begin{aligned} \mathrm{d}\boldsymbol{e}_1 &= -\varpi \boldsymbol{e}_2 - \omega^1 \boldsymbol{e}_3, \\ \mathrm{d}\boldsymbol{e}_2 &= \ \ \varpi \boldsymbol{e}_1 - \omega^2 \boldsymbol{e}_3, \\ \mathrm{d}\boldsymbol{e}_3 &= \omega^1 \boldsymbol{e}_1 + \omega^2 \boldsymbol{e}_2. \end{aligned} \tag{2.83}$$

Equation (2.74) is written as

$$\begin{aligned} \mathrm{d}(\mathrm{d}\boldsymbol{x}) &= \mathrm{d}(\boldsymbol{\sigma e}) = \mathrm{d}\boldsymbol{\sigma e} - \boldsymbol{\sigma} \wedge \mathrm{d}\boldsymbol{e} \\ &= \mathrm{d}\boldsymbol{\sigma e} - \boldsymbol{\sigma} \wedge \Omega \boldsymbol{e} = (\mathrm{d}\boldsymbol{\sigma} - \boldsymbol{\sigma} \wedge \Omega)\boldsymbol{e} = 0. \end{aligned} \tag{2.84}$$

Thus, Eqs. (2.75) and (2.76) are represented by a single expression,

$$\mathrm{d}\boldsymbol{\sigma} - \boldsymbol{\sigma} \wedge \Omega = 0. \tag{2.85}$$

Equations (2.82) and (2.85) are called the *structure equations*. Similarly, from $\mathrm{d}(\mathrm{d}\boldsymbol{e}) = 0$, we have

$$\begin{aligned} \mathrm{d}(\mathrm{d}\boldsymbol{e}) &= \mathrm{d}(\Omega \boldsymbol{e}) = \mathrm{d}\Omega \boldsymbol{e} - \Omega \wedge \mathrm{d}\boldsymbol{e} \\ &= \mathrm{d}\Omega \boldsymbol{e} - \Omega \wedge \Omega \boldsymbol{e} = (\mathrm{d}\Omega - \Omega \wedge \Omega)\boldsymbol{e} = 0. \end{aligned} \tag{2.86}$$

Thus, we obtain the second integrability equation,

$$\mathrm{d}\Omega - \Omega \wedge \Omega = 0. \tag{2.87}$$

In components, it is written as $\mathrm{d}\omega_i^k + \omega_j^k \wedge \omega_i^j = 0$ $(i, k = 1, 2, 3)$, or as

$$\mathrm{d}\varpi - \omega^1 \wedge \omega^2 = 0, \quad \mathrm{d}\omega^1 + \varpi \wedge \omega^2 = 0, \quad \mathrm{d}\omega^2 - \varpi \wedge \omega^1 = 0. \tag{2.88}$$

Thus, from two vectorial structure equations (2.81) and (2.82), we have obtained six integrability conditions (2.75), (2.76) (equivalent to (2.85)) and (2.88). Almost all of local surface theory is contained in these equations.

In particular, from the first equation of (2.88), we obtain

$$\mathrm{d}\varpi = \omega^1 \wedge \omega^2 = K\sigma^1 \wedge \sigma^2, \tag{2.89}$$

$$K = b_{11}b_{22} - b_{12}b_{21}, \tag{2.90}$$

where (2.77) and (2.79) are used. This is often called the Cartan's second structure equation. From (2.56), it is seen that the coefficient K is the Gaussian curvature since $\det(g_{\alpha\beta}) = 1$.

It is reminded that Eq. (2.75) of the first integrability condition are

$$\mathrm{d}\sigma^1 = \sigma^2 \wedge \varpi, \quad \mathrm{d}\sigma^2 = \varpi \wedge \sigma^1, \tag{2.91}$$

while Eq. (2.89) is deduced from the second integrability condition.

The first equation (2.89) implies that the Gaussian curvature K is given once we know σ^1, σ^2 and ϖ. When σ^1 and σ^2 are given, Eqs. (2.75) suffice to determine ϖ. In fact, if the two equations of (2.75) give $\mathrm{d}\sigma^1 = A\sigma^1 \wedge \sigma^2$ and $\mathrm{d}\sigma^2 = B\sigma^1 \wedge \sigma^2$, we must have $\varpi = -A\sigma^1 - B\sigma^2$. Thus, K is completely determined from σ^1 and σ^2. General consideration of this aspect will be given in §3.5.

2.7.3. *Geodesic equation*

On a geodesic curve $\gamma(s)$, its tangent vector T is translated parallel along itself. This is represented by $\bar{\nabla}_T T = 0$. From (2.36), this is written as

$$(\mathrm{d}T^\alpha)\boldsymbol{e}_\alpha + T^\alpha \bar{\mathrm{d}}\boldsymbol{e}_\alpha = 0.$$

From (2.83), we have $\bar{\mathrm{d}}\boldsymbol{e}_1 = -\varpi \boldsymbol{e}_2$ and $\bar{\mathrm{d}}\boldsymbol{e}_2 = \varpi \boldsymbol{e}_1$. Therefore we obtain

$$\mathrm{d}T^1 + T^2\varpi = 0, \quad \mathrm{d}T^2 - T^1\varpi = 0. \tag{2.92}$$

If the parameter s is taken as the arc-length of the curve, $T = \mathrm{d}\gamma/\mathrm{d}s$ is a unit tangent vector and can be represented as

$$T^1 = \cos\varphi(s), \quad T^2 = \sin\varphi(s), \tag{2.93}$$

where $\varphi(s)$ is the angle of the tangent T with respect to the axis $\boldsymbol{e}_1$ at a point s (Fig. 2.12). It is easily shown that $T^1\mathrm{d}T^2 - T^2\mathrm{d}T^1 = \mathrm{d}\varphi$. From (2.92), we also obtain $T^1\mathrm{d}T^2 - T^2\mathrm{d}T^1 = \varpi$. Thus, it is found that

$$\varpi = \mathrm{d}\varphi. \tag{2.94}$$

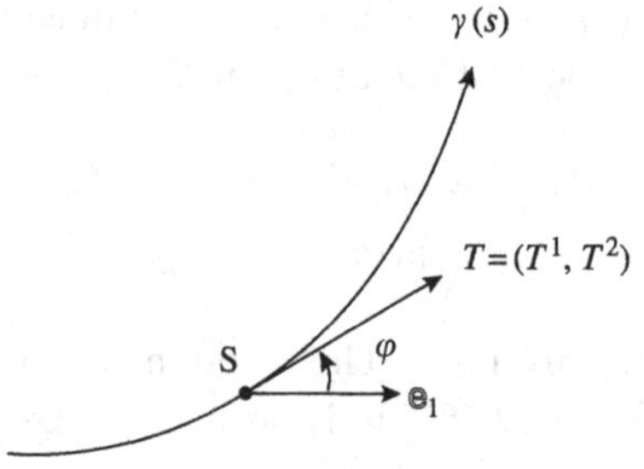

Fig. 2.12. Angle $\phi(s)$.

2.8. Gauss Spherical Map

In Appendix D.2, the curvature κ of a plane curve is interpreted by the Gauss map G, which indicates that the κ is given by the ratio of the arc-length over the Gauss' unit circle with respect to the arc-length along which a point $\boldsymbol{x}$ moves. This can be generalized to the surface in $\mathbb{R}^3$ [Kob77].

It is recalled that $\sigma^1 \wedge \sigma^2$ is a surface element on Σ^2. As a point $\boldsymbol{x}$ moves over Σ^2, $\boldsymbol{e}_3$ moves over a region on a unit sphere S^2. This is a map G, called the *Gauss spherical map* or *spherical image*, i.e. $G : \boldsymbol{x}(p) \mapsto \boldsymbol{e}_3(p)$ for $p \in \Sigma^2$. The two unit vectors $\boldsymbol{e}_1$ and $\boldsymbol{e}_2$ orthogonal to $\boldsymbol{e}_3$ lie in the tangent plane to the spherical image as well as in Σ^2, and form an orthonormal frame on the sphere S^2. From the equation $\mathrm{d}\boldsymbol{x} = \sigma^1\boldsymbol{e}_1 + \sigma^2\boldsymbol{e}_2$, it is understood that the 2-form $\sigma^1 \wedge \sigma^2$ represents a surface element of Σ^2. Analogously, from the equation $\mathrm{d}\boldsymbol{e}_3 = \omega^1\boldsymbol{e}_1 + \omega^2\boldsymbol{e}_2$, it is read that the 2-form $\omega^1 \wedge \omega^2$ represents a surface element of the spherical image S^2 (Fig. 2.13).

Because there exists only one linearly independent 2-form on the two-dimensional manifold,[9] the two 2-forms, $\sigma^1 \wedge \sigma^2$ and $\omega^1 \wedge \omega^2$, are linearly dependent, and one can write the connecting relation in terms of a scalar K' in the following form:

$$\omega^1 \wedge \omega^2 = K'\sigma^1 \wedge \sigma^2. \tag{2.95}$$

Equations (2.89) and (2.62) show that the scalar coefficient K' is nothing but the Gaussian curvature $K = \kappa_1\kappa_2$. This is analogous to the curvature interpretation of a plane curve by the Gauss map in Appendix D.2 where

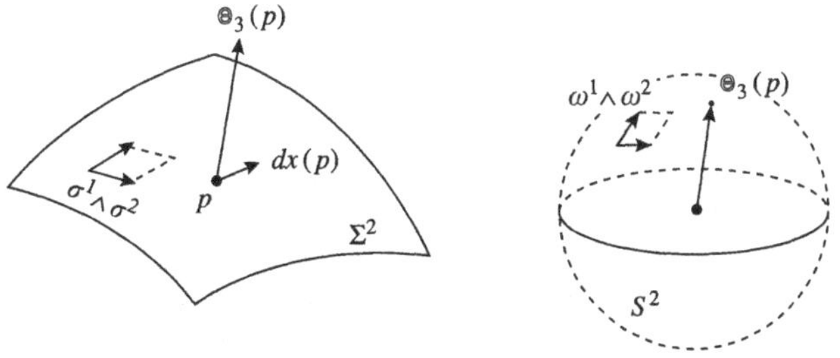

Fig. 2.13. Gauss spherical map.

[9]Dimension of 2-forms on vectors in $\mathbb{R}^2$ is 1, since the formula (B.6) gives $\binom{2}{2} = 1$ with $n = 2$ and $k = 2$.

the line element $\Delta \boldsymbol{p}$ of a plane curve corresponds to the surface element $\sigma^1 \wedge \sigma^2$ of Σ^2.

Similarly, we have a 2-form $\sigma^1 \wedge \omega^2 - \sigma^2 \wedge \omega^1$ on Σ^2, which can be represented as

$$\sigma^1 \wedge \omega^2 - \sigma^2 \wedge \omega^1 = (b_{11} + b_{22})\sigma^1 \wedge \sigma^2 = 2H\sigma^1 \wedge \sigma^2,$$

where the scalar coefficient H is the mean curvature of Σ^2 since $2H = b_{11} + b_{22} = \kappa_1 + \kappa_2$ due to $g_{ij} = \delta_{ij}$ (see the footnote of §2.5.4).

2.9. Gauss–Bonnet Theorem I

Consider a subdomain A on a surface Σ^2 with a boundary ∂A. Let us integrate Eq. (2.89) over A and transform it by using the Stokes theorem (B.46), then we obtain

$$\int_A K\sigma^1 \wedge \sigma^2 = \int_A \mathrm{d}\varpi = \int_{\partial A} \varpi. \tag{2.96}$$

Let us consider the integral over a *geodesic triangle* $A^{(3)}$, that is a triangle whose three sides are geodesics. The triangle is assumed to enclose a simply connected area such that the curvature K has the same sign within or on the triangle. Such a triangle is shown in Fig. 2.14(a). Three oriented sides are denoted by ∂A_1, ∂A_2 and ∂A_3, and three exterior angles are denoted by ϵ_1, ϵ_2, ϵ_3.

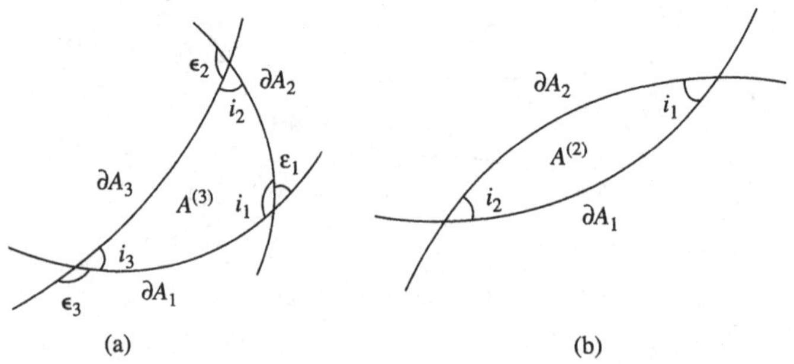

Fig. 2.14. (a) Geodesic triangle, (b) Geodesic di-angle.

Then from (2.94), we have

$$\int_{\partial A^{(3)}} \varpi = \int_{\partial A^{(3)}} \mathrm{d}\varphi = \sum_{k=1}^{3} \int_{\partial A_k} \mathrm{d}\varphi, \tag{2.97}$$

for the piecewise-geodesics $\partial A^{(3)} = \partial A_1 + \partial A_2 + \partial A_3$. On the other hand, going around the closed curve consisting of the three piecewise-geodesics, we have

$$\sum_{k=1}^{3} \int_{\partial A_k} \mathrm{d}\varphi + \sum_{k=1}^{3} \epsilon_k = 2\pi. \tag{2.98}$$

Combining the above three equations, we find the following equation,

$$\int_{A^{(3)}} K\sigma^1 \wedge \sigma^2 = 2\pi - \sum_{k=1}^{3} \epsilon_k. \tag{2.99}$$

This is the **Gauss–Bonnet Theorem** for a geodesic triangle. This is immediately generalized to a geodesic n-polygon $A^{(n)}$ by replacing $\sum_{k=1}^{3} \epsilon_k$ by $\sum_{k=1}^{n} \epsilon_k$ in the above equation.

Since the exterior angle ϵ_k is given as $\pi - i_k$ in terms of the interior angle i_k, the Gauss–Bonnet Theorem for the geodesic n-polygon is given by

$$\int_{A^{(n)}} K\sigma^1 \wedge \sigma^2 = (2-n)\pi + \sum_{k=1}^{n} i_k. \tag{2.100}$$

For a geodesic triangle $n = 3$, we obtain

$$\int_{A^{(3)}} K\sigma^1 \wedge \sigma^2 = -\pi + (i_1 + i_2 + i_3). \tag{2.101}$$

It follows that *the sum of the interior angles is greater or smaller than* π *according to the Gaussian curvature being positive or negative respectively*, and that the sum is equal to π for a triangle on a flat plane ($K = 0$) as is well-known in the euclidean geometry.

For a geodesic di-angle ($n = 2$, Fig. 2.14(b)), Eq. (2.100) reduces to

$$\int_{A^{(2)}} K\sigma^1 \wedge \sigma^2 = i_1 + i_2.$$

This makes sense for $A^{(2)}$ of positive K, while it does not for $A^{(2)}$ of negative K since $i_1 + i_2$ should be positive if it exists. As a consequence, it follows that *two geodesics on a surface of negative curvature cannot meet at two points and cannot enclose a simply connected area.*

2.10. Gauss–Bonnet Theorem II

Suppose that M_0^2 is a closed submanifold of $\mathbb{R}^3$. The *total curvature* of M_0 is given by

$$\int_{M_0} K\mathrm{d}A, \quad \mathrm{d}A = \sigma^1 \wedge \sigma^2. \tag{2.102}$$

Then, the (Brower) **degree**, Deg(Gn), of the Gauss normal map (Gn), $M^2 \to S^2$, is defined as

$$\mathrm{Deg}(M_0) := \frac{1}{4\pi}\int_{M_0} K\mathrm{d}A = \frac{1}{4\pi}\int_{S^2} \mathrm{d}S^2 = 1, \tag{2.103}$$

by (2.95), where $\mathrm{d}S^2 = \omega^1 \wedge \omega^2$, since the area of the unit sphere S^2 is 4π. If we smoothly deform M_0, the curvature K will change smoothly and likewise the area form $\mathrm{d}A$, yet the total curvature normalized by 4π remains constant to be the integer 1. This implies that the degree Deg is a topological invariant.

For a surface M_g of *genus* g, i.e. the surface of a multihole doughnut with g-holes (Fig. 2.15), we have

$$\mathrm{Deg}(M_g) := \frac{1}{4\pi}\int_{M_g} K\mathrm{d}A = 1 - g. \tag{2.104}$$

This is another form of the **Gauss–Bonnet Theorem**, and gives a measure of the genus of the surface. For example, the degree of a single-hole doughnut, that is T^2, is $\mathrm{Deg}(T^2) = 0$.

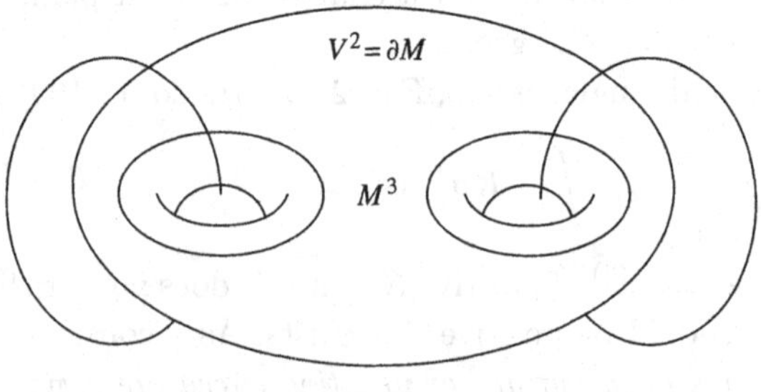

Fig. 2.15. A surface ∂M of genus $g = 2$ of a manifold M^3.

For a closed manifold M^2, the **Euler characteristic** $\chi(M^2)$ is also defined by

$$\chi(M^2) = \#(\text{vertices}) - \#(\text{edges}) + \#(\text{faces}), \tag{2.105}$$

for all triangulations of M^2 (breaking M^2 up into a number of triangle-*simplexes*), where #(A's) denotes the number of A's, and $\chi(M^2)$ is independent of the triangulation. It can be shown that

$$\chi(M^2) = (1/2\pi) \int_{M^2} K \mathrm{d}A = 2 - 2g \tag{2.106}$$

from the above Gauss–Bonnet Theorem [Fra97, Ch. 16, 17, 22.3]. Euler characteristic of a single-hole doughnut T^2 is $\chi(T^2) = 2 - 2 = 0$, whereas for a sphere (without a hole), $\chi(S^2) = 2$.

Regarding a circular disk D^2, we obtain $\chi(D^2) = 1$ (Fig. 2.16). This is obtained with a simple-minded argument that a sphere S^2 may be collapsed into two disks (top and bottom). Topologically, the unit disk (with two *antipodal* points on the unit circle identified) is equivalent to the real projective space $\mathbb{R}P^2$, whose Euler characteristic is half of S^2 [Fra97, §16.2b]. Likewise, a ring which is defined by a circular disk with a circular hole inside has the Euler characteristic $\chi = 0$, a half of $\chi(T^2) = 0$.

An example of triangulation of a torus T^2 is made as in Fig. 2.17. In T^2, the edge P_1R_1 is identified with P_4R_4, and likewise, P_1P_4 with R_4R_4 (*Example: Torus* of §2.1). Namely, the two vertical edges are brought together by bending and then sewn together, and moreover the two horizontal edges are brought together by bending and then sewn together. Therefore, the number of vertices, $P_1, P_2, P_3, Q_1, Q_2, Q_3$, is six, whereas the number of edges is 18, and the number of faces is 12. Thus, the Euler characteristic of T^2 is $\chi(T^2) = 6 - 18 + 12 = 0$, consistent with the above, which is independent of the triangulation.

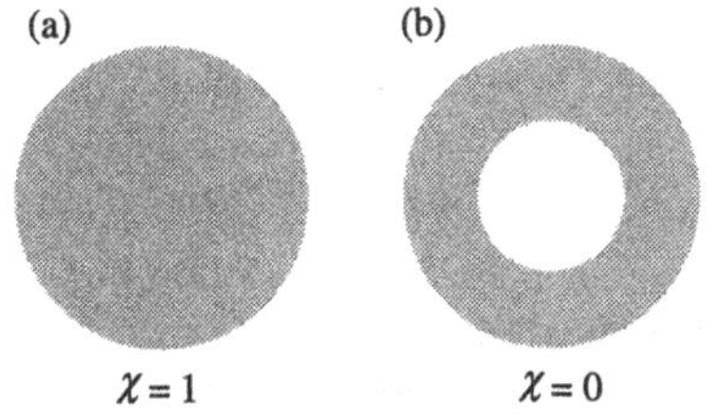

Fig. 2.16. (a) Circular disk, (b) circular annulas.

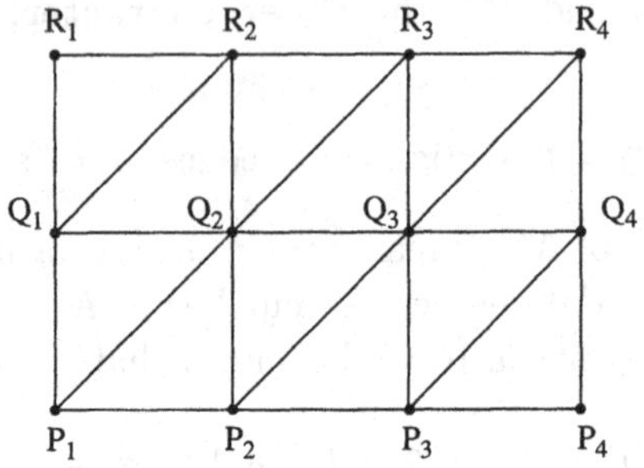

Fig. 2.17. A triangulation of torus T^2.

2.11. Uniqueness: First and Second Fundamental Tensors

The first and second fundamental tensors $g_{\alpha\beta}$ and $b_{\alpha\beta}$ of a surface satisfy the Gauss equation (2.48) and the Mainardi–Codazzi equation (2.49). In this section, we inquire conversely whether the two tensors satisfying these equations denote in fact the fundamental tensors of a certain surface. This means that we investigate integrability of the Weingarten equation (2.19) and the Gauss's surface equation (2.30).

In order to answer this question, we introduce the tangent vector $\boldsymbol{p}$ defined by

$$\boldsymbol{p}_\alpha = (p^i_\alpha) := \frac{\partial x^i}{\partial u^\alpha} = \boldsymbol{x}_\alpha, \quad (i = 1, 2, 3), \tag{2.107}$$

and recall Eqs. (2.3) and (2.18), from which we have

$$\langle \boldsymbol{p}_\alpha, \boldsymbol{p}_\beta \rangle = g_{\alpha\beta}, \quad \langle \boldsymbol{N}, \boldsymbol{N} \rangle = 1, \quad \langle \boldsymbol{N}, \boldsymbol{p}_\alpha \rangle = 0. \tag{2.108}$$

The Weingarten equation and the surface equation are reproduced here,

$$\left.\begin{aligned} \boldsymbol{N}_\alpha &= -b_{\alpha\beta}\, g^{\beta\gamma} \boldsymbol{p}_\gamma, \\ (\boldsymbol{p}_\alpha)_\beta &= b_{\alpha\beta} \boldsymbol{N} + \Gamma^\gamma_{\alpha\beta}\, \boldsymbol{p}_\gamma, \end{aligned}\right\} \tag{2.109}$$

respectively, for $\boldsymbol{N} = (N^i), \boldsymbol{p}_\alpha = (p^i_\alpha) \in \mathbb{R}^3$ and $\alpha, \beta, \gamma = 1, 2$.

The nine equations (2.109) (for $\boldsymbol{N}, \boldsymbol{p}_1, \boldsymbol{p}_2$) and six functional equations (2.108) constitute a *mixed* system of a first order partial differential equations for N^i and p^i_α, [Eis47, §23] where N^i, p^i_α $(i = 1, 2, 3; \alpha = 1, 2)$ are nine dependent variables, whereas Eq. (2.108) constitutes six constraint conditions on such nine variables, when the tensors $g_{\alpha\beta}$ and $b_{\alpha\beta}$ are given.

The existence conditions of an integral surface are given by the Gauss equation (2.48) and the Mainardi–Codazzi equation (2.49), which are *presupposed* to be satisfied. When Eqs. (2.108) are differentiated with respect to u^α, the resulting equations are satisfied in consequence of (2.109), and consequently we have three independent equations among the equations (2.109). According to the theorem [23.2] of [Eis47, §23], the solution to Eq. (2.109) under (2.108) involves three arbitrary constants (say a^1, a^2, a^3) associated with initial values.

When such a solution is given, the equations of the surface are given by the quadratures:

$$x^i = \int p^i_\alpha \mathrm{d}u^\alpha + b^i,$$

by (2.107), where b^i are three additional constants. The vector $\boldsymbol{N}$ is a unit normal to this surface due to the second and third of (2.108).

We have now six arbitrary constants (a^i, b^i) involved in the present problem for a surface in $\mathbb{R}^3$. This arbitrariness is interpreted on the basis of the following observation. Namely, suppose that $\boldsymbol{x} = (x^i)$ and $\boldsymbol{N} = (N^i)$ constitute a solution of Eqs. (2.107)–(2.109). Consider a pair of new quantities $\bar{\boldsymbol{x}} = (\bar{x}^i)$ and $\bar{\boldsymbol{N}} = (\bar{N}^i)$ defined by

$$\bar{\boldsymbol{x}} = A\boldsymbol{x} + (b^i), \quad \bar{\boldsymbol{N}} = A\boldsymbol{N}, \tag{2.110}$$

(and $\bar{\boldsymbol{p}} = A\boldsymbol{p}$), where b^i are constants, and $A = (a^i_k)$ is an element of $SO(3)$, i.e. a constant 3×3 tensor, representing rotational transformation (§1.8.2). The position vector $\bar{\boldsymbol{x}}$ is obtained by rotating $\boldsymbol{x}$ with the orthogonal matrix A and translating the origin by (b^i), while the vector $\bar{\boldsymbol{N}}$ is obtained by rotating the normal $\boldsymbol{N}$ with the matrix A. The matrix A is subject to the following orthogonality condition:

$$AA^T (= A^T A) = I, \quad (\det A = 1). \tag{2.111}$$

This condition keeps the inner product of (2.108) invariant, and the equations of (2.109) are unchanged for the overbar variables too. From the first equation of (2.110), the two surfaces defined by $\boldsymbol{x}$ and $\bar{\boldsymbol{x}}$ may be obtained from one another by a rotation and a translation, that is, a *rigid motion* without change of form. This transformation of rigid-motion involves arbitrary constants whose number is 6 since the number of b^i is 3 and the number of matrix elements of A is 9, with the total 12. However the constraint conditions (2.111) are 6. Therefore we have 6 arbitrary constants.

Thus, the transformation of rigid-motion involves the same number of arbitrary constants as the general solution of the present problem of determining a surface in $\mathbb{R}^3$ for a given set of fundamental tensors satisfying (2.48) and (2.49). It follows that *any two surfaces with the same tensors $g_{\alpha\beta}$ and $b_{\alpha\beta}$ which satisfy the equations of Gauss and Mainardi–Codazzi are transformed into one another by a rigid-motion in space.*

Chapter 3

Riemannian Geometry

We consider the "inner" geometry of a manifold which is not a part of an euclidean space. We consider only tangential vectors, and any vector normal to the manifold is not available. We presuppose that each tangent bundle possesses an inner product depending on points of its base space smoothly. The space is curved in general. The Riemannian curvature of a manifold governs the behavior of geodesics on it and corresponding dynamical system. Dimension of the manifold is not always finite.

3.1. Tangent Space

3.1.1. *Tangent vectors and inner product*

If a manifold under consideration were a part of an euclidean space, it would inherit a local euclidean geometry (such as the length) from the enveloping euclidean space, as is the case of surfaces in $\mathbb{R}^3$ considered in §2.1–2.3. What we consider here is not a part of an euclidean space, so the existence of a local geometry must be postulated.

Let M^n be an n-dimensional manifold. The problem is how to define a tangent vector X when we are constrained to the manifold M^n. According to the theory of manifolds in Chapter 1, we introduced a local coordinate frame $(x^1, \ldots, x^n)$. In §1.2, guided by the experience of a flow in an euclidean space, we defined a tangent vector $X \in T_xM^n$ by

$$X = X^i \frac{\partial}{\partial x^i} = X^i \partial_i,$$

where $\partial/\partial x^1, \ldots, \partial/\partial x^n$ is a natural frame associated with the coordinate system. Furthermore, in the language of the differential form (§1.10.3(ii)), we defined a vector-valued 1-form,[1]

$$\omega = \partial_i \otimes \mathrm{d}x^i.$$

From the calculus of differential forms, we have

$$\omega[X] = \partial_i \otimes \mathrm{d}x^i[X] = X^i \partial_i = X.$$

This is interpreted as follows. The 1-form ω *yields* the same vector X by *eating* a vector X.

We consider the intrinsic (inner) geometry of the manifold M^n. It is supposed that an inner product $\langle\cdot,\cdot\rangle$ is given in the tangent space $T_x M^n$ at each point x of M^n. If X and Y are two smooth tangent vector fields of the tangent bundle TM^n (see §1.3.1, 1.4.2), then $\langle X, Y\rangle$ is a smooth real function on M^n.

Every inner product space has an orthonormal basis ([Fla63], §2.5). Let us introduce the natural coordinate frame of the orthonormal vectors, $\boldsymbol{e}_1, \ldots, \boldsymbol{e}_n$ for $T_x M^n$, where $\partial_i = A_i^j \boldsymbol{e}_j$. Then, one can write a tangent vector as $\mathrm{d}\boldsymbol{x} = \partial_i \otimes \mathrm{d}x^i = \boldsymbol{e}_j \otimes \sigma^j$, where σ^j are 1-forms represented in the form, $\sigma^j = A_i^j \mathrm{d}x^i$.

3.1.2. *Riemannian metric*

On a Riemannian manifold M^n, a positive definite inner product $\langle\cdot,\cdot\rangle$ is defined on the tangent space $T_x M^n$ at $x = (x^1, \ldots, x^n) \in M$ and assumed to be differentiable. For two tangent fields $X = X^i(x)\partial_i, Y = Y^j(x)\partial_j \in TM^n$ (tangent bundle), the *Riemannian metric* is given by

$$\langle X, Y\rangle(x) = g_{ij} X^i(x) Y^j(x),$$

as already defined in (1.28),[2] where the metric tensor, $g_{ij}(x) = \langle \partial_i, \partial_j\rangle = g_{ji}(x)$, is symmetric and differentiable with respect to x^i. This bilinear quadratic form is called the *first fundamental form*. In terms of differential

[1]This definition is independent of the local coordinate system. Indeed, if $u^1, \ldots, u^n$ is another system, then $\mathrm{d}u^i = (\partial u^i/\partial x^k)\mathrm{d}x^k$ and $(\partial/\partial u^i) = (\partial x^l/\partial u^i)(\partial/\partial x^l)$. Hence, we have $\mathrm{d}u^i(\partial/\partial u^i) = (\partial u^i/\partial x^k)(\partial x^l/\partial u^i)\mathrm{d}x^k(\partial/\partial x^l) = \delta_k^l \mathrm{d}x^k(\partial/\partial x^l) = \mathrm{d}x^k(\partial/\partial x^k)$.

[2]If the inner product is only nondegenerate rather than positive definite like that in Minkowski space, the resulting structure on M^n is called a *pseudo*-Riemannian.

1-forms dx^i, this is equivalent to

$$\mathrm{I} := g_{ij} dx^i \otimes dx^j.$$

Eating two vectors $X = X^i(x)\partial_i$ and $Y = Y^j(x)\partial_j$, this yields

$$\mathrm{I}(X, Y) = g_{ij} dx^i(X) dx^j(Y) = g_{ij} X^i Y^j. \tag{3.1}$$

The inner product is said to be *nondegenerate*,

$$\text{if} \quad \langle X, Y \rangle = 0, \quad {}^\forall Y \in TM^n, \quad \text{only when } X = 0. \tag{3.2}$$

3.1.3. *Examples of metric tensor*

(a) *Finite dimensions*
Consider a dynamical system of N degrees of freedom in a *gravitational* field with the potential $V(\bar{q})$ and the kinetic energy,

$$T = \frac{1}{2} a_{ij} \dot{q}^i \dot{q}^j, \quad \text{where} \quad \bar{q} = (q^i), \quad (i = 1, \ldots, N).$$

The metric is defined by $g_{ij} \dot{q}^i \dot{q}^j$, where, for an energy constant E,

$$g_{ij} = g_{ij}^J(\bar{q}) := 2(E - V(\bar{q})) a_{ij}, \quad \text{for } i, j = 1, \ldots, N, \tag{3.3}$$

is the *Jacobi's* metric tensor [Ptt93]. In Chapter 6, we consider another metric called the Eisenhart metric g_{ij}^E.

(b) *Infinite dimensions*
For two tangent fields $X = u(x)\partial_x, Y = v(x)\partial_x$ on the tangent space $T_{id}D(S^1)$, an inner product is defined by

$$\langle X, Y \rangle := \int_{S^1} u(x) v(x) \, dx.$$

Correspondingly, a right-invariant metric on the group $\mathcal{D}(S^1)$ of diffeomorphisms (§1.8) is defined in the following way:

$$\langle U, V \rangle_\xi := \int_{S^1} (U_\xi \circ \xi^{-1}, V_\xi \circ \xi^{-1})_x \, dx = \int_{S^1} (u, v)_x \, dx = \langle X, Y \rangle, \tag{3.4}$$

(see also §7.2), where U_ξ and V_ξ are right-invariant fields defined by $U_\xi(x) = u \circ \xi(x)$ and $V_\xi(x) = v \circ \xi(x)$ for $\xi \in \mathcal{D}(S^1)$, and $(\cdot, \cdot)_x$ denotes the scalar product pointwisely at $x \in S^1$.

This metric is *right-invariant*, and we have $\langle U, V\rangle_\xi = \langle U, V\rangle_{id} = \langle X, Y\rangle$. We will see a left-invariant metric in Chapter 4, and right-invariant metrics in Chapters 5 and 8.

3.2. Covariant Derivative (Connection)

We are going to introduce an additional structure to a manifold that allows to form a covariant derivative, *taking vector fields into second-rank tensor fields.*

3.2.1. *Definition*

Here, a general definition is given to a covariant derivative, called a connection in mathematical literature, on a Riemannian *curved* manifold M^n. Let two vector fields X, Y be defined near a point $p \in M^n$ and two vectors U and V be defined at p. A *covariant derivative* (or *connection*) is an operator ∇. The operator ∇ assigns a vector $\nabla_U X$ at p to each pair (U, X) and satisfies the following relations:

$$\left.\begin{array}{ll}\text{(i)} & \nabla_U(aX+bY) = a\nabla_U X + b\nabla_U Y,\\ \text{(ii)} & \nabla_{aU+bV}X = a\nabla_U X + b\nabla_V X,\\ \text{(iii)} & \nabla_U(f(x)X) = (Uf)X + f(x)\nabla_U X,\end{array}\right\} \tag{3.5}$$

for a smooth function $f(x)$ and $a, b \in \mathbb{R}$, where $Uf = \mathrm{d}f[U] = U^j\partial_j f$ (§1.4.1), and $U = U^j\partial_j$. Using the representations,

$$X = X^i\partial_i, \qquad Y = Y^j\partial_j,$$

and applying the above properties (i)–(iii), we obtain

$$\nabla_X Y = \nabla_{X^i\partial_i}(Y^j\partial_j) = X^i\nabla_{\partial_i}(Y^j\partial_j)$$
$$= (X^i\partial_i Y^k)\partial_k + X^i Y^j \Gamma^k_{ij}\partial_k = (\nabla_X Y)^k\partial_k, \tag{3.6}$$
$$\nabla_{\partial_i}\partial_j := \Gamma^k_{ij}\partial_k, \tag{3.7}$$

where Γ^k_{ij} is called the *Christoffel symbol.* The ith component of $\nabla_X Y$ is

$$(\nabla_X Y)^i = X^j\frac{\partial Y^i}{\partial x^j} + \Gamma^i_{jk}X^j Y^k = X^j\nabla_j Y^i \tag{3.8}$$
$$= \mathrm{d}Y^i(X) + (\Gamma^i_{jk}Y^k)\mathrm{d}x^j(X) := \nabla Y^i(X) \tag{3.9}$$
$$\nabla Y^i = \mathrm{d}Y^i + \Gamma^i_{jk}Y^k\mathrm{d}x^j, \qquad \nabla_j Y^i = \partial_j Y^i + \Gamma^i_{jk}Y^k, \tag{3.10}$$

where ∇Y^i is called a *connection* 1-*form.*

On a manifold M^n, an *affine frame* consists of n vector fields $e_k = \partial_k$ ($k = 1, \ldots, n$), which are linearly independent and furnish a basis of the tangent space at each point p. Writing (3.7) and (3.9) in the form of vector-valued 1-forms, we have

$$\nabla e_j = \omega_j^k e_k, \qquad \nabla Y = (\mathrm{d}Y^k)e_k + \omega_j^k Y^j e_k, \tag{3.11}$$

respectively, where $\omega_j^k = \Gamma_{ij}^k \mathrm{d}x^i$. The operator ∇ is called the *affine connection*, and we have the following representation,

$$\nabla Y(X) = \nabla_X Y. \tag{3.12}$$

3.2.2. *Time-dependent case*

Most dynamical systems are *time dependent* and every tangent vector is written in the form, $\tilde{X} = \tilde{X}^i \partial_i = \partial_t + X^\alpha \partial_\alpha$, where $x^0 = t$ (time), and α denotes the indices of the spatial part, $\alpha = 1, \ldots, n$ (see (1.13), roman indices denote $0, 1, \ldots, n$). Correspondingly, the connection is written as

$$\nabla_{\tilde{X}} \tilde{Y} = \nabla_{\tilde{X}^i \partial_i} \tilde{Y}^j \partial_j = \nabla_{\partial_t} \tilde{Y}^j \partial_j + X^\alpha \nabla_{\partial_\alpha} (\tilde{Y}^j \partial_j),$$

where $\tilde{Y} = \partial_t + Y^\alpha \partial_\alpha$. We assume that the time axis is *straight*, which is represented by[3]

$$\nabla_{\partial_t} \partial_k = 0, \qquad \nabla_{\partial_k} \partial_t = 0. \tag{3.13}$$

The first property is equivalent to (3.15) below, and the second means $\nabla_{\partial_k} \tilde{Y} = \nabla_{\partial_k} Y$. Namely, the time part vanishes identically. Corresponding to (3.13), we obtain

$$\Gamma_{0k}^i = \Gamma_{k0}^i = 0 \qquad (i, k = 0, \ldots, n). \tag{3.14}$$

Writing the spatial part of $\tilde{X}$ and $\tilde{Y}$ as $X = X^\alpha \partial_\alpha$ and $Y = Y^\alpha \partial_\alpha$, respectively, we obtain

$$\nabla_{\partial_t} \tilde{Y} = \partial_t Y := \frac{\partial Y^\alpha}{\partial t} \partial_\alpha, \tag{3.15}$$

$$\nabla_{\tilde{X}} \tilde{Y} = \partial_t Y + \nabla_X Y. \tag{3.16}$$

[3]This is drawn cartoon-like in Fig. 1.9.

3.3. Riemannian Connection

There is one connection that is of special significance, having the property that parallel displacement *preserves inner products of vectors, and the connection is symmetric.*

3.3.1. *Definition*

There is a unique connection ∇ on a Riemannian manifold M called the *Riemannian connection* or *Levi–Civita* connection that satisfies

$$\text{(i)} \quad Z\langle X, Y\rangle = \langle \nabla_Z X, Y\rangle + \langle X, \nabla_Z Y\rangle \tag{3.17}$$

$$\text{(ii)} \quad \nabla_X Y - \nabla_Y X = [X, Y] \qquad \text{(torsion free)}, \tag{3.18}$$

for vector fields $X, Y, Z \in TM$, where $Z\langle\cdot,\cdot\rangle = Z^j\partial_j\langle\cdot,\cdot\rangle$. The property (i) is a compatibility condition with the metric. The torsion-free property (ii) requires the following symmetry, $\Gamma^k_{ij} = \Gamma^k_{ji}$. In fact, writing $X = X^i\partial_i$ and $Y = Y^j\partial_j$, the definitive expression (3.6) leads to

$$(\nabla_X Y - \nabla_Y X)^k = (XY - YX)^k + (\Gamma^k_{ij} - \Gamma^k_{ji})X^iY^j, \tag{3.19}$$

where $XY = X^i\partial_i(Y^k\partial_k) = X^i\partial_i Y^k\partial_k + X^iY^k\partial_i\partial_k$. One consequence of the first compatibility with metric will be given at the end of the next section. (See also [Fra97; Mil63].)

Owing to the above two properties, the Riemannian connection satisfies the following identity,

$$\begin{aligned} 2\langle \nabla_X Y, Z\rangle = {} & X\langle Y, Z\rangle + Y\langle Z, X\rangle - Z\langle X, Y\rangle \\ & + \langle [X, Y], Z\rangle - \langle [Y, Z], X\rangle + \langle [Z, X], Y\rangle. \end{aligned} \tag{3.20}$$

This equation *defines the connection* ∇ *by means of the inner product* $\langle\cdot,\cdot\rangle$ *and the commutator* $[\cdot,\cdot]$.[4]

[4]It is postulated that this formula would be applied to a system of infinite dimensions as well if an inner product and a commutator are defined consistently, under the restriction that other properties do not contradict with those of a finite dimensional system.

3.3.2. *Christoffel symbol*

The Christoffel symbol Γ^k_{ij} can be represented in terms of the metric tensor $g = (g_{ij})$ by the following formula:

$$\Gamma^k_{ij} = g^{k\alpha}\Gamma_{ij,\alpha}, \quad \Gamma_{ij,\alpha} = \frac{1}{2}(\partial_i g_{j\alpha} + \partial_j g_{\alpha i} - \partial_\alpha g_{ij}), \tag{3.21}$$

where $g^{k\alpha}$ denotes the component of the inverse g^{-1}, that is $g^{k\alpha} = (g^{-1})^{k\alpha}$, satisfying the relations $g^{k\alpha}g_{\alpha l} = g_{l\alpha}g^{\alpha k} = \delta^k_l$. The symmetry $\Gamma^k_{ij} = \Gamma^k_{ji}$ with respect to i and j follows immediately from (3.21) and $g_{ij} = g_{ji}$.

The formula (3.21) can be verified by using (3.17), $g_{ij}(x) = \langle \partial_i, \partial_j \rangle$ and $\nabla_{\partial_i}\partial_j = \Gamma^k_{ij}\partial_k$, and noting that

$$\partial_m g_{ij} = \partial_m \langle \partial_i, \partial_j \rangle = \langle \nabla_{\partial_m}\partial_i, \partial_j \rangle + \langle \partial_i, \nabla_{\partial_m}\partial_j \rangle \tag{3.22}$$

$$= \Gamma^k_{mi} g_{kj} + \Gamma^k_{mj} g_{ki}, \tag{3.23}$$

and that $\partial_i g_{jm} + \partial_j g_{mi} - \partial_m g_{ij} = 2g_{km}\Gamma^k_{ij}$.

3.4. Covariant Derivative along a Curve

3.4.1. *Derivative along a parameterized curve*

Consider a curve $x(t)$ on M^n passing through a point p whose tangent at p is given by

$$T = T^k\partial_k = \frac{\mathrm{d}x}{\mathrm{d}t} = \dot{x} = \dot{x}^k\partial_k,$$

and let Y be a tangent vector field defined along the curve $x(t)$. According to (3.6) or (3.9), the covariant derivative $\nabla_T Y$ is given by

$$\nabla_T Y := \frac{\nabla Y}{\mathrm{d}t} = [\mathrm{d}Y^i(T) + \Gamma^i_{kj}T^kY^j]\partial_i \tag{3.24}$$

$$= \left[\frac{\mathrm{d}}{\mathrm{d}t}Y^i + \Gamma^i_{kj}\dot{x}^kY^j\right]\partial_i, \tag{3.25}$$

since $T^k = \dot{x}^k$. When Y^i is a function of $x^k(t)$, then $(\mathrm{d}/\mathrm{d}t)Y^i = \dot{x}^k(\partial Y^i/\partial x^k)$. The expression $\nabla Y/\mathrm{d}t$ emphasizes the derivative along the curve $x(t)$ parameterized with t.

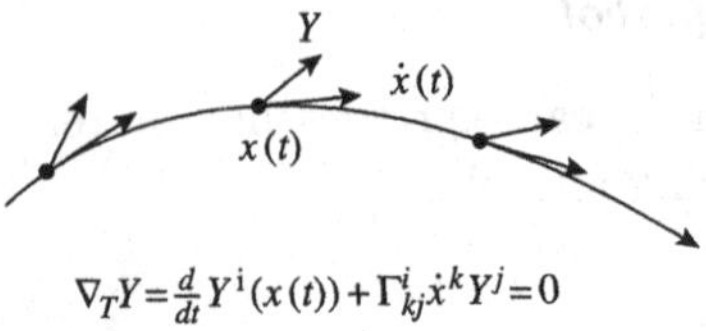

Fig. 3.1. Parallel translation.

3.4.2. *Parallel translation*

On the manifold M^n endowed with the connection ∇, one can define *parallel displacement* of a vector Y along a parameterized curve $x(t)$ (Fig. 3.1). Geometrical interpretation of the parallel displacement will be given in §3.7.1. Mathematically, this is defined by

$$\frac{\nabla Y}{dt} = \nabla_T Y = 0. \tag{3.26}$$

Thus, $Y = Y^i \partial_i$ is translated parallel along the curve $x(t)$ when[5]

$$\nabla Y = (\mathrm{d}Y^i + \omega^i_j Y^j)\partial_i = 0, \tag{3.27}$$

by using the connection-form representation of (3.11), or more precisely when $\dot{x}^k(\partial Y^i/\partial x^k) + \Gamma^i_{kj}\dot{x}^k Y^j = 0$.

For two vector fields X and Y translated parallel along the curve, we obtain

$$\langle X, Y\rangle = \text{constant} \quad \text{(under parallel translation)}, \tag{3.28}$$

because the scalar product is invariant:

$$T\langle X, Y\rangle = \langle \nabla_T X, Y\rangle + \langle X, \nabla_T Y\rangle = 0, \tag{3.29}$$

by (3.17), where each term vanishes due to (3.26).

3.4.3. *Dynamical system of an invariant metric*

As for the metric of right-invariant fields defined in §3.1.3(b), the scalar product $\langle X, Y\rangle$ is *unchanged* by right-translation. Moreover, in most dynamical systems to be studied below (Chapters 4, 5 and 8), the metrics are kept constant by the flows determined by tangent fields. In other

[5]For another interpretation of the covariant derivative, see §3.7.2.

words, the physical system evolves with time such that every metric for any pair of tangent fields is kept invariant.

Provided that the scalar products are constant along every flow, the first three terms on the right-hand side of (3.20) vanish identically.[6] Hence on the Riemannian manifold of *left-invariant* (or *right-invariant*) vector fields with an invariant metric, Eq. (3.20) reduces to

$$2\langle\nabla_X Y, Z\rangle = \langle[X, Y], Z\rangle - \langle[Y, Z], X\rangle + \langle[Z, X], Y\rangle. \tag{3.30}$$

We will examine this formula in §3.7.3 and confirm its consistency. In fact, the covariant derivative $\nabla_X Y$ determined by this formula assures that the metric is conserved, in which the combination of two terms in (3.29) vanishes, in contrast with the parallel translation.

3.5. Structure Equations

We consider reformulation of the theory on the basis of the differential forms and structure equations. As a simplest example of a two-dimensional Riemannian space, a surface of negative constant curvature will be considered at the end. A manifold of constant positive Gaussian curvature is called a sphere, while a manifold of negative constant curvature is called a pseudosphere.

3.5.1. *Structure equations and connection forms*

We investigate the geometry determined only by the first fundamental form, i.e. the *intrinsic geometry*. In §3.1, we introduced a structure equation of M^n already for local Riemannian geometry given by

$$\mathrm{d}\boldsymbol{x} = \mathrm{d}x^i \otimes \partial_i = \sigma^1 \boldsymbol{e}_1 + \cdots + \sigma^n \boldsymbol{e}_n, \tag{3.31}$$

which is a mixed form that assigns to each vector the same vector, where $\boldsymbol{e}_1, \ldots, \boldsymbol{e}_n$ are the orthonormal basis vectors for $T_x M^n$ and $\sigma^1, \ldots, \sigma^n$ are its dual form-basis taking the value $\sigma^i[x^k \boldsymbol{e}_k] = x^i$. The first fundamental form is given by

$$\mathrm{I} = \langle \mathrm{d}\boldsymbol{x}, \mathrm{d}\boldsymbol{x}\rangle = \sigma^1\sigma^1 + \cdots + \sigma^n\sigma^n, \tag{3.32}$$

where $g_{\alpha\beta} = \delta_{\alpha\beta}$. These are analogous to (2.68) (equivalently (2.81)) and (2.70) of §2.7. However, there is an essential departure from it for the structure equations now defined analogously to (2.82), since we are constrained

[6] It would be sufficient to say that we consider such vector fields only.

to the manifold M^n which is not a part of an euclidean space. Here, we can consider only the "tangential" component, and no "normal" component is available.

We define the connection form in the following way,

$$\nabla e_k = \omega^i_k e_i, \tag{3.33}$$

which is a vector-valued 1-form,[7] and try to find the connection 1-forms ω^k_i which are consistent with the following two conditions:

$$\langle \nabla e_i, e_j \rangle + \langle e_i, \nabla e_j \rangle = 0, \qquad \mathrm{d}(\mathrm{d}\boldsymbol{x}) = 0. \tag{3.34}$$

The first condition is associated with the orthonormality, $\langle e_i, e_j \rangle = \delta_{ij}$, and the second is the euclidean analogue. The expressions (3.33) and (3.34) are consistent with the first of (3.11) of §3.2 and (3.22) for $g_{ij} = \delta_{ij}$ (constant). From the above equations, we obtain

$$\omega^j_i + \omega^i_j = 0, \qquad \mathrm{d}\sigma^i - \sigma^j \wedge \omega^i_j = 0, \tag{3.35}$$

respectively, which are generalization of Eqs. (2.72) and (2.75) for $T_x\Sigma^2$ of §2.7.1. It can be verified that this problem has exactly one solution, represented as

$$\omega^k_j = \Gamma^k_{ij}\sigma^i, \tag{3.36}$$

([Fla63], §8.3), where ω^k_j are called *connection* 1-forms, and the coefficients Γ^k_{ji} the Christoffel symbols (connection coefficients). Using the property that ω^k_i is a 1-form, we have from (3.33) and (3.36),

$$\nabla e_j[e_i] = e_k \omega^k_j[e_i] = \Gamma^k_{ij} e_k.$$

The left-hand side corresponds to $\nabla_{e_i} e_j = \nabla_{\partial_i}\partial_j$ in §3.2.

Introducing the matrix notation,

$$e = \begin{pmatrix} e_1 \\ \vdots \\ e_n \end{pmatrix}, \quad \boldsymbol{\sigma} = (\sigma^1, \ldots, \sigma^n), \quad \Omega = \begin{pmatrix} 0 & \ldots & \omega^n_1 \\ \vdots & \ddots & \vdots \\ \omega^1_n & \ldots & 0 \end{pmatrix}, \tag{3.37}$$

[7] The symbol ∇ is used in order to avoid confusion with an ordinary exterior derivative d, and is consistent with the connection defined in §3.2. The equation $\mathrm{d}(\mathrm{d}e_i) = 0$ of §2.7 held since it is for surfaces Σ^2 *in the space* $\mathbb{R}^3$. Here, it is not assumed.

the *structure equations* are summarized as

$$d\boldsymbol{x} = \boldsymbol{\sigma e}, \qquad \text{(a vector-valued 1-form)} \tag{3.38}$$

$$\nabla \boldsymbol{e} = \Omega \boldsymbol{e}, \qquad \Omega + \Omega^T = 0, \tag{3.39}$$

$$d\boldsymbol{\sigma} = \boldsymbol{\sigma} \wedge \Omega, \tag{3.40}$$

where $d\boldsymbol{x}$ is the form that assigns to each vector the same vector, Ω is the *connection 1-form*, and the last equation is a condition of integrability (second of (3.35)) without any torsion form. Equations (3.39) and (3.40) are *Cartan's structural equations.*

For a tangent vector $v = v^j \boldsymbol{e}_j$, the covariant derivative of v in the direction X is $\nabla_X v$, which is written as

$$\nabla v(X) = \boldsymbol{e}_j dv^j(X) + v^j \nabla \boldsymbol{e}_j(X) = \boldsymbol{e}_k(dv^k + v^j \omega_j^k)(X), \tag{3.41}$$

from (3.12) and (3.11).

There is no reason for believing $d(\nabla \boldsymbol{e}) = 0$, which holds only for surfaces in euclidean space $\mathbb{R}^n$. Here, we have

$$d(\nabla \boldsymbol{e}) = d(\Omega \boldsymbol{e}) = (d\Omega)\boldsymbol{e} - \Omega \wedge \Omega \boldsymbol{e} = \Theta \boldsymbol{e},$$

where we defined the *curvature 2-form* Θ by

$$\Theta := d\Omega - \Omega \wedge \Omega. \tag{3.42}$$

In the euclidean space $\mathbb{R}^n$, we have the *flat* connection,

$$d\Omega - \Omega \wedge \Omega = 0. \tag{3.43}$$

In general spaces, writing (3.42) with components, we have

$$\theta_j^i = d\omega_j^i - \omega_j^k \wedge \omega_k^i, \tag{3.44}$$

where $\Theta = (\theta_j^i)$. Each 2-form entry θ_j^i is skew-symmetric, since $\omega_j^i = -\omega_i^j$ and $\omega_j^k \wedge \omega_k^i = -\omega_i^k \wedge \omega_k^j$. Equation (3.44) may be written as

$$\theta_j^i = \frac{1}{2} R^i{}_{jkl} \sigma^k \wedge \sigma^l, \tag{3.45}$$

by using the tensor coefficient $R^i{}_{jkl} = -R^j{}_{ikl} = -R^i{}_{jlk}$, called the *Riemannian curvature tensor* (see §3.9.2).

3.5.2. *Two-dimensional surface M^2*

Let us consider a simpler two-dimensional Riemannian manifold M^2, where $g_{\alpha\beta} = \delta_{\alpha\beta}$. The vector-valued 1-form (3.38) that assigns to each vector the same vector is represented by

$$\mathrm{d}\boldsymbol{x} = \sigma^1 \boldsymbol{e}_1 + \sigma^2 \boldsymbol{e}_2, \tag{3.46}$$

and the skew-symmetric connection form is given by

$$\Omega = \begin{pmatrix} 0 & -\varpi \\ \varpi & 0 \end{pmatrix}, \tag{3.47}$$

where $\varpi = \omega_2^1 = -\omega_1^2$. Hence, Ω is completely given by a single entry ϖ. The same is true of the curvature 2-form Θ. Since $\Omega \wedge \Omega = 0$ from (3.47), we obtain $\Theta = \mathrm{d}\Omega$. Hence, it is found that the curvature 2-form θ_1^2 is *exact*,[8] that is,

$$\theta_2^1 = \mathrm{d}\omega_2^1 = \mathrm{d}\varpi. \tag{3.48}$$

The second of the structure equations, (3.39), is written as

$$\nabla \boldsymbol{e}_1 = -\varpi \boldsymbol{e}_2, \qquad \nabla \boldsymbol{e}_2 = \varpi \boldsymbol{e}_1, \tag{3.49}$$

and the third structure equation (3.40) is

$$(\mathrm{d}\sigma^1, \mathrm{d}\sigma^2) = (\sigma^1, \sigma^2) \wedge \begin{pmatrix} 0 & -\varpi \\ \varpi & 0 \end{pmatrix} \tag{3.50}$$

(Fig. 3.2). Equation (3.45) of the *curvature form* reduces to

$$\theta_2^1 = \frac{1}{2} R^1{}_{2kl} \sigma^k \wedge \sigma^l = R^1{}_{212} \sigma^1 \wedge \sigma^2 := K \sigma^1 \wedge \sigma^2. \tag{3.51}$$

Note that the tensor $R^1{}_{212} = g_{2\alpha} R^{1\alpha}{}_{12} = R^{12}{}_{12}$ is the Gaussian curvature K for the present Riemannian metric of $g_{22} = 1$ and $g_{21} = 0$. As noted in the footnote, the Gaussian curvature at a point p gives the angle of rotation under parallel translation of vectors along an infinitely small closed parallelogram around p.

It is interesting to observe close similarity between the present expressions (3.40), (3.51) and Eqs. (2.85) and (2.89) of §2.7.2, respectively. In

[8] The curvature 2-form on a pair of tangent vectors is equal to the angle of rotation under parallel translation (§3.7.1) of vectors along an infinitely small closed parallelogram determined by these vectors [Arn78, App. 1].

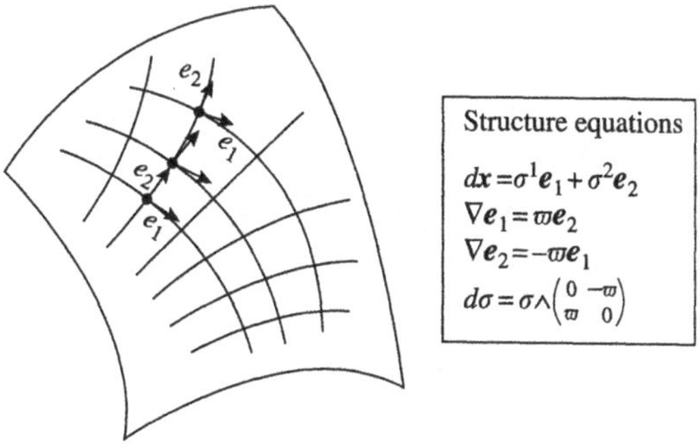

Fig. 3.2. Riemannian surface M^2.

particular, the 1-forms ϖ in both cases are playing the same role. Equations (3.50) and (3.51) with (3.48) are rewritten as follows:

$$\left.\begin{aligned} \mathrm{d}\sigma^1 &= \sigma^2 \wedge \varpi, \\ \mathrm{d}\sigma^2 &= \varpi \wedge \sigma^1, \\ \mathrm{d}\varpi &= K\sigma^1 \wedge \sigma^2. \end{aligned}\right\} \tag{3.52}$$

Remarkably, these are equivalent to the equations of integrability (2.91) and (2.89) in §2.7.

3.5.3. *Example: Poincaré surface (I)*

Let us consider an example of M^2, which is the upper half-plane $(y > 0, -\infty < x < \infty)$ equipped with the first fundamental form defined by

$$\mathrm{I} = \frac{1}{y^2}(\mathrm{d}x)^2 + \frac{1}{y^2}(\mathrm{d}y)^2, \tag{3.53}$$

called the *Poincaré metric*. This implies that all small line-elements whose length (in the (x, y) plane) is proportional to their y-coordinate (with a common proportional constant) is regarded as having the same magnitude (horizontal arrows along the vertical line L in Fig. 3.3). Corresponding to (3.32), we obtain the 1-form basis,

$$\sigma^1 = \frac{\mathrm{d}x}{y}, \qquad \sigma^2 = \frac{\mathrm{d}y}{y}. \tag{3.54}$$

The vector-valued 1-form (3.31), or (3.38), that assigns to each vector the same vector is represented as follows:

$$\mathrm{d}\boldsymbol{x} = \mathrm{d}x \otimes \partial_x + \mathrm{d}y \otimes \partial_y = \sigma^1 \boldsymbol{e}_1 + \sigma^2 \boldsymbol{e}_2, \tag{3.55}$$

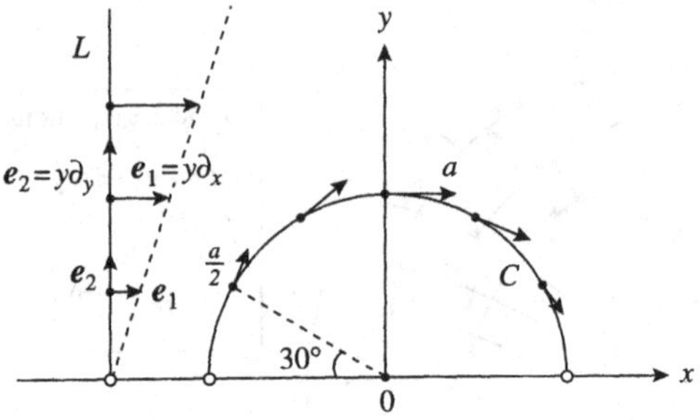

Fig. 3.3. Parallel translation on Poincaré surface.

so that the orthonormal basis vectors are given by $e_1 = y\partial_x$ and $e_2 = y\partial_y$. From the expressions of σ^1 and σ^2, we obtain

$$\mathrm{d}\sigma^1 = \frac{1}{y^2}\mathrm{d}x \wedge \mathrm{d}y = \sigma^1 \wedge \sigma^2, \qquad \mathrm{d}\sigma^2 = 0. \tag{3.56}$$

Hence, we obtain a structure equation,

$$\mathrm{d}\boldsymbol{\sigma} = (\mathrm{d}\sigma^1, \mathrm{d}\sigma^2) = (\sigma^1, \sigma^2) \wedge \begin{pmatrix} 0 & \sigma^1 \\ -\sigma^1 & 0 \end{pmatrix} \tag{3.57}$$

which corresponds to the third structure equation (3.40). Comparing with (3.50), it is found that $-\varpi = \sigma^1 = \mathrm{d}x/y$. Therefore, the second structure equation (3.49) is

$$\nabla e_1 = \frac{\mathrm{d}x}{y} e_2, \qquad \nabla e_2 = -\frac{\mathrm{d}x}{y} e_1. \tag{3.58}$$

Furthermore, comparing the first equation of (3.56) with (3.51) implies the remarkable result,

$$K = -1. \tag{3.59}$$

Namely, the Poincaré surface has a constant negative curvature $K = -1$, often called a *pseudosphere*.

Consider a vector defined by $u = e_1$, i.e. $u^1 = 1$ and $u^2 = 0$ (Fig. 3.3). Covariant derivative of u in the direction e_2 is

$$\nabla u(e_2) = \nabla e_1(e_2) = e_2 \frac{\mathrm{d}x(e_2)}{y} = 0,$$

since $\mathrm{d}x(\boldsymbol{e}_2) = 0$. This says that the *horizontal* vector $\boldsymbol{e}_1 = y\partial_x$ of same magnitude (proportional to y) is parallel when translated along the direction $\boldsymbol{e}_2 = y\partial_y$, i.e. along the line L (Fig. 3.3).

Next, consider another vector $v = \boldsymbol{e}_1 \sin\theta + \boldsymbol{e}_2 \cos\theta = a\sin\theta(\sin\theta\, \partial_x + \cos\theta\, \partial_y)$ defined at the point $p = (-a\cos\theta, a\sin\theta)$ located on the circle C of radius a in the Fig. 3.3 where θ is the angle from the negative x-axis. The vector v is tangent to the circle C with its length equal to $y = a\sin\theta$. Let us take the covariant derivative (defined by (3.41)) of v along itself v. Then we have

$$\begin{aligned}\nabla v(v) &= \nabla(\boldsymbol{e}_1 \sin\theta)(v) + \nabla(\boldsymbol{e}_2 \cos\theta)(v) \\ &= \boldsymbol{e}_1 \mathrm{d}(\sin\theta)(v) + \sin\theta\, \nabla\boldsymbol{e}_1(v) + \boldsymbol{e}_2 \mathrm{d}(\cos\theta)(v) + \cos\theta\, \nabla\boldsymbol{e}_2(v) \\ &= \boldsymbol{e}_1 \left(\sin\theta\cos\theta - \cos\theta \frac{\mathrm{d}x(v)}{y}\right) + \boldsymbol{e}_2 \left(-(\sin\theta)^2 + \sin\theta \frac{\mathrm{d}x(v)}{y}\right) = 0,\end{aligned}$$

since $\mathrm{d}x(v) = y\sin\theta$, where $\mathrm{d}f(\theta)(v) = y(\sin\theta\, \partial_x + \cos\theta\, \partial_y)f = \sin\theta\, \partial_\theta f$.

Thus it is found that the vector v tangent to the curve C is *parallel-translated* along itself, i.e. $Y = T$ in (3.26) (Fig. 3.3). This means that the semi-circle C is a *geodesic curve* (see (3.60) below).

In §3.6.3, it will be shown that both the semi-straight line L and the semi-circle C are geodesic curves on the Poincaré surface.

3.6. Geodesic Equation

One curve of special significance in a curved space is the geodesic curve, whose tangent vector is displaced parallel along itself locally.

3.6.1. *Local coordinate representation*

A curve $\gamma(t)$ on a Riemannian manifold M^n is said to be *geodesic* if its tangent $T = \mathrm{d}\gamma/\mathrm{d}t$ is displaced parallel along the curve $\gamma(t)$, i.e. if

$$\nabla_T T = \frac{\nabla T}{\mathrm{d}t} = \frac{\nabla}{\mathrm{d}t}\left(\frac{\mathrm{d}\gamma}{\mathrm{d}t}\right) = 0. \tag{3.60}$$

In local coordinates $\gamma(t) = (x^i(t))$, we have $\mathrm{d}\gamma/\mathrm{d}t = T = T^i\partial_i = (\mathrm{d}x^i/\mathrm{d}t)\partial_i$. By setting $Y = T$ in (3.24) and (3.25), we obtain

$$\nabla_T T = \left[\frac{\mathrm{d}T^i}{\mathrm{d}t} + \Gamma^i_{jk} T^j T^k\right] \partial_i = 0. \tag{3.61}$$

Thus the *geodesic equation* is

$$\frac{dT^i}{dt} + \Gamma^i_{jk} T^j T^k = 0, \tag{3.62}$$

$$\text{or} \qquad \frac{d^2 x^i}{dt^2} + \Gamma^i_{jk} \frac{dx^j}{dt} \frac{dx^k}{dt} = 0, \qquad (T^i = dx^i/dt). \tag{3.63}$$

It is observed that the geodesic equations (2.64) or (2.65) for Σ^2 are equivalent in form with (3.62) and (3.63), respectively.

3.6.2. *Group-theoretic representation*

On the Riemannian manifold of invariant metric considered in §3.4.3, another formulation of the geodesic equation is possible, because most dynamical systems considered below are equipped with invariant metrics (with respect to either right- or left translation). In such cases, the following derivation would be useful. Using the adjoint operator $ad_X Z = [X, Z]$ introduced in (1.63), let us define the *coadjoint* operator ad^* by

$$\langle ad^*_X Y, Z \rangle := \langle Y, ad_X Z \rangle = \langle Y, [X, Z] \rangle. \tag{3.64}$$

Then Eq. (3.30) is transformed to

$$2\langle \nabla_X Y, Z \rangle = \langle ad_X Y, Z \rangle - \langle ad^*_Y X, Z \rangle - \langle ad^*_X Y, Z \rangle.$$

The nondegeneracy of the inner product (see (3.2)) leads to

$$\nabla_X Y = \frac{1}{2}(ad_X Y - ad^*_X Y - ad^*_Y X). \tag{3.65}$$

Geodesic curve $\gamma(t)$ is a curve whose tangent vector, say X, is displaced parallel along itself, i.e. X satisfies $\nabla_X X = 0$. From (3.65), another form of the geodesic equation is given by

$$\nabla_X X = -ad^*_X X = 0, \tag{3.66}$$

since $ad_X X = [X, X] = 0$.

In a time-dependent problem, the covariant derivative is represented by (3.16) as $\nabla_{\tilde{Y}} \tilde{X} = \partial_t X + \nabla_Y X$ where the time part vanishes identically (so that $\tilde{X}$ is replaced by X), and

$$\tilde{X} = \partial_t + X^\alpha \partial_\alpha, \qquad \tilde{Y} = \partial_t + Y^\alpha \partial_\alpha.$$

Thus, the geodesic equation of a time-dependent problem is given by

$$\nabla_{\tilde{X}} \tilde{X} = \partial_t X + \nabla_X X = \partial_t X - ad^*_X X = 0. \tag{3.67}$$

Remark. It should be remarked that there is a difference in the sign $\pm$ for the expression of the commutator of the Lie algebra depending on the time evolution as described by *left*-translation or *right*-translation, as illustrated in §1.8.1 and the footnote there.

When the time evolution is described by *left*-translation $(L_\gamma)_*$ as in the case of the rotation group of §1.6.1 or Chapter 4, the negative sign should be taken for both $ad_X Y$ and $[\cdot,\cdot]$ to obtain the time evolution equation. In this regard, it is instructive to see the negative sign in front of the term $t(\mathbf{ab} - \mathbf{ba})$ of (1.65) (and the footnote to (1.63)). This requires that $\nabla^{(\mathrm{L})}_X Y$ should be defined by using the $[\cdot,\cdot]^{(\mathrm{L})}$ of (1.66) in place of $[\cdot,\cdot]$.

In the case of the *right*-translation $(R_\gamma)_*$ for the time evolution, the commutator is the Poisson bracket $\{X, Y\}$ of (1.77) and the time evolution is represented by (3.67).

See §3.7.3 for more details.

3.6.3. *Example: Poincaré surface (II)*

In §3.5.3, we considered the metric and structure equations of the Poincaré surface, and found that it is a pseudosphere with a constant negative curvature. Here we are going to derive its geodesic equation, which can be carried out in two ways, and obtain geodesic curves by solving it.

(a) *Direct method.* The line element is represented as (from (3.53))

$$(\mathrm{d}s)^2 = \frac{1}{y^2}(\mathrm{d}x)^2 + \frac{1}{y^2}(\mathrm{d}y)^2. \tag{3.68}$$

Therefore the metric tensor g is given by $g_{xx} = 1/y^2, g_{xy} = 0, g_{yy} = 1/y^2$, and its inverse g^{-1} is given by $g^{xx} = y^2, g^{xy} = 0, g^{yy} = y^2$ from (2.26).

The Christoffel symbols $\Gamma_{ij,x}$ and $\Gamma_{ij,y}$ are calculated by (3.21), as was done in §2.4 for the torus. The result is

$$\Gamma_{ij,x} = \begin{pmatrix} 0 & -y^{-3} \\ -y^{-3} & 0 \end{pmatrix}, \quad \Gamma_{ij,y} = \begin{pmatrix} y^{-3} & 0 \\ 0 & -y^{-3} \end{pmatrix}. \tag{3.69}$$

Next, using the inverse g^{-1}, we obtain $\Gamma^x_{ij} = g^{xx}\Gamma_{ij,x} + g^{xy}\Gamma_{ij,y} = g^{xx}\Gamma_{ij,x}$ since $g^{xy} = 0$. Similarly, we have $\Gamma^y_{ij} = g^{yy}\Gamma_{ij,y}$. Thus,

$$\Gamma^x_{ij} = \begin{pmatrix} 0 & -y^{-1} \\ -y^{-1} & 0 \end{pmatrix}, \quad \Gamma^y_{ij} = \begin{pmatrix} y^{-1} & 0 \\ 0 & -y^{-1} \end{pmatrix}. \tag{3.70}$$

The tangent vector in the (x, y)-frame is $T = (x', y')$ where $x' = \mathrm{d}x/\mathrm{d}s$. Thus, the geodesic equation (3.62) is written down as follows:

$$\frac{\mathrm{d}}{\mathrm{d}s}x' - \frac{2}{y}x'y' = 0, \tag{3.71}$$

$$\frac{\mathrm{d}}{\mathrm{d}s}y' + \frac{1}{y}(x')^2 - \frac{1}{y}(y')^2 = 0. \tag{3.72}$$

(b) *Structure equations in differential forms.* We considered the structure equations of the Poincaré surface in §3.5.3 (Eqs. (3.54)–(3.58)). The geodesic equation is an equation of parallel translation of a tangent vector $X = X^k \boldsymbol{e}_k$ in the local orthonormal-frame representation,

$$\nabla X(X) = \frac{\mathrm{d}X^k}{\mathrm{d}s}\boldsymbol{e}_k + X^k \nabla \boldsymbol{e}_k(X) = 0, \tag{3.73}$$

(see §3.5.1 for $\nabla X(X)$), where $X^1 = \sigma^1(X) = x'/y$ and $X^2 = \sigma^2(X) = y'/y$ since $\sigma^1 = \mathrm{d}x/y$ and $\sigma^2 = \mathrm{d}y/y$ according to (3.54) and (3.55). The connection form $\nabla \boldsymbol{e}_k$ is given by (3.58), which is now written as $\nabla \boldsymbol{e}_1 = \sigma^1 \boldsymbol{e}_2$ and $\nabla \boldsymbol{e}_2 = -\sigma^1 \boldsymbol{e}_1$. Using these and writing $X = X^1$ and $Y = X^2$, the above geodesic equation reads

$$\frac{\mathrm{d}}{\mathrm{d}s}X - XY = 0, \tag{3.74}$$

$$\frac{\mathrm{d}}{\mathrm{d}s}Y + XX = 0 \tag{3.75}$$

[Kob77]. It is readily seen that Eqs. (3.74) and (3.75) reduce to (3.71) and (3.72) respectively by using $X = x'/y$ and $Y = y'/y$.

(c) *Geodesic curves (Solutions).* We can show that geodesic curves in the (x, y)-plane are upper semi-circles of any radius centered at any point on the x-axis, and upper-half straight lines parallel to the y-axis (Fig. 3.4).

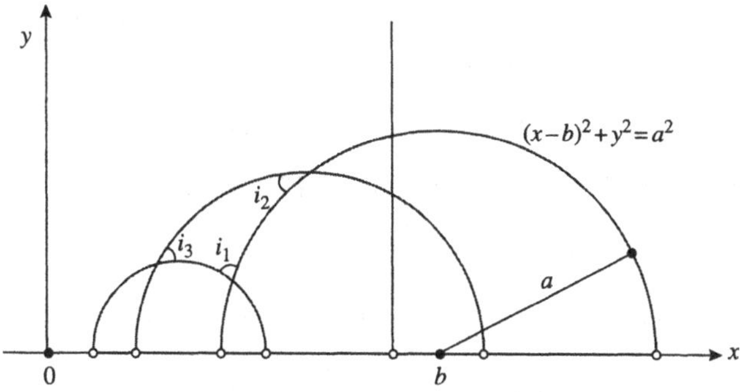

Fig. 3.4. Geodesic curves on Poincaré surface.

Multiplying (3.74) with X and (3.75) with Y, and taking their sum, we immediately obtain $XX' + YY' = 0$. Integrating this, we obtain

$$X^2 + Y^2 = 1, \tag{3.76}$$

where the integration constant on the right-hand side must be unity because of Eq. (3.68). Hence, we can represent X and Y in terms of a parameter t as

$$X = \frac{x'}{y} = \sin t, \qquad Y = \frac{y'}{y} = \cos t. \tag{3.77}$$

Next, Eq. (3.74) is rewritten as

$$\frac{X'}{X} = Y = \frac{y'}{y}.$$

This can be integrated immediately, which leads to the relation $y = aX$ for a nonzero constant a if $X \neq 0$. The right-hand side is $aX = a \sin t$ from the first of (3.77). Therefore, we obtain

$$y = a \sin t \ (> 0). \tag{3.78}$$

Accordingly we assume $a > 0$ and $0 < t < \pi$. Substitution of $X = y/a$ in (3.76) leads to

$$(y')^2 = y^2(1 - (y/a)^2) = a^2 \sin^2 t \cos^2 t.$$

Since $y' = a \cos t(\mathrm{d}t/\mathrm{d}s)$ from (3.78), we obtain

$$\mathrm{d}s = \pm \mathrm{d}t / \sin t.$$

Since $y\,y = y\,aX = a(\mathrm{d}x/\mathrm{d}s) = \pm a \sin t(\mathrm{d}x/\mathrm{d}t)$, we obtain $(\mathrm{d}x/\mathrm{d}t) = \pm a \sin t$. Integrating this, we have

$$x = \mp a \cos t + b, \qquad (b : \text{a constant}). \tag{3.79}$$

Thus, eliminating t between (3.78) and (3.79), we find

$$(x-b)^2 + y^2 = a^2, \qquad (y > 0),$$

which represents an upper semicircle of radius a centered at $(b, 0)$.

If $X = 0$, the system of Eqs. (3.74) and (3.75) results in $Y = y'(s)/y(s) = c$ (constant). Then we have the solution $(x(s), y(s)) = (x_0, y_0 e^{cs})$ for constants x_0 and y_0 (> 0). This represents upper-half straight lines parallel to the y-axis.

From the Gauss–Bonnet theorem (§2.9), it was verified for a negative-curvature surface M^2 such as the Poincaré surface that two geodesic curves do not intersect more than once (as inferred from the curves of Fig. 3.4). Furthermore, the sum of three inner angles $i_1 + i_2 + i_3$ of a geodesic triangle is less than π on the negative-curvature surface.

3.7. Covariant Derivative and Parallel Translation

A geodesic curve is characterized by vanishing of the covariant derivative of its tangent vector along itself, i.e. parallel translation of its tangent vector. How is a vector translated parallel in a curved space?

3.7.1. *Parallel translation again*

Parallel translation of a tangent vector X along a geodesic $\gamma(s)$ with unit tangent T is defined by (3.26) as $\nabla_T X = 0$. By setting $Y = Z = T$ in the second property (3.17) of the Riemannian connection, we obtain

$$\frac{\mathrm{d}}{\mathrm{d}s}\langle X, T\rangle = T\langle X, T\rangle = \langle \nabla_T X, T\rangle, \tag{3.80}$$

since $\nabla_T T = 0$ by the definition of a geodesic. Hence, the inner product $\langle X, T\rangle$ is kept constant by the parallel translation ($\nabla_T X = 0$).

Let us define the angle θ between two vectors X and T by

$$\cos\theta = \frac{\langle X, T\rangle}{\|X\|\,\|T\|} = \frac{\langle X, T\rangle}{\|X\|}, \quad \|X\| = \langle X, X\rangle^{1/2}, \tag{3.81}$$

as before (see (2.15)), where $\|T\| = \|\mathrm{d}\gamma/\mathrm{d}s\| = 1$. First, the parallel translation along the geodesic γ is carried out such that the vector X translates

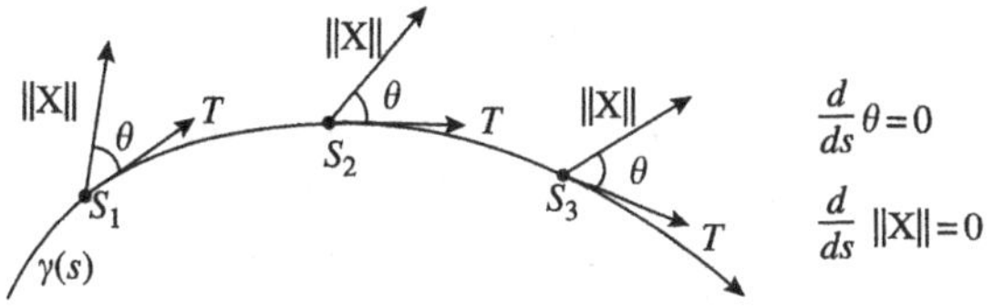

Fig. 3.5. Parallel translation in M^2.

along γ smoothly by keeping its angle θ and its magnitude $\|X\|$ invariant. Then, differentiating (3.81) with respect to the arc-length s and using (3.80), we obtain

$$\frac{\mathrm{d}}{\mathrm{d}s}\langle X, T\rangle = \langle \nabla_T X, T\rangle = \frac{\mathrm{d}}{\mathrm{d}s}(\|X\|\cos\theta) = 0. \tag{3.82}$$

This can define the parallel translation on two-dimensional Riemannian manifold M^2 uniquely (Fig. 3.5). However, in general, this is not sufficient for the parallel translation, because in higher dimensions (than two) the direction of the translated vector can rotate around T and it is not determined uniquely by θ only. To fix that, a surface including the geodesic $\gamma(s)$ must be chosen for the parallel translation. This can be done as follows.

At the initial point p of the geodesic, a plane Σ_0 spanned by X and T is defined. We consider all geodesics starting from p with their tangents lying in Σ_0. The set of all such geodesics close to p forms a smooth surface S_0 containing $\gamma(s)$. At a small distance Δ from p, a new tangent plane Σ_1 is defined so as to be tangent to the surface S_0 and contain $\gamma(s)$ at the new point p_1. Next, we take p_1 as the initial point and use the tangent plane Σ_1 to construct a new geodesic surface S_1. Moving along the $\gamma(s)$ again by Δ and so on, we repeat the construction successively. As $\Delta \to 0$, we obtain a field of two-dimensional tangent planes Σ_X along the geodesic $\gamma(s)$ in the limit [Arn78, App. 1].

Thus the *parallel translation* along a geodesic is defined such that the vector X must remain in the tangent plane field Σ_X, keeping its magnitude and the angle θ invariant. By this construction, we obtain a vector field X_s of parallel translation of the vector $X_0 = X$ at $s = 0$. This is a linear map P_0^s from 0 to s, where $X_s = P_0^s X_0$. Thus, we obtain $\nabla_T X = 0$ from this and (3.82).

Parallel translation *along any smooth curve* is defined by a limiting construction. Namely the curve is approximated by polygons consisting of geodesic arcs, and then the above procedure is applied to each geodesic arc.

3.7.2. *Covariant derivative again*

Writing a geodesic curve as $\gamma_t = \gamma(t)$ with t the time parameter, one can give another interpretation of the covariant derivative $\nabla_T X$, i.e. it is the time derivative of the vector $P_t^0 X_t$ at γ_0. The vector $P_t^0 X_t$ is defined as a vector obtained by parallel-translating the vector $X_t = X(\gamma_t)$ back to γ_0 along the geodesic curve γ_t (Fig. 3.6). This is verified by using the property $\nabla_T T = 0$ of the geodesic curve γ_t and the invariance of the scalar product of parallel-translated vector fields along γ_t in the following way.

The left-hand side of (3.80) is understood as the time derivative by replacing s with t. It is rewritten, by using $T_t = T(\gamma_t)$, as follows:

$$T\langle X, T\rangle|_{\gamma_0} = \lim_{t\to 0} \frac{1}{t}(\langle X_t, T_t\rangle - \langle X_0, T_0\rangle) = \left\langle \frac{\mathrm{d}}{\mathrm{d}t} P_t^0 X_t|_{\gamma_0}, T_0 \right\rangle \tag{3.83}$$

since $\langle X_t, T_t\rangle = \langle P_t^0 X_t, T_0\rangle$ (Fig. 3.7). Comparing with the right-hand side of (3.80) at γ_0, we find that

$$\nabla_T X = \frac{\mathrm{d}}{\mathrm{d}t} P_t^0 X_t. \tag{3.84}$$

3.7.3. *A formula of covariant derivative*

In §3.6.2, we obtained an expression of the covariant derivative (3.65), which is reproduced here:

$$\nabla_X Y = \frac{1}{2}([X, Y] - (ad_X^* Y + ad_Y^* X)), \tag{3.85}$$

where $[X, Y] = ad_X Y$ from (1.63). Comparing (3.85) with (3.84), rewriting it as $\nabla_X Y = (\mathrm{d}/\mathrm{d}t) P_t^0 Y_t$, it can be shown that the term $\frac{1}{2}[X, Y]$ came from the t-derivative of the factor Y_t.

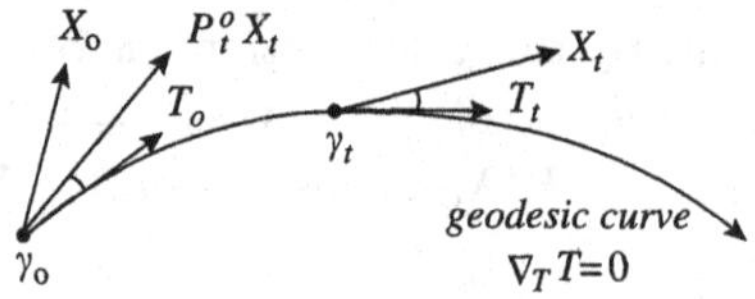

Fig. 3.6. Covariant derivative.

To show this, it is useful to note that the following identity holds for any tangent vectors A and B (likewise matrices):

$$(\exp A)(\exp B) = \exp\left(A + B + \frac{1}{2}(AB - BA) + O(A^2, B^2)\right).$$

Substituting $A = Xt$ and $B = Ys$, differentiating with s and setting $s = 0$, we obtain $Y(\gamma_t) := (\mathrm{d}/\mathrm{d}s)\gamma_t\eta_s|_{s=0}$, which is given by

$$\exp(Xt)\left(Y + \frac{1}{2}[X, Y]t + O(t^2)\right) := \exp(Xt)Y_t, \tag{3.86}$$

where $[X, Y] = XY - YX = [X, Y]^{(\mathrm{L})}$ by definition (1.66) of left-invariant fields. The γ_t and η_s denote flows generated by X and Y, respectively. Writing $Y(\gamma_t) = \exp(Xt)Y_t$, we have $Y_t = Y + \frac{1}{2}[X, Y]t + O(t^2)$.

As for the right-invariant fields, one can regard Eq. (1.74) to be an expansion of

$$\eta_s\gamma_t = \exp\left(tX + sY + \frac{1}{2}[tX, sY]\right),$$

where $X = X^k\partial_k$ and $Y = Y^k\partial_k$, and $[X, Y] = [X, Y]^{(\mathrm{R})} := \{X, Y\}$ defined by (1.76) and (1.77). Differentiating with s and setting $s = 0$, we obtain

$$Y(\gamma_t) = \left(Y + \frac{1}{2}[X, Y]t + O(t^2)\right)\exp(Xt) = Y_t \exp(Xt), \tag{3.87}$$

with the same expression for Y_t as (3.86) with $[X, Y]^{(\mathrm{R})}$. Thus the above statement that $\frac{1}{2}[X, Y]$ came from the t-derivative of the factor Y_t has been verified.

Likewise, the terms $ad_X^*Y + ad_Y^*X$ came from the t-derivative of the operation P_t^0. For its verification, we just refer to the paper [Arn66], where ad_X^*Y is written as $B(Y, X)$.

It can be shown by the covariant derivative (3.85) that the *right-invariant metric $\langle Y, Z\rangle$ of right-invariant fields $Y(\gamma_t)$ and $Z(\gamma_t)$ is constant* along the flow γ_t (generated by X). In fact, taking the derivative along γ_t, we find

$$X\langle Y, Z\rangle = \langle\nabla_X Y, Z\rangle + \langle Y, \nabla_X Z\rangle = 0. \tag{3.88}$$

The first equality is the definition relation of (3.17). Substituting the formula (3.85), the first term is

$$\langle\nabla_X Y, Z\rangle = \frac{1}{2}(\langle[X, Y], Z\rangle - \langle Y, [X, Z]\rangle - \langle X, [Y, Z]\rangle)$$

by using (3.64). The second term is obtained by interchanging Y and Z in the above expression. Thus, it is seen that summation of both terms vanishes [Arn66]. This shows that the formula (3.85) for $\nabla_X Y$ is consistent with the assumption made in §3.4.3 to derive the formula (3.30). The same is true for the left-invariant case.

The parallel translation conserves the scalar product as well (§3.4.2). In this case, each term of (3.88) vanishes.

3.8. Arc-Length

A geodesic curve denotes a path of shortest distance connecting two nearby points, or globally of an extremum distance.

A geodesic curve is characterized by the property that the first variation of arc-length vanishes for all variations with fixed end points. Let $C_0 : \gamma_0(s)$ be a geodesic curve with the length parameter $s \in [0, L]$. We consider the first variation of arc-length as we vary the curve. A varied curve is denoted by $C_\alpha : \gamma(s,\alpha)$ with $\gamma(s,0) = \gamma_0(s)$ (Fig. 3.7), where $\alpha \in (-\varepsilon, +\varepsilon)$ is a variation parameter ($\varepsilon > 0$) and s is the arc-length for $\gamma_0(s)$.

The arc-length of the curve C_α is defined by

$$L(\alpha) = \int_0^L \left\| \frac{\partial \gamma(s,\alpha)}{\partial s} \right\| \mathrm{d}s = \int_0^L \left\langle \frac{\partial \gamma(s,\alpha)}{\partial s}, \frac{\partial \gamma(s,\alpha)}{\partial s} \right\rangle^{1/2} \mathrm{d}s$$

$$= \int_0^L \langle T(s,\alpha), T(s,\alpha)\rangle^{1/2} \mathrm{d}s, \quad T = \frac{\partial \gamma}{\partial s}.$$

Its variation is given by

$$L'(\alpha) = \int_0^L \frac{\partial}{\partial \alpha} \left\langle \frac{\partial \gamma}{\partial s}, \frac{\partial \gamma}{\partial s} \right\rangle^{1/2} \mathrm{d}s = \int_0^L \left\| \frac{\partial \gamma(s,\alpha)}{\partial s} \right\|^{-1} \left\langle \frac{\nabla}{\partial \alpha} \frac{\partial \gamma}{\partial s}, \frac{\partial \gamma}{\partial s} \right\rangle \mathrm{d}s.$$

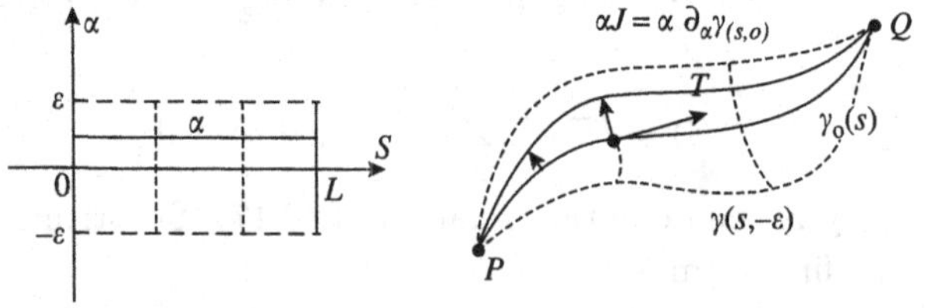

Fig. 3.7. Varied curves.

For $\alpha = 0$, we have $\|\partial\gamma(s,0)/\partial s\| = 1$ by definition, and we have

$$L'(0) = \int_0^L \left\langle \frac{\nabla}{\partial\alpha}\partial_s\gamma, \partial_s\gamma \right\rangle \mathrm{d}s \tag{3.89}$$

$$= \int_0^L \frac{\partial}{\partial s}\langle \partial_\alpha\gamma, \partial_s\gamma\rangle \mathrm{d}s - \int_0^L \left\langle \partial_\alpha\gamma, \frac{\nabla}{\partial s}\partial_s\gamma \right\rangle \mathrm{d}s, \tag{3.90}$$

where $\partial_s\gamma = \partial\gamma/\partial s$ and $\partial_\alpha\gamma = \partial\gamma/\partial\alpha$. To obtain the last expression, we used the following identity:

$$\left(\frac{\nabla}{\partial\alpha}\right)\partial_s\gamma = \left(\frac{\nabla}{\partial s}\right)\partial_\alpha\gamma. \tag{3.91}$$

This is verified as follows. Using local coordinate $(x^1, \ldots, x^n)$ for M^n, we represent $\partial_\alpha\gamma$ and $\partial_s\gamma$ as $(\partial x^i/\partial\alpha)\partial_i$ and $(\partial x^i/\partial s)\partial_i$ respectively. Then, we take the covariant derivative of $\partial_\alpha\gamma$ along the s-curve with α fixed:

$$\frac{\nabla}{\partial s}(\partial_\alpha\gamma) = \frac{\nabla}{\partial s}\left(\frac{\partial x^i}{\partial\alpha}\partial_i\right) = \left(\frac{\partial^2 x^i}{\partial s\partial\alpha}\right)\partial_i + \left(\frac{\partial x^i}{\partial\alpha}\right)\nabla_{\partial_s}\partial_i$$

$$= \left(\frac{\partial^2 x^i}{\partial s\partial\alpha}\right)\partial_i + \left(\frac{\partial x^i}{\partial\alpha}\right)\left(\frac{\partial x^j}{\partial s}\right)\Gamma^k_{ij}\partial_k.$$

The last expression is symmetric with respect to α and s, thus equal to $(\nabla/\partial\alpha)(\partial x^i/\partial s)\partial_i = (\nabla/\partial\alpha)\partial_s\gamma$, which verifies (3.91).

The vectors $T = \partial_s\gamma$ and $\partial_\alpha\gamma$ are denoted as T and J and termed the tangent field and Jacobi field, respectively in §3.10.1. Equation (3.91) is shown to be equivalent to the equation $\mathcal{L}_T J = 0$ in (8.53) of §8.4, and interpreted as the equation of J-field *frozen* to the flow generated by T.

Thus, the first variation $L'(0)$ of arc-length is given by

$$L'(0) = \langle J, T\rangle_{\mathrm{Q}} - \langle J, T\rangle_{\mathrm{P}} - \int_0^L \left\langle J, \frac{\nabla}{\partial s}T \right\rangle \mathrm{d}s, \tag{3.92}$$

where $T = \partial_s\gamma(s,0)$, $J = \partial_\alpha\gamma(s,0)$ and P$= \gamma(0,0)$, Q$= \gamma(L,0)$.

Suppose that all variations vanish at the endpoints P and Q. For such variations, we have $J = 0$ at P and Q for all α. Then we have

$$\left\langle J, \frac{\nabla}{\partial s}T \right\rangle = 0 \quad \text{for } 0 < s < L, \tag{3.93}$$

for every vector J tangent to M along the geodesic C_0. Thus the vector $\nabla T/\partial s = \nabla_T T = 0$ must vanish at all $s \in (0, L)$ by the nondegeneracy of the metric. This is necessary, and also sufficient for vanishing of the first variation $L'(0)$.

Thus, it is found that *the geodesic curve described by* $\nabla_T T = 0$ *is characterized by the extremum* $L'(0) = 0$ *of the arc-length among nearby curves having common endpoints.*

If the endpoints are sufficiently near, the geodesic curve denotes a path of shortest distance connecting the two nearby points.

3.9. Curvature Tensor and Curvature Transformation

Parallel translation in a curved space results in a curvature transformation represented by curvature tensors. The curvature tensors are given once Christoffel symbols are known.

3.9.1. *Curvature transformation*

Let us consider parallel translation of a vector along a small closed path C. Take any vector Z in the tangent space T_pM at a point $p \in C \subset M$. After making one turn along C from p to p, the vector does not necessarily return back to the original one in a curved space, but to a different vector of the same length. This is considered a map of the tangent space to itself, which represents small rotational transformation of vectors, in other words, an orthogonal transformation (close to the identity e). Any operator g of an orthogonal transformation group (or a group $SO(n)$) near e can be written in the form, $g(A) = \exp[A] = e + A + (A^2/2!) + \cdots$, where e is an identity operator and A is a small skew-symmetric operator (see Appendix C).

Let X and Y be two tangent vectors in T_pM. We construct a small curvilinear parallelogram Π_ε, in which the sides of Π_ε are given by εX and εY emanating from p, where ε is a small parameter. We carry out a parallel translation of Z along the sides of Π_ε, making a circuit C^* from p along the side εY first and returning back to p along the side εX (Fig. 3.8).[9]

The parallel translation results in an orthogonal transformation of T_pM close to the indentity e, which can be represented in the following form:

$$g_\varepsilon(X, Y) = e + \varepsilon^2 R(X, Y) + O(\varepsilon^3), \tag{3.94}$$

[9]The sense of circuit is opposite to that of [Arn78, App. 1.E], but consistent with the definition of the covariant derivative (3.84).

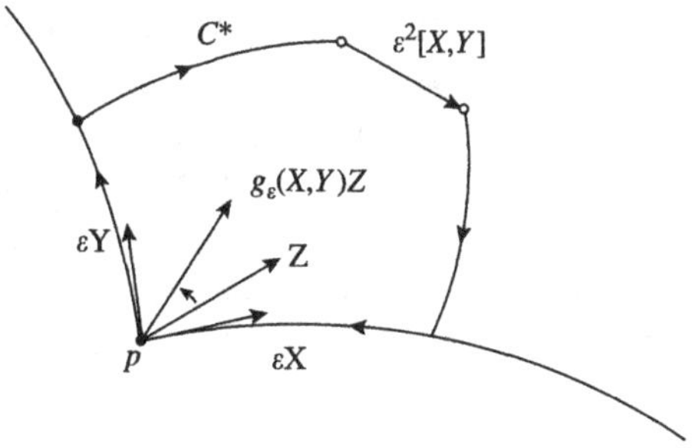

Fig. 3.8. Curvature transformation.

where R is a skew-symmetric operator depending on X and Y. Thus, the R is defined by

$$R(X,Y) = \lim_{\varepsilon \to 0} \frac{g_\varepsilon(X,Y) - e}{\varepsilon^2}. \tag{3.95}$$

The function R takes a real value for a pair of vectors X and Y in T_pM, namely a 2-form. The 2-form $R(X,Y)$ is called a *curvature 2-form*, *curvature tensor*, or *curvature transformation*. The curvature transformation describes an infinitesimal rotational transformation in the tangent space, obtained by parallel translation around an infinitely small parallelogram. The explicit representation is given by (E.8) in Appendix E. In order to derive the formula, the circuit must be closed by appending a line-segment of $\varepsilon^2[X,Y]$ to fill in the gap in the incomplete curvilinear parallelogram (hence the circuit is five-sided). It is found that the curvature transformation $R(X,Y)$ is given by

$$R(X,Y) = \nabla_X \nabla_Y - \nabla_Y \nabla_X - \nabla_{[X,Y]}. \tag{3.96}$$

3.9.2. *Curvature tensor*

The curvature transformation is represented in terms of curvature tensors. Namely, for a vector field $Z \in TM$, the transformation $Z \to R(X,Y)Z$ is defined by

$$R(X,Y)Z := \nabla_X(\nabla_Y Z) - \nabla_Y(\nabla_X Z) - \nabla_{[X,Y]} Z \tag{3.97}$$

$$= (R^\alpha{}_{kij} Z^k X^i Y^j)\partial_\alpha, \tag{3.98}$$

$$R^\alpha{}_{kij} := \partial_i \Gamma^\alpha_{jk} - \partial_j \Gamma^\alpha_{ik} + \Gamma^m_{jk}\Gamma^\alpha_{im} - \Gamma^m_{ik}\Gamma^\alpha_{jm}, \tag{3.99}$$

for the vector fields $X, Y \in TM$, where $X = X^i \partial_i, Y = Y^j \partial_j, Z = Z^k \partial_k$. This describes a linear transformation $T_x M \to T_x M$, i.e. $Z^\alpha \partial_\alpha \to (R^\alpha{}_{kij} Z^k X^i Y^j)\partial_\alpha$, where $R^\alpha{}_{kij}$ is the *Riemannian curvature tensor* and the tri-linearity (with respect to each one of X, Y, Z) is clearly seen. This is verified by using the definition (3.6) for the covariant derivative repeatedly and using $[X, Y] = \{X, Y\}^k \partial_k$ for $\nabla_{[X,Y]}$ (see (1.76) and (1.77)).

In fact, writing $\nabla_Y Z = U^l \partial_l$, we have $U^l = Y^j \partial_j Z^l + Y^j Z^k \Gamma^l_{jk}$ from (3.6). Then, we obtain

$$\nabla_X(\nabla_Y Z) = X^i \nabla_{\partial_i}(U^l \partial_l) = X^i \partial_i U^\alpha + X^i U^l \Gamma^\alpha_{il}.$$

Hence, by using the definition, $\nabla_X Z = V^l \partial_l = (X^j \partial_j Z^l + X^j Z^k \Gamma^l_{jk})\partial_l$,

$$\begin{aligned}\nabla_X(\nabla_Y Z) - \nabla_Y(\nabla_X Z) &= (X^i \partial_i U^\alpha + X^i U^l \Gamma^\alpha_{il} - Y^i \partial_i V^\alpha - X^i V^l \Gamma^\alpha_{il})\partial_\alpha \\ &= [(X^i \partial_i Y^j - Y^i \partial_i X^j)(\partial_j Z^\alpha + Z^k \Gamma^\alpha_{jk}) \\ &\quad + X^i Y^j Z^k (\partial_i \Gamma^\alpha_{jk} - \partial_j \Gamma^\alpha_{ik} \\ &\quad + \Gamma^m_{jk} \Gamma^\alpha_{im} - \Gamma^m_{ik} \Gamma^\alpha_{jm})]\partial_\alpha. \end{aligned} \tag{3.100}$$

On the other hand, we have

$$\nabla_{[X,Y]} Z = \{X, Y\}^j \nabla_{\partial_j}(Z^k \partial_k) = (X^i \partial_i Y^j - Y^i \partial_i X^j)(\partial_j Z^\alpha + Z^k \Gamma^\alpha_{jk}).$$

Thus the equality of (3.97) and (3.98) is verified. All the derivative terms of X^i, Y^j, Z^k cancel out and only the nonderivative terms remain, resulting in the expressions (3.98) with the definition (3.99) of the Riemann tensors $R^\alpha_{\ kij}$. The expression (3.99) can also be derived compactly as follows. Using the particular representation, $X = \partial_i, Y = \partial_j$ and $Z = \partial_k$, we have

$$R(\partial_i, \partial_j)\partial_k = \nabla_{\partial_i}(\nabla_{\partial_j} \partial_k) - \nabla_{\partial_j}(\nabla_{\partial_i} \partial_k) = R^\alpha_{kij} \partial_\alpha, \tag{3.101}$$

where the third term $\nabla_{[\partial_i, \partial_j]}$ does not appear because $[\partial_i, \partial_j] = 0$. The definitive equation $\nabla_{\partial_i} \partial_j = \Gamma^k_{ij} \partial_k$ leads to the representation (3.99), defining $R^\alpha{}_{kij}$ in terms of Γ^k_{ij} only. It can be shown that the tensor $R^i_{\ jkl}$ in (3.45) is equivalent to the present curvature tensor$R^\alpha_{\ kij}$, by using (3.36) and (3.44).

From the definition (3.96), one may write

$$R(X,Y) = [\nabla_X, \nabla_Y] - \nabla_{[X,Y]}, \tag{3.102}$$

The anti-symmetry with respect to X and Y is obvious:

$$R(X,Y) = -R(Y,X), \quad \text{or} \quad R^{\alpha}_{kij} = -R^{\alpha}_{kji}. \tag{3.103}$$

Taking the inner product of $R(X,Y)Z$ with $W = W^{\alpha}\partial_{\alpha} \in TM$, we have

$$\langle W, R(X,Y)Z \rangle = \langle \partial_{\alpha}, \partial_m \rangle R^m{}_{\beta ij} W^{\alpha} Z^{\beta} X^i Y^j = R_{\alpha\beta ij} W^{\alpha} Z^{\beta} X^i Y^j, \tag{3.104}$$

where $R_{\alpha\beta ij} = g_{m\alpha} R^m{}_{\beta ij}$ and $g_{m\alpha} = \langle \partial_m, \partial_{\alpha} \rangle$. In addition to the anti-symmetry

$$\langle W, R(X,Y)Z \rangle + \langle W, R(Y,X)Z \rangle = 0 \tag{3.105}$$

due to (3.103), one can verify the following anti-symmetry [Mil63],

$$\langle W, R(X,Y)Z \rangle + \langle Z, R(X,Y)W \rangle = 0. \tag{3.106}$$

In fact, using (3.17) repeatedly, we obtain

$$\begin{aligned}\langle W, \nabla_X \nabla_Y Z \rangle = &-\langle \nabla_X \nabla_Y W, Z \rangle + XY \langle W, Z \rangle \\ &- \langle \nabla_X W, \nabla_Y Z \rangle - \langle \nabla_Y W, \nabla_X Z \rangle,\end{aligned}$$

and a similar expression for $\langle W, \nabla_Y \nabla_X Z \rangle$. Noting that $(XY - YX)\langle W, Z \rangle = [X,Y]\langle W, Z \rangle = \langle \nabla_{[X,Y]} W, Z \rangle + \langle W, \nabla_{[X,Y]} Z \rangle$, we obtain (3.106). Thus, we find the following anti-symmetry with respect to (α, β) from (3.106), in addition to (i,j) from (3.103):

$$R_{\alpha\beta ij} = -R_{\beta\alpha ij}. \tag{3.107}$$

Finally, another useful expression is given as follows:

$$\begin{aligned}\langle W, R(X,Y)Z \rangle = &X\langle W, \nabla_Y Z \rangle - Y\langle W, \nabla_X Z \rangle - \langle \nabla_X W, \nabla_Y Z \rangle \\ &+ \langle \nabla_Y W, \nabla_X Z \rangle - \langle W, \nabla_{[X,Y]} Z \rangle.\end{aligned} \tag{3.108}$$

3.9.3. *Sectional curvature*

Consider a two-dimensional subspace Σ in the tangent space T_pM, and suppose that geodesics pass through the point p in all directions in Σ. These geodesics form a smooth two-dimensional surface lying in the Riemannian manifold M. One can define a Riemannian curvature at p of the surface

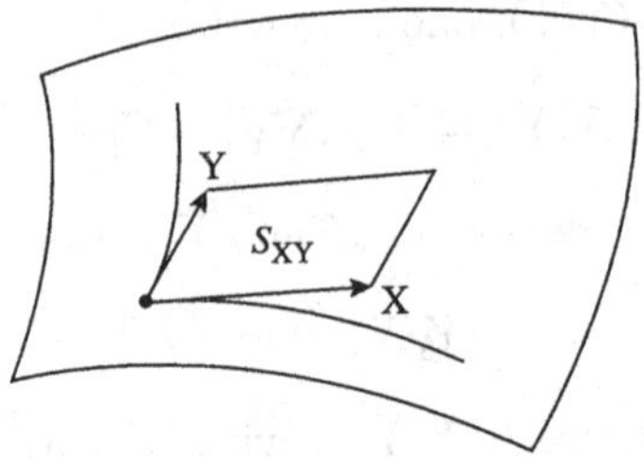

Fig. 3.9. Sectional curvature, $\hat{K}(X,Y) = \langle R(X,Y)Y, X\rangle / S_{XY}$.

thus obtained. The curvature in the two-dimensional section determined by a pair of tangent vectors X and Y (Fig. 3.9) can be expressed in terms of the curvature tensor R by

$$\hat{K}(X,Y) = \frac{\langle R(X,Y)Y, X\rangle}{S_{XY}}, \tag{3.109}$$

$$S_{XY} := \|X\|^2 \, \|Y\|^2 - \langle X, Y\rangle^2, \tag{3.110}$$

where S_{XY} denotes square of the area of the parallelogram spanned by X and Y. The $\hat{K}(X,Y)$ is called the Riemannian *sectional curvature.* If X and Y are orthogonal unit vectors, then $\hat{K}(X,Y)$ is given simply by $\langle R(X,Y)Y, X\rangle$ since $S_{XY} = 1$ in this case.[10] Using (3.108), we have

$$\begin{aligned} K(X,Y) := \langle R(X,Y)Y, X\rangle &= X\langle \nabla_Y Y, X\rangle - Y\langle \nabla_X Y, X\rangle \\ &\quad - \langle \nabla_X X, \nabla_Y Y\rangle + \langle \nabla_X Y, \nabla_Y X\rangle - \langle \nabla_{[X,Y]} Y, X\rangle \qquad (3.111) \\ = \langle R(Y,X)X, Y\rangle &= Y\langle \nabla_X X, Y\rangle - X\langle \nabla_Y X, Y\rangle \\ &\quad - \langle \nabla_Y Y, \nabla_X X\rangle + \langle \nabla_Y X, \nabla_X Y\rangle - \langle \nabla_{[Y,X]} X, Y\rangle. \qquad (3.112) \end{aligned}$$

Equality of (3.111) and (3.112) is obvious since

$$XY\langle X, Y\rangle - YX\langle X, Y\rangle = [X,Y]\langle X, Y\rangle$$

and due to (3.18). In general, because of the two anti-symmetries (3.105) and (3.106), we have $K(X,Y) = K(Y,X)$ and $\hat{K}(X,Y) = \hat{K}(Y,X)$.

---

[10] According to the definition, the expression of (3.96) is opposite in sign to the definition of [Arn78]. However, the sectional curvature $\hat{K}(X,Y)$ becomes same in both formulations, owing to the difference of respective definitions.

3.9.4. *Ricci tensor and scalar curvature*

A certain average of the sectional curvatures are sometimes useful. The *Ricci tensor* is defined by contracting the first and third indices of the Riemann tensor:

$$R_{ij} := R^k{}_{ikj}. \tag{3.113}$$

Suppose that a vector X is given at a point $p \in M^n$. Let us introduce an orthonormal frame of T_pM^n: $(\boldsymbol{e}_1, \ldots, \boldsymbol{e}_{n-1}, \boldsymbol{e}_n)$, where $X = X^n \boldsymbol{e}_n$. A sectional curvature for the plane spanned by $\boldsymbol{e}_\alpha$ and $\boldsymbol{e}_\beta$ is

$$\begin{aligned} K(\boldsymbol{e}_\alpha, \boldsymbol{e}_\beta) &= \langle R(\boldsymbol{e}_\alpha, \boldsymbol{e}_\beta)\boldsymbol{e}_\beta, \boldsymbol{e}_\alpha \rangle \\ &= g_{lm} R^l{}_{ikj} (\boldsymbol{e}_\beta)^i (\boldsymbol{e}_\alpha)^k (\boldsymbol{e}_\beta)^j (\boldsymbol{e}_\alpha)^m = R_{\alpha\beta\alpha\beta}. \end{aligned} \tag{3.114}$$

Then, one can define the Ricci curvature K_R at p specified by the direction $X = X^n \boldsymbol{e}_n$ as the sum of sectional curvatures in the following way:

$$\begin{aligned} K_R(p, X) &:= \sum_{\alpha=1}^{n-1} K(\boldsymbol{e}_\alpha, X) = \sum_\alpha g_{lm} R^l{}_{ikj} X^i X^j (\boldsymbol{e}_\alpha)^k (\boldsymbol{e}_\alpha)^m \\ &= \sum_\alpha R^\alpha{}_{i\alpha j} X^i X^j = R_{ij} X^i X^j = R_{nn} X^n X^n. \end{aligned} \tag{3.115}$$

The *scalar curvature* $\mathcal{R}$ is defined by the trace of Ricci tensor:

$$\mathcal{R} := R^i_i = g^{ik} R_{ki}. \tag{3.116}$$

In an *isotropic* manifold, i.e. all sectional curvatures being equal to a *constant* K, the Riemann curvature tensors have remarkably simple form,

$$R^\alpha{}_{jkl} = K(\delta^\alpha_k g_{jl} - \delta^\alpha_l g_{jk}), \quad R_{ijkl} = K(g_{ik} g_{jl} - g_{il} g_{jk}), \tag{3.117}$$

where K is a constant (equal to the sectional curvature, $R_{\alpha\beta\alpha\beta} = K$, for orthonormal basis $\boldsymbol{e}_\alpha$). Then, the Ricci tensors are

$$R_{ij} = R^k{}_{ikj} = K_R \, g_{ij} \tag{3.118}$$

where $K_R = (n-1)K$. The scalar curvature is

$$\mathcal{R} = g^{ik} R_{ki} = n K_R = n(n-1)K. \tag{3.119}$$

3.10. Jacobi Equation

Behavior of a family of neighboring geodesic curves describes stability of a system considered.

3.10.1. *Derivation*

Let $C_0 : \gamma_0(s)$ be a geodesic curve with the length parameter $s \in [0, L]$, and $C_\alpha : \gamma(s, \alpha)$ a varied geodesic curve where $\alpha \in (-\varepsilon, +\varepsilon)$ is a variation parameter ($\varepsilon > 0$) and $\gamma_0(s) = \gamma(s, 0)$ with s being the arc-length for $\alpha = 0$.[11] Because $\gamma(s, \alpha)$ is a geodesic, we have $\nabla(\partial_s \gamma)/\partial s = 0$ for all α. An example is the family of great circles originating from the north pole on the sphere S^2. The function $\gamma(s, \alpha)$ is a differentiable map $\gamma : U \subset \mathbb{R}^2 \to S$ (a surface) $\subset M^n$ with the property $[\partial/\partial s, \partial/\partial \alpha] = 0$ on S, since $\alpha =$ const and $s =$ const are considered as coordinate curves on the surface S (see (1.79)). Under these circumstances, the following identity is useful,

$$\frac{\nabla}{\partial s}\left(\frac{\nabla Z}{\partial \alpha}\right) - \frac{\nabla}{\partial \alpha}\left(\frac{\nabla Z}{\partial s}\right) = R(\partial_s \gamma, \partial_\alpha \gamma) Z \qquad (3.120)$$

[Fra97], where $Z(s, \alpha)$ is a vector field *defined along* S.

This is verified as follows. Using local coordinate $(x^1, \ldots, x^n)$ on M^n, we represent $\partial_\alpha \gamma$ and $\partial_s \gamma$ as $(\partial x^i/\partial \alpha)\partial_i$ and $(\partial x^i/\partial s)\partial_i$, and write $Z = z^i(s, \alpha)\partial_i$. Then we have the double covariant derivative,

$$\frac{\nabla}{\partial s}\left(\frac{\nabla Z}{\partial \alpha}\right) = \left(\frac{\partial^2 z^i}{\partial s \partial \alpha}\right)\partial_i + \left(\frac{\partial z^i}{\partial \alpha}\right)\frac{\nabla \partial_i}{\partial s} + \left(\frac{\partial z^i}{\partial s}\right)\frac{\nabla \partial_i}{\partial \alpha} + z^i \frac{\nabla}{\partial s}\left(\frac{\nabla \partial_i}{\partial \alpha}\right),$$

which is obtained by carrying out covariant derivatives $\nabla/\partial \alpha$ and $\nabla/\partial s$ consecutively, where we note

$$\frac{\nabla \partial_i}{\partial \alpha} = \nabla_{\partial_\alpha \gamma}\partial_i = \left(\frac{\partial x^j}{\partial \alpha}\right)\nabla_{\partial_j}\partial_i \qquad \left(\partial_\alpha \gamma = \frac{\partial x^j}{\partial \alpha}\partial_j\right),$$

and furthermore,

$$\frac{\nabla}{\partial s}\left(\frac{\nabla \partial_i}{\partial \alpha}\right) = \left(\frac{\partial^2 x^j}{\partial s \partial \alpha}\right)\nabla_{\partial_j}\partial_i + \left(\frac{\partial x^j}{\partial \alpha}\right)\left(\frac{\partial x^k}{\partial s}\right)\nabla_{\partial_k}\nabla_{\partial_j}\partial_i. \qquad (3.121)$$

[11] In this section, the variable s is used in the sense of *length parameter* instead of t.

Changing the order of α and s, we obtain a similar expression of $(\nabla/\partial\alpha)(\nabla/\partial s)Z$. Taking their subtraction, we obtain

$$\frac{\nabla}{\partial s}\left(\frac{\nabla Z}{\partial \alpha}\right) - \frac{\nabla}{\partial \alpha}\left(\frac{\nabla Z}{\partial s}\right) = z^i \left(\frac{\nabla}{\partial s}\left(\frac{\nabla \partial_i}{\partial \alpha}\right) - \frac{\nabla}{\partial \alpha}\left(\frac{\nabla \partial_i}{\partial s}\right)\right). \tag{3.122}$$

Thus, using (3.121), this equation reduces to

$$\begin{aligned}
&\frac{\nabla}{\partial s}\left(\frac{\nabla Z}{\partial \alpha}\right) - \frac{\nabla}{\partial \alpha}\left(\frac{\nabla Z}{\partial s}\right) \\
&\quad = z^i \left(\left(\frac{\partial x^j}{\partial \alpha}\right)\left(\frac{\partial x^k}{\partial s}\right) - \left(\frac{\partial x^j}{\partial s}\right)\left(\frac{\partial x^k}{\partial \alpha}\right)\right) \nabla_{\partial_k}\nabla_{\partial_j}\partial_i \\
&\quad = z^i \left(\frac{\partial x^j}{\partial \alpha}\right)\left(\frac{\partial x^k}{\partial s}\right)(\nabla_{\partial_k}\nabla_{\partial_j}\partial_i - \nabla_{\partial_j}\nabla_{\partial_k}\partial_i) \\
&\quad = z^i \left(\frac{\partial x^j}{\partial \alpha}\right)\left(\frac{\partial x^k}{\partial s}\right) R(\partial_k, \partial_j)\partial_i \\
&\quad = R\left(\left(\frac{\partial x^k}{\partial s}\right)\partial_k, \left(\frac{\partial x^j}{\partial \alpha}\right)\partial_j\right)(z^i\partial_i) \\
&\quad = R(\partial_s\gamma, \partial_\alpha\gamma)Z.
\end{aligned}$$

See (3.97), (3.98) and (3.101) for the last three equalities. The last shows (3.120).

Along the reference geodesic $\gamma_0(s)$, let us use the notation $T = \partial_s\gamma$ $(\alpha = 0)$ for the tangent to the geodesic and

$$J = \partial_\alpha \gamma(s,\alpha)|_{\alpha=0} \tag{3.123}$$

for the variation vector. The geodesic variation is $\gamma(s,\alpha) - \gamma(s,0)$, which is approximated linearly as

$$\gamma(s,\alpha) - \gamma(s,0) \approx \alpha\partial_\alpha\gamma(s,\alpha)|_{\alpha=0} = \alpha J. \tag{3.124}$$

Setting $Z = T$ in (3.120) and using $\nabla T/\partial s = 0$ and (3.91), we have

$$\begin{aligned}
0 = \frac{\nabla}{\partial \alpha}\frac{\nabla T}{\partial s} &= \frac{\nabla}{\partial s}\frac{\nabla T}{\partial \alpha} - R(T,J)T = \frac{\nabla}{\partial s}\frac{\nabla}{\partial \alpha}\partial_s\gamma + R(J,T)T \\
&= \frac{\nabla}{\partial s}\frac{\nabla}{\partial s}\partial_\alpha\gamma + R(J,T)T = \frac{\nabla}{\partial s}\frac{\nabla}{\partial s}J + R(J,T)T,
\end{aligned}$$

where we used the anti-symmetry $R(T,J) = -R(J,T)$.

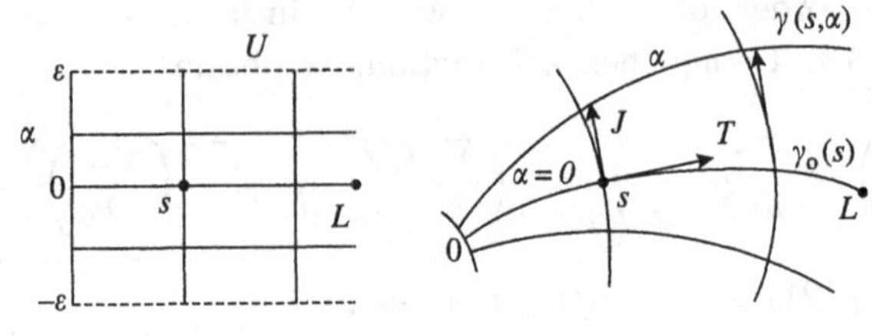

Fig. 3.10. Jacobi field.

Thus we have obtained the *Jacobi equation* for the geodesic variation J,

$$\frac{\nabla}{\partial s}\frac{\nabla}{\partial s}J + R(J,T)T = 0. \tag{3.125}$$

The variation vector field J is called the *Jacobi field* (Fig. 3.10). Note that *the Jacobi equation has been derived on the basis of the geodesic equation* $\nabla T/\partial s = 0$, *the equation of frozen field* (3.91), *and the definition of the curvature tensor* (3.97).

We are allowed to restrict ourselves to the study of the component $J_\perp$ normal to the tangent T. In fact, writing $J = J_\perp + cT$ (c: constant) satisfying $\langle J_\perp, T\rangle = 0$, we immediately find that Eq. (3.125) reduces to

$$\frac{\nabla}{\partial s}\frac{\nabla}{\partial s}J_\perp + R(J_\perp,T)T = 0, \tag{3.126}$$

since $R(J,T)T = R(J_\perp,T)T + cR(T,T)T$ and $R(T,T) = 0$, and $(\nabla T/\partial s) = 0$.

Defining $\|J\|^2 := \langle J, J\rangle$ and differentiating it two times with respect to s and using (3.125) and (3.17), we obtain

$$\frac{\mathrm{d}^2}{\mathrm{d}s^2}\frac{1}{2}\|J\|^2 = \|\nabla_T J\|^2 - K(T,J), \tag{3.127}$$

where $(\mathrm{d}/\mathrm{d}s)\frac{1}{2}\|J\|^2 = \langle \nabla_T J, J\rangle$, $\nabla_T J = \nabla J/\partial s$, and

$$K(T,J) := \langle R(J,T)T, J\rangle = R_{ijkl}J^iT^jJ^kT^l \tag{3.128}$$

is a sectional curvature factor associated with the two-dimensional section spanned by J and T. The proper *sectional curvature* $\hat{K}(T,J)$ is defined by (3.109), which reduces to $K(T,J)$ when T and J are orthonormal, hence $S_{TJ} = 1$.

Writing $J = \|J\| e_J$ where $\|e_J\| = 1$, Eq. (3.127) is transformed to

$$\frac{\mathrm{d}^2}{\mathrm{d}s^2}\|J\| = (\|\nabla_T e_J\|^2 - K(T, e_J))\|J\|. \tag{3.129}$$

3.10.2. *Initial behavior of Jacobi field*

Jacobi field $J(s)$ is defined by a C^∞ vector field along the geodesic $\gamma_0(s)$ satisfying Eq. (3.125). Such a Jacobi field is uniquely determined by its value and the value of $\nabla_T J$ at a point (at the origin e, say) on the geodesic [Hic65, §10], where $T = \partial_s$ is the unit tangent vector to γ_0. This is a consequence of the uniqueness theorem for solutions of the second order differential equation (3.125).

Suppose that the initial values are such that $J(0) = 0$ and $J'(0) = A_0 \neq 0$. This corresponds to considering a family of geodesics emanating from the origin e radially outward on a curved manifold M. Let us write $J = sA$. Then, we obtain

$$\begin{aligned} T|J| = T|sA| &= T(\langle sA, sA\rangle)^{1/2} = \langle \nabla_T sA, sA\rangle/|sA| \\ &= |A| + s\langle \nabla_T A, A\rangle/|A|, \end{aligned} \tag{3.130}$$

where $|sA| = \langle sA, sA\rangle^{1/2}$ and $\nabla_T sA = A + s\nabla_T A$. Applying T again, we have

$$\begin{aligned} T^2|sA| &= (\langle \nabla_T\nabla_T sA, sA\rangle + \langle \nabla_T sA, \nabla_T sA\rangle)/|sA| - \langle \nabla_T sA, sA\rangle^2/|sA|^3 \\ &= -\frac{\langle R(sA,T)T, sA\rangle}{|sA|} + \frac{|\nabla_T sA|^2}{|sA|} - \frac{\langle \nabla_T sA, sA\rangle^2}{|sA|^3} \\ &= -|sA|\kappa(s) + H(s), \end{aligned} \tag{3.131}$$

where

$$\begin{aligned} \kappa(s) &= \frac{\langle R(A,T)T, A\rangle}{|A|^2} = \frac{\langle R(J,T)T, J\rangle}{|J|^2}. \qquad (3.132) \\ H(s) &= |\nabla_T sA|^2/|sA| - \langle \nabla_T sA, sA\rangle^2/|sA|^3 \\ &= \frac{s}{|A|^3}(|\nabla_T A|^2|A|^2 - \langle \nabla_T A, A\rangle^2). \end{aligned}$$

For $J = sA$, Eq. (3.125) gives $\nabla_T\nabla_T(sA) = 0$ at $s = 0$, whereas we have $\nabla_T\nabla_T(sA) = 2\nabla_T A + s\nabla_T\nabla_T A$. Hence, we obtain $\nabla_T A = 0$ at $s = 0$. Therefore, we have $H'(0) = 0$ as well as $H(0) = 0$.

In view of $T = \partial_s$ and $J = sA$, the above results give

$$|J|(0) = 0, \quad |J|'(0) = A_0, \quad |J|''(0) = 0, \quad |J|'''(0) = -A_0\, \kappa(0).$$

Therefore, we have

$$|J|(s)/A_0 = s - \kappa(0)\frac{s^3}{3!} + O(s^4) \tag{3.133}$$

[Hic65]. Using the definition $\cos\theta = \langle J, T\rangle / |J|\,|T|$, we have

$$\kappa(0) = \frac{\langle R(J,T)T, J\rangle}{|J|^2}\bigg|_{s=0} = \frac{|J|^2|T|^2 - \langle J,T\rangle^2}{|J|^2}\hat{K}(J,T)$$

$$= |T|^2 \sin^2\theta\, \hat{K}(J,T), \tag{3.134}$$

$$\hat{K}(J,T) = \frac{\langle R(J,T)T, J\rangle}{|J|^2|T|^2 - \langle J,T\rangle^2}. \tag{3.135}$$

Thus it is found that initial development of magnitude of the Jacobi field is controlled by the sectional curvature $\hat{K}(J,T)$.

3.10.3. *Time-dependent problem*

In the following chapters, we consider various dynamical systems in which field variables are *time-dependent.* A tangent vector to the geodesic curve is represented in the form, $\tilde{T} = \tilde{T}^i\partial_i = \partial_t + T^\alpha\partial_\alpha$ according to §3.2.2, where $x^0 = t$ (time) and α denotes the indices of the spatial part. The covariant derivative is given by (3.16) as $\nabla_{\tilde{T}}\tilde{J} = \partial_t J + \nabla_T J$ where $J = J^\alpha\partial_\alpha$. Thus, using the time t instead of s, we have

$$\frac{\nabla \tilde{J}}{\partial t} = \partial_t J + \nabla_T J, \tag{3.136}$$

where T and J are the spatial parts.

Note that the curvature tensor in the Jacobi equation is unchanged, i.e. $R(\tilde{J},\tilde{T})\tilde{T} = R(J,T)T$ which does not include any ∂_t component, because the curvature tensor R^α_{kij} of (3.98) vanishes, when one of k, i, j takes 0, owing to the definition (3.99) using (3.14), under the reasonable assumption that the metric tensor and the Christoffel symbols do not depend on t. Thus, the Jacobi equation (3.125) is replaced by

$$\partial_t^2 J + \partial_t(\nabla_T J) + \nabla_T \partial_t J + \nabla_T \nabla_T J + R(J,T)T = 0. \tag{3.137}$$

This equation provides the link between the stability of geodesic curves and the Riemannian curvature, and one of the basic elements for the geometrical description of dynamical systems.

3.10.4. *Two-dimensional problem*

On a two-dimensional Riemannian surface M^2, Eq. (3.125) is much simplified. We introduce a unit vector field $\boldsymbol{e}$ along $\gamma_0(s)$ that is orthogonal to its tangent $T = \partial_s \gamma_0$. The unit tangent T is displaced parallel along $\gamma_0(s)$, likewise the unit normal $\boldsymbol{e}$ is also displaced parallel along $\gamma_0(s)$ because $\langle \boldsymbol{e}, T \rangle = 0$ is satisfied along $\gamma_0(s)$ (§3.4). Hence we have $(\nabla \boldsymbol{e}/\partial s) = 0$ as well as $(\nabla T/\partial s) = 0$. Let us represent the Jacobi field $J(s)$ as

$$J(s) = x(s)\,T(s) + y(s)\boldsymbol{e}(s),$$

where $x(s)$ and $y(s)$ are the tangential and normal components. Then, we obtain from (3.125)

$$\left(\frac{\nabla}{\partial s}\right)^2 J = \frac{\mathrm{d}^2 x}{\mathrm{d}s^2} T + \frac{\mathrm{d}^2 y}{\mathrm{d}s^2} \boldsymbol{e} = -R(xT + y\boldsymbol{e}, T)T = -yR(\boldsymbol{e}, T)T.$$

Taking a scalar product with $\boldsymbol{e}$, we obtain

$$\frac{\mathrm{d}^2 y}{\mathrm{d}s^2} = -y\langle R(\boldsymbol{e}, T)T, \boldsymbol{e}\rangle.$$

Representing vectors and tensors with respect to the orthonormal frame $(\boldsymbol{e}_1, \boldsymbol{e}_2) = (T, \boldsymbol{e})$ along $\gamma_0(s)$, we have

$$\begin{aligned}\langle R(\boldsymbol{e}, T)T, \boldsymbol{e}\rangle &= \langle R(\boldsymbol{e}_2, \boldsymbol{e}_1)\boldsymbol{e}_1, \boldsymbol{e}_2\rangle = R_{2121} = R^1{}_{212} \\ &= K(\boldsymbol{e}_2, \boldsymbol{e}_1) = K(\boldsymbol{e}_1, \boldsymbol{e}_2) := K(s),\end{aligned}$$

which is the only nonzero sectional curvature on M^2. Thus, the Jacobi equation becomes

$$\frac{\mathrm{d}^2 y}{\mathrm{d}s^2} + K(s)y = 0. \tag{3.138}$$

On the Poincaré surface considered in §3.5.3, we found that $K = R^1{}_{212} = -1$. Therefore, the Jacobi field on the Poincaré surface is represented as a linear combination of e^s and e^{-s}:

$$y(s) = A\mathrm{e}^s + B\mathrm{e}^{-s}.$$

3.10.5. *Isotropic space*

For an isotropic manifold, the curvature tensor is given by (3.117). Then we have

$$R(J_\perp, T)T = R^\alpha_{ikj} T^i J^k_\perp T^j = K J^\alpha_\perp \langle T, T\rangle - K T^\alpha \langle J_\perp, T\rangle = K J^\alpha_\perp.$$

Thus, the Jacobi equation (3.126) reduces to

$$\left(\frac{\nabla}{\partial s}\right)^2 J^\alpha_\perp + K J^\alpha_\perp = 0. \tag{3.139}$$

Choosing an orthonormal frame $(e_1, \ldots, e_n)$, the covariant derivative becomes ordinary derivative, i.e. $\nabla/\partial s = \mathrm{d}/\mathrm{d}s$, since the orthonormal frame can be transported parallel along the geodesic (see the previous section and §3.4).

3.11. Differentiation of Tensors

In §1.8.3, we learned the Lie derivatives of a scalar function and a vector field. Here, we consider Lie derivative of 1-form and covariant derivative of tensors.

3.11.1. *Lie derivatives of 1-form and metric tensor*

The value of 1-form $\omega \in (TM^n)^*$ for a vector field $W(x) \in TM^n$ is a function $\omega(W)(x) = \omega_i W^i \in \mathbb{R}$. Lie derivative of $\omega(W)$ along a tangent vector $U \in TM^n$ is

$$\begin{aligned} \mathcal{L}_U[\omega(W)] &= U^j \partial_j(\omega_i W^i) = \omega_i (U^j \partial_j W^i) + (U^j \partial_j \omega_i) W^i \\ &:= \omega_i (\mathcal{L}_U W)^i + (\mathcal{L}_U \omega)_i W^i, \end{aligned} \tag{3.140}$$

where

$$(\mathcal{L}_U W)^i := [U, W]^i = U^j \partial_j W^i - W^j \partial_j U^i, \tag{3.141}$$
$$(\mathcal{L}_U \omega)_i := U^k \partial_k \omega_i + \omega_k \partial_i U^k. \tag{3.142}$$

In terms of a metric tensor g_{ij}, 1-form may be written as $\omega_i(V) = g_{ij}V^j$ ($V \in TM^n$). Then,

$$(\mathcal{L}_U \omega)_i(V) = U^k(\partial_k g_{ij})V^j + g_{ij}U^k \partial_k V^j + g_{kj}V^j \partial_i U^k \tag{3.143}$$
$$= (\mathcal{L}_U g)_{ij} V^j + g_{ij}(\mathcal{L}_U V)^j, \tag{3.144}$$

where the Lie derivative of the metric tensor g is defined by

$$(\mathcal{L}_U g)_{ij} := U^k \partial_k g_{ij} + g_{ik}\partial_j U^k + g_{kj}\partial_i U^k. \tag{3.145}$$

Thus, the Lie derivative of $\omega(W) = \omega_i W^i = g_{ij}V^j W^i = \langle V, W\rangle$ is given by

$$\mathcal{L}_U[g_{ij}V^iW^j] = (\mathcal{L}_U g)_{ij}V^iW^j + g_{ij}(\mathcal{L}_U V)^i W^j + g_{ij}V^i(\mathcal{L}_U W)^j$$
$$= (\mathcal{L}_U g)_{ij}V^iW^j + \langle \mathcal{L}_U V, W\rangle + \langle V, \mathcal{L}_U W\rangle. \tag{3.146}$$

3.11.2. *Riemannian connection* ∇

On a Riemannian manifold with the metric tensor g_{ij} and Riemannian connection ∇, time derivative d/dt of the inner product $\langle Y, Z\rangle$ ($Y, Z \in TM^n$), along a parameterized curve ϕ_t generated by a vector field $V = V^k\partial_k \in TM^n$, is given by

$$\frac{\mathrm{d}}{\mathrm{d}t}\langle Y, Z\rangle = \langle \nabla_V Y, Z\rangle + \langle Y, \nabla_V Z,\rangle \tag{3.147}$$

(see §3.4). In components, this is rewritten as

$$V^k\partial_k(g_{ij}Y^iZ^j) = g_{ij}V^k\left(\frac{\partial Y^i}{\partial x^k} + \Gamma^i_{kl}Y^l\right)Z^j + g_{ij}Y^iV^k\left(\frac{\partial Z^j}{\partial x^k} + \Gamma^j_{kl}Z^l\right).$$

Since this holds for all Y and Z ($\in TM^n$), it is concluded that the Riemannian metric tensor g_{ij} must satisfy

$$\frac{\partial g_{ij}}{\partial x^k} - g_{lj}\Gamma^l_{ki} - g_{il}\Gamma^l_{kj} = 0. \tag{3.148}$$

This is understood to mean that covariant derivative of the metric tensor vanishes (see next section).

3.11.3. *Covariant derivative of tensors*

We have already defined the covariant derivative of a *vector* field $v = v^i \partial_i$ as

$$v^i_{;k} := \nabla_k v^i = \nabla v^i(\partial_k) = \partial_k v^i + \Gamma^i_{kl} v^l. \tag{3.149}$$

We define the covariant derivative of a *covector* field $\omega = \omega_i dx^i$ such that the following "rule of derivative" holds:

$$\partial_k(\omega_i v^i) = \omega_i v^i_{;k} + \omega_{i;k} v^i. \tag{3.150}$$

Using (3.149), we obtain

$$\omega_i \partial_k(v^i) + (\partial_k \omega_i) v^i = \omega_i(\partial_k v^i + \Gamma^i_{kl} v^l) + (\partial_k \omega_i - \Gamma^l_{ki}\omega_l) v^i.$$

So, the covariant derivative of ω_i is defined by

$$\nabla_k \omega_i = \omega_{i;k} := \partial_k \omega_i - \Gamma^l_{ki}\omega_l. \tag{3.151}$$

We generalize the above rules regarding general *tensors*. For a *mixed* tensor of the type (p, q), $M^{i_1 \cdots i_q}_{j_1 \cdots j_p}$ (see §1.10.3(v)), we define

$$\begin{aligned} \nabla_k M^{i_1 \cdots i_q}_{j_1 \cdots j_p} = M^{i_1 \ldots i_q}_{j_1 \ldots j_p;k} &:= \partial_k M^{i_1 \cdots i_q}_{j_1 \cdots j_p} + \Gamma^{i_1}_{kr} M^{r i_2 \cdots i_q}_{j_1 \cdots j_p} + \Gamma^{i_2}_{kr} M^{i_1 r \cdots i_q}_{j_1 \cdots j_p} + \cdots \\ &\quad - \Gamma^r_{kj_1} M^{i_1 \cdots i_q}_{r j_2 \cdots j_p} - \Gamma^r_{kj_2} M^{i_1 \cdots i_q}_{j_1 r \cdots j_p} - \cdots . \end{aligned} \tag{3.152}$$

This is obtained by using the rules (3.149) and (3.151) repeatedly for each covariant and each contravariant index of $M^{i_1 \cdots i_q}_{j_1 \cdots j_p}$. The covariant derivative of the mixed tensor M^i_j is

$$M^i_{j;k} = \partial_k M^i_j + \Gamma^i_{kr} M^r_j - \Gamma^r_{kj} M^i_r, \tag{3.153}$$

while the covariant derivative of the tensor M_{ij} of $(0, 2)$ type is

$$M_{ij;k} = \partial_k M_{ij} - \Gamma^r_{ki} M_{rj} - \Gamma^r_{kj} M_{ir}. \tag{3.154}$$

Hence, Eq. (3.148) means that

$$g_{ij;k} = 0. \tag{3.155}$$

Using contraction on i and j in M^i_j in (3.153), we have

$$M^i_{i;k} = \partial_k M^i_i + \Gamma^i_{kr} M^r_i - \Gamma^r_{ki} M^i_r = \partial_k M^i_i, \tag{3.156}$$

which is consistent with (3.150). Therefore, the covariant differentiation commutes with contraction, that is, contraction of the covariant derivative of M^i_j is equal to the covariant derivative of contracted tensor (scalar) M^i_i, i.e. $(\partial/\partial x^k) M^i_i$.

3.12. Killing Fields

We consider Killing vector field, Killing tensor fields and associated invariants.

3.12.1. *Killing vector field X*

A Killing field (after the mathematician Killing) is *defined* to be a vector field X such that the Lie derivative of the metric tensor g along it vanishes:

$$(\mathcal{L}_X g)_{ij} = X^k \partial_k g_{ij} + g_{ik} \partial_j X^k + g_{kj} \partial_i X^k = 0, \tag{3.157}$$

from (3.145). Using the property (3.148) of the Riemannian metric tensor g_{ij}, this is rewritten as

$$\begin{aligned} 0 = (\mathcal{L}_X g)_{ij} &= X^k (g_{lj} \Gamma^l_{ki} + g_{il} \Gamma^l_{kj}) + g_{ik} \partial_j X^k + g_{kj} \partial_i X^k \\ &= g_{ik} (\partial_j X^k + \Gamma^k_{lj} X^l) + g_{jk} (\partial_i X^k + \Gamma^k_{li} X^l) \\ &= g_{ik} X^k_{;j} + g_{jk} X^k_{;i} = (g_{ik} X^k)_{;j} + (g_{jk} X^k)_{;i} \\ &:= X_{i;j} + X_{j;i}, \end{aligned} \tag{3.158}$$

where $X_i = g_{ik} X^k$, since $g_{ik;j} = 0$ from (3.155).[12] Thus, we have found another relation equivalent to (3.157),

$$(\mathcal{L}_X g)_{ij} = X_{i;j} + X_{j;i} = 0, \tag{3.159}$$

which is called the **Killing's equation**. This equation implies

$$(\mathcal{L}_X g)_{ij} Y^i Z^j = \langle \nabla_Y X, Z \rangle + \langle Y, \nabla_Z X \rangle = 0. \tag{3.160}$$

In problems of infinite dimensions considered in Chapters 5 and 7, the inner product is defined by an integral, and the Killing field X is required to satisfy identically

$$\langle \nabla_Y X, Z \rangle + \langle Y, \nabla_Z X \rangle = 0, \quad \text{for } {}^\forall Y, Z \in TM. \tag{3.161}$$

Using the definition relation (3.30) for the connection, this is transformed to

$$\langle [Y, X], Z \rangle + \langle Y, [Z, X] \rangle = 0, \quad \text{for } {}^\forall Y, Z \in TM. \tag{3.162}$$

---

[12]Note that $X^j_{;i}$ is equivalent to $\nabla_i X^j = \nabla X^k(\partial_j)$ defined by (3.10) in §3.2.

3.12.2. *Isometry*

A Killing field X generates a one-parameter group $\phi_t = e^{tX}$ of isometry. If Y and Z are fields that are invariant under the flow (see Remark of §1.8.3), the inner product $\langle Y, Z\rangle = g_{ij}Y^iZ^j$ is independent of t along the flow ϕ_t. From (3.147), invariance of $\langle Y, Z\rangle$ is given by

$$\frac{\mathrm{d}}{\mathrm{d}t}\langle Y, Z\rangle = \langle \nabla_X Y, Z\rangle + \langle Y, \nabla_X Z\rangle = 0. \tag{3.163}$$

This is satisfied if the vector fields Y and Z are invariant, i.e. $Y\phi_t = \phi_t Y$ and $Z\phi_t = \phi_t Z$. Such a field as Y or Z is also called a *frozen field* (frozen to the flow ϕ_t). Then, we have from (1.81),

$$\mathcal{L}_X Y = \nabla_X Y - \nabla_Y X = 0, \quad \mathcal{L}_X Z = \nabla_X Z - \nabla_Z X = 0.$$

Thus, it is found that the two Eqs (3.160) and (3.163) are equivalent.

Using (3.146) and the property of Killing field (3.157), we obtain another expression for the invariance,

$$\frac{\mathrm{d}}{\mathrm{d}t}\langle Y, Z\rangle = \mathcal{L}_X\langle Y, Z\rangle = \langle \mathcal{L}_X Y, Z\rangle + \langle Y, \mathcal{L}_X Z\rangle = 0. \tag{3.164}$$

In an unsteady problem, $\nabla_X Y$ should be replaced by $\partial_t Y + \nabla_X Y$. Hence the equation $\nabla_X Y = \nabla_Y X$, representing invariance of Y along ϕ_t, is replaced by[13]

$$\partial_t Y + \nabla_X Y = \partial_t X + \nabla_Y X. \tag{3.165}$$

3.12.3. *Positive curvature and simplified Jacobi equation*

The sectional curvature K_X in the section spanned by a Killing field X and an arbitrary variation field J can be shown to be *positive*. The curvature K_X is given by the formula (3.111)[14]:

$$\begin{aligned} K_X(X, J) = \langle R(X, J)J, X\rangle &= -\langle \nabla_X X, \nabla_J J\rangle \\ &\quad + \langle \nabla_J X, \nabla_X J\rangle + \langle \nabla_{[X,J]}X, J\rangle. \end{aligned} \tag{3.166}$$

[13] Usually Killing field X is stationary: $\partial_t X = 0$. Examples will be seen in §5.5 and 9.6.
[14] The formula $U\langle Y, Z\rangle = \langle \nabla_U Y, Z\rangle + \langle Y, \nabla_U Z\rangle = 0$ is used repeatedly since the metric is invariant.

The first term vanishes because the Killing field X satisfies the geodesic equation $\nabla_X X = 0$. The second term is rewritten by using the Killing equation (3.161) for X as

$$\langle \nabla_J X, \nabla_X J \rangle = -\langle \nabla_{(\nabla_X J)} X, J \rangle,$$

where Y and Z are replaced by J and $\nabla_X J$, respectively. To the third term, we apply the two properties, the torsion-free $[X, J] = \nabla_X J - \nabla_J X$ and the property (ii) of (3.5), $\nabla_{(U-V)} X = \nabla_U X - \nabla_V X$, and obtain

$$\langle \nabla_{[X,J]} X, J \rangle = \langle \nabla_{(\nabla_X J)} X, J \rangle - \langle \nabla_{(\nabla_J X)} X, J \rangle.$$

Substituting these into (3.166), we find

$$\begin{aligned} K_X(X, J) &= -\langle \nabla_{(\nabla_J X)} X, J \rangle = \langle \nabla_J X, \nabla_J X \rangle \\ &= \|\nabla_J X\|^2 = \|\nabla_X J\|^2, \end{aligned} \tag{3.167}$$

by using (3.161) again. The last equality holds when J is a Jacobi field and the equation of frozen field $\nabla_J X = \nabla_X J$ is satisfied. Thus, it is found that *the sectional curvature K_X between a Killing field X and an arbitrary variation field J is positive.*

In this case, the Jacobi equation (3.127) reduces to

$$\frac{\mathrm{d}^2}{\mathrm{d}s^2} \frac{1}{2} \|J\|^2 = \|\nabla_X J\|^2 - K(X, J) = 0. \tag{3.168}$$

Hence, the Jacobi field grows only linearly with s (does not grow exponentially with s): $\|J\|^2 = as + b$ (a, b: constants). Thus, *stability of Killing field is verified.*

3.12.4. *Conservation of $\langle X, T \rangle$ along $\gamma(s)$*

If the curve $\gamma(s)$ is a geodesic with its tangent $\mathrm{d}\gamma/\mathrm{d}s = T = T^i \partial_i$ and in addition X is a Killing vector field, then we have

$$\begin{aligned} \frac{\mathrm{d}}{\mathrm{d}s} \langle X, T \rangle &= \langle \nabla_T X, T \rangle + \langle X, \nabla_T T \rangle = \langle \nabla_T X, T \rangle \\ &= g_{ik} (\nabla_T X)^i T^k = g_{ik} X^i_{;j} T^j T^k \\ &= \frac{1}{2} (X_{k;j} + X_{j;k}) T^j T^k = 0, \end{aligned} \tag{3.169}$$

where $\nabla_T T = 0$ is used. Thus, the following inner product,

$$\langle X, T\rangle = X_i T^i, \tag{3.170}$$

is *conserved* along any flow $\gamma(s)$, where $X_i = g_{ik}X^k$ is the Killing covector.

3.12.5. *Killing tensor field*

Equation (3.169) is rewritten as

$$\frac{\mathrm{d}}{\mathrm{d}s}(X_i v^i) = \frac{1}{2}(X_{i;j} + X_{j;i})v^i v^j = 0.$$

This states that the quantity $I = X_i v^i$ is conserved along a geodesic flow $g_t(v)$ generated by $^\forall v = v^i\partial_i \in TM^n$. A generalization can be made to a tensor field $X_{k_1 k_2\cdots k_p}$, where $I = X_{k_1k_2\cdots k_p}v^{k_1}v^{k_2}\cdots v^{k_p}$ is conserved along any geodesic flow. According to [ClP02], we look for the condition that assures

$$\frac{\mathrm{d}}{\mathrm{d}s}(X_{k_1k_2\cdots k_p}v^{k_1}v^{k_2}\cdots v^{k_p}) = v^j\nabla_j(X_{k_1,k_2,\ldots,k_p}v^{k_1}v^{k_2}\cdots v^{k_p}) = 0$$

for $^\forall v = v^i\partial_i \in TM^n$. Using (3.152) for the covariant derivative of the tensor $X_{k_1k_2\cdots k_p}$ and in addition (3.149) for the covariant derivative of the vector v^k (which vanishes because $v^j\nabla_j v^k = 0$), we obtain

$$\begin{aligned}\frac{\mathrm{d}}{\mathrm{d}s}(X_{k_1k_2\cdots k_p}v^{k_1}v^{k_2}\cdots v^{k_p}) &= v^{k_1}v^{k_2}\cdots v^{k_p}v^j\nabla_j X_{k_1k_2\cdots k_p}\\ &= \frac{1}{p+1}v^{k_1}v^{k_2}\cdots v^{k_p}v^j\nabla_{(j)}X_{(K)_p},\end{aligned} \tag{3.171}$$

after deleting terms of the form $v^i\nabla_j v^k$ which must vanish by the geodesic condition. The second equality is due to the permutation invariance of $k_1, k_2, \ldots, k_p$, and j, where

$$\nabla_{(j)}X_{(K)_p} := \nabla_j X_{k_1k_2\cdots k_p} + \nabla_{k_1}X_{jk_2\cdots k_p} + \cdots + \nabla_{k_p}X_{k_1k_2\cdots j}.$$

Thus, the invariance of $I = X_{k_1k_2\cdots k_p}v^{k_1}v^{k_2}\cdots v^{k_p}$ along the geodesic flow $g_t(v)$ is guaranteed by the tensor field $X_{k_1k_2\cdots k_p}$ fulfilling the conditions (for fixed value of p),

$$\nabla_{(j)}X_{(K)_p} = 0, \tag{3.172}$$

where each one of $(j, k_1, k_2, \ldots, k_p)$ takes values, $1, 2, \ldots, n$. Hence the number of equations is n^{p+1}, whereas the number of unknown variables is $(n+p-1)!/p!(n-1)!$. These overdetermined equations generalize the

Killing's equation (3.159) for the Killing vector field ($p = 1$). Such tensors $X_{k_1 k_2 \cdots k_p}$ are termed the *Killing tensor fields* [ClP02], whose existence is rather exceptional.

3.13. Induced Connection and Second Fundamental Form

Space of volume-preserving flows is a subspace embedded in a space of general (compressible) fluid flows. An induced connection for the subspace can be defined analogously to the case of a curved surface in $\mathbb{R}^3$. Here, we consider such a case with finite-dimensional spaces.

Let V^r be a submanifold of a Riemannian manifold M^n equipped with a metric g_{ij}. Let us consider the restriction of the Riemannian metric g_{ij} of M^n to the space tangent to V^r. This induces a Riemannian metric (an induced metric) for V^r. An arbitrary vector field Z of M^n defined along V^r can be decomposed into two orthogonal components[15]: $Z(p) = Z_V + Z_N$, where $Z_V = \mathtt{P}\{Z\}$ is the projected component to T_pV^r at a point $p \in V^r$ and $Z_N = \mathtt{Q}\{Z\}$ is the component perpendicular to T_pV^r. The symbols $\mathtt{P}$ and $\mathtt{Q}$ denote the orthogonal projections onto the space V^r and the space orthogonal to it, respectively. Let ∇^M be the Riemannian connection for M^n, and define a new connection ∇^V for V^r ($r < n$) as follows. Consider a vector X tangent to V^r and a vector field Z in M^n defined along V^r where Z is not necessarily tangent to V^r. Then, the ∇^V is defined by

$$\nabla_X^V Z(p) := \mathtt{P}\{\nabla_X^M Z\} = \nabla_X^M Z - \mathtt{Q}\{\nabla_X^M Z\}, \tag{3.173}$$

where the right-hand side is the projection of $\nabla_X^M Z$ onto the tangent space of T_pV^r. It can be checked that the operator ∇^V satisfies the properties (3.5) and an induced connection. Suppose that X, Y and Z are tangent to V^r, then one has $\mathtt{Q}\{X\} = 0, \mathtt{Q}\{Y\} = 0$ and $\mathtt{Q}\{Z\} = 0$, and furthermore $\mathtt{Q}\{[X,Y]\} = 0$.

This is shown as follows. Extending the vectors X and Y to the vectors in M^n, which is accomplished just by adding 0 components in the space perpendicular to V^r, we consider $[X,Y]$ in M^n. By the torsion-free of the Riemannian connection ∇^M, one has

$$\mathtt{Q}\{[X,Y]\} = \mathtt{Q}\{\nabla_X^M Y - \nabla_Y^M X\} = 0, \tag{3.174}$$

---

[15]See Appendix F for the Helmholtz decomposition of vector fields.

which is verified by using the expression (3.6). In fact, all the terms including the terms Γ^k_{ij} cancel out with the symmetry $\Gamma^k_{ij} = \Gamma^k_{ji}$ and the remaining terms are within the space V^r. Hence, $\nabla^M_X Y - \nabla^M_Y X = \mathrm{P}\{\nabla^M_X Y - \nabla^M_Y X\} = \nabla^V_X Y - \nabla^V_Y X$. Thus, it is found that the connection ∇^V is also torsion-free:

$$\nabla^V_X Y - \nabla^V_Y X = [X, Y]. \tag{3.175}$$

Therefore the connection ∇^V is also Riemannian. The second condition (3.18) is satisfied by (3.175). The first condition (3.17) is also valid for ∇^V.

In §2.3, we considered a relation between the connection ∇ in the enveloping $\mathbb{R}^3$ and the induced connection $\bar{\nabla}$ of a curved surface Σ^2, which is represented by the Gauss' s surface equation (2.35) including the second fundamental form.

Analogously for the case $r = n - 1$, taking the second fundamental form as $S(X,Y)$ instead of $\mathrm{II}(X,Y)\boldsymbol{N}$, the corresponding Gauss' formula is given by

$$\nabla^M_X Y = \nabla^V_X Y + S(X,Y), \qquad X, Y \in TV^{n-1}. \tag{3.176}$$

This equation can be generalized for the case $n = r + p$ as follows (Fig. 3.11):

$$\nabla^M_X Y := \nabla^V_X Y + S(X,Y), \tag{3.177}$$

$$S(X,Y) = \sum_a \langle \nabla^M_X Y, \boldsymbol{N}_a \rangle \boldsymbol{N}_a,$$

where $\boldsymbol{N}_a$ $(a = 1, \ldots, p)$ are p normal vector fields along V^r that are orthonomal. This is the *surface equation* generalizing the Gauss's equation (2.35). According to (3.174), it is not difficult to see that the function

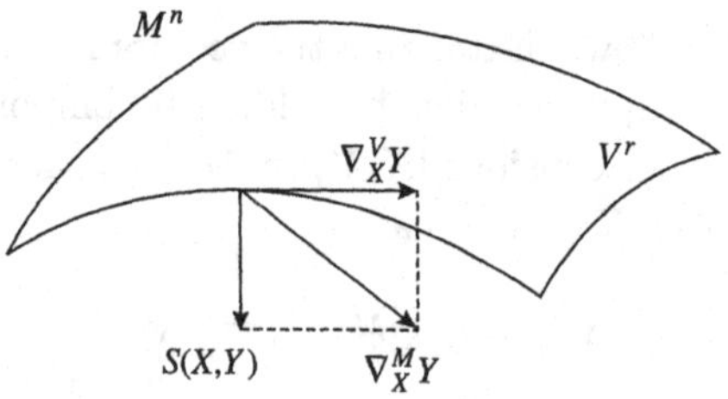

Fig. 3.11. Gauss's surface equation.

$S(X,Y)$ satisfies the following relation,

$$\begin{aligned} S(X,Y) &:= \nabla^M_X Y - \nabla^V_X Y \\ &= \mathtt{Q}\{\nabla^M_X Y\} = \mathtt{Q}\{\nabla^M_Y X\} = S(Y,X). \end{aligned} \tag{3.178}$$

Thus, $S(X,Y)$ is found to be symmetric with respect to X and Y.

Corresponding to ∇^M and ∇^V, we have two kinds of curvature tensors, $R^M(X,Y)Z$ and $R^V(X,Y)Z$, respectively. Using the definition (3.97) of $R(X,Y)Z$ and the above relations (3.177) and (3.178), one can show the following formula:

$$\begin{aligned} \langle W, R^M(X,Y)Z\rangle &= \langle W, R^V(X,Y)Z\rangle \\ &\quad + \langle S(X,Z), S(Y,W)\rangle - \langle S(X,W), S(Y,Z)\rangle, \end{aligned} \tag{3.179}$$

where $X, Y, Z, W \in TV^r$.

This can be verified by using the definition (3.173) repeatedly. For example, we have

$$\nabla^V_X \nabla^V_Y Z = \nabla^M_X (\nabla^M_Y Z - \mathtt{Q}\{\nabla^M_Y Z\}) - \mathtt{Q}\{\nabla^M_X (\nabla^M_Y Z - \mathtt{Q}\{\nabla^M_Y Z\})\}.$$

Taking the scalar product with $W \in TV^r$, we obtain

$$\langle W, \nabla^V_X \nabla^V_Y Z\rangle = \langle W, \nabla^M_X \nabla^M_Y Z\rangle - \langle W, \nabla^M_X S(Y,Z)\rangle \tag{3.180}$$

$$= \langle W, \nabla^M_X \nabla^M_Y Z\rangle + \langle S(X,W), S(Y,Z)\rangle. \tag{3.181}$$

The last equality can be shown by using

(i) $\langle W, \nabla^M_X S(Y,Z)\rangle + \langle \nabla^M_X W, S(Y,Z)\rangle = X\langle W, S(Y,Z)\rangle = 0$,

(ii) $W \perp S(Y,Z)$, and

(iii) $\mathtt{Q}\{\nabla^M_X W\} = S(X,W)$.

Similar expressions can be derived for the other terms. Using those expressions, one verifies (3.179).

Part II
Dynamical Systems

Chapter 4

Free Rotation of a Rigid Body

We now consider a physical problem and try to apply the geometrical theory formulated in Part I to one of the simplest dynamical systems: *Euler's top*, i.e. free rotation of a rigid body (free from external torque). We begin with this simplest system in order to illustrate the underlying geometrical ideas and show how powerful is the method. The basic philosophy is that the governing equation is derived as a geodesic equation over a manifold of a *symmetry group*, i.e. the rotation group $SO(3)$ (a Lie group) in three-dimensional space. The equation thus obtained describes rotational motions of a rigid body. A highlight of this chapter is the metric bi-invariance on the group SO(3) and associated integrability. Some new analysis on the stability of regular precession is presented, in addition to the basic formulation according to [LL76; Arn78; Kam98; SWK98].

4.1. Physical Background

4.1.1. *Free rotation and Euler's top*

A rigid body has six degrees of freedom in general, and equations of motion can be put in a form which gives time derivatives of momentum P and angular momentum M of the body as

$$\frac{\mathrm{d}P}{\mathrm{d}t} = F, \qquad \frac{\mathrm{d}M}{\mathrm{d}t} = N,$$

where F is the total external force acting on the body, and N is the total torque, i.e. the sum of the moments of all the external forces about a reference point O. Correspondingly, the angular momentum M is defined as the

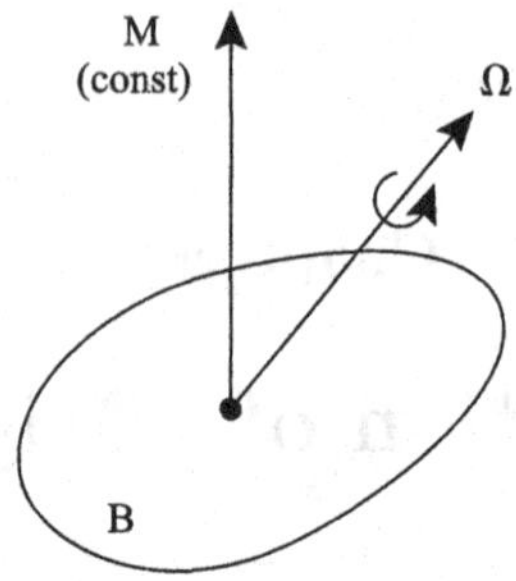

Fig. 4.1. Free rotation in fixed frame F (*inertial* frame).

one about the point O. If the body is free from any external force, we obtain the conservation of total momentum and angular momentum: $P = \text{const}$ and $M = \text{const}$.

When the rigid body is fixed at the point O but the external torque N about O is zero, then the equation of angular momentum is again given by $\mathrm{d}M/\mathrm{d}t = 0$. This situation reduces to the problem of *free* rotation of a *top*, termed the *Euler's top* (Fig. 4.1).

In either case of the Euler's top or the free rotation (with no external force and the point O coinciding with the center of mass), the equation of angular momentum is given by the same equation,

$$\frac{\mathrm{d}M}{\mathrm{d}t} = 0. \tag{4.1}$$

However, in order to describe detailed rotational motion of the body, it is simpler to consider it in the body frame (i.e. in the frame of reference fixed to the moving body), which is *noninertial*. Equation (4.1) in the inertial frame is transformed into the following *Euler's equations* in the rotating frame,

$$\begin{aligned} &J_1\frac{\mathrm{d}\Omega^1}{\mathrm{d}t} - (J_2 - J_3)\Omega^2\Omega^3 = 0, \quad J_2\frac{\mathrm{d}\Omega^2}{\mathrm{d}t} - (J_3 - J_1)\Omega^3\Omega^1 = 0, \\ &J_3\frac{\mathrm{d}\Omega^3}{\mathrm{d}t} - (J_1 - J_2)\Omega^1\Omega^2 = 0, \end{aligned} \tag{4.2}$$

where $\Omega = (\Omega^1, \Omega^2, \Omega^3)$ is the angular velocity (a tangent vector) relative to the body frame, and (J_1, J_2, J_3) are principal values of the moment of inertia of the body B. In general, the *moment of inertia* is a second order symmetric tensor defined by

$$J_{\alpha\beta} = \int_B (|\boldsymbol{x}|^2\delta_{\alpha\beta} - x^\alpha x^\beta)\rho \mathrm{d}^3\boldsymbol{x}, \quad |\boldsymbol{x}|^2 = x^\alpha x^\alpha, \tag{4.3}$$

(which is also called the *inertia tensor* J), where ρ is the body's mass density, assumed constant. Like any symmetric tensor of rank 2, the inertia tensor can be reduced to a diagonal form by an appropriate choice of coordinate frame for the body, called a *principal frame* (x^1, x^2, x^3), in which the inertia tensor is represented by a diagonal form $J = \mathrm{diag}(J_1, J_2, J_3)$.[1]

Relative to the instantaneous principal frame, the angular momentum is given by the following (cotangent) vector,

$$M = (M_\alpha) = J\Omega = (J_{\alpha\beta}\Omega^\beta) = (J_1\Omega^1, J_2\Omega^2, J_3\Omega^3). \tag{4.4}$$

The kinetic energy K is given by the following expression,

$$K = \frac{1}{2}M_\alpha\Omega^\alpha = \frac{1}{2}(J\Omega, \Omega)_s, \tag{4.5}$$

where

$$(M, T)_s = M_\alpha T^\alpha = M_1T^1 + M_2T^2 + M_3T^3 \tag{4.6}$$

is the scalar product, i.e. a scalar pairing of a tangent vector T and a cotangent vector M (see (1.55)). The kinetic energy K is a *scalar*, that is, invariant with respect to the frame transformation from a fixed inertial frame to the moving frame fixed instantaneously to the body.

It is an advantage that the moments of inertia (J_1, J_2, J_3) are fixed to be constant in the frame relative to the moving body although Eqs. (4.2) became nonlinear, while the inertia tensors were time-dependent in the fixed system, i.e. in the nonrotating inertial frame, where Eq. (4.1) is much simpler.

Using the angular momentum $M = J\Omega$, the Euler's equation (4.2) is converted into a vectorial equation:

$$\frac{\mathrm{d}}{\mathrm{d}t}M = M \times \Omega. \tag{4.7}$$

---

[1] From the definition (4.3), we obtain the following inequality in the principal frame: $J_2 + J_3 = \int([(x^3)^2 + (x^1)^2] + [(x^1)^2 + (x^2)^2])\rho \mathrm{d}^3\boldsymbol{x} \geq J_1$, and its cyclic permutation for the indices $(1, 2, 3)$.

4.1.2. *Integrals of motion*

Two integrals of the Euler's equation (4.2), or (4.7), are known:

$$\frac{1}{2}(J_1^{-1}M_1^2 + J_2^{-1}M_2^2 + J_3^{-1}M_3^2) = E, \tag{4.8}$$

$$M_1^2 + M_2^2 + M_3^2 = |M|^2. \tag{4.9}$$

It is not difficult to see from (4.2) and (4.4) that E and $|M|$ are invariants of the motion. The first represents the conservation of the energy $\frac{1}{2}(M, \Omega)$ and the second describes the conservation of the magnitude $|M|$ of the angular momentum.

From these equations, one can draw a useful picture concerning the orbit of $M(t)$. In the space of angular momentum (M_1, M_2, M_3), the vector $M(t)$ moves over the sphere of radius $|M|$ given by (4.9), and simultaneously it must lie over the surface of the ellipsoid of semiaxes $\sqrt{2J_1E}$, $\sqrt{2J_2E}$, and $\sqrt{2J_3E}$ (the energy surface (4.8) corresponds to an ellipsoid in the angular momentum space). Hence the vector $M(t)$ moves along the curve of intersection of the two surfaces (Fig. 4.2). It is almost obvious that the solutions are closed curves (i.e. periodic), or fixed points, or heteroclinic orbits (connecting different unstable fixed points). Thus it is found that the system of equations (4.2) is *completely integrable.* In fact, the solutions are represented in terms of elliptic functions (see e.g. [LL76, §37]).

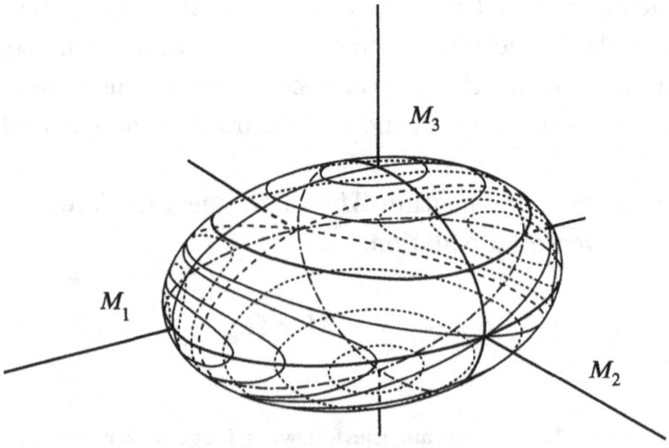

Fig. 4.2. Orbits over an energy ellipsoid in M-space for $J_1 < J_2 < J_3$.

4.1.3. *Lie–Poisson bracket and Hamilton's equation*

The Euler's equation (4.7) can be interpreted as a Hamilton's equation, which is written in the form,

$$\frac{\mathrm{d}}{\mathrm{d}t}M_\alpha = \{M_\alpha, H\}, \tag{4.10}$$

where $M = M_\alpha \boldsymbol{e}_\alpha$, and $\boldsymbol{e}_\alpha$ ($\alpha = 1, 2, 3$) is the orthonormal base vectors. The Hamiltonian function H is given by (4.5):

$$H = \frac{1}{2}\left(\frac{M_1^2}{J_1} + \frac{M_2^2}{J_2} + \frac{M_3^2}{J_3}\right), \tag{4.11}$$

and the bracket $\{\cdot, \cdot\}$ is defined by the following rigid-body Poisson bracket (a kind of the Lie–Poisson bracket[2]),

$$\{A, H\} := -M \cdot (\nabla_M A \times \nabla_M H), \tag{4.12}$$

where $\nabla_M = (\partial/\partial M_1, \partial/\partial M_2, \partial/\partial M_3)$. We have

$$\nabla_M H = (M_1/J_1, M_2/J_2, M_3/J_3) = \Omega, \quad \nabla_M M_\alpha = \boldsymbol{e}_\alpha.$$

Then, we obtain

$$\frac{\mathrm{d}}{\mathrm{d}t}M_\alpha = \{M_\alpha, H\} = -M \cdot (\boldsymbol{e}_\alpha \times \Omega) = \boldsymbol{e}_\alpha \cdot (M \times \Omega). \tag{4.13}$$

This is nothing but the αth component of Eq. (4.7).

In the following sections, the system of governing equations (4.2) will be rederived from a geometrical point of view, and stability of the motion will be investigated by deriving the Jacobi equation for the geodesic variation.

4.2. Transformations (Rotations) by $SO(3)$

Rotation of a rigid body is regarded as a smooth sequence of transformations of the body, i.e. transformations of the frame fixed to the body with

---

[2] A general definition of ($\pm$) Lie–Poisson brackets is given by $\{F, G\}_\pm(\mu) := \pm\left\langle \mu, \left[\frac{\delta F}{\delta \mu}, \frac{\delta G}{\delta \mu}\right]\right\rangle$, where $\delta F/\delta\mu$ is a *functional derivative* and $[\cdot, \cdot]$ is the Lie bracket [MR94; HMR98].

respect to the nonrotating *fixed* space F (an inertial space). The transformations are represented by matrix elements in the group of *special orthogonal* transformation in three-dimensional space **SO(3)** (see Appendix C.3).[3]

4.2.1. *Transformation of reference frames*

Let us take a point $\boldsymbol{y}_b$ fixed to the body which is located at $\boldsymbol{x} = X^1\boldsymbol{e}_1 + X^2\boldsymbol{e}_2 + X^3\boldsymbol{e}_3$ at $t = 0$, where $\boldsymbol{e}_i$ $(i = 1, 2, 3)$ are orthonormal basis fixed to the space F. By a transformation matrix $A = (A^i_j) \in SO(3)$, the initial point $\boldsymbol{x} = (X^i)$ is mapped to the current point $\boldsymbol{y}_b$ at a time t:

$$\boldsymbol{y}_b(t) = A(t)\boldsymbol{x} = y^i(t)\boldsymbol{e}_i, \qquad y^i(t) = A^i_j(t)X^j. \tag{4.14}$$

In terms of the group $G = SO(3)$, this transformation is understood such that an element $g_t(= A(t))$ of the group G represents a position of the body at a time t attained by its motion over G from the initial position $\mathbf{e}$ (represented by the unit tensor I).

On the other hand, relative to the body frame F_B which is the frame *instantaneously* fixed to the moving body, the same point $\boldsymbol{y}_b(t)$ fixed to the body is expressed as

$$Y(= \boldsymbol{y}_b) = Y^1\boldsymbol{b}_1 + Y^2\boldsymbol{b}_2 + Y^3\boldsymbol{b}_3,$$

where $\boldsymbol{b}_i$ $(i = 1, 2, 3)$ are orthonormal basis fixed to the body which coincided with $\boldsymbol{c}_i$ $(i = 1, 2, 3)$ at $t = 0$ (Fig. 4.3). From the property of a rigid

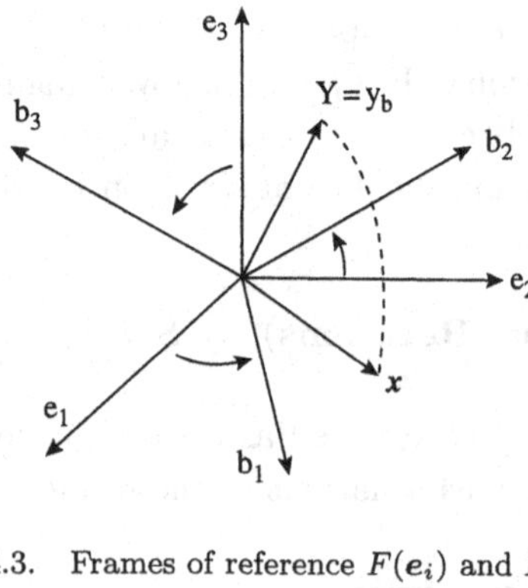

Fig. 4.3. Frames of reference $F(\boldsymbol{e}_i)$ and $F_B(\boldsymbol{b}_i)$.

[3]The group $G = SO(3)$ is a Lie group and consists of all orientation-preserving rotations, i.e. $A^T = A^{-1}$, $\det A = +1$ for $A \in G$, where A^T is transpose of A.

body, it is required that $Y^i = X^i$ which do not change with t. According to §1.5.1(b), the basis-transformation is written as (see (1.40))

$$\boldsymbol{b}_k(t) = B^i_k(t)\boldsymbol{e}_i, \quad \text{hence} \quad Y = Y^k\boldsymbol{b}_k = Y^k B^i_k \boldsymbol{e}_i, \tag{4.15}$$

where $B = (B^i_k) \in SO(3)$, i.e. $BB^T = I$. The transformation matrices B and A are identical, i.e. $A = B$.

This is because Y of (4.15) must be equal to $\boldsymbol{y}_b$ of (4.14), whence we have

$$y^i = A^i_j X^j = B^i_k Y^k \quad (X^i = Y^i).$$

Note that (4.14) is an example of §1.6.1(a).

In mechanics, the configuration of a rotating rigid body is represented by a set of three angles called the *Eulerian angles*, that is, the configuration space of a rotating rigid body is three-dimensional. In the present formulation, this is described by the manifold of the group $SO(3)$, whose dimension is also three (Appendix C.1).

4.2.2. *Right-invariance and left-invariance*

Rotational motion of a rigid body is described by a curve $\mathrm{C} : t \to g_t$ on the manifold of the group G with t the time parameter [Arn66; Arn78]:

$$g_t = A(t), \qquad g_0 = I, \quad \text{where} \quad A(t) \in SO(3), \quad t \in \mathbb{R}.$$

For an infinitesimal time increment δt, motion of the body from the position g_t is described by the left-translation $g_{t+\delta t} = g_{\delta t} g_t$. This is interpreted as an infinitesimal rotation $\delta t\, \bar{\Omega}$ at g_t represented by a matrix $\bar{\Omega}$ defined by

$$\begin{aligned} A(t+\delta t) - A(t) &= A(\delta t)A(t) - A(t) = \delta t\, \bar{\Omega} \cdot A(t), \\ A(t+\delta t)^i_k &= (A(\delta t)A(t))^i_k = A(\delta t)^i_l A(t)^l_k. \end{aligned} \tag{4.16}$$

In the language of the geometrical theory, the matrix $\bar{\Omega} = (A(\delta t) - I)/\delta t$ is called a *tangent vector* at the identity e. This is because the tangent vector $\dot{g}_t$ is defined by

$$\dot{g}_t := \frac{\mathrm{d}g_t}{\mathrm{d}t} = \frac{\mathrm{d}A}{\mathrm{d}t} = \bar{\Omega} g_t, \quad \bar{\Omega} = \dot{g}_t \circ (g_t)^{-1}, \tag{4.17}$$

namely $\bar{\Omega}$ is obtained from the tangent vector $\dot{g}_t$ by the *right-translation* with $(g_t)^{-1}$, since $A(t)$ evolves by the left translation according to (4.16) (see §1.7). The matrix $\bar{\Omega}$ can be shown to be skew-symmetric (Appendix C.3).

It is instructive to see that the $\bar{\Omega}$ is identified as an *axial* vector $\hat{\Omega}$ of the angular velocity of the body rotation relative to the *fixed* space. In fact, operating g_t on $\boldsymbol{x}$, we have $g_t\boldsymbol{x} = \boldsymbol{y}(t)$ and $g_{t+\delta t}\boldsymbol{x} = \boldsymbol{y}(t+\delta t)$, hence

$$\boldsymbol{y}(t+\delta t) = g_t\boldsymbol{x} + \delta t\bar{\Omega} g_t\boldsymbol{x} = \boldsymbol{y}(t) + \delta t\hat{\Omega} \times \boldsymbol{y}(t) \tag{4.18}$$

(see Eq. (C.15) in Appendix C). Differentiating,

$$\boldsymbol{v}_y = \mathrm{d}\boldsymbol{y}/\mathrm{d}t = \dot{g}_t\boldsymbol{x} = \bar{\Omega} \cdot g_t\boldsymbol{x} = \hat{\Omega} \times \boldsymbol{y}. \tag{4.19}$$

This means that the point $\boldsymbol{y}(t)$ is moving with the velocity $\boldsymbol{v}_y = \hat{\Omega} \times \boldsymbol{y}$, i.e. $\hat{\Omega}$ is the angular velocity in the fixed space.

Corresponding to the right-invariant expression $\dot{g}_t = \bar{\Omega} g_t$ of the tangent vector, the angular momentum is similarly written in the right-invariant way as

$$M_t = J\dot{g}_t = \bar{M} g_t, \quad \bar{M} = J\bar{\Omega}, \tag{4.20}$$

where J is an inertia operator and time-dependent in the fixed space. In mechanics [LL76], the angular momentum is defined by $M_t = \int(\boldsymbol{y} \times \boldsymbol{v}_y)\rho \mathrm{d}^3\boldsymbol{y} = \int(g_t\boldsymbol{x} \times (\bar{\Omega} g_t\boldsymbol{x}))g_t(\rho \mathrm{d}^3\boldsymbol{x})$. This enables us the definition, $M_t := \bar{M} g_t$.

Relative to the body frame F_B, the same velocity (relative to the fixed space) is represented as $\mathrm{d}\boldsymbol{y}_b/\mathrm{d}t = v_b^i\boldsymbol{b}_i$, as shown below, where

$$\boldsymbol{v}_b = \hat{\Omega}_b \times Y \tag{4.21}$$

(see (4.26)), and $\hat{\Omega}_b$ is the angular velocity relative to the body frame. The matrix version of $\hat{\Omega}_b$ is derived by *left-translation* of the tangent vector $\dot{g}_t$ with $(g_t)^{-1}$ as

$$\bar{\Omega}_b = g_t^{-1} \circ \dot{g}_t = g_t^{-1}\bar{\Omega} g_t, \tag{4.22}$$

(verified below). An important point is that the tangent vector $\dot{g}_t$ is represented by the *left-invariant* form (see §1.6 and 1.7)

$$\dot{g}_t = g_t\Omega_b, \tag{4.23}$$

by the Ω_b at e, whereas it is also represented by the *right-invariant* form, $\dot{g}_t = \bar{\Omega} g_t$ with $\bar{\Omega}$ at e in (4.17) (Fig. 4.4).

Proof of $\Omega_b = g_t^{-1}\bar{\Omega} g_t$. By the motion $\boldsymbol{y}(t) = Y^k\boldsymbol{b}_k$, the basis is transformed by (4.15) as $\boldsymbol{b}_i(t) = \boldsymbol{e}_j B_i^j(t)$. Therefore,

$$\begin{aligned} \boldsymbol{b}_i(t+\delta t) = B_i^k(t+\delta t)\boldsymbol{e}_k &= B_i^l(\delta t)B(t)_l^k\boldsymbol{e}_k \\ &= B(t)_l^k B_i^l(\delta t)\boldsymbol{e}_k, \end{aligned} \tag{4.24}$$

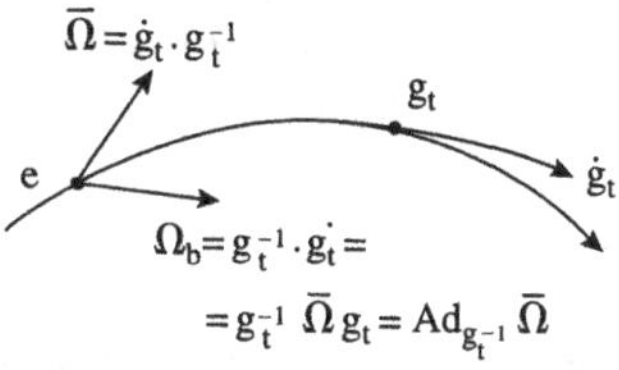

Fig. 4.4. $\bar{\Omega}$ and Ω_b.

namely, the element $B(t+\delta t)$ is obtained from $B(t)$ by the right-translation with $B(\delta t)$. Taking $\boldsymbol{y}_b(t) = y_b^j \boldsymbol{e}_j$ which is given by $Y^i B_i^j(t)\boldsymbol{e}_j$, we have the transformation law for the components y_b^j:

$$y_b^j = B_i^j(t) Y^i, \quad y_b = BY, \tag{4.25}$$

where Y is a constant vector. Owing to the right-translation property of $B(t)$, one may write $\mathrm{d}B/\mathrm{d}t = B\bar{\Omega}_b$, and

$$\frac{\mathrm{d}}{\mathrm{d}t}\boldsymbol{y}_b = \frac{\mathrm{d}}{\mathrm{d}t}(\boldsymbol{e}B(t)Y) = \boldsymbol{e}B\bar{\Omega}_b Y = \boldsymbol{b}\bar{\Omega}_b Y.$$

Writing $\mathrm{d}\boldsymbol{y}_b/\mathrm{d}t = \boldsymbol{b}_i v_b^i$, we find that the rotation velocity $\boldsymbol{v}_b = (v_b^i)$ relative to the instantaneous frame F_B is given by

$$\boldsymbol{v}_b = \bar{\Omega}_b Y = \hat{\Omega}_b \times Y, \tag{4.26}$$

since $\bar{\Omega}_b$ is skew-symmetric. Now we have $A(t) = B(t)$, however $\partial_t A = \bar{\Omega} A$ and $\partial_t B = B\bar{\Omega}_b$. Thus, we obtain

$$\Omega_b = A^{-1}\bar{\Omega}A = g_t^{-1}\bar{\Omega}g_t.$$

This verifies the expression of (4.22).

4.3. Commutator and Riemannian Metric

Aiming at geometrical formulation of rotational motion of a rigid body, we define the commutation rule for the Lie algebra **so**(3) (already considered in §1.8.2), and introduce a metric for the tangent bundle $TSO(3)$.

A *tangent* vector at the identity e of the group $G = SO(3)$ is said to be an element of the Lie algebra **so**(3). The space of such vectors is denoted by $T_e G = \mathbf{so}(3)$. It is useful to replace each element of skew-symmetric tensor $\bar{\Omega}_b \in T_e G$ with an equivalent axial vector denoted by $\hat{\Omega}_b$. According to (1.66) and (1.67) in §1.8.2, the *commutator* $[\cdot,\cdot]^{(\mathrm{L})}$ for the left-invariant field such as (4.22) (or (4.23)) is given by the vector product in $\mathbb{R}^3$:

$$[X,Y]^{(\mathrm{L})} := \hat{X} \times \hat{Y}, \quad \text{for} \quad X,Y \in T_eG = \mathbf{so}(3). \tag{4.27}$$

Kinetic energy K is given by the scalar pairing of a tangent vector $\dot{g}_t = \bar{\Omega} g_t$ and a corresponding cotangent vector $M_t = \bar{M} g_t$ (see (4.20)), both of which are right-invariant fields. Thus, the kinetic energy K is *defined* by the following scalar product in the right-invariant way (see (3.4)):

$$K := \frac{1}{2}(M_t(g_t)^{-1}, \quad \dot{g}_t(g_t)^{-1})_s. \tag{4.28}$$

The factor g_t on the right of $\dot{g}_t = \bar{\Omega} g_t$ indicates that the tangent vector is represented in terms of the original position $\boldsymbol{x}$, whereas its right-translation $\dot{g}_t(g_t)^{-1} = \bar{\Omega}$ denotes the tangent vector at the current position $\boldsymbol{y}(t) = g_t\boldsymbol{x}$ (see (4.19)).[4] The same is true for $M_t(g_t)^{-1} = \bar{M}$ (see (4.20)). This expression (4.28) can be rewritten in terms of $\hat{\Omega}$ and the angular momentum $\bar{M} = J\hat{\Omega}$ as

$$K = \frac{1}{2}(\bar{M}, \bar{\Omega})_s = \frac{1}{2}(J\hat{\Omega}, \hat{\Omega})_s, \tag{4.29}$$

where J is the inertia tensor (called an inertia operator) relative to the fixed space.

The kinetic energy is a frame-independent scalar. In other words, it is invariant with respect to the transformation from a fixed inertial frame $(\boldsymbol{e}_i)$ to the instantaneous frame $(\boldsymbol{b}_i)$ fixed to the body. In this case, the tangent vector was given by the left-invariant form $\dot{g}_t = g_t\Omega_b$ in the previous section.[5] Correspondingly, the angular momentum is written as $M_t = g_t M_b$, where $M_b = (g_t)^{-1}\bar{M} g_t$ and $\bar{M} = J\bar{\Omega}$. Hence, left-invariance on the group results in the following. The energy K of (4.28) is given by

$$K = \frac{1}{2}(g_t M_b (g_t)^{-1}, g_t \Omega_b (g_t)^{-1})_s \tag{4.30}$$

$$= \frac{1}{2}(J g_t \hat{\Omega}_b, g_t \hat{\Omega}_b)_s = \frac{1}{2}(J_b \hat{\Omega}_b, \hat{\Omega}_b)_s, \tag{4.31}$$

---

[4]Equation (4.28) is a symbolic expression in the sense that each entry should be provided with an equivalent vector to obtain the scalar product $(\,,\,)_s$, as in (4.29).

[5]There, the body frame $(\boldsymbol{b}_i)$ is transformed by the right-translation (4.24), which results in the left-invariant vector field $\dot{g}_t = (g_t)_*\Omega_b$. See §1.7.

where the second equality is due to the right-invariance and

$$g_t M_b = \bar{M} g_t = J\bar{\Omega} g_t = J g_t \Omega_b,$$

and in the last expression, $J_b = (g_t)^{-1} J g_t$ since $(g_t)^T = (g_t)^{-1}$ for $g_t \in SO(3)$. Equivalence of (4.29) and (4.31) is due to the fact that the kinetic energy K is frame-independent between the $\boldsymbol{e}_i$-frame and $\boldsymbol{b}_i$-frame. The symmetric inertia tensor J_b can be made a diagonal matrix (relative to the principal axes) with positive elements $J_\alpha(>0)$, owing to the definition (4.3). Then, the expression (4.31) is equivalent to (4.5).

Now, one can define the metric $\langle \cdot, \cdot \rangle$ on $T_e G$ by

$$\langle X, Y \rangle := (J\hat{X}, \hat{Y})_s \equiv J\hat{X} \cdot \hat{Y}, \quad \text{for} \quad X, Y \in T_e G, \tag{4.32}$$

where $(\cdot, \cdot)_s$ is defined by (4.6). Then the kinetic energy is given by $K = \frac{1}{2}\langle \Omega, \Omega \rangle$ for $\Omega \in T_e G$. Thus, the group $G = SO(3)$ is a Riemannian manifold endowed with the left-invariant metric (4.32) (which is also right-invariant in a trivial way as given by (4.28)).

4.4. Geodesic Equation

4.4.1. *Left-invariant dynamics*

Let us consider the geodesic equation on the manifold $SO(3)$. We have already introduced the commutator (4.27) and the metric (4.32). Furthermore, the metric is left-invariant. In such a case, the connection $\nabla_X Y$ satisfies Eq. (3.30), where $X, YZ \in \mathbf{so}(3)$. In a time-dependent problem such as in the present case, the geodesic equation is given by the form (3.67):

$$\partial_t X + \nabla_X X = 0. \tag{4.33}$$

In terms of the operators ad and ad^* of §3.6.2, we have the expression (3.65) for the connection $\nabla_X Y$:

$$\nabla_X Y = \frac{1}{2}(ad_X Y - ad^*_X Y - ad^*_Y X). \tag{4.34}$$

The expression of $ad^*_X Y$ is obtained by using the present commutator $[X, Y]^{(\mathrm{L})} = \hat{X} \times \hat{Y}$ and the definition $\langle ad^*_X Y, Z \rangle = \langle Y, [X, Z]^{(\mathrm{L})} \rangle$, which leads to (by using (4.32))

$$\langle ad^*_X Y, Z \rangle = (J\hat{Y}, \hat{X} \times \hat{Z})_s = (J\hat{Y} \times \hat{X}, \hat{Z})_s = \langle J^{-1}(J\hat{Y} \times \hat{X}), Z \rangle. \tag{4.35}$$

Hence, the nondegeneracy of the metric yields

$$ad^*_X Y = J^{-1}(J\hat{Y} \times \hat{X}). \tag{4.36}$$

Thus it is found from the above that

$$\begin{aligned}\nabla_X Y &= \frac{1}{2} J^{-1}(J(\hat{X} \times \hat{Y}) - (J\hat{X}) \times \hat{Y} - (J\hat{Y}) \times \hat{X}) \\ &= \frac{1}{2} J^{-1}(\tilde{K}\hat{X} \times \hat{Y}),\end{aligned} \tag{4.37}$$

[Kam98; SWK98], where $\tilde{K}$ is a diagonal matrix with the diagonal elements of

$$\tilde{K}_\alpha := -J_\alpha + J_\beta + J_\gamma \tag{4.38}$$

for $(\alpha, \beta, \gamma) = (1, 2, 3)$ and its cyclic permutation (all $\tilde{K}_\alpha > 0$ according to the definition (4.3) and the footnote there).

The tangent vector at the identity e is the angular velocity vector $\hat{\Omega}$. The geodesic equation of a time-dependent problem is given by (4.33) with X replaced by $\hat{\Omega}$ and using the ordinary time derivative $\mathrm{d}/\mathrm{d}t$ in place of ∂_t (since t is the only independent variable):

$$\frac{\mathrm{d}}{\mathrm{d}t}\hat{\Omega} - ad^*_{\hat{\Omega}}\hat{\Omega} = 0.$$

Using (4.36) and multiplying J on both sides, we obtain

$$J\frac{\mathrm{d}}{\mathrm{d}t}\hat{\Omega} - (J\hat{\Omega}) \times \hat{\Omega} = 0. \tag{4.39}$$

This is nothing but the Euler's equation (4.7) if it is represented with components relative to the body frame, i.e. instantaneously fixed to the moving body. In fact, the component with respect to the $\boldsymbol{b}_1$ axis (one of the principal axes) is written as $J_1(\mathrm{d}\Omega^1/\mathrm{d}t) - (J_2\Omega^2\Omega^3 - J_3\Omega^3\Omega^2) = 0$. In the body frame, the inertia tensor J is time-independent.

Thus, based on the framework of geometrical formulation, we have successfully recovered the equation of motion for free rotation of a rigid body well known in *mechanics*.

4.4.2. *Right-invariant dynamics*

Let us try to see this dynamics in the fixed space, where it can be verified that $(\mathrm{d}/\mathrm{d}t)J\hat{\Omega} = 0$. In this space, the inertia tensors $J = (J_{\alpha\beta})$ are time-dependent, and we have $(\mathrm{d}/\mathrm{d}t)J\hat{\Omega} = (\mathrm{d}J/\mathrm{d}t)\hat{\Omega} + J(\mathrm{d}\hat{\Omega}/\mathrm{d}t)$. In fact, from the definition (4.3), we obtain

$$\frac{\mathrm{d}}{\mathrm{d}t}J_{\alpha\beta} = \dot{J}_{\alpha\beta} = \int (x^k \dot{x}^k \delta_{\alpha\beta} - \dot{x}^\alpha x^\beta - x^\alpha \dot{x}^\beta)\rho \mathrm{d}^3\boldsymbol{x},$$

since $\mathrm{d}x^\alpha/\mathrm{d}t = \dot{x}^\alpha = (\hat{\Omega} \times \boldsymbol{x})^\alpha = \varepsilon_{\alpha jk}\Omega^j x^k$ (see (B.27) in Appendix B.4). Obviously, $x^k \dot{x}^k = \boldsymbol{x} \cdot (\hat{\Omega} \times \boldsymbol{x}) = 0$ (see B.4). Hence,

$$\dot{J}_{\alpha l} = -\varepsilon_{\alpha jk}\Omega^j \int x^k x^l \rho \mathrm{d}^3\boldsymbol{x} - \varepsilon_{ljk}\Omega^j \int x^k x^\alpha \rho \mathrm{d}^3\boldsymbol{x} = J_{\alpha jl}\Omega^j, \qquad (4.40)$$

introducing the notation $I_0 = \int |\boldsymbol{x}|^2 \rho \mathrm{d}^3\boldsymbol{x}$,

$$J_{\alpha jl} = \varepsilon_{\alpha jk}(J_{kl} - I_0\delta_{kl}) + \varepsilon_{ljk}(J_{k\alpha} - I_0\delta_{k\alpha}) = \varepsilon_{\alpha jk}J_{kl} + \varepsilon_{ljk}J_{k\alpha}. \quad (4.41)$$

The last equality is obtained since

$$\varepsilon_{\alpha jk}\delta_{kl}I_0 + \varepsilon_{ljk}\delta_{k\alpha}I_0 = I_0(\varepsilon_{\alpha jl} + \varepsilon_{lj\alpha}) = I_0(\varepsilon_{\alpha jl} + \varepsilon_{\alpha lj}) = 0$$

due to the definition (B.26) of ε_{ijk} in Appendix B.4. Then, we obtain the following expression,

$$\dot{J}_{\alpha l}\Omega^l = J_{\alpha jl}\Omega^j\Omega^l = \varepsilon_{\alpha jk}J_{kl}\Omega^j\Omega^l + \varepsilon_{ljk}J_{k\alpha}\Omega^j\Omega^l.$$

The second term vanishes because $\varepsilon_{ljk}\Omega^j\Omega^l = 0$ owing to the skew symmetry $\varepsilon_{ljk} = -\varepsilon_{jlk}$. Hence, we obtain

$$\dot{J}_{\alpha l}\Omega^l = \varepsilon_{\alpha jk}\Omega^j(J_{kl}\Omega^l) = (\hat{\Omega} \times (J\hat{\Omega}))_\alpha.$$

Therefore, it is found that, by using Eq. (4.39),

$$\frac{\mathrm{d}}{\mathrm{d}t}(J\hat{\Omega}) = \frac{\mathrm{d}J}{\mathrm{d}t}\hat{\Omega} + J\frac{\mathrm{d}\hat{\Omega}}{\mathrm{d}t} = \hat{\Omega} \times (J\hat{\Omega}) + (J\hat{\Omega}) \times \hat{\Omega} = 0. \qquad (4.42)$$

Thus, we have recovered Eq. (4.1), describing conservation of the angular momentum in the fixed inertial space.

4.5. Bi-Invariant Riemannian Metrices

There is a bi-invariant metric on every compact Lie group [Fra97, Ch. 21]. The group $SO(3)$ is a *compact* Lie group. We investigate these properties, aiming at applying it to the present problem.

4.5.1. $SO(3)$ *is compact*

The group $G = SO(3)$ is considered as the subset of real 3^2-dimensional space of 3×3 matrices, satisfying $A^T A = I$ (orthogonality), i.e. $A^T_{ij} A_{jk} = A_{ji} A_{jk} = \delta_{ik}$, and $\det A = 1$ (see C.1). In particular, its magnitude $\|A\|$ may be defined by

$$\|A\|^2 := \sum_{jk} (A_{jk})^2 = \sum_{jk} A^T_{kj} A_{jk} = \sum_{k} \delta_{kk} = 3.$$

Hence, $SO(3)$ consists of points that lie on the sphere $\|A\| = \sqrt{3}$ that satisfy $\det A = 1$. Therefore it is a bounded subset. It is also clear that the limit of a sequence of orthogonal matrices is again orthogonal. Thus, $SO(3)$ is a *closed, bounded* set, i.e. a compact set. The compactness can be generalized to $SO(n)$ for any integer n without any difficulty.

4.5.2. *Ad-invariance and bi-invariant metrices*

In the vector space of the Lie algebra $\mathfrak{g} = \mathbf{so}(3)$, a scalar product is defined by

$$\langle \boldsymbol{a}, \boldsymbol{b} \rangle_{\mathbf{so}(3)} := -\frac{1}{2} \mathrm{tr}(\boldsymbol{ab}), \tag{4.43}$$

for $\boldsymbol{a}, \boldsymbol{b} \in \mathfrak{g} = \mathbf{so}(3)$ (see Appendix C.4). Using equivalent axial vectors $\hat{\boldsymbol{a}} = (\hat{a}^i), \hat{\boldsymbol{b}} = (\hat{b}^i)$, this scalar product is

$$\langle \boldsymbol{a}, \boldsymbol{b} \rangle_{\mathbf{so}(3)} = (\hat{\boldsymbol{a}}, \hat{\boldsymbol{b}}_s) = \delta_{ij} \hat{a}^i \hat{b}^j, \tag{4.44}$$

i.e. the metric tensor is euclidean. The scalar product $\langle \boldsymbol{a}, \boldsymbol{b} \rangle_{\mathbf{so}(3)}$ is *invariant* under the adjoint action of $G = SO(3)$ on $\mathfrak{g}$. In fact, for $g \in G$, we have $Ad_g \boldsymbol{a} = g\boldsymbol{a}g^{-1}$ (see (1.62)), and

$$\langle g\boldsymbol{a}g^{-1}, g\boldsymbol{b}g^{-1} \rangle_{\mathbf{so}(3)} = -\frac{1}{2} \mathrm{tr}(g\boldsymbol{a}g^{-1} g\boldsymbol{b}g^{-1}) = -\frac{1}{2} \mathrm{tr}(\boldsymbol{ab}) = \langle \boldsymbol{a}, \boldsymbol{b} \rangle_{\mathbf{so}(3)}$$

since $g^{-1} = g^T$. This is called the *ad-invariance.*

Suppose that g is represented as $\mathrm{e}^{t\boldsymbol{a}}$ with a parameter t. From the Ad-invariance for $\boldsymbol{a}, \boldsymbol{b}, \boldsymbol{c} \in \mathfrak{g} = \mathbf{so}(3)$, we have

$$\langle \mathrm{e}^{t\boldsymbol{a}} \boldsymbol{b} \mathrm{e}^{-t\boldsymbol{a}}, \mathrm{e}^{t\boldsymbol{a}} \boldsymbol{c} \mathrm{e}^{-t\boldsymbol{a}} \rangle = \langle \boldsymbol{b}, \boldsymbol{c} \rangle. \tag{4.45}$$

Differentiating with t and putting $t = 0$, we obtain

$$\langle [\boldsymbol{a}, \boldsymbol{b}], \boldsymbol{c} \rangle + \langle \boldsymbol{b}, [\boldsymbol{a}, \boldsymbol{c}] \rangle = 0, \tag{4.46}$$

which is useful to obtain a formula of sectional curvature below.

Let us define a Riemannian metric on the group G by the right-invariant form,

$$\langle \boldsymbol{a}_g, \boldsymbol{b}_g \rangle := \langle \boldsymbol{a}_g g^{-1}, \boldsymbol{b}_g g^{-1} \rangle_e.$$

Then, we claim that it is also left-invariant. This is verified as follows. From the Ad-invariance, we have,

$$\langle \boldsymbol{a}_e, \boldsymbol{b}_e \rangle = \langle g\boldsymbol{a}_e g^{-1}, g\boldsymbol{b}_e g^{-1} \rangle = \langle g\boldsymbol{a}_e, g\boldsymbol{b}_e \rangle.$$

The second equality is due to the right-invariance (where $\boldsymbol{a}_g = g\boldsymbol{a}_e$ and $\boldsymbol{b}_g = g\boldsymbol{b}_e$). The above shows the left-invariance. Thus, we have obtained the *bi-invariance* of the metric $\langle , \rangle$. This can be generalized to $SO(n)$ for any integer n.

For any bi-invariant metric $\langle , \rangle$ on a group G, *one-parameter subgroups are geodesics.* This is verified as follows.

Let $\boldsymbol{a}_g$ be a left-invariant field represented as $g\boldsymbol{a}$ where $\boldsymbol{a} \in \mathfrak{g}$. The tangent vector $\boldsymbol{a}$ at e generates a one-parameter subgroup $A_t = \mathrm{e}^{t\boldsymbol{a}}$ of right translation. Let γ_t be a geodesic through e with the bi-invariant metric that is tangent to $g\boldsymbol{a} \equiv \boldsymbol{a}_t$ at $g = \gamma_t$ and $T_t = T(\gamma_t) = \gamma_t T_0$ be the unit tangent to γ_t there. Consider the scalar product $\langle \boldsymbol{a}_t, T_t \rangle$ along γ_t. By the left invariance, we have

$$\langle \boldsymbol{a}_t, T_t \rangle = \langle \boldsymbol{a}_0, T_0 \rangle. \tag{4.47}$$

This infers that $\nabla_T \boldsymbol{a} = 0$, from (3.83) and (3.84), that represents parallel translation of the tangent vector $\boldsymbol{a}_t$ along γ_t with its magnitude unchanged due to (4.47). Therefore the curve $A_t = \mathrm{e}^{t\boldsymbol{a}}$ coincides with the geodesic γ_t with T_0 parallel to $\boldsymbol{a} = \boldsymbol{a}_0$.

The flow $A_t = \mathrm{e}^{t\boldsymbol{a}}$ is a one-parameter group of isometry. In fact, Eq. (4.45) holds for any pair of $\boldsymbol{b}, \boldsymbol{c} \in \mathbf{so}(3)$, which results in (4.46). Using the scalar product (4.44) and the commutator (4.27), Eq. (4.46) becomes

$$(\hat{\boldsymbol{a}} \times \hat{\boldsymbol{b}}) \cdot \hat{\boldsymbol{c}} + \hat{\boldsymbol{b}} \cdot (\hat{\boldsymbol{a}} \times \hat{\boldsymbol{c}}) = (\hat{\boldsymbol{a}} \times \hat{\boldsymbol{b}}) \cdot \hat{\boldsymbol{c}} + (\hat{\boldsymbol{b}} \times \hat{\boldsymbol{a}}) \cdot \hat{\boldsymbol{c}} = 0.$$

This is satisfied identically for any pair of $\hat{\boldsymbol{b}}, \hat{\boldsymbol{c}}$. Thus, it is found that $\boldsymbol{a}$ is a *Killing field* (see §3.12), because the above equation is equivalent to (3.161): $\langle \nabla_{\hat{b}} \hat{a}, \hat{c} \rangle + \langle \nabla_{\hat{c}} \hat{a}, \hat{b} \rangle = 0$ since $\nabla_{\hat{b}} \hat{a} = \frac{1}{2} \hat{b} \times \hat{a} = -\frac{1}{2} \hat{a} \times \hat{b}$. This means that any element of $\mathbf{so}(3)$ generates a Killing field.

The scalar product $\langle \boldsymbol{a}, \boldsymbol{b} \rangle_{\mathbf{so}(3)}$ of (4.43) is equivalent to the metric $(J\hat{X}, \hat{Y})_s$ of (4.32) if the inertia tensor J is the unit tensor I (i.e. *euclidean*) that corresponds to a spherical rigid body, called a *spherical top*. In this case, the angular momentum $J_b \hat{\Omega}_b$ reduces to $\hat{\Omega}_b$. Generally for rigid bodies, the inertia tensor J is not cI (c: a constant), but a general symmetric tensor which is regarded as a *Riemannian* metric tensor. Then, the equality of (4.30) and (4.31) of scalar products represents the ad-invariance. Thus, it is seen that the rotation of a rigid body is a bi-invariant metric system as well. The formulation in §4.3 describes an *extension* to such a general metric tensor J. It would be interesting to recall the property of *complete-integrability* of the free rotation of a rigid body mentioned in §4.1.

4.5.3. *Connection and curvature tensor*

It is verified in the previous subsection that integral curves of a left-invariant field $\boldsymbol{a}_g = g\boldsymbol{a}$ for $\boldsymbol{a} \in \mathfrak{g}$ are geodesics in the bi-invariant metric. Hence we have $\nabla_{\boldsymbol{a}} \boldsymbol{a} = 0$. Likewise we have $\nabla_{(\boldsymbol{a}+\boldsymbol{b})}(\boldsymbol{a}+\boldsymbol{b}) = 0$ for $\boldsymbol{a}, \boldsymbol{b} \in \mathfrak{g}$. Since $\nabla_{\boldsymbol{a}} \boldsymbol{a} = 0$ and $\nabla_{\boldsymbol{b}} \boldsymbol{b} = 0$,

$$\nabla_{(\boldsymbol{a}+\boldsymbol{b})}(\boldsymbol{a}+\boldsymbol{b}) = \nabla_{\boldsymbol{a}} \boldsymbol{b} + \nabla_{\boldsymbol{b}} \boldsymbol{a} = 0. \tag{4.48}$$

Therefore, we obtain the Riemannian connection given by

$$2\nabla_{\boldsymbol{a}} \boldsymbol{b} = \nabla_{\boldsymbol{a}} \boldsymbol{b} - \nabla_{\boldsymbol{b}} \boldsymbol{a} = [\boldsymbol{a}, \boldsymbol{b}], \tag{4.49}$$

by the torsion-free property (3.18).

In this case, the curvature tensor (3.97) takes a particularly simple form. First note, e.g. $\nabla_{\boldsymbol{a}}(\nabla_{\boldsymbol{b}} \boldsymbol{c}) = \frac{1}{4}[\boldsymbol{a}, [\boldsymbol{b}, \boldsymbol{c}]]$, by (4.49). Then, we obtain

$$\begin{aligned} R(\boldsymbol{a}, \boldsymbol{b})\boldsymbol{c} &= \nabla_{\boldsymbol{a}}(\nabla_{\boldsymbol{b}} \boldsymbol{c}) - \nabla_{\boldsymbol{b}}(\nabla_{\boldsymbol{a}} \boldsymbol{c}) - \nabla_{[\boldsymbol{a}, \boldsymbol{b}]} \boldsymbol{c} \\ &= \frac{1}{4}([\boldsymbol{a}, [\boldsymbol{b}, \boldsymbol{c}]] - [\boldsymbol{b}, [\boldsymbol{a}, \boldsymbol{c}]] - 2[[\boldsymbol{a}, \boldsymbol{b}], \boldsymbol{c}]) = -\frac{1}{4}[[\boldsymbol{a}, \boldsymbol{b}], \boldsymbol{c}], \end{aligned} \tag{4.50}$$

by using the Jacobi identity (1.60). For the *sectional curvature* defined by

$$K(\boldsymbol{a}, \boldsymbol{b}) := \langle R(\boldsymbol{a}, \boldsymbol{b})\boldsymbol{b}, \boldsymbol{a} \rangle, \tag{4.51}$$

we obtain,

$$\begin{aligned} 4K(\boldsymbol{a}, \boldsymbol{b}) &= -\langle [[\boldsymbol{a}, \boldsymbol{b}]\boldsymbol{b}], \boldsymbol{a} \rangle = \langle [\boldsymbol{b}, [[\boldsymbol{a}, \boldsymbol{b}], \boldsymbol{a}] \rangle \\ &= -\langle [\boldsymbol{b}, [\boldsymbol{a}, [\boldsymbol{a}, \boldsymbol{b}]]\rangle = \langle [[\boldsymbol{a}, \boldsymbol{b}], [\boldsymbol{a}, \boldsymbol{b}]] \rangle, \end{aligned}$$

where the formula (4.46) is used repeatedly. Hence,

$$K(\boldsymbol{a}, \boldsymbol{b}) = \frac{1}{4}\|[\boldsymbol{a}, \boldsymbol{b}]\|^2. \tag{4.52}$$

Thus, we have found *non-negativeness* of the sectional curvature, $K(\boldsymbol{a}, \boldsymbol{b}) \geq 0$.

4.6. Rotating Top as a Bi-Invariant System

Rotating motion of a rigid body is regarded as an extension of the bi-invariant system (considered in the previous section) in the sense that the tangent vectors are time-dependent (i.e. there is an additional dimension t) and the metric tensor is not the euclidean δ_{ij}, but given by a general symmetric tensor J.

4.6.1. *A spherical top (euclidean metric)*

A spherical top is characterized by the isotropic inertia tensor $J = cI$ where I is the unit tensor with c a constant. Then, the metric is defined by (4.32) with J replaced by I:

$$\langle X, Y \rangle = \hat{X} \cdot \hat{Y}, \quad \text{for} \quad X, Y \in \mathbf{so}(3). \tag{4.53}$$

In order to apply the formulae of the bi-invariance to the time-dependent problem of a spherical top, every tangent vector such as $\boldsymbol{a}, \boldsymbol{b}$, etc. should be replaced with a vector of the form, $\tilde{X} = \partial_t + X^\alpha \partial_\alpha$, where $X = (X^\alpha) \in \mathbf{so}(3)$. For example, $\nabla_{\boldsymbol{a}} \boldsymbol{a} = 0$ (in the previous section) must be replaced by the form,

$$\nabla_{\tilde{X}} \tilde{X} = \partial_t X + \nabla_X X = 0. \tag{4.54}$$

The equation $\nabla_{(\boldsymbol{a}+\boldsymbol{b})}(\boldsymbol{a} + \boldsymbol{b}) = 0$ is replaced by $\nabla_{\tilde{X}+\tilde{Y}}(\tilde{X} + \tilde{Y}) = 0$. Corresponding to (4.48), we have

$$\nabla_X Y + \nabla_Y X = 0.$$

This is rewritten as $2\nabla_X Y = [X, Y]$, where $[X, Y] = \nabla_X Y - \nabla_Y X$ by the torsion-free property,[6] and the commutator $[X, Y]$ is given by $X \times Y$.

[6] $[\tilde{X}, \tilde{Y}] = [X, Y] + \partial_t Y - \partial_t X$, and $\nabla_{\tilde{X}} \tilde{Y} - \nabla_{\tilde{Y}} \tilde{X} = [\tilde{X}, \tilde{Y}]$ reduces to $\nabla_X Y - \nabla_Y X = [X, Y]$.

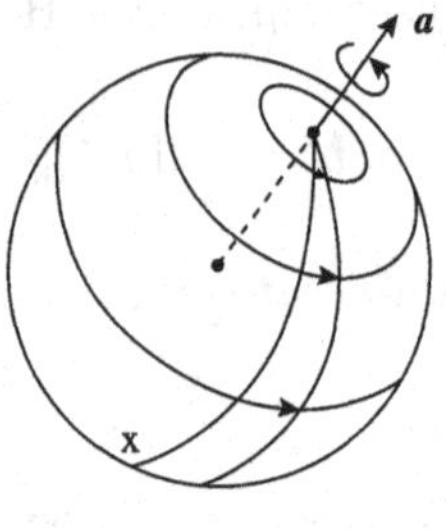

Fig. 4.5. Spherical top.

The connection $\nabla_X Y$ is given by (4.37). When $J = cI$, this reduces to the following form just obtained:

$$\nabla_X Y = \frac{1}{2} X \times Y. \tag{4.55}$$

The geodesic equation is given by $\partial_t X + X \times X = \partial_t X = 0$, i.e. X is time-independent. The one-parameter subgroup e^{tX} is a geodesic and X is a Killing field, as mentioned in §4.5.2. Namely, the general solution of free rotation of a spherical top is *steady* rotation described by any element $\boldsymbol{a} \in \mathbf{so}(3)$ (Fig. 4.5).

Note that, even in the time-dependent problem, the curvature tensor is unchanged, i.e. $R(\tilde{X}, \tilde{Y})\tilde{Z} = R(X, Y)Z$, as explained in §3.10.3. This is also verified directly by applying the expression $\nabla_{\tilde{Y}} \tilde{Z} = \partial_t Z + \nabla_Y Z$ repeatedly and using $\nabla_{[\tilde{X}, \tilde{Y}]} \tilde{Z} = \nabla_{[X,Y]} Z + \nabla_{(\partial_t Y - \partial_t X)} Z$ (the time component vanishes identically on the right-hand side). Thus, we obtain

$$R(\tilde{X}, \tilde{Y})\tilde{Z} = -\frac{1}{4}(X \times Y) \times Z, \tag{4.56}$$

$$K(\tilde{X}, \tilde{Y}) = \frac{1}{4}\|X \times Y\|^2 = \|\nabla_X Y\|^2, \tag{4.57}$$

from (4.50), (4.52) and (4.55). It is found that the sectional curvature $K(\tilde{X}, \tilde{Y})$ is *positive* if X and Y are not parallel. This characterizes the free steady rotation of a spherical top.

4.6.2. *An asymmetrical top (Riemannian metric)*

The case of free rotation of a general asymmetrical top is already studied in §4.3 and 4.4. The commutator is given by (4.27), and the metric is (4.32), which is reproduced here:

$$\langle X, Y \rangle = J\hat{X} \cdot \hat{Y}, \quad \text{for} \quad X, Y \in \mathbf{so}(3). \tag{4.58}$$

The geodesic equation is given by (4.39). Here, we consider some aspects of its stability. The variables, say $\hat{X}$ or $\hat{Y}$, will be written simply as X or Y. In particular, the connection is given by (4.37):

$$\nabla_X Y = \frac{1}{2} J^{-1}(\tilde{K}X \times Y). \tag{4.59}$$

Relation between the stability of a geodesic curve $\gamma(t)$ and Riemannian curvature tensors is described by the Jacobi equation (§3.10). The equation of the *Jacobi* field B, defined by (3.123) (using B instead of J), along the geodesic generated by $\tilde{T}$ is given by (3.137), which is reproduced here:

$$\partial_t^2 B + \partial_t(\nabla_T B) + \nabla_T \partial_t B + \nabla_T \nabla_T B + R(B,T)T = 0, \tag{4.60}$$

where T is the tangent vector to the geodesic $\gamma(t)$.

To calculate the curvature tensor $R(B,T)T$, we apply the expression of the connection ∇ of (4.59) repeatedly in the formula (3.97) together with the definition of the commutator (4.27). Finally it is found that

$$R(X,Y)Z = -\frac{1}{4J_1J_2J_3}(\tilde{\kappa}(X \times Y)) \times JZ, \tag{4.61}$$

where $\tilde{\kappa} = \mathrm{diag}(\kappa_1, \kappa_2, \kappa_3)$ is a diagonal matrix of third order with the diagonal elements,

$$\kappa_\alpha = -3J_\alpha^2 + (J_\beta - J_\gamma)^2 + 2J_\alpha(J_\beta + J_\gamma)$$

with $(\alpha, \beta, \gamma) = (1,2,3)$ and its cyclic permutation. It is readily seen that the right-hand side of (4.61) reduces to that of (4.56) when J is replaced by cI and also to that of (4.50) if the bracket is replaced by a vector product.

In the steady rotation of a spherical top, the Jacobi equation is given by (3.127):

$$\frac{\mathrm{d}^2}{\mathrm{d}t^2}\frac{\| B \|^2}{2} = \| \nabla_T B \|^2 - K(T,B), \tag{4.62}$$

along $\gamma(t) = \mathrm{e}^{tT}$. The right-hand side vanishes because $\nabla_T B = \frac{1}{2}T \times B$ and $K(T,B) = \frac{1}{4}\|T \times B\|^2$. Thus, we find a linear growth: $\|B(t)\| = at + b$ (a, b: constants), and the stability of the geodesic e^{tT} is *neutral*, because $\|B\|$ exhibits neither exponential growth, nor exponential decay.

Even in the asymmetrical top, there are solutions of steady rotation. In fact, if we substitute $(\Omega^1, 0, 0)$ for $(\Omega^1, \Omega^2, \Omega^3)$ for the Euler's equation (4.2), we immediately obtain $\mathrm{d}\Omega^1/\mathrm{d}t = 0$. Hence, the rotation about the principal axis $\boldsymbol{e}_1$ is a steady solution of (4.2). The same is true for rotation about the other two axes. Thus, each one of the three, $(\Omega^1, 0, 0) = \Omega^1 e_1$, $(0, \Omega^2, 0) = \Omega^2 e_2$ or $(0, 0, \Omega^3) = \Omega^3 e_3$, is a steady solution.

Using the unit basis vectors (e_1, e_2, e_3) in the principal frame and defining the sectional curvature by $K(e_i, e_j) = \langle R(e_i, e_j)e_j, e_i\rangle$, one obtains from (4.61) that $K(e_i, e_i) = 0$ (for $i = 1, 2, 3$), and that

$$K(e_1, e_2) = \frac{\kappa_3}{4J_3}, \quad K(e_2, e_3) = \frac{\kappa_1}{4J_1}, \quad K(e_3, e_1) = \frac{\kappa_2}{4J_2}. \tag{4.63}$$

It can be readily checked that $K(e_i, e_j)$ reduces to $\frac{1}{4}\|e_i \times e_j\|^2$ of (4.57) when J is an isotropic tensor cI. For general vectors $X = X^i e_i$ and $Y = Y^i e_i$ and with general J, we obtain

$$K(X, Y) = \langle R(X, Y)Y, X\rangle = R_{ijkl} X^i Y^j X^k Y^l, \tag{4.64}$$

where $R_{ijkl} = \langle R(e_k, e_l)e_j, e_i\rangle$ (see (3.101) and (3.104)).

4.6.3. *Symmetrical top and its stability*

(a) *Regular precession*

If two of J_i are equal (say, $J_2 = J_3 := J_\perp$ and $J_1 \neq J_\perp$),[7] we have a *symmetrical top*. If the symmetry axis is e_1, the Euler's equation (4.2) can be solved immediately, yielding a solution,

$$\Omega^1 = \beta, \qquad \Omega^2(t) + i\Omega^3(t) = \alpha \exp(i\omega_p t) \tag{4.65}$$

where $\omega_p = \Omega^1(J_1 - J_\perp)/J_\perp$ (nonzero constant) and α, β are constants. With respect to the total angular velocity $\hat{\Omega} = (\Omega^1, \Omega^2, \Omega^3)$, we may write $\beta = |\hat{\Omega}| \cos\theta$ and $\alpha = |\hat{\Omega}| \sin\theta$, where θ denotes the constant polar angle of $\hat{\Omega}$ from the pole e_1.

The steady rotation $\hat{X} = (X^1, 0, 0)$ is a *Killing vector*, because the condition (3.161) is satisfied. In fact, the covariant derivative is, from (4.37),

$$\nabla_Y X = \frac{1}{2} J^{-1}(\tilde{K}\hat{Y} \times \hat{X}) = \frac{1}{2} J^{-1}(0, K_3 Y^3 X^1, -K_2 Y^2 X^1),$$

where $J = \mathrm{diag}(J_1, J_\perp, J_\perp)$ and $\tilde{K} = \mathrm{diag}(-J_1 + 2J_\perp, J_1, J_1)$ from (4.38). Then, for $\hat{Y} = (Y^i)$ and $\hat{Z} = (Z^i)$, we have

$$\begin{aligned}\langle \nabla_Y X, Z\rangle &= \frac{1}{2}(\tilde{K}\hat{Y} \times \hat{X}) \cdot \hat{Z} = \frac{1}{2}(\hat{Z} \times \tilde{K}\hat{Y}) \cdot \hat{X} \\ &= \frac{1}{2} J_1 (Z^2 Y^3 - Z^3 Y^2) X^1.\end{aligned}$$

---

[7]In this case, we have $\kappa_1 = (4 - 3k)J_1 J_\perp$, and $\kappa_2 = \kappa_3 = k^2$ of (4.63).

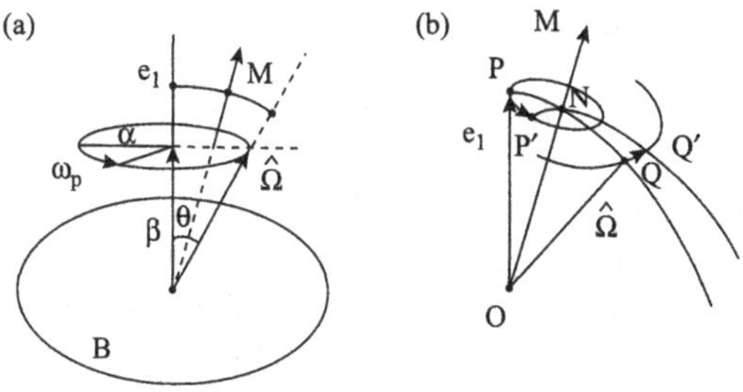

Fig. 4.6. Regular precession in (a) F_B and (b) F, for $J_1 > J_\perp$.

Thus, it is obvious that we have the indentity equation of (3.161), $\langle \nabla_Y X, Z\rangle + \langle \nabla_Z X, Y\rangle = 0$ with any $Y, Z \in \mathbf{so}(3)$, for the Killing vector $\hat{X} = (X^1, 0, 0)$. This implies existence of a *conserved quantity* (§3.12.4), which is given by

$$\langle X, Y\rangle = \hat{X} \cdot J\hat{Y} = X^1 J_1 Y^1 = \text{const} \tag{4.66}$$

along the geodesic flow generated by Y. Because X^1 is a constant, this means that the first component $J_1 Y^1$ of the angular momentum M is conserved, which interprets the first of (4.65) since J_1 is a constant.

Relative to the body (top) frame F_B, the angular momentum vector $M = (J_1\Omega^1, J_\perp\Omega^2, J_\perp\Omega^3)$ rotates with the constant angular velocity ω_p about the symmetry axis $\boldsymbol{e}_1$ with its component $J_1\Omega^1$ being constant and the magnitude $J_\perp\alpha$ perpendicular to $\boldsymbol{e}_1$ also constant. On the other hand, relative to the fixed space F, the angular momentum M is a constant vector. Relatively speaking, the symmetry axis $\boldsymbol{e}_1$ of the top rotates about M, simultaneously the top itself rotates uniformly with β about its axis $\boldsymbol{e}_1$. This motion is called the *regular precession* [LL76] (Fig. 4.6, where the points O, P, N, Q are in a plane and the same for O, P′, N, Q′ in (b)).

With respect to the regular precession about the axis $\boldsymbol{e}_1$, one can solve the Jacobi equation (4.60) where

$$T = \Omega^1\boldsymbol{e}_1 + \Omega^2(t)\boldsymbol{e}_2 + \Omega^3(t)\boldsymbol{e}_3, \qquad B = B^1\boldsymbol{e}_1 + B^2\boldsymbol{e}_2 + B^3\boldsymbol{e}_3.$$

The following is a new analytical study of stability of the regular precession (4.65).

(b) *Stability of regular precession*

In order to make the equations formally compact, we employ complex representations such as $\Omega_*(t) = \Omega^2 + i\Omega^3 = \alpha\exp(i\omega_p t)$ and $B_*(t) = B^2 + iB^3$, and introduce the variables defined by

$$(P, Q, R)(t) := (B^1, \bar{\Omega}_* B_*, \Omega_* \bar{B}_*).$$

where $R = \bar{Q}$ and the *overbar* symbol denotes complex conjugate. Then, after some nontrivial calculations given in §4.6.5, the Jacobi equation (4.60) reduces to the following three second order differential equations with constant coefficients (obtained from (4.81) and (4.82)):

$$\left.\begin{aligned} \ddot{P} - \frac{1}{2}i\dot{Q} + \frac{1}{2}i\dot{R} = 0, \\ \ddot{Q} + ik\beta\dot{Q} - ik\alpha^2\dot{P} + \gamma Q - \gamma R = 0, \\ \ddot{R} - ik\beta\dot{R} + ik\alpha^2\dot{P} + \gamma R - \gamma Q = 0, \end{aligned}\right\} \tag{4.67}$$

where $\dot{\Omega}_* = i\omega_p\Omega_*$ and $\omega_p = \beta(k-1)$ were used, together with the definitions,[8]

$$k = \frac{J_1}{J_\perp}, \qquad \alpha^2 = |\Omega_*|^2, \qquad \beta = \Omega^1, \qquad \gamma = \frac{1}{2}(1-k)\alpha^2.$$

Setting $(P, Q, R) = (P_0, Q_0, R_0)e^{ipt}$ for an exponent p (where P_0, Q_0, R_0 are time-independent complex constants), one can obtain a system of linear homogeneous equations for the amplitudes (P_0, Q_0, R_0). Nontrivial solution of (P_0, Q_0, R_0) is obtained when the determinant of their coefficients vanishes:

$$\begin{vmatrix} -2p^2 & p & -p \\ -q & F_+(p) & \gamma \\ q & \gamma & F_-(p) \end{vmatrix} = 0, \tag{4.68}$$

where $F_\pm(p) = p^2 \pm k\beta p - \gamma$ and $q = k\alpha^2 p$. This reduces to

$$2p^2(F_+F_- - \gamma^2) - pq(F_+ + F_- + 2\gamma) = 2p^4(p^2 - \alpha^2 - k^2\beta^2) = 0.$$

Hence, we have the roots of the eigenvalue equation (4.68):

$$\text{(a) } p = 0; \quad \text{(b) } p = \pm\lambda,$$

where $\lambda = \sqrt{\alpha^2 + k^2\beta^2}$, and $p = 0$ is a quadruple root. Thus, it is found that all the six roots are real.

[8] From the inequality $J_2 + J_3 \geq J_1$, in the footnote of §4.1.1, we have $2 \geq k \geq 0$.

Case (a) $p = 0$: Equations (4.67) result in $\ddot{P} = 0$ and $Q = R = \text{const}$ (real). Then the Jacobi field is given by

$$B^1 = c_1 t + c_0, \qquad B_* = B_0 e^{i\omega_p t}.$$

This shows neutral stability of the basic state (4.65) since the Jacobi field does not grow exponentially.

Case (b) $p = \pm\lambda$: For these eigenvalues, we obtain

$$B^1 = B_0^1 e^{\pm i\lambda t}, \qquad B_* = B_0 e^{i(\omega_p \pm \lambda)t}.$$

The Jacobi fields are represented by periodic functions, hence we have the neutral stability again. In general, the fields $B = (B^1, B^2, B^3)$ are represented in terms of three different frequencies, λ and $(\omega_p \pm \lambda)$ which are incommensurate to each other, therefore the Jacobi fields are quasi-periodic.

Thus, the present geometrical formulation has enabled us to study the stability of the regular precession of a symmetrical top, and the Jacobi equation predicts that the motion is *neutrally stable* in the sense that the Jacobi field does not grow exponentially.

4.6.4. *Stability and instability of an asymmetrical top*

It is shown in §4.6.2 that each of $\Omega^1 \mathbf{e}_1$, $\Omega^2 \mathbf{e}_2$, or $\Omega^3 \mathbf{e}_3$ is a steady solution of the Euler's equation (4.2). Let us try a *classical* stability analysis to see whether they are stable or not. Suppose that an infinitesimal perturbation $\omega = (\omega^1, \omega^2, \omega^3)$ is superimposed on the steady rotation $\Omega = (\Omega^1, 0, 0)$. Substituting $\Omega + \omega$ into the Euler's equation (4.2) and linearizing it with respect to the perturbation ω, we obtain the following perturbation equations: $\dot{\omega}^1 = 0$, and

$$J_2 \dot{\omega}^2 - (J_3 - J_1)\Omega^1 \omega^3 = 0,$$
$$J_3 \dot{\omega}^3 - (J_1 - J_2)\Omega^1 \omega^2 = 0.$$

Assuming the normal form $(\omega^2, \omega^3) = (a, b)e^{i\lambda t}$, we obtain an eigenvalue equation for λ:

$$\lambda^2 = \frac{(J_1 - J_2)(J_1 - J_3)}{J_2 J_3}.$$

If J_1 is the highest out of the three (J_1, J_2, J_3), then $(J_1 - J_2)(J_1 - J_3)$ is positive and λ is real, meaning *stability* of the steady rotation $\Omega^1 \mathbf{e}_1$. If J_1 is the lowest among the three, the same is true. However, if J_1 is middle among the three, λ is pure imaginary. Then, the steady rotation $\Omega^1 \boldsymbol{e}_1$ is *unstable.* Same reasoning applies to the other two steady rotations.

The analysis of the Jacobi field made for a symmetrical top in the previous section clarified that the regular precession is neutrally stable. Here are some considerations for an asymmetrical top in geometrical terms on the basis of numerical analysis [SWK98] for the sectional curvature,

$$K(\Omega^1 \boldsymbol{e}_1, B) = K(\boldsymbol{e}_1, \boldsymbol{e}_2)(\Omega^1 B^2)^2 + K(\boldsymbol{e}_1, \boldsymbol{e}_3)(\Omega^1 B^3)^2 \tag{4.69}$$

(see (4.63) and (4.64)) defined for the section spanned by the steady rotation $\Omega^1 \boldsymbol{e}_1$ and a Jacobi field $B = B^2 \boldsymbol{e}^2 + B^3 \boldsymbol{e}^3$. For the stable motion (with J_1 being the highest or lowest), it is found that the sectional curvature $K(\Omega^1 \boldsymbol{e}_1, B)$ defined by (4.69) takes either positive values always, or both positive and negative values in oscillatory manner, depending on the inertia tensor J. However, it is found that the time average $\bar{K}$ is always positive for any J in the linearly stable case. On the other hand, in the case of linear instability of the middle J_1 value, there exist some inertia tensors J which make $\bar{K}$ negative.

As a whole, the geometrical analysis is consistent with the known properties of rotating rigid bodies in *mechanics.*

4.6.5. *Supplementary notes to §4.6.3*

With respect to the regular precession about the axis $\boldsymbol{e}_1$ (considered in §4.6.3), let us write down the Jacobi equation (4.60),

$$\partial_t^2 B + \nabla_T(\partial_t B) + \partial_t(\nabla_T B) + \nabla_T \nabla_T B + R(B,T)T = 0, \tag{4.70}$$

explicitly for a general Jacobi field $B = B^1 \boldsymbol{e}_1 + B^2 \boldsymbol{e}_2 + B^3 \boldsymbol{e}_3$ along the geodesic of the regular precession with a periodic tangent vector $T = \Omega^1 \boldsymbol{e}_1 + \Omega^2(t)\boldsymbol{e}_2 + \Omega^3(t)\boldsymbol{e}_3$ given by (4.65), where $\Omega_* := \Omega^2 + i\Omega^3 = \alpha \exp(i\omega_p t)$ and $\omega_p = (k-1)\Omega^1$.

The first term is

$$\partial_t^2 B = \ddot{B}_1 \boldsymbol{e}_1 + \ddot{B}_2 \boldsymbol{e}_2 + \ddot{B}_3 \boldsymbol{e}_3, \tag{4.71}$$

where a dot denotes $\mathrm{d}/\mathrm{d}t$. Using the covariant derivative (4.59), we have

$$\begin{aligned} \nabla_T B = \Omega^k \nabla_{\boldsymbol{e}_k} B = & \frac{1}{2J_1}(\tilde{K}_2 \Omega^2 B^3 - \tilde{K}_3 \Omega^3 B^2)\boldsymbol{e}_1 \\ & + \frac{1}{2J_2}(\tilde{K}_3 \Omega^3 B^1 - \tilde{K}_1 \Omega^1 B^3)\boldsymbol{e}_2 \\ & + \frac{1}{2J_3}(\tilde{K}_1 \Omega^1 B^2 - \tilde{K}_2 \Omega^2 B^1)\boldsymbol{e}_3, \end{aligned} \tag{4.72}$$

where $J_2 = J_3 = J_\perp$, and $\tilde{K}_1 = -J_1 + 2J_\perp$, $\tilde{K}_2 = J_1$, $\tilde{K}_3 = J_1$ from (4.38). From this, we obtain

$$\nabla_T(\partial_t B) = \frac{1}{2}(\Omega^2 \dot{B}^3 - \Omega^3 \dot{B}^2)e_1 + \frac{1}{2}(k\Omega^3 \dot{B}^1 - (2-k)\Omega^1 \dot{B}^3)e_2 + \frac{1}{2}((2-k)\Omega^1 \dot{B}^2 - k\Omega^2 \dot{B}^1)e_3, \tag{4.73}$$

where B^k of (4.72) is replaced simply by $\dot{B}^k$, and $k = J_1/J_\perp$ is used. This is the second term of (4.70). The third term is obtained by differentiating (4.72):

$$\partial_t(\nabla_T B) = \nabla_T(\partial_t B) + \frac{1}{2}(\dot{\Omega}^2 B^3 - \dot{\Omega}^3 B^2)e_1 + \frac{1}{2}k(\dot{\Omega}^3 B^1 e_2 - \dot{\Omega}^2 B^1 e_3). \tag{4.74}$$

The fourth term is obtained by operating ∇_T on (4.72) again:

$$\begin{aligned}
&\nabla_T(\nabla_T B) \\
&= \frac{1}{4}[-k((\Omega^2)^2 + (\Omega^3)^2)B^1 + (2-k)\Omega^1\Omega^2 B^2 + (2-k)\Omega^1\Omega^3 B^3]e_1 \\
&+ \frac{1}{4}[k(2-k)\Omega^1\Omega^2 B^1 - k(\Omega^3)^2 B^2 - (2-k)^2(\Omega^1)^2 B^2 + k\Omega^2\Omega^3 B^3]e_2 \\
&+ \frac{1}{4}[k(2-k)\Omega^1\Omega^3 B^1 + k\Omega^2\Omega^3 B^2 - k(\Omega^2)^2 B^3 - (2-k)^2(\Omega^1)^2 B^3]e_3.
\end{aligned} \tag{4.75}$$

Regarding the last term, the curvature tensor $R(B,T)T$ is decomposed into several components as

$$R(B,T)T = B^i \Omega^j R(e_i, e_j)T, \tag{4.76}$$

by using the property of tri-linearity of general curvature tensor $R(X,Y)Z$ with respect to X, Y, Z (see (3.98)) as well as (4.61). Actually, the number of independent components are only three: $R(e_1, e_2)T$, $R(e_1, e_3)T$ and $R(e_2, e_3)T$, since $R(e_i, e_j)T$ is anti-symmetric with respect to i and j. Using (4.61), we obtain

$$\begin{aligned}
R(e_1, e_2)T &= \frac{1}{4}k\Omega^2 e_1 - \frac{1}{4}k^2\Omega^1 e_2, \\
R(e_2, e_3)T &= \frac{1}{4}(4-3k)(\Omega^3 e_2 - \Omega^2 e_3), \\
R(e_3, e_1)T &= \frac{1}{4}k^2\Omega^1 e_3 - \frac{1}{4}k\Omega^3 e_1,
\end{aligned}$$

where $\kappa_1 = (4-3k)J_1J_\perp$, and $\kappa_2 = \kappa_3 = kJ_1J_\perp$ are used for (4.61). Substituting these into (4.76), we find

$$
\begin{aligned}
&R(B,T)T \\
&= \frac{1}{4}[k((\Omega^2)^2+(\Omega^3)^2)B^1 - k\Omega^1(\Omega^2B^2+\Omega^3B^3)]e_1 + \frac{1}{4}[-k^2\Omega^1\Omega^2B^1 \\
&\quad + k^2(\Omega^1)^2B^2 + (4-3k)(\Omega^3)^2B^2 - (4-3k)\Omega^2\Omega^3B^3]e_2 + \frac{1}{4}[-k^2\Omega^1\Omega^3B^1 \\
&\quad - (4-3k)\Omega^2\Omega^3B^2 + k^2(\Omega^1)^2B^3 + (4-3k)(\Omega^2)^2B^3.
\end{aligned} \tag{4.77}
$$

Collecting all the terms (4.71), (4.73)–(4.75) and (4.77), we find that each component of the Jacobi equation (4.70) with respect to e_1, e_2 and e_3 is written down in the following way:

$$
\begin{aligned}
&\ddot{B}^1 - \Omega^3\dot{B}^2 + \Omega^2\dot{B}^3 + \frac{1}{2}(-\dot{\Omega}^3 + (1-k)\Omega^1\Omega^2)B^2 \\
&\quad + \frac{1}{2}(\dot{\Omega}^2B^2 + (1-k)\Omega^1\Omega^3)B^3 = 0,
\end{aligned} \tag{4.78}
$$

$$
\begin{aligned}
&\ddot{B}^2 + k\Omega^3\dot{B}^1 + (k-2)\Omega^1\dot{B}^3 + \frac{1}{2}k(\dot{\Omega}^3 + (1-k)\Omega^1\Omega^2)B^1 \\
&\quad - (1-k)((\Omega^1)^2 - (\Omega^3)^2)B^2 - (1-k)\Omega^2\Omega^3B^3 = 0,
\end{aligned} \tag{4.79}
$$

$$
\begin{aligned}
&\ddot{B}^3 - k\Omega^2\dot{B}^1 - (k-2)\Omega^1\dot{B}^2 + \frac{1}{2}k(-\dot{\Omega}^2 + (1-k)\Omega^1\Omega^3)B^1 \\
&\quad - (1-k)\Omega^2\Omega^3B^2 - (1-k)((\Omega^1)^2 - (\Omega^2)^2)B^2 = 0.
\end{aligned} \tag{4.80}
$$

Using complex representations such as $\Omega_*(t) = \Omega^2 + i\Omega^3 = \alpha\exp(i\omega_pt)$ and $B_*(t) = B^2 + iB^3$, these are rewritten as

$$
\ddot{B}^1 + \frac{1}{2}i(\Omega_*\dot{\bar{B}}_* - \bar{\Omega}_*\dot{B}_*) - \frac{1}{2}\omega_p(\Omega_*\bar{B}_* - \bar{\Omega}_*B_*) = 0, \tag{4.81}
$$

$$
\begin{aligned}
&\ddot{B}_* - ik\Omega_*\dot{B}^1 - i(k-2)\Omega^1\dot{B}_* - \frac{1}{2}ik\dot{\Omega}_*B^1 + \frac{1}{2}k(1-k)\Omega^1B^1\Omega_* \\
&\quad - (1-k)(\Omega^1)^2B_* + \frac{1}{2}(1-k)[|\Omega_*|^2B_* - (\Omega_*)^2\bar{B}_*] = 0,
\end{aligned} \tag{4.82}
$$

where the *overbar* symbol denotes complex conjugate. The first is obtained from (4.78) by using $\dot{\Omega}_* = i\omega_p\Omega_*$. The second is derived from (4.79) and (4.80). One more equation is obtained from them, which is found to be equivalent to the complex conjugate of (4.82) where B^1 and Ω^1 are real.

Chapter 5

Water Waves and KdV Equation

We consider the second class of dynamical systems. Physically, the equations are regarded to describe nonlinear waves, which are familiar as surface waves in shallow water, and are studied extensively in fluid mechanics. Mathematically, this is reformulated as a problem of smooth mappings of a circle S^1 along itself. This corresponds to problems of nonlinear waves under space-periodic condition. A smooth sequence of diffeomorphisms is a mathematical concept of a flow and the unit circle S^1 is one of the simplest base manifolds for physical fields. The Virasoro algebra on S^1 is considered as a fundamental problem in physics.

The manifold S^1 is spatially one-dimensional, but its diffeomorphism has infinite degrees of freedom because pointwise mapping generates arbitrary deformation of the circle. Collection of all smooth orientation-preserving maps constitutes a group $\mathcal{D}(S^1)$ of diffeomorphisms of S^1, as noted in §1.9. Two problems are considered here: the first is the geodesic equation over a manifold of the group $D(S^1)$, describing a simple diffeomorphic flow on S^1, and the second is the KdV equation, which is the geodesic equation over an extended group $\hat{D}(S^1)$,[1] obtained by the central extension of $D(S^1)$. A highlight of this chapter is the dynamical effect of the central extension, i.e. a phase shift enabling wave propagation.

In the first section, we review the physical background of long waves in shallow water, and then consider infinite-dimensional Lie groups, $D(S^1)$ and $\hat{D}(S^1)$, including an infinite-dimensional algebra called the Virasoro algebra [AzIz95]. This chapter is based on [OK87; Mis97; Kam98], and some

[1]The derivation on $D(S^1)$ (or $\hat{D}(S^1)$) in the present chapter can be made more accurate to the group $D^s(\mathbb{R}^1)$ (or $\hat{D}^s(\mathbb{R}^1)$) of diffeomorphisms of Sobolev class $H^s(\mathbb{R})$ (see §8.1.1 and Appendix F with $M = \mathbb{R}^1$). Cauchy problem is known to be well posed for a nonperiodic case as well in the Sobolev space $H^s(\mathbb{R})$ for any $s > 3/2$ [Kat83], [HM01].

additional consideration is given on a Killing field. In addition, formulae of Riemannian curvatures are given.

5.1. Physical Background: Long Waves in Shallow Water

Before considering the geometrical theory of diffeomorphic flow and the KdV equation in the subsequent sections, it would be helpful first to present its physical aspect by reviewing the historical development of the theory of water waves.

(a) Long waves of *infinitesimal* amplitudes (of wavelength λ) in shallow water of undisturbed depth h_* (Fig. 5.1) is described by a wave equation,

$$\partial_t^2 u - c_*^2 \partial_x^2 u = 0,$$

[LL87, §12], where $\lambda/h_* \gg 1$, $a/h_* \ll 1$ with a the wave amplitude. The parameter $c_* = \sqrt{gh_*}$ (g: the acceleration of gravity) denotes the phase velocity of the wave. The function $u(x,t)$ denotes the velocity of water particles along the horizontal axis x (or the surface elevation from undisturbed horizontal level h_*).

(b) Next, approximation that takes into account the finite-amplitude *nonlinear* effect is described by the following set of equations,

$$[\partial_t + (u \pm c)\partial_x](u \pm 2c) = 0 \tag{5.1}$$

[Ach90, §3.9], where $c(x,t) = \sqrt{gh(x,t)}$, and $h(x,t)$ denotes the total depth, the surface elevation given by $\zeta := h(x,t) - h_*$. Equation (5.1) represents two equations corrresponding to the upper and lower signs. This system of equations state a remarkable fact that *the two variables* $u \pm 2c$ *are constant along the two systems of characteristic curves determined by* $\mathrm{d}x/\mathrm{d}t = u \pm c$. This property can be used in solving the problem of obtaining two variables $u(x,t)$ and $c(x,t)$.

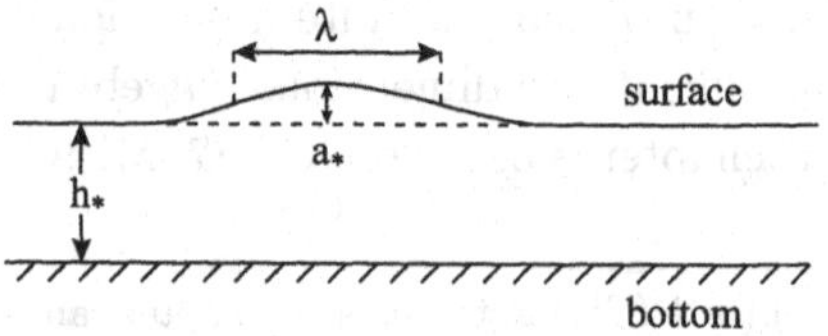

Fig. 5.1. Long waves in shallow water.

Consider a particular case with the following initial condition,

$$t = 0: \quad \begin{cases} u = 0, & c = c_* = \sqrt{gh_*} \quad \text{for } x > 0, \\ u = U > 0, & c = c_1 = \sqrt{gh_1} > c_* \quad \text{for } x < 0 \end{cases}$$

($h_1 > h_*$). Then along the characteristic curves ($\mathrm{d}x/\mathrm{d}t = u - c$) emanating from the undisturbed region of $x > 0$ where $u = 0$, we have $u - 2c = -2c_*$ from the above property. Substituting $c = c_* + \frac{1}{2}u$, the equation of the upper sign of (5.1) reduces to

$$\partial_t u + \left(\frac{3}{2}u + c_*\right)\partial_x u = 0. \tag{5.2}$$

Multiplying by $\frac{3}{2}$, this is rewritten simply as (since c_* is a constant),

$$\partial_t v + v\,\partial_x v = 0, \tag{5.3}$$

where $v = \frac{3}{2}u + c_*$. The general solution of this equation is

$$v = f(x - vt), \tag{5.4}$$

for an arbitrary differentiable function $f(x)$.

A wave profile such as that in Fig. 5.2 becomes more steep as time goes on because the point of the value $v(>0)$ moves faster than the point in front with smaller values than v. There will certainly come a time when the wave slope $\partial v/\partial x$ becomes infinite at some particular x_c. Beyond that time the wave will be broken. This critical time t_c is determined in the following way. Differentiating (5.4) with x, we obtain

$$\partial_x v = \frac{f'(\xi)}{1 + tf'(\xi)}, \quad \xi = x - vt.$$

So, the critical time when the derivative $\partial_x v$ first becomes infinite is

$$t_c = \min_{\xi}[-(1/f'(\xi))].$$

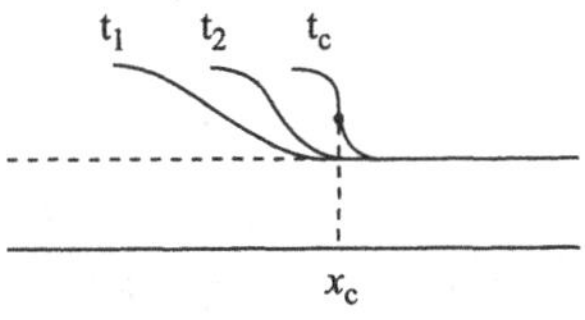

Fig. 5.2. Nonlinear wave becoming more steep.

This *finite-time* breakdown occurs at a point where $f'(\xi) < 0$, i.e. at the front part of the wave. This is understood as representing the mechanism of breakdown of waves observed in nature.

(c) However, stationary waves are also observed in nature. A well-known example is the observation by John Scott Russel in 1834, which is regarded as the first recognized observation of the solitary wave, now called the *soliton*. His observation suggested that the wave propagates with speed $\sqrt{g(h_* + a)}$ (a: wave amplitude), which was confirmed later with computation by Rayleigh (see below), although Russel's observation was exposed to criticism by his contemporaries because of the known property of breakdown described above.

Later, Korteweg and de Vries [KdV1895] succeeded in deriving an equation allowing stationary advancing waves, i.e. solutions which do not show breakdown at a finite time. In the problem of long waves in a shallow water channel (Appendix G), it is important to recognize that there are two dimensionless parameters which are small:

$$\alpha = \frac{a_*}{h_*}, \qquad \beta = \left(\frac{h_*}{\lambda}\right)^2, \tag{5.5}$$

where a_* is a normalization scale of wave amplitude and λ is a representative horizontal scale characterizing the wave width. In order to derive the equation allowing permanent waves, it is assumed that $\alpha \approx \beta \ll 1$. Performing an accurate order of magnitude estimation under such conditions, one can derive the following equation,

$$\partial_\tau u + \frac{3}{2} u\, \partial_\xi u + \frac{1}{6} \partial_\xi^3 u = 0 \tag{5.6}$$

(see Appendix G for its derivation), where

$$\xi = \left(\frac{\alpha}{\beta}\right)^{1/2} \frac{x - c_* t}{\lambda}, \qquad \tau = \left(\frac{\alpha^3}{\beta}\right)^{1/2} \frac{c_* t}{\lambda}. \tag{5.7}$$

The function $u(x,t)$ denotes not only the surface elevation normalized by a_*, but also the velocity (normalized by $g a_*/c_*$),

$$u = \mathrm{d}x_p/\mathrm{d}t, \tag{5.8}$$

of the water particle at $x = x_p(t)$. Comparing (5.6) with (5.2), it is seen that there is a new term $(1/6)\partial_\xi^3 u$. An aspect of this term is interpreted as follows. Linearizing Eq. (5.6) with respect to u, we have $\partial_\tau u + \alpha\, \partial_\xi^3 u = 0$ (where $\alpha = 1/6$). Assuming a wave form $u_w \propto \exp[i(\omega\tau - k\xi)]$ (the wave number k and frequency ω) and substituting it, we obtain a dispersion relation (i.e. a functional relation between k and ω), $\omega = -\alpha k^3$. Phase velocity of the wave is defined by $c(k) := \omega/k = -\alpha k^2$. Namely, a small amplitude wave u_w propagates with nonzero speed $c(k) = -\alpha k^2$ proportional to $\alpha = 1/6$, and the speed is different for different wavelengths $(= 2\pi/k)$. This effect is termed as *wave dispersion.* What is important is that the new term takes into account an effect of wave propagation, in addition to the particle motion $\mathrm{d}x_p/\mathrm{d}t$. (See the 'Remark' of §5.4.)

Replacing u by $v = \frac{3}{2}u$, we obtain

$$\partial_\tau v + v\,\partial_\xi v + \frac{1}{6}\partial_\xi^3 v = 0. \tag{5.9}$$

This equation is now called the **KdV equation** after Korteweg and de Vries (1895). Equation (5.9) allows *steady* wave solutions, which are called the permanent waves. Setting $v = f(\xi - b\tau)$ (b: a constant) and substituting it into (5.9), we obtain $f''' + 6ff' - 6bf' = 0$. This can be integrated twice. Choosing two integration constants appropriately, one finds two wave solutions as follows:

$$v = A\,\mathrm{sech}^2\left[\sqrt{\frac{A}{2}}\left(\xi - \frac{A}{3}\tau\right)\right] \quad \text{(solitary wave)}, \tag{5.10}$$

$$v = A\,\mathrm{cn}^2\left[\sqrt{\frac{d}{2}}\left(\xi - \frac{c}{3}\tau\right)\right], \quad c = 2A - d, \tag{5.11}$$

where $\mathrm{sech}\,x \equiv 2/(e^x + e^{-x})$ and $\mathrm{cn}\,x \equiv cn(\beta x, k)$ (Jacobi's elliptic function) with $\beta = \sqrt{d/2}$ and $k = \sqrt{a/d}$. The first solitary wave solution is obtained by setting two integration constants zero (and $b = A/3$). The

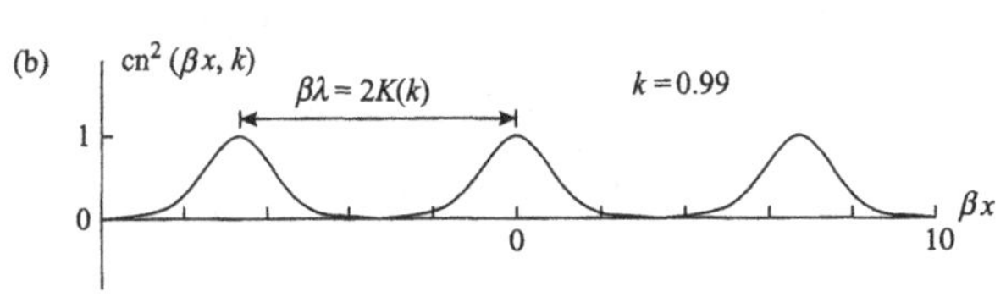

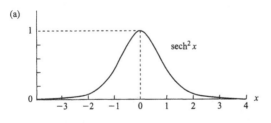

Fig. 5.3. (a) Solitary wave, (b) cnoidal wave.

second solution represents a periodic wave train called the *cnoidal wave* (Fig. 5.3).

The propagation speed of the solitary wave (5.10) is found to be consistent with Russel's observation. Namely the speed is given by $\sqrt{g(h_* + a)} \approx c_*(1 + \frac{1}{2}a/h_*)$. In fact, the argument of (5.10) is written as

$$\xi - \frac{A}{3}\tau = \left(\frac{\alpha}{\beta}\right)^{1/2} \frac{1}{\lambda}\left(x - \left(1 + \frac{1}{2}\frac{a}{h_*}\right) c_* t\right),$$

by using (5.5) and (5.7), and by noting that the amplitude of the wave is given by $a/a_* = u_{\rm amp} = (2/3)v_{\rm amp} = (2/3)A$.

5.2. Simple Diffeomorphic Flow

It was seen in the previous section that the equations of water waves describe not only the surface elevation, but also the moving velocity of the water particles. In other words, the wave propagation is regarded as continuous diffeomorphic mapping of the particle configuration. That is, the particle configuration is transformed from time to time. This observation motivates the study of a group $\mathbf{D}(\mathbf{S}^1)$ of diffeomorphisms of a circle S^1, corresponding to the space-periodic wave train (not necessarily time-periodic) in the water wave problem. For the manifold $S^1[0, 2\pi)$, every point $x + 2\pi \in \mathbb{R}$ is identified with x.

5.2.1. *Commutator and metric of* $D(S^1)$

In §1.9, we considered diffeomorphisms of the manifold S^1 (a unit circle in $\mathbb{R}^2$) by a map (Fig. 5.4),

$$g \in D(S^1) : x \in S^1 \mapsto g(x) \in S^1,$$

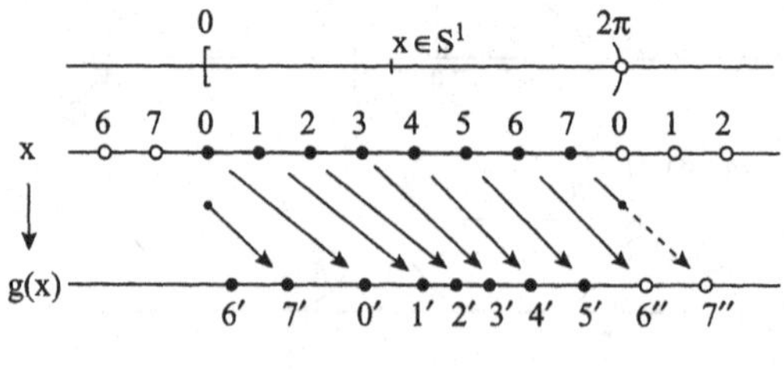

Fig. 5.4.

and defined the tangent field,

$$X(x) = u(x)\partial_x \in TS^1$$

by (1.84). There, we defined the *Lie bracket* (commutator) of two tangent fields $X = u(x)\partial_x, Y = v(x)\partial_x \in TS^1$ as

$$[X, Y] = (uv' - vu')\partial_x, \tag{5.12}$$

where $u' = \partial_x u = u_x$. Furthermore in §3.1.3(b), we introduced a *right-invariant* metric on the group $\mathcal{D}(S^1)$ defined by

$$\langle U, V\rangle_g := \int_{S^1} (U_g \circ g^{-1}, V_g \circ g^{-1})_x \mathrm{d}x,$$

for right-invariant tangent fields $U_g(x) = u \circ g(x)$ and $V_g(x) = v \circ g(x)$ with $g \in \mathcal{D}(S^1)$. This is rewritten as

$$\langle U, V\rangle_g = \int_{S^1} u(x)v(x)\mathrm{d}x = \langle X, Y\rangle_e, \tag{5.13}$$

where $X = u(x)\partial_x, Y = v(x)\partial_x \in T_eD(S^1)$ are tangent fields at the identity e. Because of this metric invariance, the Riemannian connection ∇ is given by the expression (3.65):

$$\nabla_X Y = \frac{1}{2}(ad_X Y - ad^*_Y X - ad^*_X Y). \tag{5.14}$$

Using the definition (1.63) of the *ad*-operator and the definition (5.12) of the commutator, we have $ad_X Y = [X, Y] = (uv' - vu')\partial_x$. Then by the definition (3.64) of $ad^*_X Y$ for $X = u(x)\partial_x, Y = v(x)\partial_x, Z = w(x)\partial_x$, we have

$$\langle ad^*_X Y, Z\rangle = \langle Y, ad_X Z\rangle = \int_{S^1} v(uw' - wu')\mathrm{d}x = -\int_{S^1} (uv' + 2vu')w\mathrm{d}x, \tag{5.15}$$

where integration by parts is performed with the periodic boundary conditions $u(x + 2\pi) = u(x)$, etc. Hence, one obtains

$$ad^*_X Y = -(uv' + 2vu')\partial_x.$$

From (5.14), the Riemannian connection on $D(S^1)$ is given by

$$\nabla_X Y = (2uv' + vu')\partial_x. \tag{5.16}$$

5.2.2. *Geodesic equation on* $D(S^1)$

The geodesic equation is given by (3.67):

$$\partial_t X + \nabla_X X = \partial_t X - ad_X^* X = 0.$$

Thus, the geodesic equation on the manifold $D(S^1)$ is

$$u_t + 3uu_x = 0. \tag{5.17}$$

This is equivalent to Eq. (5.3). Compared with the KdV equation (5.9) with $v = 3u$, this equation has no third order dispersion term u_{xxx}. The third order derivative term is introduced only after considering the central extension in the next section. The above equation (5.17) would be termed as the one governing a *simple diffeomorphic* flow. Its solution exhibits finite-time breakdown in general as given in §5.1.

5.2.3. *Sectional curvatures on* $D(S^1)$

Using the definition (5.16) of $\nabla_X Y$ and $\nabla_Y Y = 3vv'\partial_x$ repeatedly, we have

$$\begin{aligned}
\nabla_X(\nabla_Y Y) &= 6u(vv')' + 3vv'u', \\
\nabla_Y(\nabla_X Y) &= 2v(2uv' + vu')' + (2uv' + vu')v', \\
\nabla_{[X,Y]} Y &= 2(uv' - vu')v + v(uv' - vu')',
\end{aligned}$$

where $u' = \partial_x u$, etc. Therefore,

$$\begin{aligned}
R(X,Y)Y &= \nabla_X(\nabla_Y Y) - \nabla_Y(\nabla_X Y) - \nabla_{[X,Y]} Y \\
&= 2v'(uv' - vu')v + v(uv' - vu')'.
\end{aligned}$$

Thus, we obtain the sectional curvature,

$$K(X,Y) = \langle R(X,Y)Y, X\rangle = \int_{S^1} (uv' - vu')^2 \mathrm{d}x, \tag{5.18}$$

where integration by part is carried out for the integral $\int_{S^1} uv(uv' - vu')'\mathrm{d}x$, and the integrated term vanishes due to periodicity. It is remarkable that the sectional curvature $K(X,Y)$ is *positive*, except in the case, $u(x) = cv(x)$ for $c \in \mathbb{R}$, resulting in $K(X,Y) = 0$.

5.3. Central Extension of $D(S^1)$

An element g of the diffeomorphism group $D(S^1)$ represents a map $g : x \in S^1 \to g(x) \in S^1$. One may write $x = e^{i\phi}$ and instead consider the map $\phi \mapsto g(\phi)$ such that $g(\phi+2\pi) = g(\phi)+2\pi$. Corresponding to the map $\phi \mapsto g(\phi)$, one defines the transformation of a function $f_e := e^{i\phi} \mapsto f_g(\phi) := e^{ig(\phi)}$, where $e(\phi) = \phi$. Furthermore, associated with the group $D(S^1)$, one may define a phase shift $\eta(g) : D(S^1) \to \mathbb{R}$, which is to be introduced in a new transformed function F_g. Namely, in addition to $f_g(\phi)$, the transformation $\phi \mapsto \phi' = g(\phi)$ defines a new function,

$$F_e = e^{i\phi} \mapsto F_g(\phi') = \exp[i\eta(g)]\exp[ig(\phi)].$$

It is described in Appendix H that $f_g = e^{ig(x)}$ is a function on $D(S^1)$, whereas F_g is a function on an extended group $\hat{D}(S^1)$. Associated with this F_g, the transformation law results in the *central extension* of $D(S^1)$.

An extension of the group D is denoted by $\hat{D}$, and its elements are written as

$$\hat{f} := (f, a), \quad \hat{g} := (g, b) \in \hat{D}(S^1),$$

for $f, g \in D(S^1)$ and $a, b \in \mathbb{R}$, where $\hat{D}(S^1) = D(S^1) \oplus \mathbb{R}$. The group operation is defined by (see (H.10)):

$$\hat{g} \circ \hat{f} := (g \circ f, a + b + B(g, f)), \tag{5.19}$$

$$B(g, f) := \frac{1}{2}\int_{S^1} \ln \partial_x (g \circ f) \mathrm{d} \ln \partial_x f, \tag{5.20}$$

where $B(g, f)$ is the Bott cocycle (Appendix H). It can be readily shown that the following subgroup $\hat{D}_0$ is a *center* of the extended group $\hat{D}$, where $\hat{D}_0$ is defined by $\{\hat{f}_0 \mid \hat{f}_0 = (e, a), a \in \mathbb{R}\}$.

5.4. KdV Equation as a Geodesic Equation on $\hat{D}(S^1)$

We now consider the geodesic equation on the extended manifold $\hat{D}(S^1)$, studied by [OK87].

Nontrivial central extension of $T_e D(S^1)$ to $T_{\hat{e}} \hat{D}(S^1)$ is known as the *Virasoro algebra* [AzIz95]. A tangent field at the identity $\hat{e} := (e, 0)$ on the extended manifold $\hat{D}(S^1)$ is denoted by $\hat{u} = (u(x)\partial_x, \alpha)$. For two tangent fields at $\hat{e}$,

$$\hat{u} = (u(x,t)\partial_x, \alpha), \quad \hat{v} = (v(x,t)\partial_x, \beta) \in T_{\hat{e}}\hat{D}(S^1),$$

one can associate two flows: one is $t \mapsto \hat{\xi}_t := (\xi_t, \alpha t)$ starting at $\hat{\xi}_0 = \hat{e}$ in the direction $\hat{u} = (u(x)\partial_x, \alpha)$ and the other is $t \mapsto \hat{\eta}_t = (\eta_t, \beta t)$ in the direction $\hat{v} = (v(x)\partial_x, \beta)$, where $\alpha, \beta \in \mathbb{R}$. Then, the *commutator* is given by

$$[\hat{u}, \hat{v}] := ((u\,\partial_x v - v\,\partial_x u)\partial_x,\ c(u,v)), \tag{5.21}$$

$$c(u,v) := \int \partial_x^2 u\,\partial_x v\,dx = -c(v,u). \tag{5.22}$$

For the derivation of $c(u,v)$, see Appendix H.3. The extended component $c(u,v)$ is called the Gelfand–Fuchs cocycle [GF68].

The *metric* is defined by

$$\langle \hat{u}, \hat{v} \rangle := \int_{S^1} u(x)v(x)dx + \alpha\beta. \tag{5.23}$$

Following the procedure of §5.2, the covariant derivative is derived as

$$\nabla_{\hat{u}}\hat{v} = \left(w(u|v)\partial_x, \frac{1}{2}c(u,v) \right), \tag{5.24}$$

$$w(u|v) = 2uv_x + vu_x + \frac{1}{2}(\alpha v_{xxx} + \beta u_{xxx}). \tag{5.25}$$

The geodesic equation is written as $\partial\hat{u}/\partial t + \nabla_{\hat{u}}\hat{u} = 0$. This results in the following two equations:

$$\begin{aligned} u_t + 3uu_x + \alpha u_{xxx} &= 0, \\ \partial_t \alpha &= 0. \end{aligned} \tag{5.26}$$

The second equation follows from the property, $\int_{S^1} u_{xx}u_x dx = 0$. The first equation is of the form of *KdV equation* (5.9). The coefficient α is called the *central charge*, which was 1/6 in (5.9) for water waves (where $v = 3u$).

Remark. The central extension is associated with a phase-shift $\eta(g)$ of the transformation g describing particle rearrangement, while the central charge α represents the rate of phase-shift. The term including α induces wave motion whose phase speed is different from the speed u of particle motion. Recalling the explanation below (5.8), the term αu_{xxx} describes the wave dispersion, in other words it implies the existence of wave motion. In fact, the KdV equation describes the motion of a long wave in shallow water, where the fluid particles move translationally with speed u different from the wave speed $c(k)$.

5.5. Killing Field of KdV Equation

It may appear to be trivial that the following constant field,

$$\hat{U} = (U_* \partial_x, \alpha), \quad U_*, \alpha \in \mathbb{R} \tag{5.27}$$

is a solution of the KdV equation (5.26). This is in fact a Killing field, and it would be worth investigating from a geometrical point of view.

5.5.1. *Killing equation*

One can verify that the Killing equation is satisfied by $X = \hat{U}$. In fact, the Killing equation (3.161) reads

$$\langle \nabla_{\hat{u}} \hat{U}, \hat{v} \rangle + \langle \hat{u}, \nabla_{\hat{v}} \hat{U} \rangle = 0, \tag{5.28}$$

for any $\hat{u} = (u(x,t)\partial_x, \alpha), \hat{v} = (v(x,t)\partial_x, \alpha) \in T_{\hat{e}}\hat{D}(S^1)$. In order to show this, we apply (5.23), (5.24) and (5.25) to each of the two terms. The second term is then

$$\langle \hat{u}, \nabla_{\hat{v}} \hat{U} \rangle = \int_{S^1} \left(U_* v_x + \frac{1}{2} \alpha v_{xxx} \right) u \, dx, \tag{5.29}$$

and an analogous expression is obtained for $\langle \nabla_{\hat{u}} \hat{U}, \hat{v} \rangle$. Therefore, the left-hand side of Eq. (5.28) becomes

$$\mathcal{L}_{\hat{U}} \langle \hat{u}, \hat{v} \rangle = \int_{S^1} \left[U_* \partial_x (uv) + \frac{1}{2} \alpha \partial_x (u_{xx} v + u v_{xx} - u_x v_x) \right] dx.$$

The right-hand side can be integrated, and obviously vanishes by the periodicity of $u(x)$ and $v(x)$. Thus, the Killing equation (5.28) is satisfied, and it is seen that *the tangent field $\hat{U}$ is the Killing field.*

5.5.2. *Isometry group*

A Killing field $\hat{X}$ generates a one-parameter group of isometry $\phi_t = \mathrm{e}^{t\hat{X}}$. According to §3.12.2, along the flow ϕ_t generated by $\hat{X}$, the inner product $\langle \hat{u}, \hat{v} \rangle$ is invariant for two fields $\hat{u}$ and $\hat{v}$ that are invariant under the flow $\hat{X}$.

Suppose that $\hat{U}$ is a Killing field, then the invariance of the vector field $\hat{v} = (v(x,t)\partial_x, \alpha)$ along ϕ_t is represented by (3.170):

$$\partial_t \hat{v} + \nabla_{\hat{U}} \hat{v} = \partial_t \hat{U} + \nabla_{\hat{v}} \hat{U} \quad (\partial_t \hat{U} = 0). \tag{5.30}$$

In the time-dependent problem, one can introduce the enlarged vectors $\tilde{v} = \partial_t + \hat{v}$ and $\tilde{U} = \partial_t + \hat{U}$. Then, this is rewritten as[2]

$$\nabla_{\tilde{U}} \tilde{v} = \nabla_{\tilde{v}} \tilde{U}, = \nabla_{\hat{v}} \hat{U}. \tag{5.31}$$

Using (5.27) and the covariant derivative (5.24), the equation (5.30) (or (5.31)) takes the form,

$$v_t + 2U_* v_x + \frac{1}{2}\alpha v_{xxx} = U_* v_x + \frac{1}{2}\alpha v_{xxx}, \quad \equiv \nabla_{\tilde{v}} \tilde{U}. \tag{5.32}$$

Namely, we obtain

$$v_t + U_* v_x = 0, \qquad \text{therefore} \quad v = f(x - U_* t),$$

for an arbitrary differentiable function $f(x)$. For two such invariant fields $\hat{u}$ and $\hat{v}$, we have

$$\mathcal{L}_{\tilde{U}} \langle \tilde{u}, \tilde{v} \rangle = \langle \nabla_{\tilde{U}} \tilde{u}, \hat{v} \rangle + \langle \hat{u}, \nabla_{\tilde{U}} \hat{v} \rangle = \langle \nabla_{\hat{u}} \hat{U}, \hat{v} \rangle + \langle \hat{u}, \nabla_{\hat{v}} \hat{U} \rangle = 0, \tag{5.33}$$

(see (3.163) and (3.164)) by the Killing equation (3.161).

Thus it is found that the vector field $\hat{U}$ is the Killing field which generates *a one-parameter group of isometry*, $\phi_t = e^{t\hat{U}}$ in $\hat{D}(S^1)$, which is a stationary geodesic ($\hat{U}$ is stationary).

5.5.3. *Integral invariant*

The stationary Killing field $\hat{U}$ is analogous to the steady rotation $\hat{X} = (X^1, 0, 0)$ of a symmetrical top in §4.6.3(a). We consider an associated invariant analogous to $\langle X, Y \rangle$.

According to §3.12.4, we have the following integral invariant,

$$\langle \tilde{U}, \tilde{w} \rangle = U_* \int_{S^1} w(x) dx + \alpha^2, \tag{5.34}$$

along a geodesic generated by $\tilde{w} = \partial_t + \hat{w}$, where $\hat{w} = (w(x,t)\partial_x, \alpha)$ and $\nabla_{\tilde{w}} \tilde{w} = 0$, with the replacement of X and T by $\tilde{U}$ and $\tilde{w}$, respectively. The invariance can be verified directly, as follows, by setting $u = w$ in (5.29)

[2]The invariant field $\tilde{v}$ does not necessarily satisfy the geodesic equation $\nabla_{\tilde{v}} \tilde{v} = 0$.

since the right-hand side of the first line of (3.169) is

$$\langle \nabla_{\tilde{w}} \tilde{U}, \tilde{w} \rangle = \int_{S^1} \left(U_* w_x + \frac{1}{2}\alpha w_{xxx} \right) w dx$$
$$= \int_{S^1} \left[U_* \partial_x \left(\frac{1}{2} w^2 \right) + \frac{1}{2}\alpha \left(\partial_x (w w_{xx}) - \frac{1}{2}\partial_x ((w_x)^2) \right) \right] dx = 0. \tag{5.35}$$

5.5.4. *Sectional curvature*

Moreover, the curvatures of the two-dimensional sections spanned by the Killing field $\dot{\phi}_t (= \hat{U} \circ \phi_t)$ and any vector field $\hat{v} = (v(x,t)\partial_x, \alpha)$ are *non-negative.* The sectional curvature $K(\hat{u}, \hat{v})$ is calculated in the next section §5.6 for two arbitrary tangent fields $\hat{u}$ and $\hat{v}$. For the particular vector $\hat{u} = \hat{U}$ with constant components, the formula (5.47) is simplified to

$$K(\hat{U}, \hat{v}) = \int_{S^1} \left(U_* v_x + \frac{1}{2}\alpha v_{xxx} \right)^2 \mathrm{d}x, \tag{5.36}$$

(where $v_x = \partial_x v$, etc.). This shows positivity of $K(\hat{U}, \hat{v})$.

This can be verified by a different approach. According to the definition of sectional curvature (3.112) and using $K(\tilde{U}, \tilde{v}) = K(\hat{U}, \hat{v})$ noted in §3.10.3, the sectional curvature $K(\hat{U}, \hat{v})$ is given by

$$K(\hat{U}, \hat{v}) = \langle R(\hat{U}, \hat{v})\hat{v}, \hat{U} \rangle$$
$$= -\langle \nabla_{\hat{v}} \hat{v}, \nabla_{\hat{U}} \hat{U} \rangle + \langle \nabla_{\hat{v}} \hat{U}, \nabla_{\hat{U}} \hat{v} \rangle - \langle \nabla_{[\hat{v}, \hat{U}]} \hat{U}, \hat{v} \rangle \tag{5.37}$$

where $\hat{v}\langle \nabla_{\hat{U}} \hat{U}, \hat{v} \rangle = 0$ and $\hat{U}\langle \nabla_{\hat{v}} \hat{U}, \hat{v} \rangle = 0$.[3] Because the tangent field $\hat{U}$ generates a geodesic flow ϕ_t, we have $\nabla_{\hat{U}} \hat{U} = 0$. Hence the first term which includes $\nabla_{\hat{U}} \hat{U}$ disappears. Using the torsion-free relation (3.18) $[\tilde{U}, \tilde{v}] = \nabla_{\tilde{U}} \tilde{v} - \nabla_{\tilde{v}} \tilde{U}$, the second term can be written as

$$\langle \nabla_{\hat{v}} \hat{U}, \nabla_{\hat{U}} \hat{v} \rangle = \langle \nabla_{\hat{v}} \hat{U}, \nabla_{\hat{v}} \hat{U} \rangle + \langle \nabla_{\hat{v}} \hat{U}, [\hat{U}, \hat{v}] \rangle,$$

[3]This is because $\nabla_{\hat{U}} \hat{U} = 0$ (geodesic equation) and $\langle \nabla_{\hat{v}} \hat{U}, \hat{v} \rangle = 0$ by (5.35). The latter is also verified by $\langle \nabla_{\hat{v}} \hat{U}, \hat{v} \rangle = \hat{v}\langle \hat{U}, \hat{v} \rangle - \langle \hat{U}, \nabla_{\hat{v}} \hat{v} \rangle = 0$. The extended components of $\nabla_{\hat{U}} \hat{U}$ and $\nabla_{\hat{v}} \hat{U}$ are zero due to (5.24).

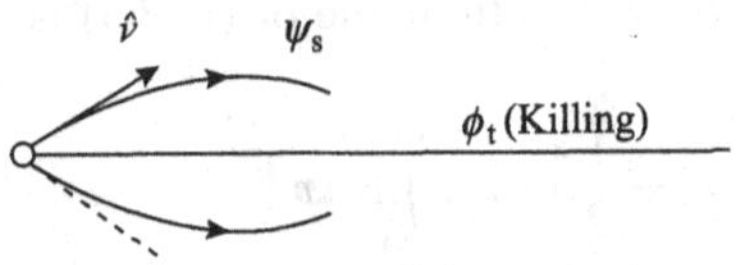

Fig. 5.5. A Killing geodesic ϕ_t and a nearby geodesic ψ_s.

where $[\hat{U}, \hat{v}] = (U_* v_x \partial_x, 0)$ by (5.21) and (5.22). Using (5.24) and (5.25), the third term is

$$-\langle \nabla_{[\hat{v},\hat{U}]} \hat{U}, \hat{v} \rangle = \int_{s1} (U_*(U_* v_x)_x + \frac{1}{2}\alpha(U_* v_x)_{xxx}) v \mathrm{d}x.$$

It is not difficult to show that this term cancels the second term $\langle \nabla_{\hat{v}} \hat{U}, [\hat{U}, \hat{v}] \rangle$. Thus we finally obtain the non-negativity of $K(\tilde{U}, \tilde{v})$:

$$K(\tilde{U}, \tilde{v}) = \langle \nabla_{\tilde{v}} \tilde{U}, \nabla_{\tilde{v}} \tilde{U} \rangle = \|\nabla_{\tilde{v}} \tilde{U}\|^2 \geq 0 \tag{5.38}$$

where $\nabla_{\tilde{v}} \tilde{U}$ is given in (5.32). This confirms the formula (5.36). This is analogous to the curvature (4.57) for the rotation of a spherical top in §4.6.1.

This positivity of $K(\hat{U}, \hat{v})$ means that the nearby geodesic ψ_s generated by $\hat{v}$, where $\dot{\psi}_s (= \hat{v} \circ \psi_s)$, will be pulled toward ϕ_t initially, according to the Jacobi equation (3.125). This could be interpreted as a kind of *stability* of the flow ϕ_t (Fig. 5.5), and furthermore investigated in the next subsection.

5.5.5. *Conjugate point*

The geodesic flow $\phi_t = \mathrm{e}^{t\hat{U}}$ in $\hat{D}(S^1)$ with initial condition $\phi_0 = (e, 0)$, generated by the Killing field $\hat{U} = (U_* \partial_x, \alpha)$, has points conjugate to $\phi_0 = (e, 0)$ along ϕ_t. This is verified, in the following way, by using the Jacobi equation (3.127) along the flow generated by $\tilde{U} = \partial_t + \hat{U}$, which is reproduced here:

$$\frac{\mathrm{d}^2}{\mathrm{d}t^2} \frac{\|J\|^2}{2} = \|\nabla_{\tilde{U}} J\|^2 - K(\hat{U}, J), \tag{5.39}$$

where the vector J is a Jacobi field,

$$J = (v(x,t)\partial_x, 0).$$

The sectional curvature $K(\hat{U}, J)$ is calculated in the previous subsection, and is given by (5.36).

As for the first term on the right-hand side, we have

$$\|\nabla_{\tilde{U}} J\|^2 = \langle \nabla_{\tilde{U}} J, \nabla_{\tilde{U}} J \rangle = \int_{S^1} \left(v_t + 2U_* v_x + \frac{1}{2}\alpha v_{xxx} \right)^2 \mathrm{d}x$$

where $v_t = \partial_t v$. The left-hand side of (5.39) is

$$\frac{\mathrm{d}^2}{\mathrm{d}t^2}\frac{\|J\|^2}{2} = \frac{1}{2}\frac{\mathrm{d}^2}{\mathrm{d}t^2}\left(\int_{S^1} v^2 \mathrm{d}x + \alpha^2\right) = \int_{S^1} (vv_{tt} + (v_t)^2)\mathrm{d}x.$$

Collecting these three terms, Eq. (5.39) becomes

$$\begin{aligned} 0 &= \int_{S^1} (vv_{tt} - 3U_*^2 v_x^2 - 4U_* v_t v_x - \alpha v_t v_{xxx} - \alpha U_* v_x v_{xxx})\mathrm{d}x \\ &= \int_{S^1} v(v_{tt} + 3U_*^2 v_{xx} + 4U_* v_{tx} + \alpha v_{txxx} + \alpha U_* v_{xxxx})\mathrm{d}x, \end{aligned} \tag{5.40}$$

where integration by parts are carried out for all the terms and the integrated terms are deleted by the periodicity.

Thus, requiring this is satisfied with non-trivial Jacobi field $v(x,t)$, we obtain

$$v_{tt} + 3U_*^2 v_{xx} + 4U_* v_{tx} + \alpha v_{txxx} + \alpha U_* v_{xxxx} = 0. \tag{5.41}$$

We have recovered the equation for the Jacobi field $v(x,t)$, obtained by [Mis97] (Eq. (3.1) in the paper).

Equation (5.41) is satisfied by the function,

$$v(x,t) = \sin(\omega_n t)\sin(nx - \nu_n t), \quad n\text{: an integer},$$

where $\omega_n = n\big(U_* - \frac{1}{2}\alpha n^2\big)$ and $\nu_n = \omega_n + U_* n$. This is orthogonal to $\hat{U} = (U_*\partial_x, \alpha)$ (the inner product vanishes), and the magnitude is

$$\|J\| = |\sin(\omega_n t)|\left(\int_{S^1} \sin^2(nx - \nu_n t)\mathrm{d}x\right)^{1/2} = \sqrt{\pi}|\sin(\omega_n t)|.$$

This magnitude $\|J\|$ vanishes at the times t_k (*conjugate points*),

$$t_k = \frac{\pi}{\omega_n}k = \frac{2\pi}{(2U_* - \alpha n^2)n}k, \quad \text{for} \quad k = 0, \pm 1, \pm 2, \ldots.$$

Thus, the neighboring geodesic ψ_s intersects with the flow ϕ_t a number of times (Fig. 5.6), which is regarded as a kind of *neutral stability* of the flow ϕ_t.

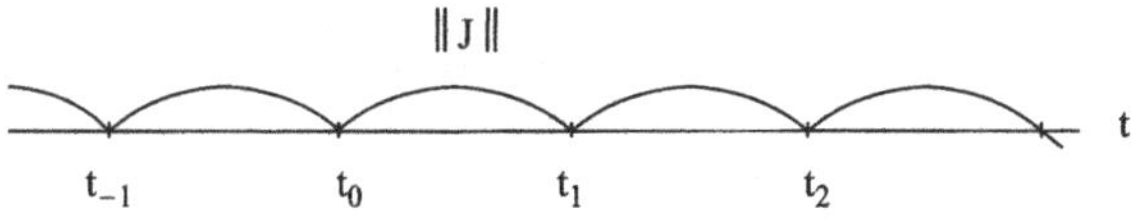

Fig. 5.6. The Jacobi field $\|J\|$ and conjugate points.

5.6. Sectional Curvatures of KdV System

By the geometrical theory formulated in Chap. 3, the *stability of geodesic curves* on a Riemannian manifold is connected with the sectional curvatures. The link is expressed by the Jacobi equation for geodesic variation J in §3.10. An evolution equation of its norm $\|J\|$ is given by Eq. (3.127), where the second term on the right-hand side $K(J,T)$ is the sectional curvature associated with the two-dimensional section spanned by J and T (the tangent to the geodesic). If $K(J,T)$ is negative, the right-hand side is positive. Then the equation predicts exponential growth of the magnitude $\|J\|$, which indicates that the geodesic is *unstable*. It was found in the previous section that the geodesic flow ϕ_t generated by the Killing field is stable. The sectional curvatures of the KdV system can be estimated according to the definition (3.111) [Mis97; Mis98; Kam98].

Expressing two tangent vectors with a common charge α as

$$\hat{u} = (u(x,t)\partial_x, \alpha), \qquad \hat{v} = (v(x,t)\partial_x, \alpha),$$

we have the sectional curvature $K(\hat{u},\hat{v}) = K(\hat{v},\hat{u})$ in the section spanned by $\hat{u}$ and $\hat{v}$ as (see Eq. (3.112), with $X = \hat{u}, Y = \hat{v}$),

$$K(\hat{u},\hat{v}) = \langle R(\hat{v},\hat{u})\hat{u},\hat{v}\rangle = -\langle \nabla_{\hat{u}}\hat{u}, \nabla_{\hat{v}}\hat{v}\rangle + \langle \nabla_{\hat{u}}\hat{v}, \nabla_{\hat{v}}\hat{u}\rangle + \langle \nabla_{[\hat{u},\hat{v}]}\hat{u}, \hat{v}\rangle. \tag{5.42}$$

In order to calculate the right-hand side,[4] we use the definition (5.24) of the covariant derivative, and obtain

$$\nabla_{\hat{u}}\hat{v} = \left(w(u|v)\partial_x, \frac{1}{2}H(u|v) \right), \quad H(u|v) = \int_{S^1} u_{xx}v_x \mathrm{d}x, \tag{5.43}$$

where $w(u|v)$ is defined by (5.25) and $u_{xx} = \partial_x^2 u$, etc. It is readily shown that $H(v|u) = -H(u|v)$ by using periodicity of functions after integration by part. Then by the definition of inner product (5.23), the second term of (5.42) is

$$\langle \nabla_{\hat{u}}\hat{v}, \nabla_{\hat{v}}\hat{u}\rangle = \int_{S^1} w(u|v)\, w(v|u)\mathrm{d}x + \frac{1}{4}H(u|v)\, H(v|u), \tag{5.44}$$

[4]The curvature tensor is represented by spatial parts $\hat{u}$ and $\hat{v}$ even in the unsteady problem, as explained in §3.10.3 and also verified directly in §4.6.1.

Similarly, the first term of (5.42) is

$$-\langle \nabla_{\hat{u}}\hat{u}, \nabla_{\hat{v}}\hat{v}\rangle = -\int_{S^1} w(u|u)w(v|v)\mathrm{d}x,$$

$$\text{since}\quad H(u|u) = \int u_{xx}u_x \mathrm{d}x = \int \partial_x\left(\frac{1}{2}(u_x)^2\right)\mathrm{d}x = 0,$$

by periodicity. Substituting the expression (5.25) for $w(u|v)$, we obtain the sum of two terms as

$$\begin{aligned} &-\langle \nabla_{\hat{u}}\hat{u}, \nabla_{\hat{v}}\hat{v}\rangle + \langle \nabla_{\hat{u}}\hat{v}, \nabla_{\hat{v}}\hat{u}\rangle \\ &\quad = -\frac{1}{4}[H(u|v)]^2 + 2\int (uv' - vu')^2 \mathrm{d}x - \alpha\int (uv' - vu')(u''' - v''')\mathrm{d}x \\ &\qquad + \frac{1}{4}\alpha^2\int (u''' - v''')^2\mathrm{d}x - 9\alpha\int u'v'(u'' + v'')\mathrm{d}x \\ &\qquad + \frac{1}{2}\alpha\int (u'' - v'')(uv'' - vu'')\mathrm{d}x, \end{aligned} \tag{5.45}$$

where integration by parts are carried out several times (when necessary) with integrated terms being deleted by the periodicity. As for the third term of (5.42), we use the definition (5.22),

$$[\hat{u}, \hat{v}] = ((uv' - vu')\partial_x, H(u|v)).$$

Then we obtain $\nabla_{[\hat{u},\hat{v}]}\hat{u} = (W(u|v)\partial_x, L(u|v))$, where

$$W(u|v) = 2u'(uv' - vu') + u(uv' - vu')' + \frac{1}{2}H(u|v)u''' + \frac{1}{2}\alpha(uv' - vu')'''$$

$$L(u|v) = -\frac{1}{2}\int (uv' - vu')'u''\mathrm{d}x,$$

according to the definition (5.24) of the covariant derivative. Finally, the last term of (5.42) is

$$\begin{aligned} \langle \nabla_{[\hat{u},\hat{v}]}\hat{u}, \hat{v}\rangle &= \frac{1}{2}H(u|v)\int u'''v\,\mathrm{d}x - \int (uv' - vu')^2\mathrm{d}x \\ &\quad - \frac{1}{2}\alpha\int (u'' - v'')(uv'' - vu'')\mathrm{d}x. \end{aligned} \tag{5.46}$$

Note that $\int u'''v\mathrm{d}x = -\int u''v'\mathrm{d}x = -H(u|v)$. Thus, collecting (5.45) and (5.46), the curvature $K(\hat{u}, \hat{v})$ of (5.42) is given by

$$K(\hat{u}, \hat{v}) = F - \frac{3}{4}H^2 - G, \tag{5.47}$$

where

$$F = \int_{S^1} \left((uv' - vu') - \frac{1}{2}\alpha(u''' - v''') \right)^2 dx, \tag{5.48}$$

$$G = 9\alpha \int_{S^1} u'v'(u'' + v'')dx, \qquad H = \int_{S^1} u''v'dx. \tag{5.49}$$

Note that the terms including the factor α and the integral $H(u|v)$ originate from the central extension. If those terms are deleted, the curvature $K(\hat{u}, \hat{v})$ is seen to reduce to $K(X, Y)$ of (5.18).

The curvatures are calculated explicitly for the sinusoidal periodic fields $\hat{u}_n = (a_n \sin nx\, \partial_x, \alpha)$ and $\hat{v}_n = (b_n \cos nx\, \partial_x, \alpha)$. For $n \geq 3$, we have

$$K(\hat{v}_1, \hat{u}_n) = \frac{\pi}{4}(\alpha^2 b_1^2 + \alpha^2 (a_n n^3)^2 + 2(b_1 a_n)^2(1 + n^2)) > 0,$$
$$K(\hat{v}_1, \hat{v}_n) = \frac{\pi}{4}(\alpha^2 b_1^2 + \alpha^2 (b_n n^3)^2 + 2(b_1 b_n)^2(1 + n^2)) > 0.$$

Therefore, both of the sectional curvatures $K(\hat{v}_1, \hat{u}_n)$ and $K(\hat{v}_1, \hat{v}_n)$ are *positive* for $n \geq 3$. Thus, most of the sectional curvatures are positive. However, there are some sections which are not always positive. In fact,

$$K(\hat{v}_1, \hat{u}_1) = \frac{\pi}{4}(a_1 b_1)^2 \left(-3\pi + 8 + \alpha^2 \frac{a_1^2 + b_1^2}{a_1^2 b_1^2} \right).$$

The term $-(3/4)(\pi a_1 b_1)^2$ was derived from the term $-(3/4)H^2$, namely from the central extension.

Similarly it can be shown that $K(\hat{v}_n, \hat{u}_n)$ is not always positive for any integer n as well.

Chapter 6

Hamiltonian Systems: Chaos, Integrability and Phase Transition

A self-interacting system of N point masses is one of the typical dynamical systems studied in the traditional analytical dynamics. Such a system is a Hamiltonian dynamical system of finite degrees of freedom. We try to investigate this class of dynamical systems on the basis of geometrical theory, mainly according to [Ptt93; CPC00], and ask whether the geometrical theory is able to provide any new characterization. A simplest nontrivial case is the Hénon–Heiles system, a two-degrees-of-freedom Hamiltonian system, which is known to give rise to chaotic trajectories. A geometrical aspect of *Hamiltonian chaos* will be considered with particular emphasis [CeP96]. Next, integrability of a generalized Hénon–Heiles system will be investigated for a special choice of parameters [ClP02].

Recently some evidence has been revealed that the phenomenon of phase transition is related to a change in the topology of configuration space characterized by the potential function [CCCP97; CPC00], which is described briefly in the last section.

Highlights of the present chapter are chaos and phase transition in certain simplified dynamical systems.

6.1. A Dynamical System with Self-Interaction

6.1.1. *Hamiltonian and metric tensor*

Consider a dynamical system described by the Lagrangian function,

$$L(\bar{q}, \dot{\bar{q}}) := E - V = \frac{1}{2} a_{ij}(\bar{q}) \dot{q}^i \dot{q}^j - V(\bar{q}), \tag{6.1}$$

where $\bar{q} := (q^1, \ldots, q^N)$ and $\dot{\bar{q}} := (\dot{q}^1, \ldots, \dot{q}^N)$ are the generalized coordinates of the configuration space and generalized velocities of the N degrees of freedom system, respectively, and $V(\bar{q})$ is a potential function of self-interaction or gravity. The first term $K = (1/2)a_{ij}\dot{q}^i\dot{q}^j$ is the kinetic energy, with a_{ij} $(i, j = 1, \ldots, N)$ being the mass tensor. We consider only the case $a_{ij} = \delta_{ij}$ (Kronecker's delta). The Hamiltonian H is

$$H := p_i\dot{q}^i - L(\bar{q}, \dot{\bar{q}}) = (1/2)a^{ij}p_ip_j + V(\bar{q}) = K + V, \qquad (6.2)$$

where $p_i = a_{ij}\dot{q}^j$ is the generalized momentum, and (a^{ij}) is the inverse of $(a_{ij}) = \underline{\underline{a}}$, i.e. $(a^{ij}) = \underline{\underline{a}}^{-1}$.

In order to make geometrical formulation, one can define an enlarged Riemannian manifold equipped with the Eisenhart metric $g^E(M^n \times \mathbb{R}^2)$ by introducing two additional coordinates $q^0(= t)$ and q^{N+1} [Eis29; Ptt93].[1] Introducing an enlarged generalized coordinate $Q := (q^\alpha) = (q^0, \bar{q}, q^{N+1})$, the arc-length $\mathrm{d}s$ is given by

$$\mathrm{d}s^2 = g^E_{\alpha\beta}(Q)\mathrm{d}Q^\alpha \mathrm{d}Q^\beta = a_{ij}\mathrm{d}q^i\mathrm{d}q^j - 2V(\bar{q})\mathrm{d}q^0\mathrm{d}q^0 + 2\mathrm{d}q^0\mathrm{d}q^{N+1},$$

$(q^0 = t)$, where the metric tensor $g^E = g^E_{\alpha\beta}(Q)$ is represented by, for $\alpha, \beta = 0, \ldots, N+1$,[2]

$$g^E = ((g^E)_{\alpha\beta}) = \begin{pmatrix} -2V(\bar{q}) & \underline{0} & 1 \\ \underline{0}^T & \underline{\underline{a}} & \underline{0}^T \\ 1 & \underline{0} & 0 \end{pmatrix}, \qquad (6.3)$$

$$(g^E)^{-1} = ((g^E)^{\alpha\beta}) = \begin{pmatrix} 0 & \underline{0} & 1 \\ \underline{0}^T & \underline{\underline{a}}^{-1} & \underline{0}^T \\ 1 & \underline{0} & 2V(\bar{q}) \end{pmatrix}, \qquad (6.4)$$

where $\underline{\underline{a}} := (a_{ij})$, $\underline{0}$ is the null row vector and $\underline{0}^T$ is its transpose.

The Christoffel symbols Γ^k_{ij} are given in (3.21) with the metric tensors g^E. Since the matrix elements $a_{ij}(= \delta_{ij})$ are constant, it is straightforward to see that only nonvanishing Γ^k_{ij}'s are

$$\Gamma^i_{00} = g^{il}\frac{\partial V}{\partial q^l} = \partial_i V, \quad \Gamma^{N+1}_{0i} = \Gamma^{N+1}_{i0} = -g^{0N+1}\frac{\partial V}{\partial q^i} = -\partial_i V \qquad (6.5)$$

[CPC00]. The natural motion is obtained as the projection on the space-time configuration space $(t, \bar{q})$ and given by the geodesics satisfying $\mathrm{d}s^2 = k^2\mathrm{d}t^2$ and $\mathrm{d}q^{N+1} = (k^2/2 - L(\bar{q}, \dot{\bar{q}}))\mathrm{d}t$. The constant k can be always set as $k = 1$, resulting in $\mathrm{d}s^2 = \mathrm{d}t^2$.

[1]The same system can be formulated in terms of the Jacobi metric of (3.3) as well.
[2]Greek indices such α, β run for $0, 1, \ldots, N, N+1$, whereas Roman indices such as i, j run for $1, \ldots, N$, in this chapter.

6.1.2. *Geodesic equation*

The geodesic equation is given by (3.63):

$$\frac{\mathrm{d}^2 x^i}{\mathrm{d}t^2} + \Gamma^i_{jk}\frac{\mathrm{d}x^j}{\mathrm{d}t}\frac{\mathrm{d}x^k}{\mathrm{d}t} = 0.$$

Using (6.5) and $\mathrm{d}s = \mathrm{d}t$ (and $\underline{\underline{a}} = \underline{\underline{a}}^{-1} = (\delta_{ij})$), we obtain

$$\frac{\mathrm{d}^2}{\mathrm{d}t^2}q^i = -\frac{\partial V}{\partial q^i} \quad (i = 1, \ldots, N), \qquad \frac{\mathrm{d}q^0}{\mathrm{d}t} = 1, \tag{6.6}$$
$$\frac{\mathrm{d}^2}{\mathrm{d}t^2}q^{N+1} = 2\frac{\partial V}{\partial q^i}\frac{\mathrm{d}q^i}{\mathrm{d}t} = 2\frac{\mathrm{d}V}{\mathrm{d}t} = -\frac{\mathrm{d}L}{\mathrm{d}t} \quad \left(\text{since } \frac{\mathrm{d}K}{\mathrm{d}t} = -\frac{\mathrm{d}V}{\mathrm{d}t}\right).$$

Choosing arbitrary constants appropriately, we have $q^0 = t$ and $\mathrm{d}q^{N+1}/\mathrm{d}t = 1/2 - L$. Alternatively, we may define $\mathrm{d}q^{N+1}/\mathrm{d}t = 2V(\bar{q})$. Equation (6.6) is the Newton's equation of motion, and we have the following energy conservation from it,

$$\frac{\mathrm{d}H}{\mathrm{d}t} = \frac{\mathrm{d}K}{\mathrm{d}t} + \frac{\mathrm{d}V}{\mathrm{d}t} = \dot{q}^i\ddot{q}^i + \dot{q}^i\partial_{q^i}V = 0.$$

An enlarged velocity vector $\hat{v}$ is written as

$$\hat{v} = \dot{Q} = (1, v, 2V) = (1, v^1, \ldots, v^N, 2V(\bar{q})), \quad v = \dot{\bar{q}}. \tag{6.7}$$

Thus, it is found that the geometric machinery works for the present dynamical system too. The Eisenhart metric (regarded as a Newtonian limit metric of the general relativity) is chosen here as can be seen immediately below to have very simple curvature properties, although there is another metric known as the Jacobi metric (§3.1.3(a)), which is useful as well [CeP95; ClP02].

6.1.3. *Jacobi equation*

The link between the stability of trajectories and the geometrical characterization of the manifold $(M(\bar{q}) \otimes \mathbb{R}^2, g^E)$ is expressed by the Jacobi equation (3.125) (rewritten):

$$\left(\frac{\nabla}{\mathrm{d}s}\right)^2 J + R(J, \dot{Q})\dot{Q} = 0, \tag{6.8}$$

for the Jacobi field J, i.e. a geodesic variation vector. From (3.99), it is found that the nonvanishing components of the curvature tensors are

$$R^i_{0j0} = -R^i_{00j} = \partial_i \partial_j V, \quad \text{for } i, j = 1, \ldots, N. \tag{6.9}$$

The *Ricci tensor* (see §3.9.4), defined by $R_{kj} := R^l_{klj}$, has only a nonzero component $R_{00} = R^l_{0l0} = \Delta V$. The *scalar curvature*, defined by $R := g^{ij} R_{ij} = g^{00} R_{00}$, *vanishes* identically since $g^{00} = 0$ from (6.4). Note that the curvature tensor is

$$R(J, \dot{Q})\dot{Q} = R^i_{\alpha\beta\gamma} \dot{Q}^\alpha J^\beta \dot{Q}^\gamma = (\partial_i \partial_j V) J^j, \tag{6.10}$$

where the Jacobi vector J is defined by

$$J = (0, J^i, 0), \tag{6.11}$$

in view of the definition (3.123).

It is interesting to find that the Jacobi equation (6.8) is equivalent to the equation of tangent dynamics, that is, the evolution equation of infinitesimal variation vector $\xi(t)$ along the reference trajectory $\bar{q}_0(t)$. In fact, writing the perturbed trajectory as $q^i(t) = q^i_0(t) + \xi^i(t)$ and substituting it to the equation of motion, $\mathrm{d}^2 q^i/\mathrm{d}t^2 = -\partial V/\partial q^i$, a linearized perturbation equation with respect to $\xi(t)$ results in

$$\frac{\mathrm{d}^2}{\mathrm{d}t^2} \xi^i = -\left(\frac{\partial^2 V(\bar{q})}{\partial q^i \partial q^j} \right)_{\bar{q}=\bar{q}_0(t)} \xi^j.$$

This is equivalent to the Jacobi equation (6.8) by using (3.25), (6.5) and (6.10) because, noting $J^0 = 0$ and $\dot{Q}^0 = 1$, one has

$$\left(\frac{\nabla J}{\mathrm{d}s} \right)^i = \frac{\mathrm{d}J^i}{\mathrm{d}t} + \Gamma^i_{00} J^0 \dot{Q}^0 = \frac{\mathrm{d}J^i}{\mathrm{d}t}.$$

6.1.4. *Metric and covariant derivative*

Introducing the velocity vectors defined by

$$\hat{u} = (u^\alpha) = (1, u^k, 2V), \qquad \hat{v} = (1, v^k, 2V), \tag{6.12}$$

and using the metric tensor g^E of (6.3), we obtain the corresponding covectors $\hat{U} = (U_\alpha)$ and $\hat{V} = (V_\alpha)$ represented as

$$\hat{U} = (g^E)_{\alpha\beta} u^\beta = (0, u^k, 1), \qquad \hat{V} = (0, v^k, 1). \tag{6.13}$$

The metric is defined by

$$\langle \hat{u}, \hat{v} \rangle = (g^E)_{00} u^0 v^0 + u^k v^k + u^0 v^{N+1} + u^{N+1} v^0 \tag{6.14}$$

$$= u^k v^k + 2V(\bar{q}), \tag{6.15}$$

$$\text{or} \quad \langle \hat{u}, \hat{v} \rangle = U_\alpha v^\alpha = u^k v^k + 2V(\bar{q}).$$

This definition is reasonable because we obtain $\langle \hat{u}, \hat{u} \rangle = 2K + 2V = 2H$, which is an invariant of motion.

Covariant derivative is defined by $(\nabla_{\hat{u}} \hat{v})^\alpha = \mathrm{d}v^\alpha(\hat{u}) + \Gamma^\alpha_{\beta\gamma} u^\beta v^\gamma$. Using the Christoffel symbols defined by (6.5), we obtain

$$(\nabla_{\hat{u}} \hat{v})^0 = \mathrm{d}v^0 + \Gamma^0_{\beta\gamma} u^\beta v^\gamma = 0,$$

$$(\nabla_{\hat{u}} \hat{v})^i = \mathrm{d}v^i(\hat{u}) + \partial_i V, \tag{6.16}$$

$$(\nabla_{\hat{u}} \hat{v})^{N+1} = 2\mathrm{d}V(\hat{u}) - (u^k \partial_k + v^k \partial_k) V = u^k \partial_k V - v^k \partial_k V. \tag{6.17}$$

It is quite natural that the geodesic equation $\nabla_{\hat{v}} \hat{v} = 0$ is consistent with (6.6), since $\mathrm{d}v^i(\hat{v}) = \mathrm{d}v^i/\mathrm{d}t$ and $\mathrm{d}V(\hat{v}) = (\mathrm{d}/\mathrm{d}t)V(\bar{q}) = v^k \partial_k V$ (see (3.25)).

6.2. Two Degrees of Freedom

6.2.1. *Potentials*

All information of dynamical behavior, either regular or chaotic, is included in the geometrical formulation in the previous section. In order to see this, let us investigate a two-degrees-of-freedom system described by the Lagrangian,

$$L = \frac{1}{2}((\dot{q_1})^2 + (\dot{q_2})^2) - V(q_1, q_2),$$

and consider two particular model systems: one is known to be integrable and the other (Hénon–Heiles system) to have chaotic dynamical trajectories. In this chapter (only), we denote the coordinates by the lower suffices such as q_1 (or J_2), in order to have a concise notation of its square, e.g. $(q_1)^2 = q_1^2$. The enlarged coordinate and velocity are

$$Q = (t, q_1, q_2, q_3), \qquad \dot{Q} = (1, \dot{q}_1, \dot{q}_2, 2V(q_1, q_2)).$$

To begin with, let us introduce a potential of *generalized* Hénon–Heiles model, which is defined by

$$V(q_1, q_2) = \frac{1}{2}(q_1^2 + q_2^2) + A q_1^2 q_2 - \frac{1}{3} B q_2^3, \tag{6.18}$$

where the constant parameters A and B are

$$A = 1, \qquad B = 1, \tag{6.19}$$

for the **Hénon–Heiles** model, which is known to yield chaotic trajectories.

With special choices of parameters, this system is known to be globally integrable [CTW82]: one is the case of $A = 0$ and $B = 0$, and the other is the case of $A = 1$ and $B = -6$.

In the next section (§6.3), we consider the first Hénon–Heiles model and show some evidence that the trajectories are influenced by the Riemannian curvatures, according to [CeP96]. In §6.5, we will investigate the latter integrable cases and try to find Killing fields and associated invariants of motion.

6.2.2. *Sectional curvature*

In order to study stability or instability of geodesic flows, i.e. existence or nonexistence of chaotic orbits, the Jacobi equation (6.8) for a geodesic variation J is useful, in which the sectional curvature is an important factor, to be studied in §6.3.

The geodesic variation vector from (6.11) is given as $J = (J_0 = 0, J_1, J_2, 0)$. On the constant energy surface and along the geodesic flow $Q(t)$, one can always assume that J is orthogonal to $\dot{Q}$, i.e. $\langle J, \dot{Q}\rangle = g_{ij} J_i \dot{Q}_j = 0$. In fact, expressing as $J = J_\perp + c\dot{Q}$ and substituting it in Eq. (6.8), it is readily seen that the terms related to the parallel component $c\dot{Q}(= cT)$ drop out and the equation is nothing only for the orthogonal component $J_\perp$ (see (3.126)).

The equation for the norm of geodesic variation $\|J\|$ is given by Eq. (3.127). The Jacobi vector is chosen as $J = (0, \dot{q}_2, -\dot{q}_1, 0)$, hence $\langle J, \dot{Q}\rangle = 0$ ([CeP96]). Then the sectional curvature normalized by $\|J\|^2$ is given by

$$\begin{aligned}\hat{K}(\dot{Q}, Q) &:= \frac{K(J, \dot{Q})}{\|J\|^2} \\ &= \frac{1}{2(E_t - V(\bar{q}))}\left(\frac{\partial^2 V}{\partial q_1^2}\dot{q}_2^2 - 2\frac{\partial^2 V}{\partial q_1 \partial q_2}\dot{q}_1\dot{q}_2 + \frac{\partial^2 V}{\partial q_2^2}\dot{q}_1^2\right),\end{aligned}$$

($E_t = K + V$, total energy), which can be computed on the surface S_E of constant energy where E_t =const.

In order to get the geometrical characterization of the dynamical orbits of the Hénon–Heiles model, it is useful to define the average value of negative curvature, $\hat{K}_- = \{\hat{K} : \hat{K}(\dot{Q}, Q) < 0\}$ by

$$\langle \hat{K}_- \rangle := \frac{1}{A(S_E)} \int_{S_E} \hat{K}_- \mathrm{d}\bar{q} \mathrm{d}\dot{\bar{q}},$$

where $A(S_E)$ is the area in the $(\bar{q}, \dot{\bar{q}})$-plane S_E which is accessible by the dynamical trajectories. The quantity $\langle \hat{K}_- \rangle$ was estimated at different energy values E.

6.3. Hénon–Heiles Model and Chaos

In the Hénon–Heiles (1964) model, the Hamiltonian is $H = (1/2)(p_1^2 + p_2^2) + V(q_1, q_2)$, and the potential V is chosen as

$$V(q_1, q_2) = \frac{1}{2}(q_1^2 + q_2^2) + q_1^2 q_2 - \frac{1}{3} q_2^3 = \frac{1}{2} r^2 + \frac{1}{3} r^3 \sin 3\theta,$$

where $q_1 = r \cos\theta$ and $q_2 = r \sin\theta$. Originally this was derived to describe the motion of a test star in an axisymmetric galactic mean gravitational field [HH64].

6.3.1. *Conventional method*

It was shown that the transition from order to chaos is quantitatively described by measuring, on a Poincaré section, the ratio μ of the area covered by the regular trajectories divided by the total area accessible to the motions.

At low values of the energy E_t, the whole area is practically covered by regular orbits and hence the ratio μ is almost 1. As E_t is increased but remains below $E_t \approx 0.1$, μ decreases very slowly from 1. As E_t is increased further, μ begins to drop rapidly to very small values (Fig. 6.1). At $E_t = 1/6 \approx 0.167$, the accessible area is marginal because the equipotential curve $V(q_1, q_2) = 1/6$ is an equilateral triangle (including the origin within it). Beyond $E_t = 1/6$, the equipotential curves are open, and the motions are unbounded. Thus the accessible area becomes infinite.

6.3.2. *Evidence of chaos in a geometrical aspect*

It is shown in [CeP96] that, for low values of E_t, the integral of the negative curvature is almost zero, but that, at the same E_t value (≈ 0.1) at which μ begins to drop rapidly, the value $\langle \hat{K}_- \rangle$ starts to increase rapidly (Fig. 6.1).

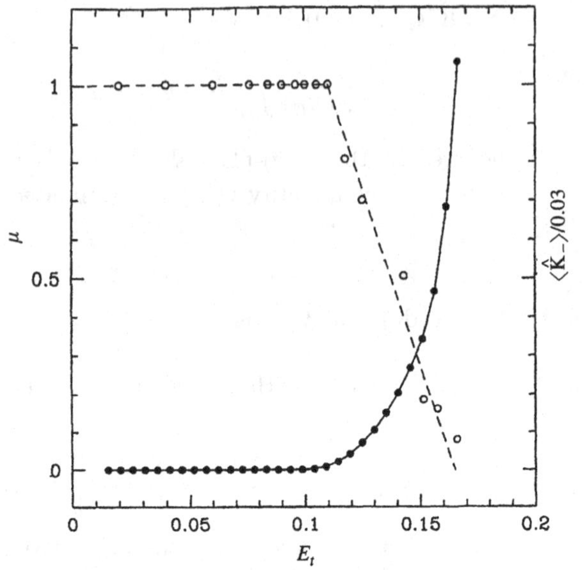

Fig. 6.1. μ (open circles versus left axis) and $\langle \hat{K}_- \rangle$ (full dots versus right axis) with respect to E_t, due to [CeP96, Fig. 8] (reprinted with permission of American Physical Society © 1996).

The exact coincidence between the critical energy levels for the μ-*decrease* below 1 and for the $\langle \hat{K}_- \rangle$-*increase* above 0 is understood to be the onset of sharp increase of chaotic domains detected by the increase of the negative curvature integral $\langle \hat{K}_- \rangle$. Along with this, the fraction of the area $A_-(S_E)$ where $\hat{K} < 0$ is also estimated as a function of E_t. The transition is again detected by this quantity as well.

6.4. Geometry and Chaos

In geodesic flows on a compact manifold with negative curvatures, nearby geodesics tend to separate exponentially. Since the geodesic flows are constrained in a bounded space, the geodesic curves are obliged to fold in due course of time. Such joint action of stretching and folding is just the mechanism yielding chaos. Ergodicity and mixing of this type of flows are investigated thoroughly [Ano67; AA68]. The particular system of Hénon–Heiles model considered in the previous section is regarded to be another example of chaos controlled by negative curvatures, but it includes positive curvatures as well.

Recently, a series of geometrical study of chaotic dynamical systems [Ptt93; CP93; CCP96; CPC00] suggested that chaos can be induced not only by negative curvatures, but also by positive curvatures, if the curvatures fluctuate stochastically along the geodesics.

In order to quantify the degree of instability of dynamical trajectories, the notion of Lyapunov exponent is introduced (e.g. [Ott93; BJPV98]). However, the Lyapunov exponents are determined only asymptotically along the trajectories, and their relation with local properties of the phase space is far from obvious. The geometrical approach described below allows us to find a quantitative link between the Largest Lyapunov exponent and the curvature fluctuations. Under suitable approximations [CPC00], a stability equation in high-dimensions can be written in the following form,

$$\left(\frac{\mathrm{d}}{\mathrm{d}t}\right)^2 J + k(t)J = 0,$$

which is similar to the Jacobi equation (3.139) for an isotropic manifold. Here however, the curvature $k(t)$ is not constant, but fluctuates as a function of time t. The fluctuation is modeled as a stochastic process. This enables us to estimate the Largest Lyapunov exponent.

Four assumptions are made before the analysis: (i) the manifold is quasi-isotropic, (ii) $k(t)$ is modeled as a stochastic process, (iii) statistics of $k(t)$ is the same as the Ricci curvature K_R (§3.9.4), and (iv) time average of K_R is replaced by a certain static phase average. After some nontrivial procedures, the following effective stability equation is derived:

$$\left(\frac{\mathrm{d}}{\mathrm{d}t}\right)^2 J + \left[\langle k_R\rangle + \langle \delta^2 k_R\rangle^{1/2}\eta(t)\right]J = 0, \qquad (6.20)$$

where J stands for any component of the Jacobi field $J_\perp$, and $\eta(t)$ is a Gaussian random process with zero mean and unit variance, and furthermore

$$\langle k_R\rangle = \frac{1}{n-1}\langle K_R(s)\rangle, \quad \langle \delta^2 k_R\rangle = \frac{1}{n-1}\langle [K_R(t) - \langle K_R\rangle]^2\rangle.$$

Equation (6.20) looks like an equation of a stochastic oscillator of frequency $\Omega = [\langle k_R\rangle + \langle \delta^2 k_R\rangle^{1/2}\eta(t)]^{1/2}$ with a Gaussian stochastic process with the average $\bar{k} = \langle k_R\rangle$ and the variance $\sigma_k = \langle \delta^2 k_R\rangle^{1/2}$.

The largest Lyapunov exponent λ is determined by the following limit [CCP96; CPC00; Ott93; BJPV98]:

$$\lambda = \lim_{t\to\infty} \frac{1}{2t} \log \frac{J^2(t) + \dot{J}^2(t)}{J^2(0) + \dot{J}^2(0)}. \tag{6.21}$$

Applying a theory of stochastic oscillator [vK76; CPC00] to (6.20), the largest Lyapunov exponent is estimated as

$$\lambda \propto \langle \delta^2 k_R \rangle \approx \frac{1}{n-1} \langle [K_R(t) - \langle K_R \rangle]^2 \rangle \tag{6.22}$$

for the case $\sigma_k \ll \langle k_R \rangle$. A formula is also given for $\sigma_k \approx O(\langle k_R \rangle)$.

A numerical exploration was carried out to compare (6.22) with (6.21) for a three-degree-of freedom system [CiP02], in which the potential was

$$V(x, y, z) = Ax^2 + By^2 + Cz^2 - axz^2 - byz^2,$$

where A, B, C, a, b are constants. This is a potential of nonlinearly coupled oscillators, which was originally derived to represent central regions of a three-axial elliptical galaxy [Cont86].

Estimate of the lyapunov exponent λ through the analytic formula (6.22) and its generalization (not shown here) shows a fairly good comparison with the largest Lyapunov exponent (6.21) obtained by solving numerically a tangent dynamics equation (equivalent to the Jacobi equation with the Eisenhart metric). Figure 6.2 shows such comparison, plotted with respect to the parameter a for fixed values of $A = 0.9, B = 0.4, C = 0.225, b = 0.3$ and $E_t = 0.00765$.

Another numerical exploration had been carried out to compare (6.22) with (6.21), for a chain of coupled nonlinear oscillators (the FPU β model) and a chain of coupled rotators (the 1-d XY model) [CP93; CCP96]. The sectional curvatures are always positive in the former model, while there is a very small probability of negative sectional curvatures in the latter model. Agreement between the relations (6.21) and (6.22) is excellent for the former case. There is some complexity in the latter case, but not inconsistent with the geometrical theory.

This series of investigations made it clear that chaos occurs in dynamical systems with positive sectional curvatures if they fluctuate stochastically, and moreover that the largest Lyapunov exponent depends on the variance of the curvature fluctuation (confirmed in general cases including $\sigma_k = O(\bar{k})$). This describes, in fact, a geometrical origin of the chaos in the models investigated.

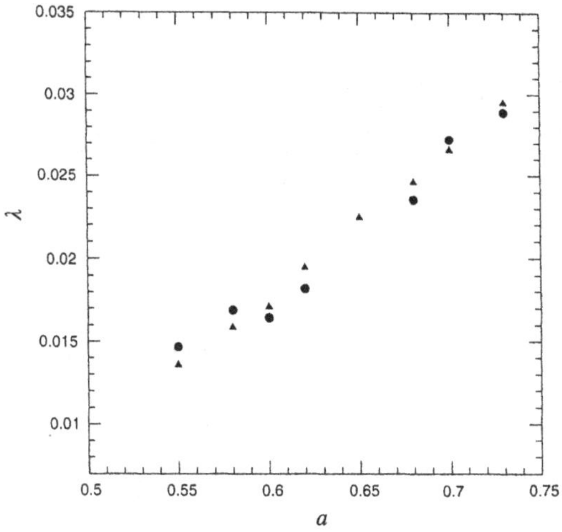

Fig. 6.2. Comparison of the largest Lyapunov exponent λ. Full circle: analytic formula (6.22) and its generalization, full triangle: (6.21) obtained by numerical computation [CiP02, Fig. 11] (with kind permission of Kluwer Academic Publishers).

6.5. Invariants in a Generalized Model

Let us investigate a class of invariants in a generalized Hénon–Heiles model, whose potential is given by

$$V(q_1, q_2) = \frac{1}{2}(q_1^2 + q_2^2) + Aq_1^2 q_2 - \frac{1}{3}Bq_2^3, \tag{6.23}$$

from (6.18). First we consider a Killing vector field and its associated invariant, and next briefly describe an invariant associated with a Killing tensor field of second order. There is a known invariant of motion, i.e. the total energy: $E = K + V$. With a special choice of parameters A and B, one can show the existence of another invariant. If there exist *two* invariants in a *two-degrees-freedom* Hamiltonian system, the system is completely integrable (*Liouville's theorem*: [Arn78, §49; Whi37]).

6.5.1. *Killing vector field*

According to the definition (6.13) of a covector, a Killing covector is denoted as $\hat{X} = (X_\alpha) = (0, X_1, X_2, 0)$, and its covariant derivative is given by

$$X_{\alpha;\beta} = \partial_\beta X_\alpha - \Gamma^\gamma_{\beta\alpha} X_\gamma, \tag{6.24}$$

from (3.151). By (3.159), we have ten Killing equations,

$$X_{\alpha;\beta} + X_{\beta;\alpha} = 0 \qquad (\alpha, \beta = 0, 1, 2, 3). \tag{6.25}$$

Using (6.24) and the Christoffel symbol Γ^k_{ij} of (6.5), and in particular, noting that $X_{0;\beta} = -X_\gamma \Gamma^\gamma_{\beta 0}$, we obtain

$$X_{0;k} = 0, \qquad X_{0;0} = -X_k \partial_k V, \quad (k = 1, 2).$$
$$X_{k;0} = \partial_t X_k.$$

Thus, the Killing equation (6.25) for $\alpha = 0$ and $\beta = k$ results in

$$X_{0;k} + X_{k;0} = \partial_t X_k = 0,$$

i.e. the Killing field must be *steady*. In addition, for $\beta = 0$,

$$X_{0;0} = -X_k \partial_k V = -X_1(q_1 + 2Aq_1q_2) - X_2(q_2 + Aq_1^2 - Bq_2^2) = 0. \tag{6.26}$$

The steady Killing field $X_1(q_1, q_2)$ and $X_2(q_1, q_2)$ must satisfy this equation. It can be shown that $X_{3;\alpha} = 0$ and $X_{\alpha;3} = 0$, hence the four equations $X_{\alpha;3} + X_{3;\alpha} = 0$ are identically satisfied. The remaining three equations are

$$X_{1;1} = 0, \qquad X_{2;2} = 0, \qquad X_{1;2} + X_{2;1} = 0. \tag{6.27}$$

Using (6.24) and (6.5), we have

$$X_{k;l} = \partial_l X_k, \qquad (k, l = 1, 2).$$

Then, the first two equations of (6.27) lead to $X_1 = X_1(q_2)$ and $X_2 = X_2(q_1)$. The third equation results in

$$X_1'(q_2) = -X_2'(q_1) = c \quad (\text{const}).$$

Hence, we have $X_1 = cq_2 + a$ and $X_2 = -cq_1 + b$. Substituting these into (6.26), it is found that the following equation,

$$0 \equiv aq_1 + bq_2 + 2aAq_1q_2 + b(Aq_1^2 - Bq_2^2) + 2cAq_1q_2^2 - cAq_1^2q_2 + cBq_2^3,$$

must hold identically. This is only satisfied by

$$a = 0, \quad b = 0; \qquad A = 0, \quad B = 0,$$

to obtain a nontrivial solution of Killing field. Thus, we find that the Killing covector (covariant vector) is given by

$$\hat{X} = (X_\alpha) = (0, cq_2, -cq_1, 0). \tag{6.28}$$

Corresponding Killing (contravariant) vector is

$$X = ((g^E)^{\alpha\beta} X_\beta) = (0, cq_2, -cq_1, 0).$$

The vector $(X^1, X^2) = (cq_2, -cq_1)$ denotes a rotation velocity in the (q_1, q_2) plane with an angular velocity $-c$.

The associated invariant $\langle X, \dot{Q} \rangle$ is given by (6.14) as

$$I_1 = \langle X, \dot{Q} \rangle = cq_2\dot{q}_1 - cq_1\dot{q}_2 = cM,$$

where $M = q_1\dot{q}_2 - q_2\dot{q}_1$ is the angular momentum. Thus, it is found that the angular momentum M is conserved. This is analogous to the invariant (4.66) of a symmetrical top. Here, the potential $V(q_1, q_2)$ of (6.23) has a rotational symmetry with respect to the origin since $A = 0$ and $B = 0$.

6.5.2. *Another integrable case*

It is shown in [ClP02] that there exists another Killing tensor field X_{ij} for the potential $V(q_1, q_2)$ of $A = 1$ and $B = -6$. The Killing tensors X_{ij} were found on the basis of the Jacobi metric (3.3) to satisfy Eq. (3.172), $\nabla_{(j)} X_{(K)_p} = 0$, in §3.12.5 with the definition of covariant derivative given in §3.11.3. The associated constant of motion is given as

$$I_2 = q_1^4 + 4q_1^2q_2^2 - \dot{q}_1^2 q_2 + 4\dot{q}_1\dot{q}_2 q_1 + 4q_1^2 q_2 + 3\dot{q}_1^2 + 3q_1^2.$$

This is consistent with that reported in [CTW82], worked out with a completely different method.

6.6. Topological Signature of Phase Transitions

Evidence has been gained recently to show that phenomena of phase transition are related to change in the topology of configuration space of the system. This is briefly described here according to [CPC00; CCP02; ACPRZ02].

There is a certain relationship between *topology change* of the configuration manifold and the dynamical behaviors on it. This is shown with a *k-trigonometric model* defined just below, which is a solvable mean-field model with a k-body interaction. According to the value of the parameter k, it is known that the system has no phase transition for $k = 1$, undergoes a second order phase transition for $k = 2$, or a first order phase transition for $k \geq 3$. This behavior is retrieved by investigation of a topological invariant, the Euler characteristics, of some submanifolds of the configuration space in terms of a Morse function defined by a potential function. However, it should be noted that not all changes of topology are related to a phase transition, but a certain characteristic topological change is identified to correspond to it.

The mean-field k-trigonometric model is defined by the following Hamiltonian for a system of N degrees of freedom,

$$H_k(q,p) = K(p) + V_k(q) = \frac{1}{2m}\sum_{i=1}^{N} p_i^2 + V_k(q), \tag{6.29}$$

where the potential energy is given by

$$V_k(q) = \frac{A}{N^{k-1}} \sum_{i_1,\ldots,i_k} (1 - \cos(\varphi_{i_1} + \cdots + \varphi_{i_k})), \tag{6.30}$$

where $q = (q_1, \ldots, q_N)$, and A is a energy scale constant, and $\varphi_i = (2\pi/L)q_i$ with L as a scale length. Note that

$$V_1 = A\sum_i (1 - \cos\varphi_i), \qquad V_2 = \frac{A}{N}\sum_{i_1,i_2}(1 - \cos(\varphi_{i_1} + \varphi_{i_2})),$$

$$V_3 = \frac{A}{N^2}\sum_{i_1,i_2,i_3}(1 - \cos(\varphi_{i_1} + \varphi_{i_2} + \varphi_{i_3})), \quad \cdots$$

6.6.1. *Morse function and Euler index*

Given a potential function $f(q)$, we define a submanifold of the configuration space M^N by

$$M_a := \{q \in M \mid f(q) \leq a\}.$$

Morse theory [Mil63] provides a way of classifying topological changes of the manifod M_a and links global topological properties with local analytical properties of a smooth function such as the potential $f(q)$ defined on M.

A point $q_c \in M$ is called a *critical point* of f if $\mathrm{d}f = 0$, i.e. if the differential of f at q_c vanishes. The function $f(q)$ is called a *Morse function* on M if its critical points are all nondegenerate, i.e. if the Hessian of f at q_c has only nonzero eigenvalues, so that the critical points q_c are isolated.

We now define the Morse function by the potential energy per particle,

$$f(q) := \bar{V}(q) = V(q)/N.$$

The submanifold is defined by $M_v = \{q \mid \bar{V} \leq v\}$. Suppose that the parameter v is increased from the minimum value v_0 where $v_0 = \min_q \bar{V}$ (if any). All the submanifolds have the same topology until a critical level l_c is crossed, where the level set is defined by $\bar{V}^{-1}(l_c) = \{q \in M \mid \bar{V}(q) = l_c\}$. At this level, the topology of M_v changes in a way completely determined by the local properties of the Morse function $\bar{V}(q)$. Full configuration space M can be constructed sequentially from the M_v by increasing v.

At any critical point q_c where $\partial\bar{V}/\partial q_i|_{q_c} = 0$ $(i = 1, \ldots, N)$, an *index* k of the critical point is defined by the number of *negative* eigenvalues of the Hessian matrix H_{ij} at q_c:

$$H_{ij} = \frac{\partial^2 \bar{V}}{\partial q_i \partial q_j}, \quad i, j = 1, \ldots, N.$$

Knowing the index of all the critical points below a given level v, one can obtain the Morse index (number) μ_k defined by the number of critical points which have index k, and furthermore obtain the *Euler characteristic* of the manifold M_v, given by

$$\chi(M_v) = \sum_{k=0}^{N} (-1)^k \mu_k(M_v),$$

[Fra97; CPC00, App. B]. The Euler characteristic is a topological invariant of the manifold M_v (see §2.10 for M^2).

6.6.2. *Signatures of phase transition*

Morse indexes μ_k are given explicitly for the potential function (6.30) [CCP02; ACPRZ02]. Thus, the Euler characteristics $\chi(M_v)$ of the manifold M_v are calculated explicitly for general k. It is found that behaviors of $\chi(M_v)$ are characteristically different between different values of $k = 1, 2$, and 3, which correspond to no phase transition, a second order phase transition, and a first order phase transition, respectively.

Based on the exact analysis [CCP02] of the case $k = 2$, called the mean-field XY model, it is found that the phase transition occurs when the Euler characteristic changes discontinuously from a big value of $O(\mathrm{e}^N)$ to zero, e.g. a large number of suddles existing on M_v disappear all of a sudden at the level of phase transition.

It is shown [ACPRZ02] that the Euler characteristic $\chi(M_v)$ shows a discontinuity in the first order derivative with respect to v at the phase transition, and the order of phase transition can be deduced from the sign of the second order derivative $\mathrm{d}^2\chi/\mathrm{d}v^2$, i.e. $\mathrm{d}^2\chi/\mathrm{d}v^2 > 0$ for the first order phase transition and $\mathrm{d}^2\chi/\mathrm{d}v^2 < 0$ for the second order transition. From this result, it is proposed that the thermodynamic entropy is closely related to a function derived from the Euler characteristic, i.e. $\sigma(v) = \lim_{N\to\infty} N^{-1} \log|\chi(v)|$ at the phase transition.

6.6.3. *Topological change in the mean-field XY model*

In the mean-field XY model, the potential energy per degree of freedom is given by

$$\bar{V}(q) = \frac{V(q)}{N} = \frac{1}{2N^2}\sum_{i,j=1}^{N}(1-\cos(\varphi_i-\varphi_j)) - \frac{h}{N}\sum_{i=1}^{N}\cos\varphi_i,$$

where $\varphi_i \in [0, 2\pi]$, and h is a strength of the external field. At the level of phase transition, it is remarkable that not only the Euler characteristic changes discontinuously from a big value of $O(\mathrm{e}^N)$ to zero, but also fluctuations of configuration-space curvature exhibit cusp-like singular behaviors at the phase transition [CPC00].

In this XY model, it is possible to project the configuration space onto a two-dimensional plane, an enormous reduction of dimensions. To this end, it is useful to define the magnetization vector $\boldsymbol{m}$ by

$$\boldsymbol{m} = (m_x, m_y) = \frac{1}{N}\left(\sum_{i=1}^{N}\cos\varphi_i, \sum_{i=1}^{N}\sin\varphi_i\right). \tag{6.31}$$

By this definition, the potential energy can be expressed as a function of m_x and m_y alone:

$$\bar{V}(\varphi_i) = \bar{V}(m_x, m_y) = \frac{1}{2}(1 - m_x^2 - m_y^2) - hm_x,$$

which is regarded as a mean-field character. The minimum energy level is at $\bar{V} = v_0 = -h$, whose configuration corresponds to $m_x^2 + m_y^2 = 1$ and $m_x = 1$ (therefore $m_y = 0$). The maximum energy level is at $\bar{V} = v_c = \frac{1}{2}(1+h^2)$, whose configuration corresponds to $m_x = -h$ and $m_y = 0$. The number of such configurations grows with N quite rapidly. The critical value v_c is isolated and tends to $\frac{1}{2}$ as $h \to 0$.

All the critical levels l_c where $d\bar{V} = 0$ are included in the interval $l_c \in [v_0, v_c]$. The submanifold M_v is defined by $\{q \mid \bar{V} \leq v\}$ as before.

First, for $v < v_0$, the manifold M_v is empty. The topological change that occurs at $v = v_0$ is that corresponding to the emergence of manifold from the empty set. Subsequently, there are many topological changes at levels $l_c \in (v_0, \frac{1}{2}]$, and there is a final topological change which corresponds to the completion of the manifold. Note that the number of critical levels l_c in the interval $[v_0, \frac{1}{2}]$ grows with N and that eventually the set of l_c becomes dense in $[v_0, \frac{1}{2}]$ in the limit $N \to \infty$. However, the maximum critical value v_c remains isolated from other critical levels l_c.

According to (6.31), the accessible configuration space in the two-dimensional (m_x, m_y)-plane is not the whole plane, but only the disk D defined by

$$D = [(m_x, m_y) : m_x^2 + m_y^2 \leq 1].$$

Instead of M_v, the submanifold D_v is defined by

$$D_v = [(m_x, m_y) \in D : \bar{V}(m_x, m_y) \leq v].$$

The sequence of topological changes undergone by D_v is very simplified in the limit $h \to 0$.

The D_v is empty as long as $v < 0$ (since $h \to 0$ is assumed). The submanifold D_v first appears as the circle $m_x^2 + m_y^2 = 1$, i.e. the boundary circle of D (Fig. 6.3). Then as v grows, D_v becomes an annulus (a ring)

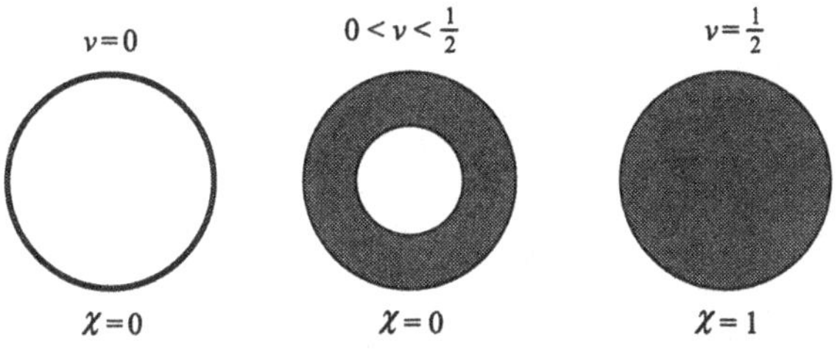

Fig. 6.3. Sequence of topological changes of D_v with increasing v (in the limit $h \to 0$). [CPC00].

bounded by two circles of radii 1 and $1 - 2v$:

$$1 - 2v \leq m_x^2 + m_y^2 \leq 1.$$

Inside it, there is a disk-hole of radius $1 - 2v$. As v continues to grow (until $\frac{1}{2}$), the hole shrinks and is eventually filled completely at $v = v_c = \frac{1}{2}$ (Fig. 6.3). In this coarse-grained two-dimensional description, all the topological changes that occur between $v = 0$ and $\frac{1}{2}$ disappear. Only two topological changes occur at $v = 0$ and $v_c = \frac{1}{2}$.

The topological change at v_c is characterized by the change of Euler characteristic χ from the value of an annulus, $\chi(\text{annulus}) = 0$, to the value of a disk, $\chi(\text{disk}) = 1$ (see §2.10). Despite the enormous reduction of dimensions to only two, there still exists a change of topology. This topological change is related to the thermodynamic *phase transition* of the mean-field XY model.

Part III

Flows of Ideal Fluids

Chapter 7

Gauge Principle and Variational Formulation

7.1. Introduction: Fluid Flows and Field Theory

Fluid mechanics is, in a sense, a theory of field of fluid flows in Newtonian mechanics. In other words, it is a *field theory of mass flow* subject to Galilei transformation.

There are various similarities between fluid mechanics and electromagnetism. For instance, the functional relation between the velocity and vorticity field is the same as the Biot–Savart law in electromagnetism between the magnetic field and electric current [Saf92]. One may ask whether the similarity is merely an analogy, or has a solid theoretical background.

In the theory of *gauge field*, a guiding principle is that laws of physics should be expressed in a form that is independent of any particular coordinate system. In §7.3, before the study of fluid flows, we review the scenario of the gauge theory in quantum field theory and particle physics [Fra97; Qui83; AH82]. According to the scenario, a free-particle Lagrangian is defined first in such a way as having an invariance under Lorenz transformation. Next, a gauge principle is applied to the Lagrangian, requiring it to have a *symmetry*, i.e. the gauge invariance. Thus, a gauge field such as the elecromagnetic field is introduced to satisfy *local* gauge invariance.

There are obvious differences between the fluid-flow field and the quantum field. Firstly, the field of fluid flow is nonquantum, which however causes no problem since the gauge principle is independent of the quantization principle. In addition, the fluid flow is subject to the Galilei transformation instead of Lorenz transformation. This is not an obstacle because the former is a limiting transformation of the latter as the relative ratio of flow velocity to the light speed tends to zero. Thirdly, relevant gauge groups should be different. Certainly, we have to find appropriate guage groups for fluid flows. A translation group and a rotation group will be shown as such groups relevant to fluid flows.

Here, we seek a scenario which has a formal equivalence with the gauge theory in the quantum field theory. We first review the Lagrangian and variational principle of fluid flows in §7.2. In order to go further over a mere analogy of the flow field to the gauge field, we define, in §7.4 and subsequent sections, a Galilei-invariant Lagrangian for fluid flows and examine whether it has gauge invariances in addition to the Galilei invariance. Then, based on the guage principle with respect to the *translation group*, we deduce the equation of motion from a variational formulation. However, the velocity field obtained by this gauge group is *irrotational*, i.e. a potential flow.

In §7.11 and subsequent sections, we consider an additional formulation with respect to the gauge group SO(3), a *rotation group* in three-dimensional space. It will be shown that the new gauge transformation introduces a *rotational* component in the velocity field (i.e. vorticity), even though the original field is irrotational. In complying with the local gauge invariance, a gauge-covariant derivative is defined by introducing a new gauge field. Galilei invariance of the covariant derivative requires that the gauge field should coincide with the *vorticity*. As a result, the covariant derivative of velocity is found to be the so-called *material* derivative of velocity, and thus the Euler's equation of motion for an ideal fluid is derived from the Hamilton's principle.

If we have a gauge invariance for the Lagrangian of a system, i.e. if we have a symmetry group of transformation, then we must have an equation of the form $\partial_\alpha J^\alpha = 0$ from the local gauge symmetry [Uti56; SS77; Qui83; Fra97], where J is a conserved current 4-vector (or tensor), i.e. a Noether current. This is called the *Noether's theorem* for local symmetry. Corresponding to a gauge invariance, the Noether's theorem leads to a conservation law. In fact, the gauge symmetry with respect to the translation group results in the conservation law of momentum, while the symmetry with respect to the rotation group results in the conservation of angular momentum. (See §7.15.5, 6.)

In addition, the Lagrangian has a symmetry with respect to particle permutation, which leads to a local law of vorticity conservation, i.e. the vorticity equation as well as the Kelvin's circulation theorem. Thus, the well-known equations in fluid mechanics are related to various symmetries of the Lagrangian.

7.2. Lagrangians and Variational Principle

7.2.1. *Galilei-invariant Lagrangian*

The field of fluid flow is subject to Galilei transformation, whereas the quantum field is subject to Lorentz transformation. Galilei transformation is considered to be a limiting transformation of the Lorentz transformation of space-time $(x^\mu) = (ct, \boldsymbol{x})$ as $v/c \to 0$.[1] In the Lorentz transformation between two frames $(t, \boldsymbol{x})$ and $(t', \boldsymbol{x}')$, a line-element of world-line is a vector represented in the form, $\mathrm{d}s = (c\mathrm{d}t, \mathrm{d}\boldsymbol{x})$, and its length $|\mathrm{d}s| = \langle \mathrm{d}s, \mathrm{d}s\rangle_{\mathrm{Mk}}^{1/2}$ is a *scalar*, namely a Lorentz-invariant:

$$\langle \mathrm{d}s, \mathrm{d}s\rangle_{\mathrm{Mk}} = -c^2\mathrm{d}t^2 + \langle \mathrm{d}\boldsymbol{x}, \mathrm{d}\boldsymbol{x}\rangle = -c^2(\mathrm{d}t')^2 + \langle \mathrm{d}\boldsymbol{x}', \mathrm{d}\boldsymbol{x}'\rangle, \tag{7.1}$$

where the light speed c is an invariant (e.g. [Chou28]). Scalar product of a 4-momentum $P = (E/c, \boldsymbol{p})$ of a particle of mass m with the line element $\mathrm{d}s$ is given by

$$\begin{aligned}\langle P, \mathrm{d}s\rangle_{\mathrm{Mk}} &= -\frac{E}{c}c\,\mathrm{d}t + \boldsymbol{p}\cdot\mathrm{d}\boldsymbol{x} = (\boldsymbol{p}\cdot\dot{\boldsymbol{x}} - E)\mathrm{d}t \qquad (= \Lambda\mathrm{d}t)\\ &= m(v^2 - c^2)\mathrm{d}t = -m_0c^2\mathrm{d}\tau,\end{aligned} \tag{7.2}$$

where m_0 is the rest mass, $\boldsymbol{v}$ and $\boldsymbol{p}(:= m\boldsymbol{v})$ are 3-velocity and 3-momentum of the particle respectively [Fra97; LL75], and

$$\begin{aligned}&E = mc^2, \quad m = \frac{m_0}{(1-\beta^2)^{1/2}}, \quad \beta = \frac{v}{c},\\ &\mathrm{d}\boldsymbol{x} = \boldsymbol{v}\mathrm{d}t, \quad \mathrm{d}\tau = (1-\beta^2)^{1/2}\mathrm{d}t \quad \text{(proper time)}.\end{aligned}$$

Either the leftmost side or rightmost side of (7.2) is obviously a scalar, i.e. an invariant with respect to the Lorentz transformation, and $\Lambda = \boldsymbol{p}\cdot\dot{\boldsymbol{x}} - E$ is what is called the Lagrangian in *Mechanics*. Hence it is found that either of the five expressions of (7.2), denoted as $\Lambda\mathrm{d}t$, might be taken as the integrand of the action $\underline{I}$ to be defined below.

Next, we consider a *Lorentz-invariant* Lagrangian $\Lambda_{\mathrm{L}}^{(0)}$ in the limit as $\beta = v/c \to 0$, and seek its appropriate counterpart Λ_{G} in the Galilei system. In this limit, the mass m and energy mc^2 are approximated by m_0 and $m_0(c^2 + \frac{1}{2}v^2 + \epsilon)$ respectively (neglecting $O(\beta^2)$ terms) in a macroscopic fluid system ([LL87], §133), where ϵ is the internal energy per unit fluid mass

[1] Spatial components, e.g. $\boldsymbol{x}$, $\boldsymbol{p}$ (called 3-vector), are denoted by bold letters, with their scalar product being written such as $\langle \boldsymbol{p}, \boldsymbol{x}\rangle$. In addition, a scalar product $\langle\cdot,\cdot\rangle_{\mathrm{Mk}}$ is defined with the Minkowski metric $g_{ij} = \mathrm{diag}(-1, 1, 1, 1)$.

which is a function of density ρ and entropy s in a single phase in general. The first expression of the second line of (7.2) is, then asymptotically,

$$\begin{aligned} mv^2 - mc^2 &\Rightarrow m_0 v^2 - m_0(c^2 + \tfrac{1}{2}v^2 + \epsilon) \\ &= (\rho \mathrm{d}^3 x)\big(\tfrac{1}{2}v^2 - \epsilon - c^2\big), \end{aligned}$$

where m_0 is replaced by $\rho(x)\mathrm{d}^3 x$. Thus, the Lagrangian $\Lambda_{\mathrm{L}}^{(0)}$ would be defined by

$$\Lambda_{\mathrm{L}}^{(0)} \mathrm{d}t = \int_M \mathrm{d}V(\boldsymbol{x})\rho(\boldsymbol{x})\Big(\tfrac{1}{2}\langle \boldsymbol{v}, \boldsymbol{v}\rangle - \epsilon - c^2\Big)\mathrm{d}t. \tag{7.3}$$

The third $-c^2\mathrm{d}t$ term is necessary so as to satisfy the Lorenz-invariance ([LL75], §87). It is obvious that the term $\langle \boldsymbol{v}, \boldsymbol{v}\rangle$ is not invariant with the Galilei transformation, $\boldsymbol{v} \mapsto \boldsymbol{v}' = \boldsymbol{v} - \boldsymbol{U}$. Using the relations $\mathrm{d}\boldsymbol{x} = \boldsymbol{v}\mathrm{d}t$ and $\mathrm{d}\boldsymbol{x}' = \boldsymbol{v}'\mathrm{d}t = (\boldsymbol{v} - \boldsymbol{U})\mathrm{d}t'$ with respect to two frames of reference moving with a relative velocity $\boldsymbol{U}$, the invariance (7.1) leads to

$$\mathrm{d}t' = \mathrm{d}t + \left(\frac{1}{c^2}\big(- \langle \boldsymbol{v}, \boldsymbol{U}\rangle + \tfrac{1}{2}U^2\big) + O(\beta^4)\right)\mathrm{d}t \tag{7.4}$$

The second term of $O(\beta^2)$ makes the Lagrangian $\Lambda_{\mathrm{L}}^{(0)}\mathrm{d}t$ Lorenz-invariant exactly in the $O(\beta^0)$ terms in the limit as $\beta \to 0$. Note that this invariance is satisfied locally at space-time, as inferred from (7.2).

When we consider a fluid flow subject to the Galilei transformation, the following prescription is adopted. Suppose that the flow is investigated in a finite domain M in space. Then the third term including c^2 in the parenthesis of (7.3) gives a constant $c^2\mathcal{M}\mathrm{d}t$, where $\mathcal{M} = \int_M \mathrm{d}^3 x \rho(x)$ is the total mass in the domain M. In carrying out variation, the total mass $\mathcal{M}$ is fixed at a constant. Next, keeping this in mind implicitly [SS77], we define the Lagrangian Λ_{G} of a fluid motion in the Galilei system by

$$\Lambda_{\mathrm{G}}\mathrm{d}t = \mathrm{d}t \int_M \mathrm{d}V(\boldsymbol{x})\rho(\boldsymbol{x})\Big(\tfrac{1}{2}\langle \boldsymbol{v}, \boldsymbol{v}\rangle - \epsilon\Big). \tag{7.5}$$

Only when we need to consider its Galilei invariance, we use the Lagrangian $\Lambda_{\mathrm{L}}^{(0)}$. In the Lagrangian formulation of subsequent sections, local conservation of mass is imposed. As a consequence, the mass is conserved globally. Thus, the use of Λ_{G} is justified (except in the case when its Galilei invariance is required).

Under Galilei transformation from one frame x to another x_* which is moving with a velocity $\boldsymbol{U}$ relative to the frame x (Fig. 7.1), the four-vectors

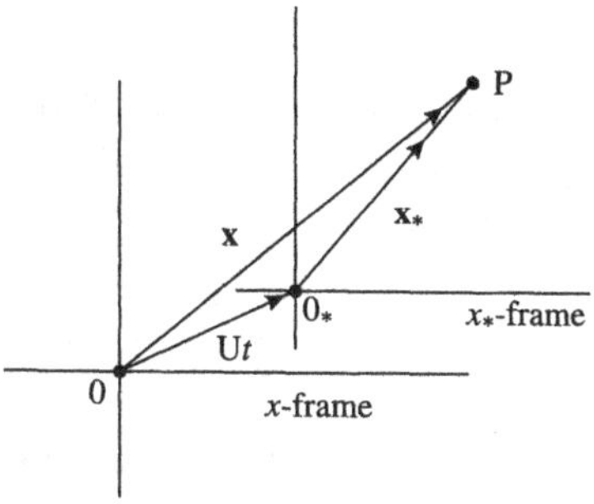

Fig. 7.1. Galilei transformation.

$x = (t, \boldsymbol{x})$ and $v = (1, \boldsymbol{v})$ are transformed as

$$x = (t, \boldsymbol{x}) \Rightarrow x_* = (t_*, \boldsymbol{x}_*) = (t, \boldsymbol{x} - \boldsymbol{U}t), \tag{7.6}$$

$$v = (1, \boldsymbol{v}) \Rightarrow v_* = (1, \boldsymbol{v}_*) = (1, \boldsymbol{v} - \boldsymbol{U}). \tag{7.7}$$

Since $\boldsymbol{v}_* = \boldsymbol{v} - \boldsymbol{U}$, the kinetic energy term $\frac{1}{2}\langle \boldsymbol{v}, \boldsymbol{v}\rangle$ is transformed as

$$\tfrac{1}{2}\langle \boldsymbol{v}_*, \boldsymbol{v}_*\rangle \Rightarrow \tfrac{1}{2}\langle \boldsymbol{v}, \boldsymbol{v}\rangle - \langle \boldsymbol{v}, \boldsymbol{U}\rangle + \tfrac{1}{2}U^2.$$

Second and third terms are written in the following form of time derivative, $(\mathrm{d}/\mathrm{d}t)[-\langle \boldsymbol{x}(t), \boldsymbol{U}\rangle + \frac{1}{2}U^2 t]$.[2] Transformation laws of derivatives are

$$\partial_t = \partial_{t_*} - \boldsymbol{U}\cdot\boldsymbol{\nabla}_*, \quad \boldsymbol{\nabla} = \boldsymbol{\nabla}_*, \tag{7.8}$$
$$\boldsymbol{\nabla} = (\partial_1, \partial_2, \partial_3), \qquad \partial_t = \partial/\partial t, \qquad \partial_k = \partial/\partial x^k.$$

Hence, we have the invariance of combined differential operators:

$$\partial_t + (\boldsymbol{v}\cdot\boldsymbol{\nabla}) = \partial_{t_*} + (\boldsymbol{v}_*\cdot\boldsymbol{\nabla}_*). \tag{7.9}$$

In the following, we reconsider the Lagrangian from the point of view of the gauge principle, and try to reconstruct the Lagrangian Λ_{G} of (7.5).

---

[2]It is understood in Newtonian mechanics that these additional terms do not play any role in the variational formulation of the action integral, where end values at two times are fixed in the variation of (7.5).

7.2.2. *Hamilton's variational formulations*

Variational principle is formulated in terms of the action functional $\underline{I}$ defined by the following integral of a Lagrangian L. The variational principle, i.e. principle of least action, is

$$\delta \underline{I} = 0, \quad \text{where } \underline{I} = \int_{t_0}^{t_1} L[\boldsymbol{v}, \rho, \cdots]\, \mathrm{d}t, \tag{7.10}$$

where the Lagrangian L is a functional depending on velocity field $\boldsymbol{v}$, density ρ, internal energy ϵ, etc.

In a system of point masses of classical mechanics, the Lagrangian L is usually defined by a sum of kinetic energies of each particle $\sum_a k_a$ and potential energy $-V$: $L = \sum_a k_a - V$, and the variational principle is called the *Hamilton's principle.*[3] However, for flows of a macroscopic continuous material such as a fluid, a certain generalization must be made. In addition, as already seen with (7.5), the Lagrangian must include a term of the internal energy ϵ in a form $\int \epsilon$, which will be given a certain gauge-theoretic meaning in the present formulation. As a consequence of the Hamilton's principle, we have the energy conservation equation.

The Hamilton's principle for an ideal fluid was formulated variously by [Heri55; Ser59; Eck60; SS77]. There are two main approaches, Lagrangian *particle* representation and Eulerian *spatial* representation (see the footnote to §7.6.1), which are reviewed by [Ser59; Bre70; Sal88]. The Lagrangian function in every case is composed of the terms of kinetic energy, internal energy and potential energy, with additional constraint conditions. (The potential energy term missing in (7.5) can be taken into account without difficulty.) However, it may be said that there are some complexity or incompleteness in those formulations made so far.

In the variational formulation with the Lagrangian particle representation, the equation of motion finally derived includes a term of particle acceleration $\mathrm{d}\boldsymbol{v}/\mathrm{d}t$ [Ser59; Sal88], where the time derivative $\mathrm{d}/\mathrm{d}t$ is replaced by $\partial_t + \boldsymbol{v} \cdot \nabla$ with an intuitive argument. When this time derivative acts on a scalar function, then it is given a definite sense. However, if it acts on vectors, we need a careful consideration. In fact, the Lie derivative of a vector $\boldsymbol{v}$ is different from $\mathrm{d}\boldsymbol{v}/\mathrm{d}t$ (§1.8.3). The derivative $\mathrm{d}\boldsymbol{v}/\mathrm{d}t = \partial_t \boldsymbol{v} + (\boldsymbol{v} \cdot \nabla)\boldsymbol{v}$

[3]Originally, Lagrange (*Mécanique Analytique*, 1788) extended the principle of virtual displacements, by applying the d'Alembert's principle, from static equilibrium systems to dynamical systems. The d'Alembert–Lagrange principle is expressed equivalently in the form of Hamilton's principle. [Ser59]

(accepted in the Euler's equation of motion) is defined as an extension of the operation on a scalar function. In the gauge theory which we are going to consider, we will arrive at $\mathrm{d}\boldsymbol{v}/\mathrm{d}t = \partial_t \boldsymbol{v} + (\boldsymbol{v}\cdot\nabla)\boldsymbol{v}$ from the gauge priciple.

On the other hand, it was pointed out by [Bre70] and [Sal88] that the connection between Hamilton's principle and the variation with the Eulerian representation is obscure. This vagueness is partly due to the fact that no mention is made of how the positions of massive particles in the physical system are related to the field variables in a standard Eulerian formulation, and that the system is described by the velocity field $\boldsymbol{u}(\boldsymbol{x}, t)$ at a fixed spatial point $\boldsymbol{x}$, regardless of which material particle is actually there. In other words, the equation of particle motion $\mathrm{d}\boldsymbol{x}_p/\mathrm{d}t = \boldsymbol{u}(\boldsymbol{x}_p(t), t)$ is not specified in the variation. Lin's constraint [Lin63; Ser59; Sal88] was introduced to bridge the gap between the two approaches, but it remains as local validity only in a neighborhood of the point under consideration, and in addition physical significance of the new fields (called potentials) thus introduced is not clear. In particular, for a homentropic fluid in which the entropy s takes a constant value everywhere, the velocity field is limited only to such a field as the vortex lines are not knotted [Bre70].

7.2.3. *Lagrange's equation*

Consider that we have a system of n point masses whose positions are denoted by $\boldsymbol{x}_1 = (q^1, q^2, q^3), \cdots, \boldsymbol{x}_n = (q^{3n-2}, q^{3n-1}, q^{3n})$, and velocities by $\boldsymbol{v}_1 = (q_t^1, q_t^2, q_t^3), \cdots, \boldsymbol{v}_n = (q_t^{3n-2}, q_t^{3n-1}, q_t^{3n})$, and that we have a Lagrangian L of the form,

$$L = L[q, q_t], \tag{7.11}$$

which depends on the coordinates $q = (q^i)$ and the velocities $q_t = (q_t^i)$ for $i = 1, 2, \cdots, N$, where $N = 3n$. It is said that the Lagrangian L describes a dynamical system of N degrees of freedom.

The action principle (7.10) may be written as

$$\delta \underline{I} = \int_{t_0}^{t_1} \delta L[q, q_t]\mathrm{d}t = 0. \tag{7.12}$$

We consider a variation to a reference curve $q(t)$, where the varied curve is written as $q'(t, \varepsilon) = q(t) + \varepsilon\xi(t)$ and $q'_t = \dot{q}(t) + \varepsilon\dot{\xi}(t)$ with an infinitesimal parameter ε by using a virtual displacement $\xi(t)$ satisfying

$\xi(t_0) = \xi(t_1) = 0$. Resulting variation of the Lagrangian is

$$\delta L = \frac{\partial L}{\partial q}\delta q + \frac{\partial L}{\partial q_t}\delta q_t = \frac{\partial L}{\partial q^i}\varepsilon\xi^i + \frac{\partial L}{\partial q_t^i}\varepsilon\dot{\xi}^i$$
$$= \varepsilon\left[\frac{\partial L}{\partial q^i} - \partial_t\left(\frac{\partial L}{\partial q_t^i}\right)\right]\xi^i + \varepsilon\partial_t\left(\frac{\partial L}{\partial q_t^i}\xi^i\right). \quad (7.13)$$

When this is substituted in (7.12), the second term vanishes because of the assumed conditions $\xi(t_0) = \xi(t_1) = 0$. Therefore, it is deduced that, for an arbitrary variation $\xi(t)$, the first term of δL must vanish at each time t. Thus, we obtain the *Euler–Lagrange equation*:

$$\partial_t\left(\frac{\partial L}{\partial q_t^i}\right) - \frac{\partial L}{\partial q^i} = 0. \quad (7.14)$$

If the Lagrangian is given by the following form of kinetic energy of the same mass m,

$$mL_{\mathrm{f}}(q_t) = \tfrac{1}{2}m\langle q_t, q_t\rangle, \quad (7.15)$$

then the above equation (7.14) describes *free motion* of point masses. In fact, Eq. (7.14) results in $\partial_t(q_t) = 0$, i.e. the velocity q_t is constant.

7.3. Conceptual Scenario of the Gauge Principle

Typical successful cases of the gauge theory are the Dirac equation or Yang–Mills equation in particle physics, which are reviewed here briefly for later purpose. A free-particle Lagrangian density Λ_{free}, e.g. for a free electron, is constructed so as to be invariant under the Lorenz transformation of space-time (x^μ) for $\mu = 0, 1, 2, 3$, where

$$\Lambda_{\mathrm{free}} = \bar{\psi}i\gamma^\mu\partial_\mu\psi - m\bar{\psi}\psi, \quad \bar{\psi} = \psi^\dagger\begin{pmatrix} I & 0 \\ 0 & -I \end{pmatrix}, \quad (7.16)$$

m is the mass, ψ is a Dirac wave function of four components for electron and positron with $\pm 1/2$ spins, $\psi^\dagger$ the hermitian conjugate of ψ, and γ^μ the Dirac matrices ($\mu = 0, 1, 2, 3$) with $x^0 = t$ (time), I being 2×2 unit matrix. In the Yang–Mills case, the wave functions are considered to represent two internal states of fermions, e.g. up and down quarks, or a lepton pair.

The above Lagrangian has a *symmetry* called a global gauge invariance. Namely, its form is invariant under the transformation of the wave function, e.g. $\psi \mapsto e^{i\alpha}\,\psi$ for an electron field. The term *global* means that the phase α

is a real constant, i.e. independent of coordinates. This keeps the probability density, $|\psi|^2$, unchanged.[4]

In addition, we should be able to have invariance under a *local* gauge transformation,

$$\psi(x) \mapsto \psi'(x) = e^{i\alpha(x)}\psi(x) := g(x)\psi(x), \tag{7.17}$$

where $\alpha = \alpha(x)$ varies with the space-time coordinates $x = (x^\mu)$. With this transformation too, the probability density $|\psi|^2$ obviously is not changed.

However, the free-particle Lagrangian Λ_{free} is not invariant under such a transformation because of the derivative operator $\partial_\mu = \partial/\partial x^\mu$ in Λ_{free}. This demands that some *background field* interacting with the particle should be taken into account, that is the *Electromagnetic field* or a *gauge field.* If a *new* gauge field term is included in the Lagrangian Λ, the local gauge invariance will be attained [Fra97, Ch.19 & 20; Qui83; AH82]. This is the *Weyl's principle of gauge invariance.*

If a proposed Lagrangian including a partial derivative of some matter field ψ is invariant under global gauge transformation as well as Lorentz transformation, but not invariant under local gauge transformation, then the Lagrangian is to be altered by replacing the partial derivative with a *covariant* derivative including a gauge field $\mathcal{A}(x)$, $\partial \to \nabla = \partial + \mathcal{A}(x)$, so that the Lagrangian Λ acquires local gauge invariance [Uti56]. The second term $\mathcal{A}(x)$ is also called *connection.* The aim of introducing the gauge field is to obtain a generalization of the gradient that transforms as

$$\nabla\psi \mapsto \nabla'\psi' = (\partial + \mathcal{A}')g(x)\psi = g(\partial + \mathcal{A})\psi = g\nabla\psi, \tag{7.18}$$

where $\psi' = g(x)\psi$. In dynamical systems which evolve with respect to the time coordinate t, the replacement $\partial \to \nabla = \partial + \mathcal{A}(x)$ is carried out only for the t derivative. This will be considered below again (§7.5.1).

Finally, the principle of least action is applied,

$$\delta \underline{I} = 0, \quad \underline{I} = \int_{M^4} \Lambda(\psi, \mathcal{A})\mathrm{d}^4 x$$

where M^4 is a certain (x^μ) space-time manifold, and $\underline{I}$ is the action functional. Let us consider two examples.

[4]The gauge invariance results in conservation of Noether current. See §7.4, 7.15.

(i) In the case of an electron field, the local gauge transformation is given by an element g of the unitary group $U(1)$,[5] i.e. $g(x) = e^{iq\alpha(x)}$ at every point x with $\alpha(x)$ a scalar function, and q a charge constant.[6] Transformation of the wave function ψ is

$$\psi'(x) = g(x)\psi(x) = e^{iq\alpha(x)}\psi(x). \tag{7.19}$$

The gauge-covariant derivative is then defined by

$$\nabla_\mu = \partial_\mu - i\,q\,\mathcal{A}_\mu(x), \tag{7.20}$$

where $\mathcal{A}_\mu = (-\varphi, \mathcal{A}_k)$ is the electromagnetic potential (4-vector potential with the electric potential φ and magnetic 3-vector potential $\mathcal{A}_k$, $k = 1, 2, 3$). The electromagnetic potential (connection term) transforms as

$$\mathcal{A}'_\mu(x) = \mathcal{A}_\mu(x) + \partial_\mu \alpha(x). \tag{7.21}$$

It is not difficult to see that this satisfies the relation (7.18), $\nabla'\psi' = g\nabla\psi$. Thus, the Dirac equation with an electromagnetic field is derived. (See Appendix I for its brief summary.)

(ii) In the second example of Yang–Mills's formulation of two-fermion field, the local gauge transformation of the form (7.17) is given with $g(x) \in SU(2)$.[7] Consider an infinitesimal gauge transformation written as

$$g(x) = \exp\left[iq\boldsymbol{\sigma}\cdot\boldsymbol{\alpha}(x)\right] = I + iq\boldsymbol{\sigma}\cdot\boldsymbol{\alpha}(x) + O(|\boldsymbol{\alpha}|^2), \quad |\boldsymbol{\alpha}| \ll 1, \tag{7.22}$$

where $\boldsymbol{\sigma}\cdot\boldsymbol{\alpha} = \sigma_1\alpha^1 + \sigma_2\alpha^2 + \sigma_3\alpha^3$ with real functions $\alpha^k(x)$ $(k = 1, 2, 3)$, and $\boldsymbol{\sigma} = (\sigma_1, \sigma_2, \sigma_3)$ are the Pauli matrices,

$$\sigma_1 = \begin{bmatrix} 0 & 1 \\ 1 & 0 \end{bmatrix}, \quad \sigma_2 = \begin{bmatrix} 0 & -i \\ i & 0 \end{bmatrix}, \quad \sigma_3 = \begin{bmatrix} 1 & 0 \\ 0 & -1 \end{bmatrix}, \tag{7.23}$$

---

[5] The unitary group $U(1)$ is a group of complex numbers $z = e^{i\theta}$ of absolute value 1.
[6] $q = e/(\hbar c)$ with c the light-speed, e the electric-charge, and $\hbar$ the Planck constant.
[7] $SU(2)$ is the special unitary group, consisting of complex 2×2 matrices $g = (g_{ij})$ with $\det g = 1$. The hermitian conjugate $g^\dagger = (g^\dagger_{ij}) = (\bar{g}_{ji})$ is equal to g^{-1} where the overbar denotes complex conjugate. Its Lie algebla $su(2)$ consists of skew hermitian matrices of trace 0.

composing a basis of the algebra $su(2)$ which is considered as a *real* three-dimensional (3D) vector space.[8] The commutation relations are given by[9]

$$[\sigma_j, \sigma_k] = 2i\epsilon_{jkl}\sigma_l, \tag{7.24}$$

The second term on the right-hand side of (7.22) is a generator of an infinitesimal gauge transformation. The gauge-covariant derivative is represented by

$$\nabla_\mu = \partial_\mu - iq\boldsymbol{\sigma} \cdot \boldsymbol{A}_\mu(x), \tag{7.25}$$

where the connection consists of three terms, $\boldsymbol{\sigma} \cdot \boldsymbol{A}_\mu = \sigma_1 A^1_\mu + \sigma_2 A^2_\mu + \sigma_3 A^3_\mu$ in accordance with the three-dimensionality of $su(2)$, and q the coupling constant of interaction. The connection $\boldsymbol{A}_\mu = (A^1_\mu, A^2_\mu, A^3_\mu)$ transforms as

$$\boldsymbol{A}'_\mu = \boldsymbol{A}_\mu - 2q\boldsymbol{\alpha} \times \boldsymbol{A}_\mu + \partial_\mu \boldsymbol{\alpha}, \tag{7.26}$$

instead of (7.21). The three gauge fields $\boldsymbol{A}^k_{\mathrm{YM}} = (A^k_0, A^k_1, A^k_2, A^k_3)$ for $k = 1, 2, 3$ (3D analogues of the electromagnetic potential, associated with three colors) are thus introduced and called the Yang–Mills gauge fields. A characteristic feature distinct from the previous electrodynamic case, compared with (7.21), is the *non-abelian* nature of the algebra $su(2)$, represented by the new second term of (7.26) arising from the *non-commutativity* of the gauge transformations, i.e. the commutation rule (7.24).

In the subsequent sections, we consider fluid flows and try to formulate the flow field on the basis of the *generalized* gauge principle.[10] It will be found that the flow fields are characterized by two gauge groups: a translation group and a rotation group. Interestingly, the former is abelian and the latter is non-abelian. So, the flow fields are governed by two different transformation laws.

---

[8]Vector space $su(2)$ is closed under multiplication by real numbers α^k, e.g. [Fra97].

[9]The structure constant ϵ_{jkl} takes 1 or -1 according to (jkl) being an even or odd permutation of (123), and 0 if (jkl) is not a permutation of (123).

[10]Gauge symmteries are often referred to as internal symmetries because gauge transformations correspond to rotation in an internal space. In the general gauge theory [Uti56], however, gauge invariance with respect to Poincaré transformation of space-time yields the gravitational field of general relativity theory. The present study of fluid flows considers gauge invariance with respect to Galilei transformation (translation invariance). See §7.5, 6, 7.

7.4. Global Gauge Transformation

Suppose that our system is characterized by a symmetry group of transformation (of a Lie group G), and the action is invariant under an associated infinitesimal transformation [Uti56]:

$$\left.\begin{aligned} &q(t) \to q(t) + \delta q \\ &\delta q^i = \xi^\alpha T^i_{\alpha j}\, q^j, \qquad\qquad \delta q_t = \partial_t(\delta q), \end{aligned}\right\} \tag{7.27}$$

where T_α ($\alpha = 1, \cdots, K$) are generators of the group G (represented in a matrix form $T^i_{\alpha_j}$), i.e. Lie algebra of dimension K (say), and ξ^α are infinitesimal variation parameters. An example of T_α is given by (7.40). In terms of the group theory (§1.8), the operators T_α are elements of the Lie algebra $\mathbf{g} = T_eG$, and in general satisfy the commutation relation,

$$[T_\alpha, T_\beta]^i_j = C^\gamma_{\alpha\beta} T^i_{\gamma\, j},$$

where $C^\gamma_{\alpha\beta}$ are the *structure constants*.

If ξ^α ($\alpha = 1, \ldots, K$) are constants, i.e. if the transformation is *global*, then invariance of I under the transformation results in

$$0 \equiv \delta L = \frac{\partial L}{\partial q^i}\delta q^i + \frac{\partial L}{\partial q^i_t}\, \xi^\alpha T^i_{\alpha\, j}\partial_t q^j \tag{7.28}$$

$$= \left[\frac{\partial L}{\partial q^i} - \partial_t\left(\frac{\partial L}{\partial q^i_t}\right)\right]\delta q^i + \partial_t\left(\frac{\partial L}{\partial q^i_t}\delta q^i\right). \tag{7.29}$$

Using δq^i of (7.27), vanishing of the right-hand side of (7.28) gives

$$\frac{\partial L}{\partial q^i} T^i_{\alpha j} q^j + \frac{\partial L}{\partial q^i_t} T^i_{\alpha j}\partial_t q^j = 0, \tag{7.30}$$

since the parameters ξ^α can be chosen arbitrarily. The first term of (7.29) vanishes owing to the Euler–Lagrange equation (7.14). Hence the second term must vanish identically, and we obtain the *Noether's theorem* for the global invariance:

$$\partial_t\left(\frac{\partial L}{\partial q^i_t}\delta q^i\right) = 0 \Rightarrow \partial_t\left(\frac{\partial L}{\partial q^i_t} T^i_{\alpha j} q^j\right) = 0. \tag{7.31}$$

7.5. Local Gauge Transformation

When we consider local gauge transformation, the physical system under consideration is to be modified. Concept of local transformation allows us to consider a continuous field, rather than the original discrete system. We replace the discrete variables q^i by continuous parameters $\boldsymbol{a} = (a^1, a^2, a^3)$ to represent continuous distribution of particles in a three-dimensional Euclidean space M. Spatial position $\boldsymbol{x} = (x^1, x^2, x^3)$ of each *mass* particle of label $\boldsymbol{a}$ (Lagrange parameter) is denoted by $x_a^k(\boldsymbol{a}, t)$, a function of $\boldsymbol{a}$ as well as the time t. Conversely, the particle locating at the point $\boldsymbol{x}$ at a time t is denoted by $a^k(\boldsymbol{x}, t)$. Functions $x_a^k(\boldsymbol{a}, t)$ or $a^k(\boldsymbol{x}, t)$ $(k = 1, 2, 3)$ may be taken as field variables.

Thus, we consider a continuous distribution of mass, i.e. *fluid*, and its motion. The variation field is represented by differentiable functions $\xi^\alpha(\boldsymbol{x}, t)$. Suppose that we have a *local* transformation expressed by

$$\left.\begin{aligned} &q = \boldsymbol{x}_a \to q' = \boldsymbol{x}_a + \delta q, && \delta q^i = \xi^\alpha T^i_{\alpha j} x^j, \\ &q_t = \partial_t \boldsymbol{x}_a \to q'_t = \partial_t \boldsymbol{x}_a + \delta q_t, && \delta q_t = \partial_t(\delta q), \\ &\xi^\alpha = \xi^\alpha(\boldsymbol{x}, t), && \end{aligned}\right\} \tag{7.32}$$

and examine local gauge invariance, i.e. gauge invariance at each point $\boldsymbol{x}$ in the space. In this case, the variation of $L[q, q_t]$ is

$$\delta L = \left[\frac{\partial L}{\partial q^i} - \partial_t\left(\frac{\partial L}{\partial q^i_t}\right)\right]\delta q^i + \partial_t\left(\frac{\partial L}{\partial q^i_t}\delta q^i\right). \tag{7.33}$$

This does not vanish owing to the arbitrary function $\xi^\alpha(\boldsymbol{x}, t)$ depending on time t. In fact, using the Euler–Lagrange equation (7.14), we have

$$\delta L = \partial_t\left(\frac{\partial L}{\partial q^i_t}\delta q^i\right) = \frac{\partial L}{\partial q^i_t}\partial_t \xi^\alpha(\boldsymbol{x}, t) T^i_{\alpha j} x^j, \tag{7.34}$$

where (7.31) is used. Note that this vanishes in the global transformation since $\partial_t \xi = 0$.

7.5.1. *Covariant derivative*

According to the gauge principle, the nonvanishing of δL is understood such that there exists some *background field* interacting with flows of a fluid, and that a *new* field $\mathcal{A}$ must be taken into account in order to achieve the local gauge invariance of the Lagrangian (i.e. in order to have vanishing of (7.34)). To that end, the partial time derivative ∂_t must be replaced by a covariant derivative D_t. The covariant derivative is defined by

$$\mathrm{D}_t = \partial_t + \mathcal{A}, \tag{7.35}$$

where $\mathcal{A}$ is a gauge-field operator. Correspondingly, the time derivatives $\partial_t \xi$ and $\partial_t q$ are replaced by

$$\mathrm{D}_t \xi = \partial_t \xi + \mathcal{A}\xi, \qquad u := \mathrm{D}_t q = \partial_t q + \mathcal{A}q. \tag{7.36}$$

In most dynamical systems like the present case, the *time derivative* is the primary concern in the analysis of local gauge transformation.

7.5.2. *Lagrangian*

The Lagrangian mL_{f} of (7.15) is replaced by

$$L_{\mathrm{F}} = \tfrac{1}{2} \int \langle \mathrm{D}_t q, \mathrm{D}_t q \rangle \, \mathrm{d}^3 \boldsymbol{a} = \tfrac{1}{2} \int \langle u, \, u \rangle \, \rho \, \mathrm{d}^3 \boldsymbol{x}, \tag{7.37}$$

where $\mathrm{d}^3 \boldsymbol{a} = \rho \, \mathrm{d}^3 \boldsymbol{x}$ denotes the mass in a volume element $\mathrm{d}^3 \boldsymbol{x}$ of the $\boldsymbol{x}$-space with ρ as the *mass-density* (see §7.8.1), and the integrand is $L'(q, q_t, \mathcal{A}) = \frac{1}{2} \langle \mathrm{D}_t q, \mathrm{D}_t q \rangle$.

In fact, assuming that the Lagrangian density is of the form $L'(q, q_t, \mathcal{A})$, the invariance postulate demands

$$\delta L' = \frac{\partial L'}{\partial q} \delta q + \frac{\partial L'}{\partial q_t} \delta q_t + \frac{\partial L'}{\partial \mathcal{A}} \delta \mathcal{A} = 0.$$

Suppose that the variations are given by the following forms:

$$\left. \begin{aligned} \delta q^i &= \xi^\alpha T_\alpha q^i, \\ \delta q^i_t &= \delta(\partial_t q)^i = \partial_t(\delta q^i) - (\mathrm{D}_t \xi)^\alpha T_\alpha q^i \\ &= \xi^\alpha T_\alpha q^i_t - (\mathcal{A}\xi)^\alpha T_\alpha q^i, \\ \delta \mathcal{A}^i &= \xi^\alpha T_\alpha \mathcal{A}^i + (\mathcal{A}\xi)^i, \end{aligned} \right\} \tag{7.38}$$

(see the footnote[11])where $\mathcal{A}$ is assumed to be represented as $\mathcal{A} = \mathcal{A}^k T_k$ and $\mathrm{D}_t \xi = \partial_t \xi + \mathcal{A}\,\xi$. It will be seen that these variations are consistent with the translational transformation (Fig. 7.2) considered in §7.7. Substituting

[11]According to the local Galilei transformation in §7.7, we have $\delta(\partial_t q) = [\partial_t - (\mathrm{D}_t\xi)^\alpha T_\alpha](q + \delta q) - \partial_t q = \partial_t(\delta q^i) - (\mathrm{D}_t \xi)^\alpha T_\alpha \, q^i + O(|\delta q|^2)$ (see (7.8)). The form of δA^i determines the character of the present gauge field A.

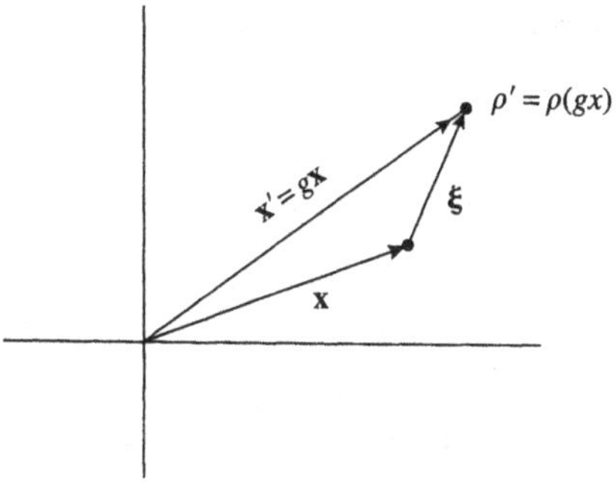

Fig. 7.2. Translational transformation:
$\boldsymbol{x} \to \boldsymbol{x}' = g\boldsymbol{x} := \boldsymbol{x} + \boldsymbol{\xi}$, $\quad \boldsymbol{u} \to \boldsymbol{u}' = \boldsymbol{u}(g\boldsymbol{x}) + \mathrm{D}_t\boldsymbol{\xi}$.
$\delta x^i = \xi^i = \xi^\alpha T_\alpha x^i$, $\quad \delta u^i = \xi^\alpha T_\alpha u^i + \mathrm{D}_t \xi^i$.

these, we obtain a variational equation for arbitrary functions of $\xi(\boldsymbol{x}, t)$ and $\mathcal{A}\xi(\boldsymbol{x}, t)$. Requiring the coefficient of $(\mathcal{A}\xi)^k$ to vanish, we have

$$-\frac{\partial L'}{\partial q_t^i} T_k q^i + \frac{\partial L'}{\partial \mathcal{A}^k} = 0. \tag{7.39}$$

From this, it is found that $\mathcal{A}$ should be contained in L' only through the combination:

$$q_t + \mathcal{A}^k T_k q = \partial_t q + \mathcal{A} q = \mathrm{D}_t q,$$

confirming the second expression of (7.36). This implies that the operators ∂_t and $\mathcal{A} = \mathcal{A}^k T_k$ are to be combined as $\mathrm{D}_t = \partial_t + \mathcal{A}^k T_k = \partial_t + \mathcal{A}$. Thus, the expression (7.35) has been found. In case that L' does not include q explicitly, the above result implies that L_{F} of (7.37) is a possible Lagrangian.

Before carrying out variational formulation with a complete Lagrangian, we must first define the variation field $\xi(\boldsymbol{x}, t)$ which should be the subject of a certain consistency condition considered in §7.6.3. Next, we determine the form of the gauge operator $\mathcal{A}$ in §7.7, and then define a Lagrangian L_A associated with the gauge-field $\mathcal{A}$ in §7.8 for translational invariance.

7.6. Symmetries of Flow Fields

It is readily seen that the Lagrangian (7.15) of point masses has symmetries with respect to two transformation groups, a translation group and rotation group. Lagrangian of flows of an ideal fluid has the same properties as the point-mass system globally. Local gauge invariance of such a system is one of primary concerns in the study of fluid flows.

7.6.1. *Translational transformation*

First, we consider *translational* transformation.[12] The coordinates q^i are regarded as the spatial coordinates x^α ($\alpha = 1, 2, 3$) where x^α are continuous variables, and q^i_t are taken as the velocity components u^α of fluid flow. The operator of translational transformation is given by $T_\alpha = \partial/\partial x^\alpha$, denoted also as ∂_α. This is rewritten in a matrix form,

$$T^i_{\alpha,j} = \delta^i_j \partial_\alpha, \tag{7.40}$$

so as to have the same notations as those in §7.4, where $T^i_{\alpha,j}$ is the (i, j) entry of the matrix operator T_α and δ^i_j the Kronecker delta. (This form can be applied to the other case as well, i.e. the rotational transformation to be considered next.) Then, variations of q^i and q^i_t (Fig. 7.2) are defined by

$$\delta x^i = \xi^\alpha T^i_{\alpha,j} x^j = \xi^\alpha T_\alpha\, x^i = (\boldsymbol{\xi} \cdot \nabla)\, x^i = \xi^i, \tag{7.41}$$

$$\delta u^i = (\boldsymbol{\xi} \cdot \nabla) u^i + \mathrm{D}_t \xi^i. \tag{7.42}$$

The last term $\mathrm{D}_t \xi^i$ is a characteristic term related to fluid flows.

The generators are commutative, i.e. the commutator is given by

$$[T_\alpha, T_\beta] = \partial_\alpha \partial_\beta - \partial_\beta \partial_\alpha = 0.$$

Hence all the structure constants $C^\gamma_{\alpha\beta}$ vanish, i.e. abelian.

7.6.2. *Rotational transformation*

When we consider local rotation of a fluid element about a reference point $\boldsymbol{x} = (x^1, x^2, x^3)$, attention is directed to relative motion at neighboring points $\boldsymbol{x} + \boldsymbol{s}$ for small $\boldsymbol{s}$. Mathematically speaking, at each point within the fluid, local rotation is represented by an element of the rotation group SO(3) (a Lie group). Local infinitesimal rotation is described by the generators of

---

[12] When the author (Kambe) discussed rotational gauge invariance of fluid flows and its noncommutative property (in China, 2002), Professor MoLin Ge (Nankai Institute of Mathematics) suggested that there would be another gauge symmetry of commutative property which is the subject of this subsection.

SO(3), i.e. Lie algebra **so**(3) of three dimensions. The basis vectors of the space of **so**(3) are denoted by (e_1, e_2, e_3):

$$e_1 = \begin{bmatrix} 0 & 0 & 0 \\ 0 & 0 & -1 \\ 0 & 1 & 0 \end{bmatrix}, \qquad e_2 = \begin{bmatrix} 0 & 0 & 1 \\ 0 & 0 & 0 \\ -1 & 0 & 0 \end{bmatrix}, \qquad e_3 = \begin{bmatrix} 0 & -1 & 0 \\ 1 & 0 & 0 \\ 0 & 0 & 0 \end{bmatrix}.$$

which satisfy the *non-commutative* relations:

$$[e_\alpha, e_\beta] = e_\alpha e_\beta - e_\beta e_\alpha = \epsilon_{\alpha\beta\gamma} e_\gamma \tag{7.43}$$

where $\epsilon_{\alpha\beta\gamma}$ is the third order completely skew-symmetric tensor.

A rotation operator is defined by $\theta = (\theta^i_j) = \theta^\alpha\, e_\alpha$ where $\hat{\theta} = (\theta^1, \theta^2, \theta^3)$ is an infinitesimal angle vector. Then, an infinitesimal rotation of the displacement vector $\boldsymbol{s} = (s^1, s^2, s^3)$ is expressed as $\theta^i_j s^j$. The same rotation is represented by a vector product as well:

$$(\hat{\theta} \times \boldsymbol{s})^i = \theta^i_j s^j = \theta^\alpha\, e^i_{\alpha,j}\, s^j. \tag{7.44}$$

The left-hand side clearly says that this represents a rotation of an infinitesimal angle $|\hat{\theta}|$ around the axis in the direction coinciding with the vector $\hat{\theta}$. Thus the operator of rotational transformation is given by $T_\alpha = e_\alpha$, where $T^i_{\alpha,j}$ is the (i,j)-th entry of the matrix operator e_α.

In the rotational transformation, the generalized coordinte q^i is replaced by the displacement s^k from the reference point $\boldsymbol{x}$, and q^i_t is the velocity at neighboring points $\boldsymbol{x} + \boldsymbol{s}$ with $q_t(\boldsymbol{s} = 0)$ being $\boldsymbol{u}(\boldsymbol{x})$. Hence, we have

$$q^i = s^i, \quad q^i_t = \mathrm{D}_t q^i(\boldsymbol{s}) = u^i(\boldsymbol{x}) + s^j\, \partial_j u^i(\boldsymbol{x}). \tag{7.45}$$

Then, variations of q^i and q^i_t are defined by

$$\delta q^i = \theta^\alpha\, e^i_{\alpha\, j}\, s^j = \hat{\theta} \times \boldsymbol{s}, \tag{7.46}$$

$$\delta q^i_t = \mathrm{D}_t(\delta q^i) = \mathrm{D}_t\big(\theta^\alpha\, e^i_{\alpha\, j}\, s^j\big). \tag{7.47}$$

7.6.3. *Relative displacement*

We consider local deformation of a fluid element about a reference point $\boldsymbol{x}$, and direct our attention to relative motion of neighboring points $\boldsymbol{x} + \boldsymbol{s}$ for small $\boldsymbol{s}$. Suppose that the point $\boldsymbol{x}$ is displaced to a new position $\boldsymbol{X}(\boldsymbol{x}) = \boldsymbol{x} + \boldsymbol{\xi}(\boldsymbol{x})$. Correspondingly, a neighboring point $\boldsymbol{x} + \boldsymbol{s}$ is displaced to $\boldsymbol{X}(\boldsymbol{x} + \boldsymbol{s})$.

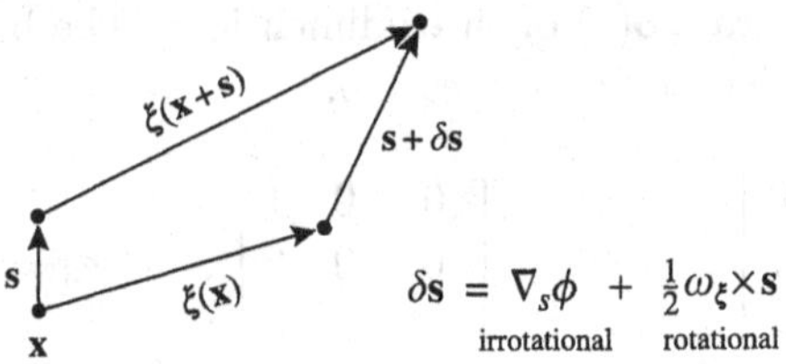

Fig. 7.3. Relative local motion.

Relative variation $\delta \boldsymbol{s}$ is defined by

$$\delta \boldsymbol{s} := \boldsymbol{X}(\boldsymbol{x}+\boldsymbol{s}) - [\boldsymbol{X}(\boldsymbol{x}) + \boldsymbol{s}] = (\boldsymbol{s}\cdot\nabla)\boldsymbol{\xi}(\boldsymbol{x}) + O(|\boldsymbol{s}|^2) \tag{7.48}$$

(Fig. 7.3). To the first order of $|\boldsymbol{s}|$, we have

$$\delta \boldsymbol{s} = \nabla_s \phi + \tfrac{1}{2}\boldsymbol{\omega}_\xi \times \boldsymbol{s}, \tag{7.49}$$

[Bat67, §2.3], where the first term (which came from the symmetric part of $\partial_j \xi^k$) is a potential part of $\delta \boldsymbol{s}$ with the potential $\phi(\boldsymbol{x},\boldsymbol{s}) = \frac{1}{2}e_{jk}(\boldsymbol{x})s^j s^k$ (where $e_{jk}(\boldsymbol{x}) = \frac{1}{2}[\partial_j \xi^k(\boldsymbol{x}) + \partial_k \xi^j(\boldsymbol{x})]$). The second term (which came from the anti-symmetric part of $\partial_j \xi^k$) represents a rigid-body rotation with an infinitesimal rotation angle $\frac{1}{2}\boldsymbol{\omega}_\xi$, where

$$\boldsymbol{\omega}_\xi(\boldsymbol{x}) = \operatorname{curl}\boldsymbol{\xi}(\boldsymbol{x})$$

i.e. curl of the displacement vector $\boldsymbol{\xi}(\boldsymbol{x})$. Requiring that the transformation should be *commutative*, we must have $\operatorname{curl}\boldsymbol{\xi}(\boldsymbol{x}) = 0$. Therefore, we have

$$\operatorname{curl}\boldsymbol{\xi}(\boldsymbol{x}) = 0, \qquad \boldsymbol{\xi}(\boldsymbol{x}) = \operatorname{grad}\varphi. \tag{7.50}$$

A vector field $\boldsymbol{\xi}(\boldsymbol{x})$ satisfying $\operatorname{curl}\boldsymbol{\xi}(\boldsymbol{x}) = 0$ is said to be *irrotational*. For such an irrotational field, there always exists a certain scalar function φ, and $\boldsymbol{\xi}$ is represented as above.

7.7. Laws of Translational Transformation

7.7.1. *Local Galilei transformation*

For the translational transformation, we represented in §7.6.1 the variations of position and velocity as $\delta x^i = \xi^i$ and $\delta u^i = (\boldsymbol{\xi}\cdot\nabla)u^i + \mathrm{D}_t \xi^i$ respectively, by using the operator $\mathrm{D}_t = \partial_t + \mathcal{A}$ according to the gauge principle. The

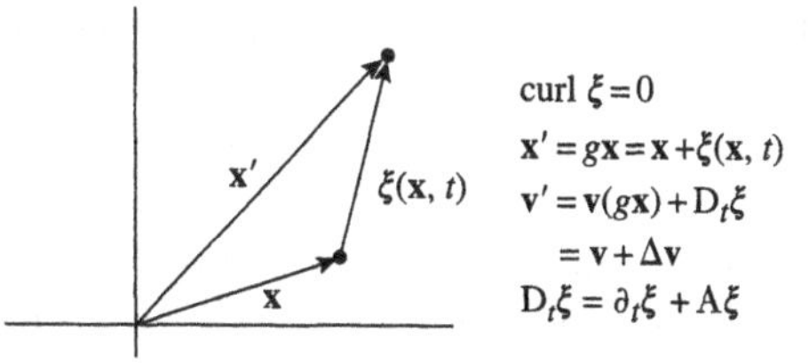

Fig. 7.4. Local translational gauge transformation.

same transformations are rewritten as

$$x' = gx := x + \boldsymbol{\xi}(x, t), \tag{7.51}$$

$$u' = u(gx) + \mathrm{D}_t\boldsymbol{\xi}, \tag{7.52}$$

(Fig. 7.4), which are supplemented with the (unchanged) transformation of spatial derivatives:

$$\nabla' = \nabla, \qquad \partial/\partial x'^{\alpha} = \partial/\partial x^{\alpha}. \tag{7.53}$$

This is, in fact, the *local* Galilei transformation. Namely, the transfomations (7.51) and (7.52) are understood to mean that the local coordinate origin is moving with the velocity $-\mathrm{D}_t\boldsymbol{\xi}$ (in accelerating motion) which corresponds to $\boldsymbol{U}$ of (7.6) if x' and u' are replaced by x_* and v_*, respectively. Under the above local transformation, the time derivative is transformed as

$$\partial_{t'} = \partial_t - (\mathrm{D}_t\boldsymbol{\xi}) \cdot \nabla.$$

7.7.2. *Determination of gauge field $\mathcal{A}$*

We require that the covariant derivative $\mathrm{D}_t u$, like u, should be transformed as follows:

$$(\mathrm{D}_t u)'(\equiv \partial_{t'}u' + \mathcal{A}'\, u') = \mathrm{D}_t u(gx) + \mathrm{D}_t(\mathrm{D}_t\boldsymbol{\xi}). \tag{7.54}$$

This requirement results in

$$[\mathcal{A}' - (\mathrm{D}_t\boldsymbol{\xi}) \cdot \nabla - \mathcal{A}]\,(u(gx) + \mathrm{D}_t\boldsymbol{\xi}) = 0.$$

As a consequence, we obtain the transformation law of $\mathcal{A}$:

$$\mathcal{A}' = \mathcal{A}(gx) + (\mathrm{D}_t\boldsymbol{\xi}) \cdot \nabla.$$

Eliminating $\mathrm{D}_t\boldsymbol{\xi}$ by (7.52) and using (7.53), this is rewritten as

$$\mathcal{A}' - \boldsymbol{u}' \cdot \nabla' = \mathcal{A}(g\boldsymbol{x}) - \boldsymbol{u}(g\boldsymbol{x}) \cdot \nabla.$$

This means that the operator $\mathcal{A} - \boldsymbol{u} \cdot \nabla$ is independent of reference frames. Denoting the frame-independent scalar, or tensor, of dimension $(\text{time})^{-1}$ by Ω, we have

$$\mathcal{A} = \boldsymbol{u} \cdot \nabla + \Omega. \tag{7.55}$$

Then, we shall have $\mathrm{D}_t\xi = \partial_t\xi + (\boldsymbol{u}\cdot\nabla)\,\xi + \Omega\xi$. However, the global invariance of translational transformation requires $\Omega = 0$, because we should have $\mathrm{D}_t\xi = \Omega\xi$ for constant ξ (if that is the case) and Eq. (7.34) does not vanish in the global transformation if ∂_t is replaced by D_t.[13] Thus, we obtain

$$\mathcal{A} = \boldsymbol{u} \cdot \nabla = \mathcal{A}^k\,\partial_k, \quad \mathcal{A}^k = u^k. \tag{7.56}$$

In this case, the covariant derivative $\mathrm{D}_t\boldsymbol{u}$ is represented by

$$\mathrm{D}_t\boldsymbol{u} = \partial_t\boldsymbol{u} + \mathcal{A}\boldsymbol{u} = \partial_t\boldsymbol{u} + (\boldsymbol{u} \cdot \nabla)\boldsymbol{u}, \tag{7.57}$$

which is usually called the time derivative of $\boldsymbol{u}$ following the particle motion, i.e. *material derivative*, or the Lagrange derivative.

7.7.3. *Irrotational fields* $\boldsymbol{\xi}(\boldsymbol{x})$ *and* $\boldsymbol{u}(\boldsymbol{x})$

Local gauge transformation $\boldsymbol{x} \to \boldsymbol{X}(\boldsymbol{x}, t) = \boldsymbol{x} + \boldsymbol{\xi}(\boldsymbol{x}, t)$ is determined by the irrotational $\boldsymbol{\xi}(\boldsymbol{x})$ (see (7.50)), i.e. the vector $\boldsymbol{\xi}(\boldsymbol{x})$ can be represented in terms of a potential $\varphi(\boldsymbol{x}, t)$:

$$\boldsymbol{\xi}(\boldsymbol{x}, t) = \operatorname{grad}\varphi(\boldsymbol{x}, t) = (\partial_i\varphi). \tag{7.58}$$

As a consequence, it can be verified that the velocity field $\boldsymbol{u}$ must be irrotational as well, i.e. we have potential flows of the form, $\boldsymbol{u} = \operatorname{grad} f$ with a certain scalar potential f, in the following way.

Transformed velocity is given by (7.52) with $\mathcal{A} = \boldsymbol{u} \cdot \nabla$:

$$\boldsymbol{u}'(\boldsymbol{x}) = \boldsymbol{u}(\boldsymbol{X}) + \partial_t\boldsymbol{\xi}(\boldsymbol{x}) + \boldsymbol{u}(\boldsymbol{X}) \cdot \nabla\boldsymbol{\xi}(\boldsymbol{x}), \tag{7.59}$$

[13] The case of nonzero tensor Ω will be considered in the next rotational transformation.

where $\boldsymbol{X}(\boldsymbol{x}) = \boldsymbol{x} + \boldsymbol{\xi}$. Suppose that $\boldsymbol{\xi}$ is infinitesimal, then we have

$$\Delta \boldsymbol{u} = \boldsymbol{u}'(\boldsymbol{x}) - \boldsymbol{u}(\boldsymbol{x}) = \partial_t \boldsymbol{\xi} + (\boldsymbol{u} \cdot \nabla)\boldsymbol{\xi} + (\boldsymbol{\xi} \cdot \nabla)\boldsymbol{u}, \tag{7.60}$$

to the first order of $|\boldsymbol{\xi}|$. The last equation can be rewritten in component form by using (7.58) as

$$\Delta u^i = \partial_t \partial_i \varphi + \partial_i (u^k \partial_k \varphi) + \partial_k \varphi (\partial_k u^i - \partial_i u^k). \tag{7.61}$$

This implies a remarkable property that if the velocity $\boldsymbol{u}$ is irrotational, i.e. if $\partial_k u^i - \partial_i u^k$ vanishes, Δu^i has a potential function Δf. Namely,

$$\Delta u^i = \partial_i \Delta f, \quad \Delta f = \partial_t \varphi + u^k \partial_k \varphi = \mathrm{D}_t\, \varphi. \tag{7.62}$$

Hence, if the original velocity field $\boldsymbol{u}(\boldsymbol{x})$ is irrotational, the transformed velocity $\boldsymbol{u}'$ will be irrotational as well.

The case of finite transformation $\boldsymbol{\xi}(\boldsymbol{x}, t)$ is verified as follows. Suppose that we have a velocity potential $f(\boldsymbol{x}, t)$, and $\boldsymbol{u}$ is given by

$$\boldsymbol{u} = \mathrm{grad}\, f. \tag{7.63}$$

Let us consider the variation of the transformed velocity $\boldsymbol{u}'$ of (7.59) about the reference point $\boldsymbol{x}_*$, i.e. we consider the variational field $\delta \boldsymbol{u}'(\boldsymbol{s}) = \boldsymbol{u}'(\boldsymbol{x}_* + \boldsymbol{s}) - \boldsymbol{u}'(\boldsymbol{x}_*)$, which is developed as

$$\begin{aligned} \delta(u')^k(\boldsymbol{s}) = \frac{\partial u_*^k}{\partial X^l}(s^l + s^m \partial_m \xi^l) + \partial_t (s^l \partial_l \xi^k) \\ + s^l \partial_l [u^m(\boldsymbol{X}_*) \partial_m \xi^k(\boldsymbol{x})], \end{aligned} \tag{7.64}$$

(see (7.48) for comparison) to the first order of $|\boldsymbol{s}|$, where $u_*^k = u^k(\boldsymbol{X}_*)$ and $\boldsymbol{X}_* = \boldsymbol{x}_* + \boldsymbol{\xi}(\boldsymbol{x}_*)$. Using the potential functions φ and f, the above equation can be written as $\delta(u')^k(\boldsymbol{s}) = D_{kl}(\boldsymbol{x}_*)\, s^l$, where[14]

$$\begin{aligned} D_{kl}(\boldsymbol{x}_*) &= \partial_k \partial_l f + \partial_k \partial_l (\partial_t \varphi) + u^m \partial_m (\partial_k \partial_l \varphi) \\ &\quad + (\partial_{*k} u_*^m)\, \partial_l \xi^m + (\partial_{*l} u_*^m)\, \partial_k \xi^m + (\partial_{*n} u_*^m)\, \partial_l \xi^n \partial_k \xi^m \\ &= D_{lk}(\boldsymbol{x}_*), \end{aligned}$$

and $\partial_k = \partial / \partial x^k$, $\partial_{*k} = \partial / \partial X_*^k$ and

$$\partial_{*k} u_*^m = \partial u^m(\boldsymbol{X}_*) / \partial X_*^k = \partial_{*k} \partial_{*m} f(\boldsymbol{X}_*) = \partial_{*m} u_*^k \quad \text{(symmetric)}.$$

---

[14] $\partial_m \xi^k = \partial_m \partial_k \varphi = \partial_k \xi^m$, $\partial_l\, u^m(\boldsymbol{X}_*) = (\partial_{*n} u_*^m)(\delta_{nm} + \partial_l \xi^m)$, and $(\partial_{*l} u_*^k) s^m (\partial_m \xi^l) = s^l (\partial_{*m} \partial_{*k} f_*)(\partial_l \partial_m \varphi) = s^l (\partial_{*k} u_*^m)(\partial_l \xi^m)$.

Thus, we obtain

$$\delta(u')^k(\boldsymbol{s}) = D_{kl}(\boldsymbol{x}_*)\, s^l = \frac{\partial}{\partial s^k}\left[\tfrac{1}{2}D_{ij}(\boldsymbol{x}_*)s^i s^j\right].$$

It is found that *the transformed field $\boldsymbol{u}'(\boldsymbol{x})$ is irrotational if both $\boldsymbol{\xi}$ and $\boldsymbol{u}$ are irrotational*, namely that the irrotational property is preserved by the (locally parallel) translational transformations.

For the potential velocity field (7.63), the gauge term is given by a potential form as well:

$$\mathcal{A}\boldsymbol{u} = (\boldsymbol{u}\cdot\nabla)\boldsymbol{u} = \mathrm{grad}\left(\tfrac{1}{2}|\boldsymbol{u}|^2\right), \tag{7.65}$$

since $u^k\partial_k u^i = (\partial_k f)\partial_k\partial_i f = (\partial_k f)\partial_i\partial_k f = \partial_i(u^2/2)$.

The covariant derivative $\mathrm{D}_t\boldsymbol{u}$ is represented as

$$\mathrm{D}_t\boldsymbol{u} = \partial_t\boldsymbol{u} + (\boldsymbol{u}\cdot\nabla)\boldsymbol{u} = \mathrm{grad}\,\partial_t f + \mathrm{grad}\left(\tfrac{1}{2}|\boldsymbol{u}|^2\right). \tag{7.66}$$

This leads to an important consequence in §7.9.4 that the equation of motion *can be integrated* once.

Remark. There exist some fluids such as the superfluid He^4 or Bose–Einstein condensates, composed of indistinguishable equivalent particles. In such a fluid, local rotation would not be captured because there should be no difference by such a local rotation that would be conceived in an ordinary fluid (i.e. nonquantum fluid). Therefore the flow should be inevitably *irrotational* [SS77; Lin63]. This is not the case when we consider the motion of an ordinary fluid composed of distinguishable particles.

7.8. Fluid Flows as Material Motion

7.8.1. *Lagrangian particle representation*

From the local gauge invariance under the local (irrotational) translation, we arrived at the covariant derivative (7.66), which is regarded as a derivative following a material particle moving with the irrotational velocity $\boldsymbol{u} = \mathrm{grad}\, f$. This suggests such a formulation which takes into account displacement of individual *mass*-particles [Eck60; Bre70]. In other words, the gauge invariance requires that laws of fluid motion should be expressed in a form equivalent to every individual particle.

Suppose that continuous distribution of mass-particles is represented by three continuous parameters, $\boldsymbol{a} = (a^i) = (a, b, c)$ in $\mathbb{R}^3$, and that each

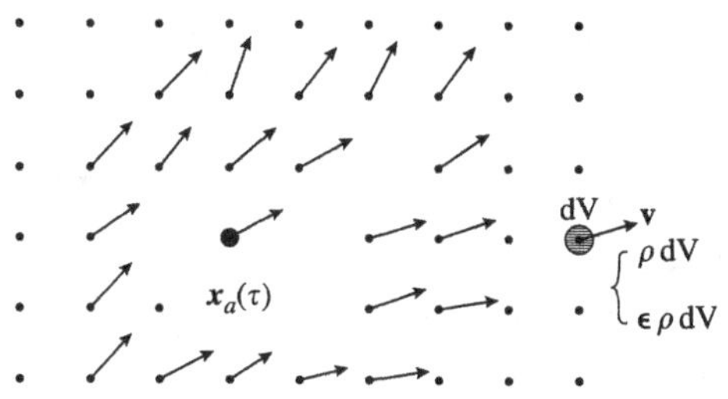

Fig. 7.5. Motion of material particles (2D case).

particle $\boldsymbol{a}$ is moving with a velocity $\boldsymbol{v}_a$. Their motion as a whole is described by a *flow* ϕ_τ which takes a particle located at $\phi_0\boldsymbol{a} \equiv \boldsymbol{a}$ when $\tau = 0$ to the position $\phi_\tau\boldsymbol{a} = \phi(\tau, \boldsymbol{a}) = \boldsymbol{x}(\tau, \boldsymbol{a})$ at a time τ.[15] The coordinate parameters $(a, b, c) = \boldsymbol{a}$ remain fixed during the motion of each paticle, so that a fluid particle is identified by $\boldsymbol{a}$ (Fig. 7.5). Then the particle velocity is given by

$$\boldsymbol{u}_a(\tau) = \boldsymbol{u}(\tau, \boldsymbol{a}) := \partial_\tau\phi(\tau, \boldsymbol{a}) = \partial_\tau\boldsymbol{x}_a, \tag{7.67}$$

where $\boldsymbol{x}_a(\tau) := \boldsymbol{x}(\tau, \boldsymbol{a})$ denotes the position of a particle $\boldsymbol{a}$ at a time τ.

Suppose that the velocity $\partial_\tau\boldsymbol{x}_a$ is defined continuously and differentiably at points in space and represented by the velocity *field* $\boldsymbol{u}(t, \boldsymbol{x})\ (= \partial_\tau\boldsymbol{x}_a)$ in terms of the Eulerian coordinates $\boldsymbol{x}$ and t. Note that

$$\partial_\tau\,|_{a=\text{const}} = \mathrm{D}_t = \partial_t + \boldsymbol{u}\cdot\nabla,$$

which is called the Lagrange derivative. By definition, we have

$$\partial_\tau\boldsymbol{a} = \mathrm{D}_t\boldsymbol{a} = \partial_t\boldsymbol{a} + \boldsymbol{u}\cdot\nabla\boldsymbol{a} = 0. \tag{7.68}$$

The covariant derivative of the potential velocity $\boldsymbol{u}$ is given by

$$\mathrm{D}_t\boldsymbol{u} = \partial_\tau\boldsymbol{u}|_{a=\text{const}} = \partial_t\boldsymbol{u} + \text{grad}\left(\tfrac{1}{2}|\boldsymbol{u}|^2\right). \tag{7.69}$$

Next, consider a small volume δV consisting of a set of material particles of total mass δm. Suppose that the motion of its center of mass from $\tau = 0$ to τ is described by the displacement from $\boldsymbol{x}_a(0) = \boldsymbol{a}$ to $\boldsymbol{x}_a(\tau)$ (Fig. 7.6). The volume of the same mass will change, and equivalently the density ρ will change. The density is expressed in two ways by $\rho_a(\tau)$ with the Lagrangian

---

[15] The configuration space of a fluid flow is represented in two ways: one is the *Lagrangian particle* coordinates $(\tau, \boldsymbol{a})$, and the other is the *Eulerian* coordinates $(t, \boldsymbol{x})$. The time coordinate used in combination with $\boldsymbol{a} = (a, b, c)$ is denoted by τ. Inverse map of the function $\boldsymbol{x} = \phi(\tau, \boldsymbol{a})$ is $\boldsymbol{a} = \boldsymbol{a}(t, \boldsymbol{x})$ and $\tau = t$, where $\boldsymbol{x} = (x, y, z)$ and t is the time.

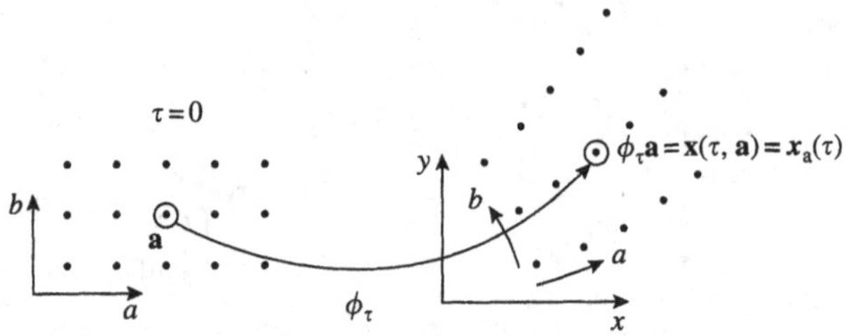

Fig. 7.6. Map $\boldsymbol{a}(a,b) \mapsto \boldsymbol{x}(x,y)$ (2D case).

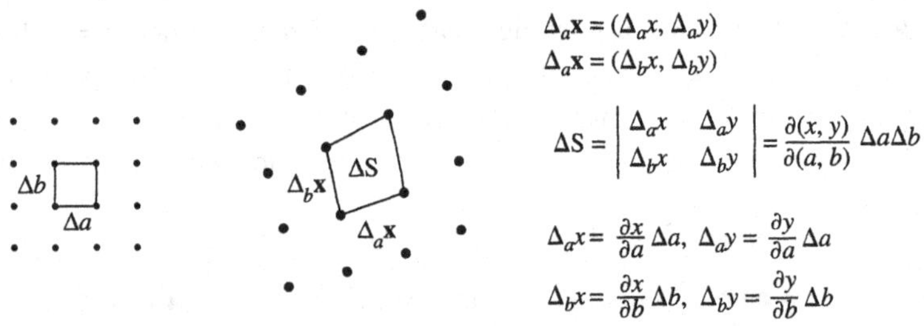

Fig. 7.7. 2D problem: change of area: $\Delta a \, \Delta b \to \Delta S$.

description and by $\rho(t,\boldsymbol{x})$ with the Eulerian one, and the invariance of mass in a volume element during the motion is represented by

$$\delta m = \rho(t,\boldsymbol{x})\mathrm{d}V = \rho_a(0)\mathrm{d}V_a,$$

where $\mathrm{d}V = \mathrm{d}x\,\mathrm{d}y\,\mathrm{d}z = \mathrm{d}^3\boldsymbol{x}$ and $\mathrm{d}V_a = \mathrm{d}a\,\mathrm{d}b\,\mathrm{d}c = \mathrm{d}^3\boldsymbol{a}$.

It is useful to normalize that $\rho_a(0) = 1$, so that the Lagrangian coordinates $\boldsymbol{a} = (a,b,c)$ represent the *mass coordinates*. Then, we have

$$\rho(t,\boldsymbol{x})\mathrm{d}^3\boldsymbol{x} = \mathrm{d}^3\boldsymbol{a}, \tag{7.70}$$

Since $\mathrm{d}^3\boldsymbol{x} = J_a \mathrm{d}^3\boldsymbol{a}$ (Fig. 7.7), where J_a is the Jacobian $J_a = \partial(\boldsymbol{x})/\partial(\boldsymbol{a})$ of the map $\boldsymbol{a} \mapsto \boldsymbol{x}$:

$$J_a = \frac{\partial(\boldsymbol{x})}{\partial(\boldsymbol{a})} = \frac{\partial(x,y,z)}{\partial(a,b,c)}, \tag{7.71}$$

we have

$$\rho(t,\boldsymbol{x}) = 1/J_a. \tag{7.72}$$

The relation $\boldsymbol{u}(t, \boldsymbol{x}) = \partial_\tau \boldsymbol{x}_a(\tau)$ has an important consequemce. Namely, this *kinematical constraint* connects the velocity fired $\boldsymbol{u}(t, \boldsymbol{x})$ with the change of volume $\mathrm{d}V = \mathrm{d}^3\boldsymbol{x}$. This is considered below.

7.8.2. *Lagrange derivative and Lie derivative*

Given a velocity field $\boldsymbol{u}(\boldsymbol{x}) = (u, v, w)$, the *divergence* of $\boldsymbol{u}$, written as $\mathrm{div}\,\boldsymbol{u}$, is defined by relative rate of change of a volume element δV during an infinitesimal time δt along the flow ϕ_τ generated by $\boldsymbol{u}(\boldsymbol{x})$, namely

$$\delta \mathrm{d}V/\mathrm{d}V = (\mathrm{div}\,\boldsymbol{u})\,\delta t \tag{7.73}$$

(Fig. 7.8), where $\mathrm{div}\,\boldsymbol{u} = \partial_x u + \partial_y v + \partial_z w$ for $\boldsymbol{x} = (x, y, z) \in M$.

In the external algebra (Appendix B), a volume element δV is represented by a volume form $\mathcal{V}^3 (= \mathrm{d}x \wedge \mathrm{d}y \wedge \mathrm{d}z)$. The rate of change of $\mathcal{V}^3$ along the flow ϕ_τ is defined by the Lie derivative $\mathcal{L}_{\boldsymbol{u}}$ in the form,

$$\mathcal{L}_{\boldsymbol{u}}\,\mathcal{V}^3 = (\mathrm{div}\,\boldsymbol{u})\,\mathcal{V}^3, \tag{7.74}$$

given by (B.44) in Appendix B.6.

The differential operator $\mathrm{D}_t = \partial_t + \boldsymbol{u} \cdot \nabla$ defines a time derivative analogous to $\mathcal{L}_{\boldsymbol{u}}$, but their difference must be remarked. The operator D_t defined by (7.57) was derived by the gauge principle (a physical principle) in §7.5, 7.7, whereas the Lie derivative $\mathcal{L}$ of forms is mathematically defined by the Cartan's formula (B.20) in Appendix B.4.[16] It is already explained

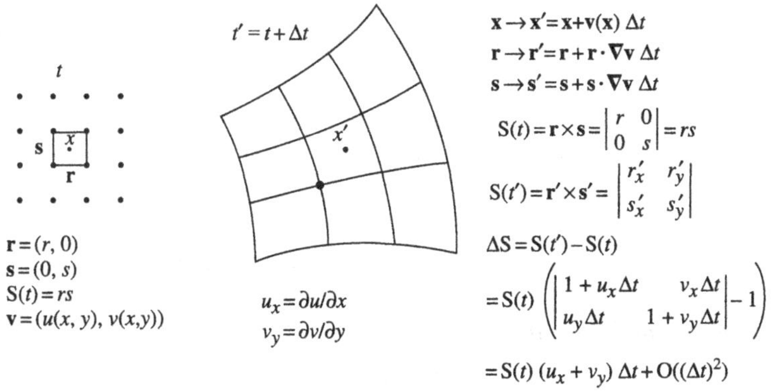

Fig. 7.8. Fluid flow and $\mathrm{div}\,\boldsymbol{u}$.

[16]§3.11.1 describes Lie derivatives of 1-forms and metric tensors.

in §1.8.3 that both D_t and $\mathcal{L}_{\boldsymbol{u}}$ yield the same result on 0-forms, but different results on vectors. Actually, we have $\mathrm{D}_t\boldsymbol{u} = \partial_t\boldsymbol{u} + (\boldsymbol{u}\cdot\nabla)\boldsymbol{u}$, while $\mathcal{L}_{\boldsymbol{u}}\boldsymbol{u} = 0$ according to (1.81) and (1.83) of §1.8.3. $\mathcal{L}_{\boldsymbol{u}}\boldsymbol{u} = 0$ because $\boldsymbol{u}$ is "frozen" to the velocity field $\boldsymbol{u}$ itself. The derivative D_t refers to the rate of change of a variable following the motion of *a moving point of interest*, whereas the derivative $\mathcal{L}_{\boldsymbol{u}}$ acting on a vector refers to the rate of *change of a vector field along a flow* generated by $\boldsymbol{u}$ (i.e. describing *deviation* from the frozen field).

Acting on a volume form, its rate of change is defined by the Lie derivative $\mathcal{L}_{\boldsymbol{u}}$, and $\mathrm{D}_t(\mathrm{d}V)$ is defined to be equivalent to $\mathcal{L}_{\boldsymbol{u}}\,\mathcal{V}^3$:

$$\mathcal{L}_{\boldsymbol{u}}\,\mathcal{V}^3 = (\mathrm{div}\,\boldsymbol{u})\,\mathcal{V}^3 \Longleftrightarrow \mathrm{D}_t(\mathrm{d}V) = (\mathrm{div}\,\boldsymbol{u})\mathrm{d}V. \tag{7.75}$$

7.8.3. *Kinematical constraint*

Rate of change of mass in a volume element $\mathrm{d}V$ along a flow generated by $X = \partial_t + \boldsymbol{u}\cdot\nabla$ is given by[17]

$$\begin{aligned}\mathrm{D}_t\big(\rho\,\mathrm{d}V\big) &= (\mathrm{D}_t\rho)\mathrm{d}V + \rho\,(\mathrm{D}_t\mathrm{d}V) = \big(\partial_t\rho + (\boldsymbol{u}\cdot\nabla)\rho\big)\ \mathrm{d}V \\ &\quad + \rho\,\mathrm{div}\,\boldsymbol{u}\mathrm{d}V \ := \Delta_t\rho\,\mathrm{d}V,\end{aligned} \tag{7.76}$$

where (7.75) is used. Invariance of the mass $\rho\,\mathrm{d}V$ along the flow is represented by $\mathrm{D}_t(\rho\,\mathrm{d}V) = 0$. Thus, from the kinematical argument in the above two subsections, we obtain the following *equation of continuity*,

$$\Delta_t\rho = \mathrm{D}_t\rho + \rho\,\mathrm{div}(\boldsymbol{u}) = \partial_t\rho + \mathrm{div}(\rho\boldsymbol{u}) = 0. \tag{7.77}$$

This must be satisfied in all the variations carried out below as a *kinematical constraint.*

7.9. Gauge-Field Lagrangian L_A (Translational Symmetry)

7.9.1. *A possible form*

In order to consider *possible* type of the Lagrangian L_A for the background gauge field with $\mathcal{A}_k = \partial_k f$ (f: a potential function), we suppose that L_A is

[17] The time part ∂_t is included since the fields are assumed to be time-dependent.

a function of $\mathcal{A}_k$ and its derivatives:

$$L_A = L_A(\mathcal{A}_k, \mathcal{A}_{k;\nu}), \quad \mathcal{A}_{k;\nu} = \partial_\nu \mathcal{A}_k = \partial \mathcal{A}_k / \partial x^\nu.$$

We take the following variations:

$$\left.\begin{aligned} \delta \boldsymbol{x} &= \boldsymbol{\xi}, \quad \text{where} \quad \xi^\alpha = \partial_\alpha \varphi, \\ \delta \mathcal{A}_k &= \xi^\alpha \partial_\alpha \mathcal{A}_k + \partial_t \xi^k + \mathcal{A}_\alpha \partial_\alpha \xi^k, \\ \delta \mathcal{A}_{k;\nu} &= \partial_\nu (\delta \mathcal{A}_k) = \partial_\nu \xi^\alpha \, \partial_\alpha \mathcal{A}_k + \xi^\alpha \, \partial_\nu \partial_\alpha \mathcal{A}_k \\ &\quad + \partial_\nu \partial_k \varphi_t + \partial_\nu \mathcal{A}_\alpha \, \partial_\alpha \xi^k + \mathcal{A}_\alpha \partial_\nu \partial_k \partial_\alpha \varphi, \quad (\varphi_t = \partial_t \varphi), \end{aligned}\right\} \tag{7.78}$$

where $\delta \mathcal{A}_k$ is taken according to (7.42) with u replaced by $\mathcal{A}$. The invariance of L_A with respect to such variations is given by

$$\delta L_A = \frac{\partial L_A}{\partial \mathcal{A}_k} \delta \mathcal{A}_k + \frac{\partial L_A}{\partial \mathcal{A}_{k;\nu}} \delta \mathcal{A}_{k;\nu} = 0. \tag{7.79}$$

Substituting the variations (7.78), we have a variational equation for arbitrary functions of ξ^k, $\partial_\alpha \xi^k$, $\partial_t \xi^k$, $\partial_\nu \partial_k \varphi_t$, $\partial_\nu \partial_k \partial_\alpha \varphi$. Each coefficient of five terms are required to vanish. From the symmetry of the coefficient of $\partial_\nu \partial_k \varphi_t (= \partial_k \partial_\nu \varphi_t)$ with respect to exchange of k and ν, we have

$$\frac{\partial L_A}{\partial \mathcal{A}_{k;\nu}} + \frac{\partial L_A}{\partial \mathcal{A}_{\nu;k}} = 0.$$

Hence, the derivative terms $\mathcal{A}_{k;\nu}$ should be contained in L_A through the combination,

$$\mathcal{A}_{[k,\nu]} := \mathcal{A}_{k;\nu} - \mathcal{A}_{\nu;k}.$$

In the present case, $\mathcal{A}_k = \partial_k f$. Then, we have $\mathcal{A}_{[k,\nu]} = \partial_\nu \partial_k f - \partial_k \partial_\nu f = 0$. Therefore, the Lagrangian L_A is not able to contain the derivative terms $\mathcal{A}_{k;\nu}$, and we have $L_A(\mathcal{A}_k)$. However, vanishing of the coefficient of $\partial_t \xi^k$ in the variational equation requires $\partial L_A / \partial \mathcal{A}_k = 0$. Hence, L_A is independent of $\mathcal{A}_k$ as well. Thus, the Lagrangian L_A can contain only a scalar function (if any, say $\epsilon(\boldsymbol{x})$) independent of $\mathcal{A} : L_A = L_A(\epsilon(\boldsymbol{x}))$.

7.9.2. *Lagrangian of background thermodynamic state*

According to the physical argument in §7.2.1 and [Kam03a; Kam03c] regarding the Lorentz invariance of a mechanical system, it is found that the scalar function (implied in the previous section) should be the internal energy $\epsilon(\rho, s)$ (per unit mass) and the Lagrangian L_A can be given as

$$L_\epsilon = -\int_M \epsilon(\rho, s)\rho\,\mathrm{d}V, \tag{7.80}$$

where ρ is the fluid density and s the entropy. Thus, the *thermodynamic state* of a fluid is regarded as the background field, which is represented by the internal energy ϵ of the fluid, given as a function of density ρ and entropy s (in a single phase) with ϵ and s defined per unit mass.

On the other hand, the Lagrangian of the fluid motion is

$$L_{\mathrm{F}} = \tfrac{1}{2}\int \langle \boldsymbol{u}, \boldsymbol{u}\rangle \mathrm{d}^3\boldsymbol{a} \tag{7.81}$$

(see (7.37)), where $\langle \boldsymbol{u}, \boldsymbol{u}\rangle \equiv u^i u^i = u^1u^1 + u^2u^2 + u^3u^3$.

7.10. Hamilton's Principle for Potential Flows

On the basis of the preliminary considerations with respect to the translational invariance made so far, now, we can derive the equation of motion on the basis of the Hamilton's principle by using the total Lagrangian L_{P} to be defined just below. To accomplish it, some constitutive or constraint conditions are required for the variations of the Lagrangian of an ideal fluid.

7.10.1. *Lagrangian*

According to the scenario of the gauge principle, the full Lagrangian is defined by

$$L_{\mathrm{P}} := L_{\mathrm{F}} + L_\epsilon = \int_M \tfrac{1}{2}\langle \boldsymbol{u}, \boldsymbol{u}\rangle\,\rho\,\mathrm{d}V - \int_M \epsilon(\rho, s)\,\rho\,\mathrm{d}V, \tag{7.82}$$

where $\boldsymbol{u} = \mathrm{D}_t\boldsymbol{x}$. The first integral is the Lagrangian of fluid flow including the interaction with background field which is represented by $\mathcal{A}\boldsymbol{x}$ in $\boldsymbol{u} = \mathrm{D}_t\boldsymbol{x}$.[18] The second is the Lagrangian of a thermodynamics state of background material. The action principle is given by $\delta \underline{I} = 0$, i.e.

$$\delta\,\underline{I} = \int_{t_0}^{t_1} \delta L_{\mathrm{P}}\,\mathrm{d}t = 0. \tag{7.83}$$

[18] $\mathrm{D}_t\boldsymbol{x} = \partial_t\boldsymbol{x} + \mathcal{A}\boldsymbol{x} = \mathcal{A}\boldsymbol{x} = (\boldsymbol{u}\cdot\nabla)\boldsymbol{x} = \boldsymbol{u}$. If $\boldsymbol{x}$ is substituted by $\boldsymbol{x}_p(t)$, $\mathrm{D}_t\boldsymbol{x} = \mathrm{d}\boldsymbol{x}_p/\mathrm{d}t$.

7.10.2. *Material variations: irrotational and isentropic*

We carry out *material* variations in the following way. All the variations are taken so as to follow particle displacements. Writing an infinitesimal variation of the particle position as $\delta \boldsymbol{x}_a = \boldsymbol{\xi}(\boldsymbol{x}, t)$ and the variation of particle velocity as $\Delta \boldsymbol{u}_a$, we have

$$\boldsymbol{x}_a \mapsto \boldsymbol{x}_a + \boldsymbol{\xi}(\boldsymbol{x}_a, t), \tag{7.84}$$

$$\boldsymbol{u}_a \mapsto \boldsymbol{u}_a + \Delta \boldsymbol{u}_a = \boldsymbol{u}(\boldsymbol{x}_a) + (\boldsymbol{\xi} \cdot \nabla)\boldsymbol{u} + \mathrm{D}_t \boldsymbol{\xi}, \tag{7.85}$$

(see (7.51), (7.52)), where $\boldsymbol{x}_a(t) := \boldsymbol{x}(t, \boldsymbol{a})$ denotes the position of the particle $\boldsymbol{a}$. The displacemnet $\boldsymbol{\xi}(\boldsymbol{x})$ must be irrotational, i.e. $\boldsymbol{\xi} = \mathrm{grad}\, \varphi$. All the variations are taken so as to satisfy the mass conservation.

Variations of density and internal energy consist of two components:

$$\Delta \rho = \boldsymbol{\xi} \cdot \nabla \rho + \delta \rho, \qquad \Delta \epsilon = \boldsymbol{\xi} \cdot \nabla \epsilon + \delta \epsilon. \tag{7.86}$$

The first terms are the changes due to the displacement $\boldsymbol{\xi}$, while the second terms are proper variations explained below. The entropy s (per unit mass) is written as $s = s(\tau, \boldsymbol{a})$, and the variation is carried out *adiabatically*,

$$\delta s = 0, \tag{7.87}$$

i.e. *isentropically*. Such a fluid is called an *ideal fluid*. Namely, in an ideal fluid, there is no mechanism of energy dissipation and the fluid motion is isentropic. Sometimes such a fluid is called an *inviscid* fluid too, because kinetic energy would be dissipated if there were viscosity.

On the other hand, variation $\delta \rho$ of the density is caused by the change of volume $\mathrm{d}V$ composed of the same material mass $\rho \mathrm{d}V$ which is fixed during the displacement $\delta \boldsymbol{x}_a = \boldsymbol{\xi}(\boldsymbol{x}, t)$. The condition of the fixed mass is given by

$$\delta(\rho \mathrm{d}V) = (\delta \rho)\mathrm{d}V + \rho(\delta \mathrm{d}V) = (\delta \rho + \rho\, \mathrm{div}\, \boldsymbol{\xi})\delta \mathrm{d}V = 0, \tag{7.88}$$

where $\delta \mathrm{d}V = \mathrm{div}\, \boldsymbol{\xi}\, \mathrm{d}V$. Hence, we have

$$\delta \rho = -\rho\, \mathrm{div}\, \boldsymbol{\xi}. \tag{7.89}$$

Invariance of the mass $\rho\, \mathrm{d}V$ along the flow is $\mathrm{D}_t(\rho, \mathrm{d}V) = 0$. From the kinematical argument (§7.8.3), we have obtained the following *equation of continuity*,

$$\mathrm{D}_t \rho + \rho\, \mathrm{div}(\boldsymbol{u}) = \partial_t \rho + \mathrm{div}(\rho \boldsymbol{u}) = 0. \tag{7.90}$$

This must be satisfied in all variations as a *kinematical constraint*.

Similarly, $\delta s = 0$ of (7.87) implies that $\partial_\tau s(\tau, \boldsymbol{a}) = 0$, i.e. the entropy s is invariant during the motion of particle $\boldsymbol{a}$. This may be represented alternatively by $\mathcal{L}_X s = \mathrm{D}_t s = 0$ for the 0-form s with $X = \partial_t + \boldsymbol{u} \cdot \nabla$:

$$\partial_\tau s = \mathrm{D}_t s = \partial_t s + \boldsymbol{u} \cdot \nabla s = 0. \tag{7.91}$$

Then, proper variation of the internal energy $\epsilon(\rho, s)$ is expressed in terms of the density variation $\delta\rho$ only by using (7.87), since

$$\delta\epsilon = \frac{\partial\epsilon}{\partial\rho}\delta\rho + \frac{\partial\epsilon}{\partial s}\delta s = \frac{p}{\rho^2}\delta\rho, \tag{7.92}$$

where p is the fluid pressure, since $\partial t/\partial p = p/\rho^2$.

The variation field $\boldsymbol{\xi}(\boldsymbol{x}, t)$ is constrained to vanish on the boundary surface S of M, as well as at both ends of time t_0, t_1 for the action $\underline{I}$:

$$\boldsymbol{\xi}(\boldsymbol{x}_S, t) = 0, \qquad \text{for any } {}^\forall t, \;\; \text{for } \boldsymbol{x}_S \in S = \partial M, \tag{7.93}$$

$$\boldsymbol{\xi}(\boldsymbol{x}, t_0) = 0, \qquad \boldsymbol{\xi}(\boldsymbol{x}, t_1) = 0, \;\; \text{for } {}^\forall \boldsymbol{x} \in M. \tag{7.94}$$

7.10.3. *Constraints for variations*

As a consequence of the material variation,[19] irrotational flows is derived under the constraints of the *continuity equation* and the *isentropic flow*. Schutz and Sorkin [SS77, §4] verified that any variational principle for an ideal fluid that leads to the Euler's equation of motion must be constrained. This is related to the property of fluid flows that the energy (including the mass energy) of a fluid *at rest* can be changed by adding particles (i.e. changing density) or by adding entropy without violating the equation of motion. (In addition, adding a uniform velocity to a uniform-flow state is again another state of uniform flow.) They proposed a minimally constrained variational principle for relativistic fluid flows that obey the conservations of particle number and entropy. However, *gauge principle* in the current formulation was out of scope in [SS77]. Actually, they showed only the momentum conservation equation derived in the Newtonian limit from their relativistic formulation which is gauge-invariant in the framework of the theory of *relativity*.

[19]In most variational problems of continuous fields, both coordinate and field are varied. It is said that a variation of coordinate is external, whereas field variation is internal. In the material variation, both variations are inter-related.

7.10.4. *Action principle for* $L_{\rm P}$

We now consider variation of each term of (7.82) separately.

(i) Variation of the first term (denoted by $L_{\rm F}$) is carried out as follows. Variation of the integrand of $L_{\rm F}$ is composed of two parts:

$$\boldsymbol{\xi}\cdot\nabla\left[\tfrac{1}{2}\langle\boldsymbol{u},\boldsymbol{u}\rangle\,\rho\,\mathrm{d}V\right]+\delta\left[\tfrac{1}{2}\langle\boldsymbol{u},\boldsymbol{u}\rangle\rho\,\mathrm{d}V\right]. \tag{7.95}$$

The first is the change of integrand due to the displacement $\boldsymbol{\xi}$ and the second is the proper change by the kinematical condition, as explained in (7.86) for ρ and ϵ separately. It is useful to note the vector identity:

$$\begin{aligned}\boldsymbol{\xi}\cdot\nabla[F(\boldsymbol{x})\mathrm{d}V]&=(\boldsymbol{\xi}\cdot\nabla F)\mathrm{d}V+F\boldsymbol{\xi}\cdot\nabla(\mathrm{d}V)\\&=(\boldsymbol{\xi}\cdot\nabla F)\mathrm{d}V+F(\mathrm{div}\,\boldsymbol{\xi})\mathrm{d}V=\mathrm{div}[F(\boldsymbol{x})\boldsymbol{\xi}]\mathrm{d}V\end{aligned}$$

If F is set to be $\frac{1}{2}\langle\boldsymbol{u},\boldsymbol{u}\rangle\rho$, this becomes the first term of (7.95). The equality $\boldsymbol{\xi}\cdot\nabla(\mathrm{d}V)=(\mathrm{div}\boldsymbol{\xi})\mathrm{d}V$ is obtained by setting $\boldsymbol{u}\delta t=\boldsymbol{\xi}$ and $\delta\mathrm{d}V=\boldsymbol{\xi}\cdot\nabla(\mathrm{d}V)$ in (7.73). Thus, the variation of $L_{\rm F}$ is given by

$$\begin{aligned}\delta L_{\rm F}&=\int_M\left(\mathrm{div}\left[\tfrac{1}{2}\langle\boldsymbol{u},\boldsymbol{u}\rangle\rho\,\boldsymbol{\xi}\right]\mathrm{d}V+\delta\left[\tfrac{1}{2}\langle\boldsymbol{u},\boldsymbol{u}\,\rangle\,\rho\,\mathrm{d}V\right]\right)\\&=\oint_S\tfrac{1}{2}\langle\boldsymbol{u},\boldsymbol{u}\rangle\rho\,\langle\boldsymbol{n},\boldsymbol{\xi}\rangle\mathrm{d}S+\int_M\langle\boldsymbol{u},\mathrm{D}_t\boldsymbol{\xi}\rangle\rho\mathrm{d}V\end{aligned} \tag{7.96}$$

since $\delta(\rho\mathrm{d}V)=0$ and $\delta\boldsymbol{u}=\mathrm{D}_t\boldsymbol{\xi}$, where $\boldsymbol{n}$ is the unit outwardly normal to S. The first term vanishes due to the boundary condition (7.93). Thus,

$$\begin{aligned}\delta L_{\rm F}=\int_M\langle\boldsymbol{u},\mathrm{D}_t\boldsymbol{\xi}\rangle\rho\,\mathrm{d}V&=\int_M\mathrm{D}_t\left[\langle\boldsymbol{u},\boldsymbol{\xi}\rangle\rho\,\mathrm{d}V\right]\\&\quad-\int_M\langle\mathrm{D}_t\boldsymbol{u},\boldsymbol{\xi}\rangle\rho\,\mathrm{d}V-\int_M\langle\boldsymbol{u},\boldsymbol{\xi}\rangle\mathrm{D}_t\left[\rho\,\mathrm{d}V\right]\end{aligned} \tag{7.97}$$

$$=\partial_t\int_M\left[\langle\boldsymbol{u},\boldsymbol{\xi}\rangle\rho\,\mathrm{d}V\right]-\int_M\langle\mathrm{D}_t\boldsymbol{u},\boldsymbol{\xi}\,\rangle\rho\,\mathrm{d}V, \tag{7.98}$$

where $\int_M$ and D_t are interchanged uniformly) and D_t is replaced with ∂_t in the last equation, because the integral $\int_M$ is a function of t only. The last term of (7.97) disappears due to the kinematical condition (7.76) with (7.90).

(ii) Variation of the second term (denoted by L_ϵ) of (7.82) is given by

$$\begin{aligned} -\delta L_\epsilon &= \int_M \left(\mathrm{div}(\epsilon\rho\,\boldsymbol{\xi})\mathrm{d}V + \delta\big[\epsilon(\rho,s)\rho(\boldsymbol{x})\mathrm{d}V\big]\right) \\ &= \oint_S \epsilon\rho\,\langle \boldsymbol{n},\boldsymbol{\xi}\rangle \mathrm{d}S + \int_M \delta\big[\epsilon(\rho,s)\big]\rho\,\mathrm{d}V, \end{aligned}$$

analogously to (7.96). The first term vanishes due to the boundary condition (7.93), and $\delta(\rho\,\mathrm{d}V) = 0$ is used for the first term. Thus, we obtain

$$\delta L_\epsilon = -\int_M \delta\epsilon(\rho,s)\rho\mathrm{d}V = -\int_M \frac{p}{\rho^2}\delta\rho\,\rho\,\mathrm{d}V,$$

from (7.92). Substituting in (7.89),

$$\delta L_\epsilon = \int_M p\,\mathrm{div}\,\boldsymbol{\xi}\,\mathrm{d}V = \oint_S p\langle \boldsymbol{n},\boldsymbol{\xi}\rangle \mathrm{d}S - \int_M \langle \mathrm{grad}\,p, \boldsymbol{\xi}\rangle \mathrm{d}V. \tag{7.99}$$

The first surface integral vanishes by the condition (7.93), but is retained here for later use.

Thus, collecting (7.98) and (7.99), the variation of the action $\underline{I}$ is given by

$$\begin{aligned} \delta\underline{I} &= \int_{t_0}^{t_1} (\delta L_{\mathrm{F}} + \delta L_\epsilon)\mathrm{d}t \\ &= \left[\int_M \langle \boldsymbol{u},\boldsymbol{\xi}\rangle \rho\,\mathrm{d}V\right]_{t_0}^{t_1} + \int_{t_0}^{t_1} \mathrm{d}t \oint_S p\langle \boldsymbol{n},\boldsymbol{\xi}\rangle \mathrm{d}S \\ &\quad - \int_{t_0}^{t_1} \mathrm{d}t \int_M \langle (\mathrm{D}_t\boldsymbol{u} + \rho^{-1}\mathrm{grad}\,p), \boldsymbol{\xi}\,\rangle \rho\,\mathrm{d}V. \end{aligned} \tag{7.100}$$

The first line on the right-hand side vanishes owing to the boundary conditions (7.93) and (7.94). Therefore, the action principle $\delta\,\underline{I} = 0$ leads to

$$\begin{aligned} &\mathrm{D}_t\boldsymbol{u} + \rho^{-1}\mathrm{grad}\,p = 0, \\ \text{or}\quad &\partial_t\boldsymbol{u} + (\boldsymbol{u}\cdot\nabla)\boldsymbol{u} + \rho^{-1}\mathrm{grad}\,p = 0, \end{aligned} \tag{7.101}$$

for arbitrary variation $\boldsymbol{\xi}$ under the conditions: $\boldsymbol{\xi} = \mathrm{grad}\,\varphi$.

Equation (7.101) must be supplemented by the equation of continuity (7.90) and the isentropic equation (7.91):

$$\partial_t \rho + \mathrm{div}(\rho \boldsymbol{u}) = 0, \tag{7.102}$$

$$\partial_t s + \boldsymbol{u} \cdot \nabla s = 0. \tag{7.103}$$

As a consequence of the isentropy, the *enthalpy* $h := \epsilon + p/\rho$ is written as

$$dh = \rho^{-1} \mathrm{d}p + T\mathrm{d}s = \rho^{-1}\mathrm{d}p. \tag{7.104}$$

Using this in (7.101), we obtain the equation of motion for potential flows:

$$\partial_t \boldsymbol{u} + \mathrm{grad}\left(\tfrac{1}{2}|\boldsymbol{u}|^2\right) = -\mathrm{grad}\, h, \tag{7.105}$$

since $(\boldsymbol{u} \cdot \nabla)\, u^i = \partial_i(\frac{1}{2}|\boldsymbol{u}|^2)$ for $\boldsymbol{u} = \mathrm{grad}\, f$. As a consequence, Eq. (7.105) ensures that we have a *potential flow* represented as $\boldsymbol{u} = \mathrm{grad}\, f$ at all times. Substituting $\boldsymbol{u} = \mathrm{grad}\, f$, we obtain an *integral* of (7.105):

$$\partial_t f + \tfrac{1}{2}|\mathrm{grad}\, f|^2 + h = \mathrm{const.} \tag{7.106}$$

It is remarkable that the equation of motion is *integrable.* It is interesting to recall that the flow of a superfluid in the degenerate state is irrotational [SS77; Lin63] (see the remark of §7.7.3).

The degenerate ground state is consistent with the Kelvin's theorem of minimum energy [Lamb32], which asserts that a potential flow has a minimum kinetic energy among all possible flows satisfying given conditions.[20]

This is not the case when we consider motion of fluids composed of distinguishable particles like ordinary fluids. Local rotation is distinguishable and there must be a formal mathematical structure to take into account the local rotation of fluid particles. The analysis in the present section should apply to the former class of fluids, whereas the latter class of ordinary fluids will be investigated next.

[20]Suppose that the velocity is represented as $\boldsymbol{u} = \nabla\phi + \boldsymbol{u}'$ with $\mathrm{div}\, \boldsymbol{u}' = 0$ and $\mathrm{div}\, \boldsymbol{u} = \nabla^2\phi$ (given), and that the normal component $\boldsymbol{n} \cdot \boldsymbol{u}'$ vanishes and $\boldsymbol{n} \cdot \boldsymbol{u} = \boldsymbol{n} \cdot \nabla\phi$ is given on the boundary surface S ($\boldsymbol{n}$ being a unit normal to S), where ϕ is a scalar function solving $\nabla^2\phi =$ *given* with values of $\boldsymbol{n} \cdot \nabla\phi$ on S. Then $\int (\nabla\phi + \boldsymbol{u}')^2 \mathrm{d}^3\boldsymbol{x} = \int (\nabla\phi)^2 \mathrm{d}^3\boldsymbol{x} + \int (v')^2 \mathrm{d}^3\boldsymbol{x} + 2\int_S \phi \boldsymbol{u}' \cdot \boldsymbol{n} \mathrm{d}S = \int (\nabla\phi)^2 \mathrm{d}^3\boldsymbol{x} + \int (v')^2 \mathrm{d}^3\boldsymbol{x} \geq \int (\nabla\phi)^2 \mathrm{d}^3\boldsymbol{x}$. (This is a proof generalized to a compressible case.)

7.11. Rotational Transformations

So far, we have investigated the translational symmetry, which is commutative. From this section, we are going to consider the rotational transformation briefly considered in §7.6.2. Related gauge group is the rotation group $SO(3)$ (Appendix C), i.e. a group of orthogonal transformations of $\mathbb{R}^3$ characterized with unit determinant, $\det R = 1$ for $R \in SO(3)$.

7.11.1. *Orthogonal transformation of velocity*

Consider the scalar product of velocity $\boldsymbol{v}$, i.e. $\langle \boldsymbol{v}, \boldsymbol{v} \rangle$ which is an integrand of the Lagrangian L_{F}. With an element $R \in SO(3)$, the transformation of a velocity vector $\boldsymbol{v}$ is represented by $\boldsymbol{v}' = R\boldsymbol{v}$. Then, the magnitude $|\boldsymbol{v}|$ is invariant, i.e. *isometric*: $|\boldsymbol{v}'|^2 \equiv \langle \boldsymbol{v}', \boldsymbol{v}' \rangle = \langle \boldsymbol{v}, \boldsymbol{v} \rangle$. This is equivalent to the definition of the orthogonal transformation (Fig. 7.9). In matrix notation $R = (R^i_j)$, a vector v^i is mapped to $(v')^i = R^i_j v^j$, and we have

$$\langle v', v' \rangle = (v')^i (v')^i = R^i_j v^j R^i_k v^k = v^k v^k = \langle v, v \rangle.$$

Therefore, the orthogonal transformation is described by

$$R^i_j R^i_k = (R^T)^j_i R^i_k = (R^T R)^j_k = \delta^j_k, \tag{7.107}$$

where R^T is the transpose of R and found to be equal to the inverse R^{-1}. Using the unit matrix $I = (\delta^j_k)$, this is rewritten as

$$R^T R = R R^T = I. \tag{7.108}$$

Taking its determinant, we have $(\det R)^2 = 1$. Note that every element R of SO(3) is defined by the property $\det R = 1$.

The Lagrangian L_{F} is invariant for a fixed R. In fact, the mass in a volume element $\mathrm{d}V$ is invariant by the rotational transformation:

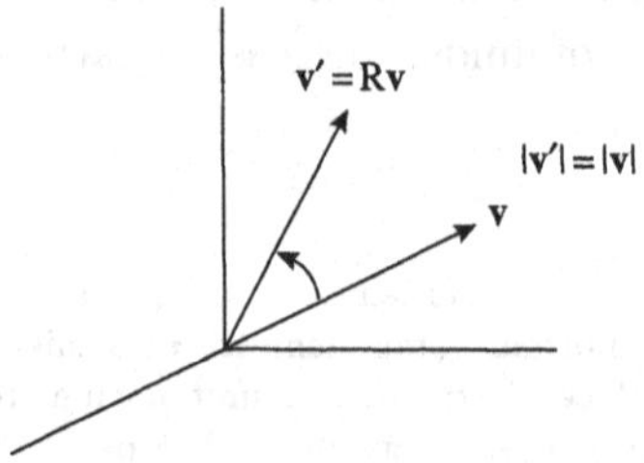

Fig. 7.9. Orthogonal transformation.

$R\,[\rho(\boldsymbol{x})\mathrm{d}V] = \rho(\boldsymbol{x})\,R[\mathrm{d}V] = \rho(\boldsymbol{x})\mathrm{d}V$, since scalar functions such as $\rho(\boldsymbol{x})$ are not influenced by the rotational transformation and the volume element is invariant. Thus, L_{F} has the global gauge invariance.

Likewise, it is not difficult to see that the Lagrangian L_{F} is *invariant* under a *local* transformation, $\boldsymbol{v}(x) \mapsto \boldsymbol{v}'(\boldsymbol{x}) = R(\boldsymbol{x})\,\boldsymbol{v}(\boldsymbol{x})$ depending on each point $\boldsymbol{x}$, where $R(\boldsymbol{x}) \in SO(3)$ at $^{\forall}\boldsymbol{x} \in M$, becasue the above invariance property of rotational transformation applies at each point.

7.11.2. *Infinitesimal transformations*

For later use, we take an element $R \in SO(3)$ and its varied element $R' = R + \delta R$ with an finitesimal variation δR. Suppose that an arbitrary vector $\boldsymbol{v}_0$ is sent to $\boldsymbol{v} = R\boldsymbol{v}_0$. We then have $\delta\boldsymbol{v} = \delta R\,\boldsymbol{v}_0 = (\delta R)R^{-1}\boldsymbol{v}$, so that $\boldsymbol{v} + \delta\boldsymbol{v}$ is represented as

$$\boldsymbol{v} \to \boldsymbol{v} + \delta\boldsymbol{v} = \big(I + (\delta R)R^{-1}\big)\,\boldsymbol{v} = \big(I + \theta\big)\,\boldsymbol{v},$$

where $\theta = (\delta R)R^{-1}$ is skew-symmetric for $R \in SO(3)$.[21] This term $\theta = (\delta R)R^{-1}$ is an element of the Lie algebra $\mathbf{so}(3)$, represented as $\theta = \theta^k e_k$ in §7.6.2 where (e_1, e_2, e_3) is the basis set of the Lie algebra $\mathbf{so}(3)$ (Appendix C.4). θ is a skew-symmetric 3×3 matrix. Analogously to (7.22), the infinitesimal gauge transformation is written as

$$\begin{aligned} R(\boldsymbol{x}) &= \exp[\theta] = I + \theta + O(|\theta|^2) \\ &= I + \big(\theta^1 e_1 + \theta^2 e_2 + \theta^3 e_3\big) + O(|\theta|^2), \end{aligned} \tag{7.109}$$

where $\theta^k \in \mathbb{R}$, $|\theta| \ll 1$. The operator $\theta = (\theta^i_j) = \theta^\alpha\, e_\alpha$ describes an infinitesimal rotation, where $\hat{\theta} := (\theta^1, \theta^2, \theta^3)$ is an infinitesimal angle vector (Fig. 7.10). Then, an infinitesimal rotation of the displacement vector $\boldsymbol{s} = (s^1, s^2, s^3)$ is given by (7.44). According to (7.49), the same local rotation is represented by $\frac{1}{2}\boldsymbol{\omega}_\xi \times \boldsymbol{s}$, where $\frac{1}{2}\boldsymbol{\omega}_\xi = \frac{1}{2}\mathrm{curl}\,\boldsymbol{\xi}$ corresponds to $\hat{\theta}$.

It is remarkable that the local rotation $R(\boldsymbol{x})$ of velocity $\boldsymbol{u}$ gives rise to a rotational component in the velocity field $\boldsymbol{v}(\boldsymbol{x})(= R(\boldsymbol{x})\boldsymbol{u}(\boldsymbol{x}))$ even though $\boldsymbol{u}(\boldsymbol{x})$ is irrotational. In fact, the velocity $\boldsymbol{v}$ is

$$\boldsymbol{u}(x) \to \boldsymbol{v} = \exp[\theta(x)]\,\boldsymbol{u} \approx \boldsymbol{u} + \theta\boldsymbol{u} = \boldsymbol{u} + \hat{\theta} \times \boldsymbol{u}. \tag{7.110}$$

---

[21] Since $RR^T = I$ and $(R + \delta R)(R + \delta R)^T = I$ from (7.108), we have $(\delta R)R^{-1} + (R^{-1})^T(\delta R)^T = 0$.

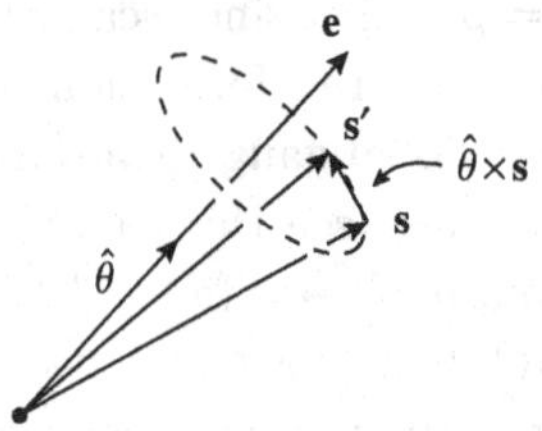

Fig. 7.10. Rotational transformation.

The transformed field $\boldsymbol{v} = R(x)\boldsymbol{u}(x)$ is *rotational*. The second term can be represented in terms of a vector potential B and a scalar potential f' as

$$\hat{\theta} \times \boldsymbol{u} = \operatorname{curl} B + \operatorname{grad} f', \tag{7.111}$$

(together with the gauge condition, $\operatorname{div} B = 0$, to fix an additional arbitrariness). Taking curl,

$$\operatorname{curl} \boldsymbol{v} = \operatorname{curl}(\hat{\theta} \times \boldsymbol{u}) = \operatorname{curl}(\operatorname{curl} B) = -\nabla^2 B,$$

which does not vanish in general, where ∇^2 is the Laplacian. The vector potential B is determined by the equation, $\nabla^2 B = -\operatorname{curl}(\hat{\theta} \times \boldsymbol{u})$. Thus, it is found that the rotational gauge transformation introduces a rotational component. Henceforth, it is understood that the vector $\boldsymbol{v}$ denotes a rotational velocity field.

This property distinguishes itself from the previous translational transformation which preserves irrotationality (§7.7.3). A gauge-covariant derivative is defined in the next section by introduing a gauge field Ω with respect to the rotational transformation.

7.12. Gauge Transformation (Rotation)

7.12.1. *Local gauge transformation*

The rotational symmetry of the flow field was considered in §7.6.2, where q^i and q^i_t are defined as $q^i = s^i$ and $q^i_t = \mathrm{D}_t q^i(\boldsymbol{s}) = u^i(\boldsymbol{x}) + s^j\,\partial_j u^i(\boldsymbol{x})$. Namely, $q(=\boldsymbol{s})$ is a local displacement vector and $q_t = \mathrm{D}_t q \approx \boldsymbol{u}(\boldsymbol{x}+\boldsymbol{s})$.

Suppose that, in order to satisfy the local rotational gauge invariance, an old form $L'(q, q_t)$ of the Lagrangian is modified to

$$L''(q, q_t, \Omega), \quad q_t = \mathrm{D}_t q, \tag{7.112}$$

where the third variable Ω is the newly introduced *guage field.* Assume that the local transformations are represented by

$$\left.\begin{aligned} q = (s^i) \to q' = q + \delta q, \qquad & \delta q^i = \theta^\beta(\boldsymbol{x}, t)\, e^i_{\beta\, j}\, q^j = (\hat{\theta} \times \boldsymbol{s})^i, \\ q_t \to q'_t = q_t + \delta q_t, \qquad & \delta q_t = \mathrm{D}_t(\delta q), \\ \Omega \to \Omega' = \Omega + \delta\Omega, \qquad & \delta\Omega^\alpha = \epsilon_{\alpha\beta\gamma}\theta^\beta\, \Omega^\gamma - \mathrm{D}_t\theta^\alpha. \end{aligned}\right\} \tag{7.113}$$

7.12.2. *Covariant derivative* ∇_t

Local invariance of L'' under the above transformation is given by

$$\delta L'' = \frac{\partial L''}{\partial q}\,\delta q + \frac{\partial L''}{\partial q_t}\,\delta q_t + \frac{\partial L''}{\partial \Omega}\,\delta\Omega = 0.$$

Substituting (7.113) and setting the coefficients of θ^β and $\mathrm{D}_t\theta^\beta$ to be zero separately, we obtain

$$0 = \frac{\partial L''}{\partial q^i}\, e^i_{\beta\, j}\, q^j + \frac{\partial L''}{\partial q^i_t}\, e^i_{\beta\, j}\, q^j_t + \frac{\partial L''}{\partial \Omega^\alpha}\,\epsilon_{\alpha\beta\gamma}\Omega^\gamma, \tag{7.114}$$

$$0 = \frac{\partial L''}{\partial q^i_t}\, e^i_{\beta\, j}\, q^j - \frac{\partial L''}{\partial \Omega^\beta}. \tag{7.115}$$

The second equation (7.115) means that the gauge field Ω is contained in L'' only through the combination expressed as

$$\mathrm{D}_t q^i + \Omega^\beta\, e^i_{\beta\, j}\, q^j = \mathrm{D}_t q^i + (\hat{\Omega} \times \boldsymbol{s})^i =: \nabla_t q^i, \tag{7.116}$$

where $\hat{\Omega} = (\Omega^\beta)$. Thus, a new velocity $\boldsymbol{v}$ is defined by $\nabla_t q^i$:

$$\boldsymbol{v} = \nabla_t q, \qquad \boldsymbol{v}(\boldsymbol{x} + \boldsymbol{s}) = \boldsymbol{u}(\boldsymbol{x} + \boldsymbol{s}) + \hat{\Omega} \times \boldsymbol{s}. \tag{7.117}$$

Variation of $\nabla_t q$ defined by (7.116) is given as

$$\delta\nabla_t q = \theta^\alpha(\boldsymbol{x}, t)\, e_\alpha\, \nabla_t q = \theta^\alpha(\boldsymbol{x}, t)\, e^i_{\alpha\, j}\, \nabla_t q^j. \tag{7.118}$$

Using $\boldsymbol{v}$, this is $\delta\boldsymbol{v} = \theta^\alpha(\boldsymbol{x}, t)\, e^i_{\alpha\, j}\, v^j = \hat{\theta} \times \boldsymbol{v}$.

7.12.3. *Gauge principle*

According to the gauge principle, the operator D_t is replaced by the covariant derivative ∇_t defined by

$$\nabla_t = \mathrm{D}_t + \Omega, \tag{7.119}$$

where the gauge-field operator Ω is characterized by

$$\Omega = \Omega^\alpha e_\alpha, \quad \Omega^i_j = \Omega^\alpha e^i_{\alpha\, j}, \quad \Omega^j_i = -\Omega^i_j.$$

According to the previous section, the Lagrangian is written as

$$L''(q,\, q_t,; \Omega) = L(q,\, \nabla_t q). \tag{7.120}$$

Using this form, we have the relations:

$$\left.\begin{aligned} \frac{\partial L''}{\partial q^i} &= \left.\frac{\partial L}{\partial q^i}\right|_{\nabla_t q: fixed} + \left.\frac{\partial L}{\partial \nabla_t q^j}\right|_{q: fixed} e^j_{\alpha\, i}\, \Omega^\alpha, \\ \frac{\partial L''}{\partial q^i_t} &= \left.\frac{\partial L}{\partial \nabla_t q^j}\right|_{q: fixed}, \qquad \frac{\partial L''}{\partial \Omega^\alpha} = \left.\frac{\partial L}{\partial \nabla_t q^j}\right|_{q: fixed} e^j_{(\alpha)\, i}\, q^i. \end{aligned}\right\} \tag{7.121}$$

In addition, global invariance of $L(q,\, \nabla_t q)$ for the variations δq and (7.118) results in

$$\frac{\partial L''}{\partial q^i} e^i_{\beta\, j}\, q^j + \left.\frac{\partial L}{\partial \nabla_t q^j}\right|_{q: fixed} e^i_{\beta\, j} (\nabla_t q)^j = 0. \tag{7.122}$$

Substituting (7.121) into (7.114) and using (7.122) and (7.116), it can be verified that Eq. (7.114) is satisfied identically.

Therefore, replacing the differential operator D_t in (7.98) with ∇_t, we obtain the variation δL_{F} of (7.98) for the rotational symmetry as

$$\delta L_{\mathrm{F}} = \partial_t \int_M [\langle \boldsymbol{v}, \boldsymbol{\xi} \rangle \rho \mathrm{d}V] - \int_M \langle \nabla_t \boldsymbol{v}, \boldsymbol{\xi} \rangle \rho \mathrm{d}V. \tag{7.123}$$

7.12.4. *Transformation law of the gauge field* Ω

We now require that not only the Lagranginan but its variational form should be *invariant* under local rotational transformations. We propose the following replacement:

$$\mathrm{D}_t \boldsymbol{v} \to \nabla_t \boldsymbol{v} := \mathrm{D}_t \boldsymbol{v} + \Omega\, \boldsymbol{v} = \partial_t \boldsymbol{v} + \mathrm{grad}\,(v^2/2) + \hat{\Omega} \times \boldsymbol{v}, \tag{7.124}$$

where $\mathrm{D}_t \boldsymbol{v} = \partial_t \boldsymbol{v} + \mathrm{grad}\,(\frac{1}{2} v^2)$ defined by (7.69), and Ω is the gauge-field operator, $\Omega \in \mathbf{so}(3)$, i.e. a 3×3 skew-symmetric matrix, and $\hat{\Omega}$ the axial vector counterpart of Ω.

Recall that local translational gauge transformation permits a term of the form Ωv for Av (§7.7.2). In the previous translational gauge invariance, this term must vanish to satisfy the global transformation. However, it will be seen that the term $\Omega\, v$ makes sense in the present rotational transformation.

According to the scenario of the gauge principle (e.g. [Qui83; FU86]), the velocity field v and the covariant derivative $\nabla_t v$ should obey the following transformation laws:

$$v \mapsto v' = \exp[\theta(t,x)]\, v \tag{7.125}$$

$$\nabla_t v \mapsto \nabla'_t v' = \exp[\theta(t,x)]\, \nabla_t v \tag{7.126}$$

where $\theta \in \mathbf{so}(3)$, i.e. θ being a skew-symmetric matrix.[22]

From the above equations (7.124)–(7.126), it is found that the gauge field operator Ω is transformed as

$$\Omega \to \Omega' = e^{\theta}\, \Omega\, e^{-\theta} - \left(\mathrm{D}_t\, e^{\theta}\right) e^{-\theta}. \tag{7.127}$$

Corresponding to the infinitesimal transformation, we have the expansion, $e^{\theta} = 1 + \theta + (|\theta|^2)$. Using $\delta\theta$ instead of θ,

$$v \to v' = \big(1 + \delta\theta\,\big)\, v = v + \delta\hat{\theta} \times v, \tag{7.128}$$

up to the first order, and the gauge field $\hat{\Omega}$ is transformed as

$$\hat{\Omega} \to \hat{\Omega}' = \hat{\Omega} + \delta\hat{\theta} \times \hat{\Omega} - \mathrm{D}_t(\delta\hat{\theta}). \tag{7.129}$$

The second term on the right-hand side came from $\delta\theta\,\Omega - \Omega\,\delta\theta$. This is equivalent to the non-abelian transformation law (7.26) for the Yang–Mills gauge field $\boldsymbol{A}_\mu$ (only for $\mu = t$) if $2q\boldsymbol{A}_\mu$ is replaced by Ω and $2q\boldsymbol{\alpha}$ by $-\delta\theta$.

7.13. Gauge-Field Lagrangian L_B (Rotational Symmetry)

Let us consider a *possible* type of the Lagrangian L_B for the gauge potential, newly defined by A_k for the rotational symmetry.[23] Its relation with the

---

[22]The property θ, $\Omega \in \mathbf{so}(3)$ means that we are considering the principal fiber bundle.

[23]Note that the new notation A_k of the gauge potential for the rotational symmetry is used by the analogy with the electromagnetic gauge potential. This is completely different from the gauge field $\mathcal{A}_k$ (no more used below) of the translational symmetry.

gauge field Ω will be considered elsewhere. Suppose that

$$L_B = L_B(A_k, A_{k;l}), \quad A_{k;l} = \partial_l A_k = \partial A_k / \partial x^l.$$

In an analogy with the electromagnetism, it is assumed that variations are assumed to be a potential-type, i.e.

$$\left.\begin{aligned} \delta A_k &= \partial_k \phi, \\ \delta A_{k;l} &= \partial_l(\delta A_k) = \partial_l \partial_k \phi, \end{aligned}\right\} \tag{7.130}$$

and in addition, it is required that the Lagrangian L_B is invariant with respect to such variations:

$$\delta L_B = \frac{\partial L_B}{\partial A_k}\,\delta A_k + \frac{\partial L_B}{\partial A_{k;l}}\,\delta A_{k;l} = 0.$$

Substituting the variations (7.130), we have a variational equation for arbitrary functions of $\partial_k \phi$, $\partial_l \partial_k \phi$. From the symmetry with respect to exchange of k and l of $\partial_l \partial_k \phi$, the vanishing of the coefficient of $\partial_l \partial_k \phi$ is

$$\frac{\partial L_B}{\partial A_{k;l}} + \frac{\partial L_B}{\partial A_{l;k}} = 0.$$

Thus, the derivative terms $A_{k;l}$ should be contained in L_B through

$$B_{kl} := \partial_k A_l - \partial_l A_k = A_{l;k} - A_{k;l}. \tag{7.131}$$

In addition, the vanishing of the coefficient of $\partial_k \phi$ is $\partial L_B / \partial A_k = 0$. Thus, the Lagrangian L_B depends only on B_{kl}: $L_B = L_B(B_{kl})$.

According to the theory of *electromagnetic field*, we define 1-form A^1 and its associated 2-form B^2 in the case of constant density ρ by

$$\left.\begin{aligned} A^1 &= A_k\,dx^k = A_1\,dx^1 + A_2\,dx^2 + A_3\,dx^3, \\ B^2 &= dA^1 = B_{12}\,dx^1 \wedge dx^2 + B_{23}\,dx^2 \wedge dx^3 + B_{31}\,dx^3 \wedge dx^1, \end{aligned}\right\} \tag{7.132}$$

where (7.131) is used. One can define a 1-form version of B^2 by

$$\left.\begin{aligned} B^1 &:= B_k\,dx^k, \quad B_i = \epsilon_{ijk}\,\partial_j A_k, \\ \boldsymbol{B} &:= \nabla \times \boldsymbol{A} = (B_i), \quad \boldsymbol{A} := (A_i). \end{aligned}\right\} \tag{7.133}$$

In addition, we introduce a vector V^1 defined by

$$V^1 = V_k\,dx^k = V_1\,dx^1 + V_2\,dx^2 + V_3\,dx^3, \tag{7.134}$$

$$\boldsymbol{B} := \rho \boldsymbol{V} = (\rho V_k), \quad (\rho = \text{const}). \tag{7.135}$$

A scalar depending on B_{kl} quadratically is defined by the following external product (since V^1 depends on B_{kl} linearly):

$$V^1 \wedge B^2 = \langle \boldsymbol{V}, \boldsymbol{B} \rangle \mathrm{d}^3\boldsymbol{x} = \langle \boldsymbol{V}, \boldsymbol{V} \rangle\, \rho\, \mathrm{d}^3\boldsymbol{x}.$$

Then, a quadratic Lagrangian is defined by

$$L_B = \tfrac{1}{2} \int \langle \boldsymbol{V}, \boldsymbol{B} \rangle \mathrm{d}^3\boldsymbol{x} = \tfrac{1}{2} \int \langle \boldsymbol{V}, \boldsymbol{V} \rangle\, \rho\, \mathrm{d}^3\boldsymbol{x}. \tag{7.136}$$

Its generalization to variable density is considered in §7.16.5 (with no formal change). Thus, it is concluded that the total Lagrangian L_{T} is defined as

$$L_{\mathrm{T}} := \tfrac{1}{2} \int_M \langle \boldsymbol{v}, \boldsymbol{v} \rangle \rho \mathrm{d}^3\boldsymbol{x} - \int_M \epsilon(\rho, s) \rho \mathrm{d}^3\boldsymbol{x} + \tfrac{1}{2} \int_M \langle \boldsymbol{V}, \boldsymbol{B} \rangle \mathrm{d}^3\boldsymbol{x}. \tag{7.137}$$

7.14. Biot–Savart's Law

7.14.1. *Vector potential of mass flux*

With respect to the newly introduced field $\boldsymbol{A}$, we try to find an equation to be derived from the total Lagrangian L_T with respect to the variation of the gauge potential $\boldsymbol{A} = (A_j)$ with other variables fixed.

Before carrying it out, we first take differential $\mathrm{d}L_T$ obtained by the difference of values of the integrands at two infinitesimally close points $\boldsymbol{x}$ and $\boldsymbol{x} + \mathrm{d}\boldsymbol{x}$. It is assumed that the density ρ is a uniform constant. Since ρ is constant, we obtain

$$\mathrm{d}L_{\mathrm{T}} = \rho \int_M \langle\, \boldsymbol{v}, \mathrm{d}\boldsymbol{v} \,\rangle\, \mathrm{d}^3\boldsymbol{x} - \rho \int_M \mathrm{d}\epsilon(\rho, s) \mathrm{d}^3\boldsymbol{x} + \rho \int_M \langle\, \boldsymbol{V}, \mathrm{d}\boldsymbol{V} \,\rangle \mathrm{d}^3\boldsymbol{x},$$

where $\rho \boldsymbol{V} = \boldsymbol{B} = \mathrm{curl}\, \boldsymbol{A}$, and

$$\mathrm{d}\boldsymbol{v} = \boldsymbol{v}(\boldsymbol{x} + \mathrm{d}\boldsymbol{x}) - \boldsymbol{v}(\boldsymbol{x}), \quad \mathrm{d}\boldsymbol{V} = \boldsymbol{V}(\boldsymbol{x} + \mathrm{d}\boldsymbol{x}) - \boldsymbol{V}(\boldsymbol{x}) = \rho^{-1}\, \mathrm{curl}\, \mathrm{d}\boldsymbol{A}.$$

Using the relation $\mathrm{d}\boldsymbol{v} = \nabla_t \mathrm{d}\boldsymbol{x}$, and carrying out integration by parts, we obtain

$$\mathrm{d}L_{\mathrm{T}} = -\rho \int_M \langle\, \nabla_t \boldsymbol{v}, \mathrm{d}\boldsymbol{x} \,\rangle\, \mathrm{d}^3\boldsymbol{x} - \rho \int_M \mathrm{d}\epsilon\; \mathrm{d}^3\boldsymbol{x} + \int_M \langle\, \Omega_V, \mathrm{d}\boldsymbol{A} \,\rangle \mathrm{d}^3\boldsymbol{x},$$

where $\nabla_t \boldsymbol{v} = \partial_t \boldsymbol{v} + \nabla\big(\tfrac{1}{2}\, v^2\big) + \hat{\Omega} \times \boldsymbol{v}$ and $\Omega_V = \nabla \times \boldsymbol{V}$.

Now, consider *variations* of the gauge fields

$$\delta \boldsymbol{A}, \quad \rho\,\delta \boldsymbol{V} = \nabla \times \delta \boldsymbol{A}, \quad \delta \Omega_V = \nabla \times \delta \boldsymbol{V},$$

with $\boldsymbol{v}(\boldsymbol{x})$, $\mathrm{d}\boldsymbol{x}$ being held *fixed* with ρ as a constant. Assuming that $\delta\Omega_V(\boldsymbol{x}) = \delta\hat{\Omega}$, and collecting nonzero terms from the first and third terms of $\mathrm{d}L_{\mathrm{T}}$ and setting $\mathrm{d}L_{\mathrm{T}} = 0$, we obtain

$$\int_M \big\langle\, \delta\hat{\Omega},\ [-\rho\,\boldsymbol{v} \times \mathrm{d}\boldsymbol{x} + 2\mathrm{d}\boldsymbol{A}]\, \big\rangle \, \mathrm{d}^3\boldsymbol{x} = 0.$$

The last equation must hold for arbitrary $\delta\hat{\Omega}$. Hence, we obtain

$$\mathrm{d}\boldsymbol{A} = \tfrac{1}{2}\rho\,\boldsymbol{v} \times \mathrm{d}\boldsymbol{x}, \qquad \mathrm{d}A_i = \tfrac{1}{2}\,\varepsilon_{ijk}\;\rho\, v^j \mathrm{d}x^k. \tag{7.138}$$

Taking curl of $\boldsymbol{A}$ (equal to $\boldsymbol{B}$), we have

$$\varepsilon_{\alpha\beta\gamma}\partial_\beta A_\gamma = \tfrac{1}{2}\varepsilon_{\alpha\beta\gamma}\varepsilon_{\gamma jk}\;\rho\, v^j\;\partial x^k/\partial x^\beta = \rho\, v_\alpha.$$

Thus, we find a connecting relation between $\boldsymbol{v}$ and $\boldsymbol{A}$:

$$\nabla \times \boldsymbol{A} = \boldsymbol{B} = \rho\,\boldsymbol{v} \qquad (\rho:\ \text{const}). \tag{7.139}$$

This means that the gauge potential $\boldsymbol{A}$ is the *vector potential* of the mass-flux field $\rho\boldsymbol{v}(\boldsymbol{x})$, which reduces to the stream function in case of two-dimensional flows. Taking curl again, we obtain

$$\hat{\Omega} \equiv \nabla \times \boldsymbol{V} = \rho^{-1}\,\nabla \times \boldsymbol{B} = \nabla \times \boldsymbol{v} \equiv \boldsymbol{\omega}, \tag{7.140}$$

where $\boldsymbol{\omega} \equiv \nabla \times \boldsymbol{v}$ is the *vorticity*. This is just the *Biot–Savart's law* between the vorticity $\boldsymbol{\omega}$ and the gauge field $\boldsymbol{V} = \boldsymbol{B}/\rho$.

This is analogous to the electromagnetic Biot–Savart's law $\boldsymbol{j} = \nabla \times \boldsymbol{B}_m$ beween the electric current density $\boldsymbol{j}$ and and the magnetic field $\boldsymbol{B}_m$ (an electromagnetic gauge field).

As a result, the covariant derivative $\nabla_t \boldsymbol{v}$ is given by

$$\nabla_t \boldsymbol{v} = \partial_t \boldsymbol{v} + \nabla\big(\tfrac{1}{2}\,v^2\big) + \boldsymbol{\omega} \times \boldsymbol{v} = \partial_t \boldsymbol{v} + (\boldsymbol{v}\cdot\nabla)\,\boldsymbol{v}. \tag{7.141}$$

The last expression denotes the *material derivative* in the rotational case, where an vector identity in $\mathbb{R}^3$ is used.[24] In terms of the particle coordinate $\boldsymbol{a}$, Eq. (7.141) is rewritten as

$$\nabla_t \boldsymbol{v} = \partial_\tau \boldsymbol{v}(\tau, \boldsymbol{a})\,|_{a=\mathrm{const}}. \tag{7.142}$$

[24] $\boldsymbol{v} \times (\nabla \times \boldsymbol{v}) = \nabla\big(\tfrac{1}{2}|\boldsymbol{v}|^2\big) - (\boldsymbol{v}\cdot\nabla)\boldsymbol{v}$.

7.14.2. *Vorticity as a gauge field*

In the previous section, we have found that the gauge operator $\hat{\Omega}$ is in fact the vorticity by the assumption of the relation $\delta\Omega_A = \delta\hat{\Omega}$. The equality $\hat{\Omega} = \nabla \times \boldsymbol{v}$ can be verified from the requirement of Galilei invariance of the covariant derivative $\nabla_t \boldsymbol{v}$ of (7.124), as follows.

Applying (7.7) and (7.8) (e.g. $\boldsymbol{v}$ is replaced by $\boldsymbol{v}_* + \boldsymbol{U}$), the covariant derivative $\nabla_t \boldsymbol{v} = \partial_t \boldsymbol{v} + \mathrm{grad}(v^2/2) + \hat{\Omega} \times \boldsymbol{v}$ is transformed to

$$\begin{aligned}(\partial_{t_*} - \boldsymbol{U}\cdot\boldsymbol{\nabla}_*)(\boldsymbol{v}_* + \boldsymbol{U}) &+ \boldsymbol{\nabla}_* \tfrac{1}{2}\,|\boldsymbol{v}_* + \boldsymbol{U}|^2 + \hat{\Omega} \times (\boldsymbol{v}_* + \boldsymbol{U}) \\ &= \partial_{t_*}\boldsymbol{v}_* + \boldsymbol{\nabla}_*\,(v_*^2/2) + \hat{\Omega} \times \boldsymbol{v}_* \\ &\quad -(\boldsymbol{U}\cdot\boldsymbol{\nabla}_*)\boldsymbol{v}_* + \hat{\Omega} \times \boldsymbol{U} + \boldsymbol{\nabla}_*(\boldsymbol{v}_* \cdot \boldsymbol{U}),\end{aligned}$$

since $\boldsymbol{U}$ is a constant vector and $\boldsymbol{\nabla}_*(U^2) = 0$. We require that the right-hand side is equal to $\partial_{t_*}\boldsymbol{v}_* + \boldsymbol{\nabla}_*(v_*^2/2) + \hat{\Omega}_* \times \boldsymbol{v}_*$, therefore

$$\begin{aligned}0 &= (\hat{\Omega} - \hat{\Omega}_*) \times \boldsymbol{v}_* - (\boldsymbol{U}\cdot\boldsymbol{\nabla}_*)\boldsymbol{v}_* + \hat{\Omega} \times \boldsymbol{U} + \boldsymbol{\nabla}_*(\boldsymbol{v}_* \cdot \boldsymbol{U}) \\ &= (\hat{\Omega} - \hat{\Omega}_*) \times \boldsymbol{v}_* + (\hat{\Omega} - \nabla_* \times \boldsymbol{v}_*) \times \boldsymbol{U},\end{aligned} \qquad (7.143)$$

where the following vector identity was used:

$$\boldsymbol{U} \times (\nabla_* \times \boldsymbol{v}_*) = -(\boldsymbol{U}\cdot\nabla_*)\boldsymbol{v}_* + \nabla_*(\boldsymbol{U}\cdot\boldsymbol{v}_*),$$

with $\boldsymbol{U}$ as a constant vector. Equation (7.143) is satisfied identically, if

$$\hat{\Omega} = \nabla \times \boldsymbol{v}, \quad \hat{\Omega}_* = \nabla_* \times \boldsymbol{v}_* = \nabla \times \boldsymbol{v} = \hat{\Omega}. \qquad (7.144)$$

The second relation holds by the Galilei transformation (7.8). Thus, the Galilei invariance of $\nabla_t \boldsymbol{v}$ results in the first relation, i.e. *the gauge field* $\hat{\Omega}$ *coincides with the vorticity*: $\boldsymbol{\omega} = \nabla \times \boldsymbol{v}$.

7.15. Hamilton's Principle for an Ideal Fluid (Rotational Flows)

We now consider the variational principle of general flows. By the requirement of local rotational gauge invariance in addition to the translational one, we have arrived at the covariant derivative (7.141), that is the *material derivative* of velocity. The variation should take into account the motion of individual particles, as carried out in §7.10 for irrotational flows. The gauge invariances require that laws of fluid motion should be expressed in a form equivalent to every individual particle.

7.15.1. *Constitutive conditions*

In order to comply with two gauge invariances (translational and rotational), we carry out again the material variations under the following three constitutive conditions.

(i) *Kinematic condition*: The $\boldsymbol{x}$-space trajectory of a material particle $\boldsymbol{a}$ is denoted by $\boldsymbol{x}_a(\tau) = \boldsymbol{x}(\tau, \boldsymbol{a})$, and the particle velocity is given by

$$\boldsymbol{v}(\tau, \boldsymbol{a}) = \partial_\tau \boldsymbol{x}(\tau, \boldsymbol{a}) \tag{7.145}$$

(Lagrangian representation). All the variations are taken with keeping $\boldsymbol{a}$ fixed, i.e. following trajectories of material particles. During the motion, the particle mass is kept constant. As a consequence, the equation of continuity must be satisfied, which is given by (7.90) with $\boldsymbol{u}$ replaced by $\boldsymbol{v}$. In the Eulerian representation, the velocity field is written by

$$\boldsymbol{v} = \boldsymbol{v}(t, \boldsymbol{x}). \tag{7.146}$$

(ii) *Ideal fluid*: An ideal fluid is characterized by the property that there is no disspative mechanism within it such as viscous dissipation or thermal conduction [LL87, §2, 49]. As a consequence, the entropy s per unit mass (i.e. specific entropy) remains constant following the motion of each material particle, i.e. *isentropic*. This is represented as $s = s(\boldsymbol{a})$, and its governing equation is given by (7.103) with $\boldsymbol{u}$ replaced by $\boldsymbol{v}$. The fluid is not necessarily homentropic, i.e. the entropy is not necessarily constant at every point.

(iii) *Gauge covariance*: All the expressions of the formulation must satisfy both global and local gauge invariance. Therefore, not only the action $\underline{I}$ defined just now, but also its varied form must be gauge-invariant, and the gauge-covariant derivative ∇_t of (7.141) must be used for the variation.

7.15.2. *Lagrangian and its variations*

Total Lagrangian of flows of an ideal fluid was given by (7.137), which is reproduced here:

$$L_{\mathrm{T}} = L_{\mathrm{F}} + L_\epsilon + L_B = \tfrac{1}{2}\int_M \langle \boldsymbol{v}, \boldsymbol{v}\rangle \rho \mathrm{d}^3\boldsymbol{x} - \int_M \epsilon(\rho, s)\rho \mathrm{d}^3\boldsymbol{x} + L_B, \tag{7.147}$$

where L_B is the Lagrangian depending on the gauge potential $\boldsymbol{A}$ which is assumed to be a function of $\boldsymbol{a}$ only and depends on $\boldsymbol{x}$ and t only through the function $\boldsymbol{a}(\boldsymbol{x},t)$. The Lagrangian L_B is invariant with respect to the material variation, and its explicit form will be given in the next section (§7.16.5).

The action principle is $\delta \underline{I} = \int_{t_0}^{t_1} \delta L_{\mathrm{T}} \, \mathrm{d}t = 0$. Variations of the Lagrangian L_{T} are carried out in two ways, with an analogy in mind of the case of quantum electrodynamics (QED), which is summarized briefly in Appendix I. In the QED case, the total Lagrangian density is composed of $\Lambda(\psi, A_\mu)$ of matter field ψ and $\Lambda_F(= -(1/16\pi)F_{\mu\nu}F^{\mu\nu})$ of the gauge field A_μ. The variation of $\Lambda(\psi, A_\mu)$ with respect to the wave function ψ (with A_μ fixed) yields the QED equation, i.e. the Dirac equation with electromagnetic field, while the variation with respect to the gauge field A_μ (with ψ fixed) yields the equations for the gauge field, i.e. Maxwell's equations of electromagnetism.

In the present case of fluid flows, the equation of motion is derived from the variation with respect to the particle position $\boldsymbol{x}_a$ in the $\boldsymbol{x}$-space by keeping $\boldsymbol{a}$ (therefore $\boldsymbol{A}(\boldsymbol{a})$) fixed. This is called the *material variation.* This will yield the Euler's equation of motion.

In §7.14.1, we carried out the variation of the gauge potential $\boldsymbol{A}$ with other variables (such as $\boldsymbol{v}(\boldsymbol{x})$, $\mathrm{d}\boldsymbol{x}$ and $\boldsymbol{a}(\boldsymbol{x})$) fixed, under the condition of constant density ρ. There, we found the relation $\mathrm{curl}\, \boldsymbol{A} \, (\equiv \boldsymbol{B}) = \rho \boldsymbol{v}$ and $\mathrm{curl}\, \boldsymbol{B} = \rho\, \boldsymbol{\omega}$ (Biot–Savart's law). This is a connecting relation between the vorticity field $\boldsymbol{\omega}(\boldsymbol{x})$ and the gauge field $\boldsymbol{B}(\boldsymbol{x})$.

Our system of fluid flows allows a third variation which is the variation with respect to the particle coordinate $\boldsymbol{a}$ in the $\boldsymbol{a}$-space by keeping $\boldsymbol{x}$, $\boldsymbol{v}(\boldsymbol{x})$ and $\rho(\boldsymbol{x})$ fixed. This is called *particle permutation.* This variation yields a gauge-field equation in the Lagrangian coordinate $\boldsymbol{a}$-space. The $\boldsymbol{a}$-variation is related to the symmetry of Lagrangian with respect to particle permutation, which leads to a local law of vorticity conservation, i.e. the vorticity equation. It is seen that there is a close analogy to the electromagnetic theory in this framework of the gauge theory (see comments in §7.14.1, 7.16.5).

In §7.15, we only consider the material variation, and the particle permutation symmetry will be considered in §7.16.

7.15.3. *Material variation: rotational and isentropic*

According to the scenario outlined in the previous subsections, we carry out an *isentropic material variation.* As a result of the variation satisfying local gauge invariance, we will obtain the Euler's equation of motion for an ideal fluid. In addition to the conservation of momentum associated with the translational invariance (considered in §7.4 briefly as a Noether's theorem), we will obtain another Noether's conservation law, i.e. the conservation of angular momentum, from the SO(3) gauge invariance.

All variations are taken so as to follow particle displacement ($\boldsymbol{a}$, $\boldsymbol{B}(\boldsymbol{a})$: fixed) under the kinematical constraint (7.90) and the isentropic condition (7.103). Associated with the variation of particle position given by

$$\boldsymbol{x}_a \mapsto \boldsymbol{x}_a + \boldsymbol{\xi}(\boldsymbol{x}_a, t), \tag{7.148}$$

the variation of particle velocity is represented by

$$\boldsymbol{v}(\boldsymbol{x}_a) \mapsto \boldsymbol{v}(\boldsymbol{x}_a + \boldsymbol{\xi}) + \mathrm{D}_t\boldsymbol{\xi}(\boldsymbol{x}_a) = \boldsymbol{v} + (\boldsymbol{\xi}\cdot\nabla)\boldsymbol{v} + \mathrm{D}_t\boldsymbol{\xi} \tag{7.149}$$

up to $O(|\boldsymbol{\xi}|)$ terms (see (7.85) and Fig. 7.11). Variations of density and internal energy consist of two components as before:

$$\Delta\rho = \boldsymbol{\xi}\cdot\nabla\rho + \delta\rho, \quad \Delta\epsilon = \boldsymbol{\xi}\cdot\nabla\epsilon + \delta\epsilon. \tag{7.150}$$

It is assumed that their variations are carried out *adiabatically*:

$$\delta s = 0. \tag{7.151}$$

The appropriate part $\delta\rho$ of the density variation is caused by the displacement $\delta\boldsymbol{x}_a = \boldsymbol{\xi}(\boldsymbol{x}, t)$. From the condition of fixed mass, we have, as before (§7.10.2),

$$\delta\rho = -\rho\,\mathrm{div}\,\boldsymbol{\xi}. \tag{7.152}$$

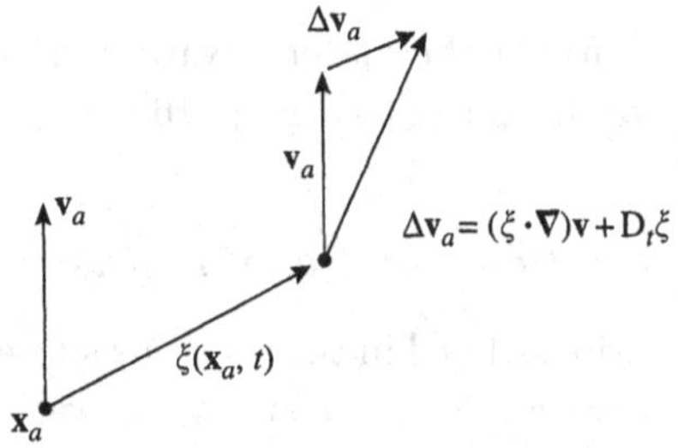

Fig. 7.11. Material variation.

Then, the proper part $\delta\epsilon(\rho, s)$ is expressed in terms of the density variation $\delta\rho$ and the pressure p:

$$\delta\epsilon = \frac{\partial\epsilon}{\partial\rho}\,\delta\rho + \frac{\partial\epsilon}{\partial s}\,\delta s = \frac{p}{\rho^2}\,\delta\rho. \tag{7.153}$$

The variation field $\boldsymbol{\xi}(\boldsymbol{x}, t)$ is constrained to vanish on the boundary surface S of M, as well as at both ends of time t_0, t_1 for the action $\underline{I}$:

$$\boldsymbol{\xi}(\boldsymbol{x}_S, t) = 0, \qquad \text{for any } {}^{\forall}t, \text{ for } \boldsymbol{x}_S \in S = \partial M, \tag{7.154}$$

$$\boldsymbol{\xi}(\boldsymbol{x}, t_0) = 0, \qquad \boldsymbol{\xi}(\boldsymbol{x}, t_1) = 0, \quad \text{for } {}^{\forall}\boldsymbol{x} \in M. \tag{7.155}$$

7.15.4. *Euler's equation of motion*

It is noted that the variations of $\boldsymbol{x}_a$, $\boldsymbol{v}$, ϵ, ρ and s, given in (7.148), (7.149)–(7.153), are different from those of irrotational case (§7.10.2) in the sense that those are extended so as to include rotational components, although their precise expressions look the same. Therefore, formal procedure of the variation of the Lagrangian L_{T} does not change. But, new consequences are to be deduced in the following sections.

Using (7.147), the variation $\delta\underline{I}$ is given by

$$\begin{aligned} \delta\,\underline{I} = \left[\int_M \langle \boldsymbol{v}, \boldsymbol{\xi}\rangle\ \rho \mathrm{d}V\right]_{t_0}^{t_1} + \int_{t_0}^{t_1} \mathrm{d}t \oint_S p\,\langle \boldsymbol{n}, \boldsymbol{\xi}\rangle \mathrm{d}S \\ - \int_{t_0}^{t_1} \mathrm{d}t \int_M \big\langle (\nabla_t \boldsymbol{v} + \rho^{-1}\mathrm{grad}\,p), \boldsymbol{\xi}\,\big\rangle\ \rho \mathrm{d}V, \end{aligned} \tag{7.156}$$

which is the same as (7.100) in the formal structure except ∇_t replaces D_t (see (7.123)). The first line on the right-hand side vanishes owing to the boundary conditions (7.154) and (7.155). Thus, the action principle $\delta\,\underline{I} = 0$ for arbitrary $\boldsymbol{\xi}$ results in

$$\nabla_t \boldsymbol{v} + \frac{1}{\rho}\,\nabla p = 0, \tag{7.157}$$

under the conditions (7.151) and (7.152). This is the **Euler's equation of motion**. In fact, using (7.141), we have

$$\partial_t \boldsymbol{v} + (\boldsymbol{v}\cdot\nabla)\,\boldsymbol{v} = -\frac{1}{\rho}\,\nabla p. \tag{7.158}$$

Using another equivalent expression of $\nabla_t \boldsymbol{v}$ in (7.141) and the thermodynamic equality $(1/\rho)\nabla p = \nabla h$ (see (7.104)), this can be rewritten as

$$\partial_t \boldsymbol{v} + \boldsymbol{\omega} \times \boldsymbol{v} + \nabla\left(\tfrac{1}{2}\, v^2\right) = -\nabla\, h. \tag{7.159}$$

This equation of motion must be supplemented by the equation of continuity and the isentropic equation:

$$\partial_t \rho + \mathrm{div}(\rho \boldsymbol{v}) = 0, \tag{7.160}$$

$$\partial_t s + \boldsymbol{v} \cdot \nabla\, s = 0. \tag{7.161}$$

Note: The form of Lagrangian L_{T} of (7.147) is compact with no constraint term, and the variation is carried out adiabatically by following particle trajectories. In the conventional variations [Ser59; Sal88], the Lagarangian has additional constraint terms which are imposed to obtain rotational component of velocity field.[25] In the conventional approaches including [Bre70], the formula (7.141) for the derivative of $\boldsymbol{v}$ is taken as an identity for the acceleration of a material particle without any further interpretation. However, in the present formulation, the expression (7.141) for $\nabla_t \boldsymbol{v}$ is the *covariant derivative*, an essential building block of gauge theory. There is another byproduct of the present gauge theory for the vorticity equation, which will be described in §7.16.

7.15.5. *Conservations of momentum and energy*

We now consider the Noether's theorem as a consequence of a symmetry of the system, which applies to every variational formulation with a Lagrangian.

We consider the global gauge transformation, $\boldsymbol{x} \mapsto \boldsymbol{x} + \boldsymbol{\xi}$ with a constant vector $\boldsymbol{\xi} \in \mathbb{R}^3$ (where $\boldsymbol{a}$ is fixed so that L_B does not change), which is a uniform translation. Variation of the Lagrangian $\delta L_T = \delta L_{\mathrm{F}} + \delta L_\epsilon$ due to this transformation is

$$\delta L_{\mathrm{T}} = \delta L_{\mathrm{F}} + \delta L_\epsilon = \partial_t \int_M \langle \boldsymbol{v}, \boldsymbol{\xi} \rangle\, \rho \mathrm{d}V + \oint_S p\, \langle \boldsymbol{n}, \boldsymbol{\xi} \rangle \mathrm{d}S$$
$$- \int_{t_0}^{t_1} \mathrm{d}t \int_M \big\langle (\nabla_t \boldsymbol{v} + \rho^{-1} \mathrm{grad}\, p), \boldsymbol{\xi} \big\rangle\, \rho \mathrm{d}V. \tag{7.162}$$

---

[25] Although the Lin's constraint yields a rotational component [Lin63; Ser59; SS77; Sal88], it is shown that the helicity of the vorticity field for a homentropic fluid in which $\mathrm{grad}\, s = 0$ vanishes [Bre70]. Such a rotational field is not general because the knotted vorticity field is excluded.

(see (7.156), (7.98) and (7.99)). This time, the second line vanishes because of the equation of motion (7.157). Thus, the Noether's theorem obtained from $\delta L_{\mathrm{T}} = 0$ is as follows:

$$\partial_t \int_M \langle \boldsymbol{v}, \boldsymbol{\xi} \rangle \, \rho \mathrm{d}V + \oint_S p \, \langle \boldsymbol{n}, \boldsymbol{\xi} \rangle \mathrm{d}S = 0,$$

for any compact space M with a bounding surface S. Since $\boldsymbol{\xi}$ is an arbitrary constant vector, this implies the following:

$$\partial_t \int_M \boldsymbol{v} \, \rho \mathrm{d}V = -\oint_S p \boldsymbol{n} \mathrm{d}S. \tag{7.163}$$

This is the *conservation law of total momentum.* In fact, the left-hand side denotes the time rate of change of total momentum within the space M, whereas the right-hand side is the resultant pressure force on the bounding surface S, where $-p\boldsymbol{n}\mathrm{d}S$ is a pressure force acting on a surface element $\mathrm{d}S$ from outside.

Equation (7.163) can be rewritten in the following way. Noting that $\partial_t \int_M = \int_M \nabla_t$, and that

$$\nabla_t(\rho \boldsymbol{v}) = \partial_t(\rho \boldsymbol{v}) + \boldsymbol{v} \cdot \nabla(\rho \boldsymbol{v}), \quad \nabla_t(\mathrm{d}V) = (\nabla \cdot \boldsymbol{v})\mathrm{d}V,$$

the left-hand side is rewritten as

$$\int_M \big[\partial_t(\rho \boldsymbol{v}) + \nabla \cdot \rho \boldsymbol{v}\boldsymbol{v}\big]\mathrm{d}V, \quad \text{where } (\nabla \cdot \rho \boldsymbol{v}\boldsymbol{v})^i = \partial_k(\rho v^k v^i).$$

The right-hand side is transformed to

$$-\int_M \nabla p \, \mathrm{d}V = -\int_M \nabla \cdot (pI) \, \mathrm{d}V$$

where $I = (\delta_{ij})$ is a unit tensor. Substituting these into (7.163), we obtain the *momentum conservation equation*[26]:

$$\partial_t(\rho \boldsymbol{v}) + \nabla \cdot \big(\rho \boldsymbol{v}\boldsymbol{v} + pI\big) = 0, \tag{7.164}$$

since Eq. (7.163) must hold with any compact M. This describes that the change of momentum density $\rho \boldsymbol{v}$ is equal to the negative divergence of the

[26]The ith component of $\nabla \cdot \big(\rho \boldsymbol{v}\boldsymbol{v} + pI\big)$ is $\partial_k(\rho v^k v^i + p\,\delta_{ki})$.

momentum flux tensor $\rho v^i v^k + p\delta_{ik}$ in the ideal fluid. Equation (7.164) is decomposed to

$$\rho\,(\partial_t \boldsymbol{v} + \boldsymbol{v}\cdot\nabla\,\boldsymbol{v} + \rho^{-1}\nabla\,p) + \boldsymbol{v}\,(\partial_t\rho + \nabla\cdot(\rho\boldsymbol{v})) = 0.$$

Because of the continuity equation (7.160), this reduces to the equation of motion (7.158).

Equation of *energy conservation* is obtained as follows. Taking scalar product of $\boldsymbol{v}$ with the equation of motion (7.158) by using the enthalpy h on the right-hand side, we obtain

$$\partial_t\big(\tfrac{1}{2}v^2\big) + (\boldsymbol{v}\cdot\nabla)\,\tfrac{1}{2}v^2 = -\boldsymbol{v}\cdot\nabla\,h.$$

Add the null-equation $\frac{1}{2}v^2(\partial_t\rho + \mathrm{div}(\rho\boldsymbol{v})) = 0$ to this equation multiplied with ρ, we obtain

$$\partial_t\big(\tfrac{1}{2}\rho v^2\big) + \nabla\cdot\big(\tfrac{1}{2}\rho v^2\boldsymbol{v}\big) = -\rho\boldsymbol{v}\cdot\nabla\,h.$$

Next, we note that $\partial_t\epsilon = (p/\rho^2)\partial_t\rho = -(p/\rho^2)\nabla\cdot(\rho\boldsymbol{v})$ by using the continuity equation. Then, we have

$$\partial_t(\rho\epsilon) = (\partial_t\rho)\,\epsilon + \rho\,(\partial_t\epsilon) = -\left(\epsilon + \frac{p}{\rho}\right)\nabla\cdot(\rho\boldsymbol{v}) = -h\,\nabla\cdot(\rho\boldsymbol{v}),$$

where $h = \epsilon + p/\rho$. Adding the last two equations, we obtain

$$\partial_t\big[\rho\big(\tfrac{1}{2}v^2 + \epsilon\big)\big] + \nabla\cdot\big[\rho\boldsymbol{v}\big(\tfrac{1}{2}v^2 + h\big)\big] = 0. \qquad (7.165)$$

This describes the change of energy density $\rho\big(\frac{1}{2}v^2 + \epsilon\big)$ under the energy flux vector $\rho\boldsymbol{v}\big(\frac{1}{2}v^2 + h\big)$ in the ideal fluid.

7.15.6. *Noether's theorem for rotations*

Next we consider a consequence of the rotational invariance. Global rotational transformation of a vector $\boldsymbol{v}$ is represented by $\boldsymbol{v}(\boldsymbol{x}) \mapsto \boldsymbol{v}'(\boldsymbol{x}) = R\,\boldsymbol{v}(\boldsymbol{x})$ with a fixed element $R \in SO(3)$ at every point $\boldsymbol{x} \in M$, which is equivalent to a uniform rotation. Let us take a rotation vector $\delta\hat{\theta} = |\delta\hat{\theta}|\,\boldsymbol{e}$ of the rotation angle $|\delta\hat{\theta}|$ about the axis $\boldsymbol{e}$. Then, an infinitesimal global transformation is defined for the position vector $\boldsymbol{x}$ as

$$\boldsymbol{x}' = \boldsymbol{x} + \delta\hat{\theta} \times \boldsymbol{x}. \tag{7.166}$$

Next, we consider an infinitesimal variation of the Lagrangian (with $\boldsymbol{A}$ and $\boldsymbol{a}$ fixed) due to this transformation. From (7.156), the variation is given by

$$\begin{aligned} \delta L_{\mathrm{T}} = & \frac{\partial}{\partial t} \int_M \langle \boldsymbol{v}, \boldsymbol{\xi} \rangle \rho \mathrm{d}V + \oint_S p \langle \boldsymbol{\xi}, \boldsymbol{n} \rangle \, \mathrm{d}S \\ & - \int_M \langle (\nabla_t \boldsymbol{v} + \frac{1}{\rho} \nabla p), \ \boldsymbol{\xi} \rangle \rho \mathrm{d}V. \end{aligned} \tag{7.167}$$

where $\boldsymbol{\xi} = \delta\hat{\theta} \times \boldsymbol{x}$ with a constant vector $\delta\hat{\theta}$, and $\boldsymbol{n}$ is a unit outward normal to S. The last term vanishes owing to the equation of motion (7.157). The Noether's theorem is given by $\delta L_{\mathrm{T}} = 0$, which reduces to

$$\frac{\partial}{\partial t} \int_M \langle \boldsymbol{v}, \boldsymbol{\xi}(\boldsymbol{x}) \rangle \rho \mathrm{d}V + \int_S p \langle \boldsymbol{\xi}(\boldsymbol{x}_S), \boldsymbol{n} \rangle \, \mathrm{d}S = 0. \tag{7.168}$$

Using $\boldsymbol{\xi} = \delta\hat{\theta} \times \boldsymbol{x}$, it is verified that this represents the conservation of total angular momentum. In fact, using the vector identity $\langle \boldsymbol{v}, \delta\hat{\theta} \times \boldsymbol{x} \rangle = \langle \delta\hat{\theta}, \boldsymbol{x} \times \boldsymbol{v} \rangle \equiv \delta\hat{\theta} \cdot (\boldsymbol{x} \times \boldsymbol{v})$ with $\delta\hat{\theta}$ being a constant vector, the first term is

$$\delta\hat{\theta} \cdot \frac{\partial}{\partial t} \int_M \boldsymbol{x} \times \boldsymbol{v} \, \rho \mathrm{d}V = \delta\hat{\theta} \cdot \frac{\partial}{\partial t} \boldsymbol{L}(M). \tag{7.169}$$

The integration term denoted by $L(M)$ is the total angular momntum of M. Similarly, the second term is

$$\delta\hat{\theta} \cdot \int_S \boldsymbol{x} \times (p\boldsymbol{n} \mathrm{d}S) = -\delta\hat{\theta} \cdot \boldsymbol{N}(S). \tag{7.170}$$

The surface integral over S is denoted by $-\boldsymbol{N}(S)$, since $\boldsymbol{N}(S)$ is the resultant moment of pressure force $-p\boldsymbol{n}\mathrm{d}S$ acting on a surface element $\mathrm{d}S$ from outside. Since $\delta\hat{\theta}$ is an arbitrary constant vector, Eq. (7.168) implies the following:

$$\frac{\partial}{\partial t} \boldsymbol{L}(M) = \boldsymbol{N}(S). \tag{7.171}$$

Thus, it is found that the Noether's theorem leads to the conservation law of total angular momentum from the SO(3) gauge invariance.

7.16. Local Symmetries in a-Space

It is instructive to formulate the action principle in terms of the particle coordinates $\boldsymbol{a} = (a^1, a^2, a^3)$. First, the equation of motion will be derived with respect to the independent variables $(\tau, \boldsymbol{a})$. Next, local rotation symmetry concerning the $\boldsymbol{a}$-coordinate results in an associated local conservation law of vorticity in the $\boldsymbol{a}$-space. Recall that $\boldsymbol{a} = (a, b, c)$ is defined as the *mass coordinate* in §7.8.1.

7.16.1. *Equation of motion in a-space*

Variation of L_{T} with respect to $\boldsymbol{x}_a$ in the $\boldsymbol{x}$-space is written as

$$\delta L_{\mathrm{T}} = \delta L_{\mathrm{F}} + \delta L_{\epsilon} = \int_M \langle \boldsymbol{v}, \delta \boldsymbol{v} \rangle \mathrm{d}^3\boldsymbol{a} - \int_M \delta\epsilon(\rho, s)\, \rho \mathrm{d}V, \tag{7.172}$$

under the condition that $\boldsymbol{A}$ and $\boldsymbol{a}$ are fixed so that $\delta L_B = 0$. In the first term (denoted by δL_{F}), $\rho \mathrm{d}V$ is replaced by $\mathrm{d}^3\boldsymbol{a}$. In the second term (denoted by δL_ϵ), we have kept the original form $\rho \mathrm{d}V$ by the reason that will become clear just below.

Following the particle motion, the variables $(a^1, a^2, a^3) = (a, b, c)$ are invariant by definition, which is expressed as

$$\mathrm{D}_t a^i = \partial_t a^i + (\boldsymbol{v} \cdot \nabla) a^i = 0. \tag{7.173}$$

In the variation $\boldsymbol{x}_a \mapsto \boldsymbol{x}_a + \delta \boldsymbol{x}_a$ of (7.148), the particle which was located originally at $\boldsymbol{x}_a$ is displaced to a new position $\boldsymbol{x}_a + \delta \boldsymbol{x}_a$ with the coordinate $\boldsymbol{a}$ fixed. Suppose that the particle which was located at $\boldsymbol{x}_a + \delta \boldsymbol{x}_a$ originally had the coordinate $\boldsymbol{a} + \delta \boldsymbol{a}$ (Fig. 7.12), then we have

$$\delta x_a^k = \frac{\partial x^k}{\partial a^i} \delta a^i, \quad \delta a^i = \frac{\partial a^i}{\partial x^k} \delta x_a^k. \tag{7.174}$$

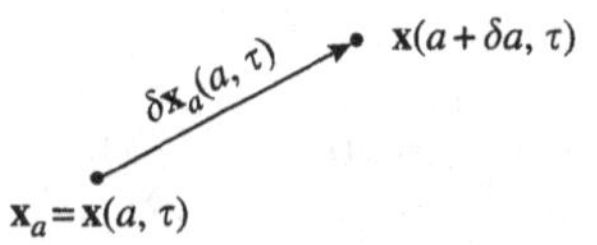

Fig. 7.12. Relation between $\delta \boldsymbol{x}_a$ and $\delta \boldsymbol{a}$, with $\delta \boldsymbol{v} = \partial_\tau(\delta \boldsymbol{x}_a)$.

As for $\delta\epsilon$, using (7.153) and (7.152), we have $\delta\epsilon = -(p/\rho)\,\mathrm{div}\,\delta\boldsymbol{x}_a$. Then, the variation of the second term $\delta L_\epsilon = -\int_M \delta\epsilon\,\rho\mathrm{d}V$ becomes

$$\begin{aligned}
\delta L_\epsilon &= \int_M p\,(\mathrm{div}\,\delta\boldsymbol{x}_a)\mathrm{d}V = \oint_S p\,\langle \boldsymbol{n}, \delta\boldsymbol{x}_a\rangle \mathrm{d}S - \int_M \delta\boldsymbol{x}_a \cdot \mathrm{grad}\,p\,\mathrm{d}V \\
&= \oint_S p\,\langle \boldsymbol{n}, \delta\boldsymbol{x}_a\rangle \mathrm{d}S - \int_M \delta a^i\,\frac{\partial x^k}{\partial a^i}\,\frac{\partial p}{\partial x^k}\,\frac{1}{\rho}\,\rho\mathrm{d}V \\
&= \oint_S p\,\langle \boldsymbol{n}, \delta\boldsymbol{x}_a\rangle \mathrm{d}S - \int_M \langle \nabla_a h, \delta\boldsymbol{a}\rangle\,\mathrm{d}^3\boldsymbol{a}, \qquad (7.175)\\
\nabla_a &:= \frac{\partial}{\partial a^i} = \frac{\partial x^k}{\partial a^i}\,\frac{\partial}{\partial x^k}, \qquad (7.176)
\end{aligned}$$

where ∇_a is the nabla operator with respect to the variables a^i, and $\nabla_a h = (1/\rho)\nabla_a p = (1/\rho)(\partial p/\partial a^i)$. Thus, we have obtained δL_ϵ represented in terms of the coordinate $\boldsymbol{a}$ only.

We seek a similar representation of the first term of (7.172). Using the relation $\delta v^k = \partial_\tau(\delta x_a^k)$, we obtain

$$\begin{aligned}
\delta L_{\mathrm{F}} &:= \int \mathrm{d}^3\boldsymbol{a}\; v^k\,\delta v^k = \int \mathrm{d}^3\boldsymbol{a}\; v^k\,\partial_\tau(\delta x_a^k) \\
&= \frac{\partial}{\partial\tau}\int \mathrm{d}^3\boldsymbol{a}\; v^k\,\delta x_a^k - \int \mathrm{d}^3\boldsymbol{a}\; \partial_\tau v^k\,\frac{\partial x^k}{\partial a^i}\,\delta a^i \qquad (7.177)\\
&= \frac{\partial}{\partial\tau}\int \mathrm{d}^3\boldsymbol{a}\; v^k\,\delta x_a^k - \int \mathrm{d}^3\boldsymbol{a}\; \langle d_\tau \boldsymbol{V}_a, \delta\boldsymbol{a}\rangle, \qquad (7.178)
\end{aligned}$$

where $v^k = \partial_\tau x^k$, and $\partial_\tau v^k$ is defined by (7.142), and

$$d_\tau \boldsymbol{V}_a := (\partial_\tau v^k)\,\nabla_a x^k = \partial_\tau \boldsymbol{V}_a - \nabla_a(v^2/2), \qquad (7.179)$$

$$\boldsymbol{V}_a := v_k\,\nabla_a x^k, \qquad (V_a)_i = v_k\,\frac{\partial x^k}{\partial a^i}. \qquad (7.180)$$

Here a covector $\boldsymbol{V} = (v_k)$ is introduced by the definition, $v_k = \delta_{kl}\,v^l = v^k$. The covector $\boldsymbol{V}_a$ is the velocity covector *transformed* to the $\boldsymbol{a}$-space. This is seen on the basis of a 1-form V^1 defined by

$$\begin{aligned}
V^1 &= v_1\mathrm{d}x^1 + v_2\mathrm{d}x^2 + v_3\mathrm{d}x^3 \ (:= \boldsymbol{V}\cdot\mathrm{d}\boldsymbol{x}) \qquad (7.181)\\
&= (V_a)_1\mathrm{d}a^1 + (V_a)_2\mathrm{d}a^2 + (V_a)_3\mathrm{d}a^3 \ (:= \boldsymbol{V}_a\cdot\mathrm{d}\boldsymbol{a}). \qquad (7.182)
\end{aligned}$$

Then, the variation of the action $\underline{I}$ with $\delta L_{\mathrm{T}} = \delta L_{\mathrm{F}} + \delta L_\epsilon$ is given by

$$\int_{t_0}^{t_1} \delta L_{\mathrm{T}}\mathrm{d}\tau = -\int \mathrm{d}\tau \int \mathrm{d}^3\boldsymbol{a}\; \big\langle [d_\tau \boldsymbol{V}_a + \nabla_a h], \delta\boldsymbol{a}\big\rangle,$$

where boundary terms are deleted by the boundary conditions (7.154) and (7.155) as before. Thus the action principle $\delta \underline{I} = 0$ results in

$$d_\tau \boldsymbol{V}_a + \nabla_a h = (\partial_\tau v^k)\, \nabla_a x^k + \nabla_a h = 0. \tag{7.183}$$

This is the *Lagrangian form* of the equation of motion [Lamb32, §13].

7.16.2. *Vorticity equation and local rotation symmetry*

We next consider a symmetry with respect to *particle permutation*. Suppose that a material particle $\boldsymbol{a}$ is replaced by a particle $\boldsymbol{a}'$ of the same mass, and that this permutation is carried out without affecting the current velocity field $\boldsymbol{v}(\boldsymbol{x})$ and carried out adiabatically, hence s being invariant. Therefore, we are going to investigate a *hidden* symmetry with respect to permutation of *equivalent* fluid particles.

To be precise, suppose that the permutation is represented by the following infinitesimal variation of particle coordinates, $\boldsymbol{a} \to \boldsymbol{a}' = \boldsymbol{a} + \delta\boldsymbol{a}(\boldsymbol{a})$. By this permutation, a fluid particle $\boldsymbol{a}$ which occupied a spatial point $\boldsymbol{x}$ shifts to a new spatial point $\boldsymbol{x} + \delta\boldsymbol{x}$ where a particle $\boldsymbol{a}' = \boldsymbol{a} + \delta\boldsymbol{a}(\tau, \boldsymbol{a})$ occupied before, and the position $\boldsymbol{x}$ where $\boldsymbol{a}$ was located before is now occupied by a particle $\boldsymbol{a}''$ (Fig. 7.13). Therefore, we have $\delta x^k = (\partial x^k / \partial a^i)\, \delta a^i$.

Since the particle $\boldsymbol{a}$ is now located at a point where the velocity is $\boldsymbol{v}(\boldsymbol{a}') = \boldsymbol{v}(\boldsymbol{a}) + \delta\boldsymbol{v}$, Eq. (7.173) is replaced by

$$\partial_t a^i + (\boldsymbol{v} + \delta\boldsymbol{v}) \cdot \nabla a^i = 0.$$

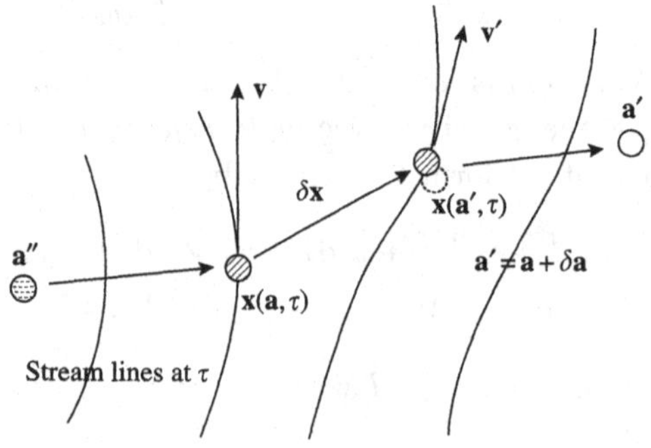

Fig. 7.13. Relation between $\delta\boldsymbol{x}_a$ and $\delta\boldsymbol{a}$, with $\boldsymbol{v}(\boldsymbol{x})$ unchanged.

Analogously, the equation for the particle $\boldsymbol{a}'' = \boldsymbol{a} - \delta\boldsymbol{a} + (|\delta\boldsymbol{a}|^2)$ (derived from the definition $\boldsymbol{a} = \boldsymbol{a}'' + \delta\boldsymbol{a}(\tau, \boldsymbol{a}''))$ is written as

$$\partial_t(a^i - \delta a^i) + \boldsymbol{v} \cdot \nabla (a^i - \delta a^i) = 0.$$

Eliminating the unvaried terms between the two equations, we obtain $\partial_t \delta a^i + (\boldsymbol{v} \cdot \nabla)\delta a^i = -(\delta\boldsymbol{v} \cdot \nabla) a^i$, which is rewritten as

$$\partial_\tau \, \delta a^i = -\frac{\partial a^i}{\partial x^k} \delta v^k,$$

where $\partial_\tau \, \delta a^i = \partial_t \delta a^i + (\boldsymbol{v} \cdot \nabla)\delta a^i$. This equation can be solved for δv^k by multiplying $\partial x^k / \partial a^i$, and we obtain

$$\delta v^k = -\frac{\partial x^k}{\partial a^i} \partial_\tau \delta a^i(\tau, \boldsymbol{a}), \tag{7.184}$$

since $(\partial x^k / \partial a^i)(\partial a^i / \partial x^j)\delta v^j = \delta^k_j \, \delta v^j = \delta v^k$.

Because the mass $\mathrm{d}a'\mathrm{d}b'\mathrm{d}c'$ at $\boldsymbol{a}' = (a', b', c')$ is replaced by the same amount of mass $\mathrm{d}a\mathrm{d}b\mathrm{d}c$ at $\boldsymbol{a} = (a, b, c)$, we have

$$\frac{\partial(a', b', c')}{\partial(a, b, c)} = 1. \tag{7.185}$$

For an infinitesimal transformation $\boldsymbol{a}' = \boldsymbol{a} + \delta\boldsymbol{a}(\boldsymbol{a}) = (a + \delta a(\boldsymbol{a}), b + \delta b(\boldsymbol{a}), c + \delta c(\boldsymbol{a}))$, Eq. (7.185) implies

$$\frac{\partial \delta a}{\partial a} + \frac{\partial \delta b}{\partial b} + \frac{\partial \delta c}{\partial c} = 0.$$

Such a divergence-free vector field $\delta\boldsymbol{a}(\tau, \boldsymbol{a})$ is represented as

$$\delta\boldsymbol{a} = \nabla_a \times \delta\Psi(\boldsymbol{a}), \tag{7.186}$$

by using a vector potential $\delta\Psi(\tau, \boldsymbol{a})$.[27]

Variation of the Lagrangian is given by (7.172) again, and variations of each term are given by (7.175) and (7.178) respectively, since the relation between δa^i and δx^i takes the same form as the previous (7.174) although their background mechanisms are different. Assuming the same boundary conditions (7.154) and (7.155) as before, both first terms of (7.175) and (7.178) vanish.

---

[27]The expression (7.185) is called *unimodular* or *measure-preserving* transformation by Eckart [Eck60]. See (7.176) for the definition of ∇_a.

Substituting (7.186) into the second term of (7.175), we have

$$\delta L_\epsilon = -\int_M \langle \nabla_a h, \nabla_a \times \delta\Psi \rangle \mathrm{d}^3\boldsymbol{a}$$
$$= -\int_M \langle \nabla_a \times \nabla_a h, \delta\Psi \rangle \mathrm{d}^3\boldsymbol{a} = 0, \qquad (7.187)$$

since $\nabla_a \times \nabla_a h \equiv 0$,[28] where partial integrations are carried out with respect to the $\boldsymbol{a}$-variables.

By the particle permutation symmetry, it is required that the action $\underline{I}$ is invariant with respect to the variation (7.186) for arbitrary vector potential $\delta\Psi(\tau, \boldsymbol{a})$. Using (7.178) and (7.187), the variation $\delta \underline{I}$ is given by

$$\delta \underline{I} = -\int \mathrm{d}\tau \int \mathrm{d}^3\boldsymbol{a}\ \langle d_\tau \boldsymbol{V}_a,\ \nabla_a \times \delta\Psi \rangle$$
$$= -\int \mathrm{d}\tau \int \mathrm{d}^3\boldsymbol{a}\ \langle \partial_\tau(\nabla_a \times \boldsymbol{V}_a),\ \delta\Psi \rangle,$$

To obtain the integrand of the last line, we used the definition (7.179) of $d_\tau \boldsymbol{V}_a$:

$$\nabla_a \times d_\tau \boldsymbol{V}_a = \nabla_a \times (\partial_\tau \boldsymbol{V}_a - \nabla_a(v^2/2)) = \partial_\tau(\nabla_a \times \boldsymbol{V}_a).$$

The variation of the vector potential $\delta\Psi$ is regarded as arbitrary, hence the action principle $\delta \underline{I} = 0$ requires

$$\partial_\tau(\nabla_a \times \boldsymbol{V}_a) = 0. \qquad (7.188)$$

This equation, found by Eckart (1960) [Eck60; Sal88], represents a conservation law of local rotation symmetry in the particle-coordinate space, and may be regarded as the *vorticity equation* in the $\boldsymbol{a}$-space. Its $\boldsymbol{x}$-space version will be given next.

7.16.3. *Vorticity equation in the x-space*

To see the meaning of (7.188) in the $\boldsymbol{x}$-space, it is useful to realize that $\nabla_a \times \boldsymbol{V}_a$ is the vorticity transformed to the $\boldsymbol{a}$-space. Curl of a vector $\boldsymbol{V}_a = (V_a, V_b, V_c)$ is denoted by $\boldsymbol{\Omega}_a = (\Omega_a, \Omega_b, \Omega_c) := \nabla_a \times \boldsymbol{V}_a$. In differential-form representation, this is described by a two-form $\Omega^2 = \mathrm{d}V^1$ (using V^1

[28] Salmon [Sal88] introduced the vector potential $\mathbf{T} = \delta\Psi(\boldsymbol{a})$ which has been later found useful, but $\delta L_\epsilon = 0$ is not mentioned explicitly.

defined by (7.182)):

$$\Omega^2 = \mathrm{d}V^1 = \Omega_a \mathrm{d}b \wedge \mathrm{d}c + \Omega_b \mathrm{d}c \wedge \mathrm{d}a + \Omega_c \mathrm{d}a \wedge \mathrm{d}b = \boldsymbol{\Omega}_a \cdot \boldsymbol{S}^2$$
$$= \omega_1 \mathrm{d}x^2 \wedge \mathrm{d}x^3 + \omega_2 \mathrm{d}x^3 \wedge \mathrm{d}x^1 + \omega_3 \mathrm{d}x^1 \wedge \mathrm{d}x^2 = \boldsymbol{\omega} \cdot \boldsymbol{s}^2, \quad (7.189)$$

where $\boldsymbol{\omega} = \nabla \times \boldsymbol{v} = (\omega_1, \omega_2, \omega_3)$ is the vorticity in the physical $\boldsymbol{x}$ space, and the 2-forms $\boldsymbol{s}^2$ and $\boldsymbol{S}^2$ are surface forms.[29] From Eqs. (7.188) and (7.189), one can conclude that

$$0 = \partial_\tau[\boldsymbol{\Omega}_a \cdot \boldsymbol{S}^2] = \partial_\tau[\Omega^2] = \partial_\tau[\boldsymbol{\omega} \cdot \boldsymbol{s}^2] = 0. \quad (7.190)$$

Let us introduce a gauge potential covector $\boldsymbol{A}_a$ of particle coordinates only: $\boldsymbol{A}_a(\boldsymbol{a}) = (A_a, A_b, A_c)$, and define a 1-form A^1 by

$$A^1 = A_a \mathrm{d}a + A_b \mathrm{d}b + A_c \mathrm{d}c. \quad (7.191)$$

The exterior product of Ω^2 and A^1 is

$$\Omega^2 \wedge A^1 = \langle \boldsymbol{\Omega}_a, \boldsymbol{A}_a \rangle \mathrm{d}^3\boldsymbol{a}, \quad \mathrm{d}^3\boldsymbol{a} = \mathrm{d}a \wedge \mathrm{d}b \wedge \mathrm{d}c. \quad (7.192)$$

In the $\boldsymbol{x} = (x, y, z)$-space, the same exterior product is given by

$$\Omega^2 \wedge A^1 = \langle \boldsymbol{\omega}, \boldsymbol{A} \rangle \mathrm{d}^3\boldsymbol{x}, \quad \mathrm{d}^3\boldsymbol{x} = \mathrm{d}x \wedge \mathrm{d}y \wedge \mathrm{d}z, \quad (7.193)$$

where $\boldsymbol{A}$ is defined by (7.132). Relation between $\boldsymbol{A}$ and $\boldsymbol{A}_a$ is the same as that of $\boldsymbol{V}$ and $\boldsymbol{V}_a$ in (7.181) and (7.182).

From the equality of (7.192) and (7.193) and $\mathrm{d}^3\boldsymbol{a} = \rho \mathrm{d}^3\boldsymbol{x}$, we obtain the following transformation law between $\boldsymbol{\omega}$ and $\boldsymbol{\Omega}_a$ [Sal88]:

$$\langle \boldsymbol{\omega}, \boldsymbol{A} \rangle = \rho \langle \boldsymbol{\Omega}_a, \boldsymbol{A}_a \rangle. \quad (7.194)$$

The derivative $\partial/\partial\tau$ is understood to be the Lie derivative $\mathcal{L}_X$ with the vector X defined by

$$X = \partial_\tau\big|_{a:\text{fixed}} = \partial_t + v^k(t, \boldsymbol{x})\, \partial_k.$$

Thus, we obtain

$$\partial_\tau[\Omega^2] = \mathcal{L}_X[\Omega^2] = \mathcal{L}_{\partial_t + v^k\partial_k}[\Omega^2] = \mathcal{L}_{\partial_t}[\Omega^2] + \mathcal{L}_{v^k\partial_k}[\Omega^2] = 0,$$

[29] $\boldsymbol{s}^2 = (\mathrm{d}x^2 \wedge \mathrm{d}x^3, \mathrm{d}x^3 \wedge \mathrm{d}x^1, \mathrm{d}x^1 \wedge \mathrm{d}x^2)$, and $\boldsymbol{S}^2 = (\mathrm{d}a^2 \wedge \mathrm{d}a^3, \mathrm{d}a^3 \wedge \mathrm{d}a^1, \mathrm{d}a^1 \wedge \mathrm{d}a^2)$. For example, the $\mathrm{d}b \wedge \mathrm{d}c$ component of $\nabla_a \times \boldsymbol{V}_a$ is given by

$$\Omega_a = \frac{\partial V_c}{\partial b} - \frac{\partial V_b}{\partial c} = \frac{\partial}{\partial b}\left(v_k \frac{\partial x^k}{\partial c}\right) - \frac{\partial}{\partial c}\left(v_k \frac{\partial x^k}{\partial b}\right).$$

from (7.190). With the $\boldsymbol{x}$-space notation, this is written as

$$\partial_t \boldsymbol{\omega} + \operatorname{curl}(\boldsymbol{\omega} \times \boldsymbol{v}) = 0. \tag{7.195}$$

To verify this, we write $\Omega^2 = \boldsymbol{\omega} \cdot \boldsymbol{s}^2$. Then,

$$\mathcal{L}_{\partial_t}(\Omega^2) = (\partial_t \boldsymbol{\omega}) \cdot \boldsymbol{s}^2.$$

Next, for $\mathcal{L}_v$ with $v = v^k \partial_k$, we use the Cartan's formula (B.20) (Appendix B.4): $\mathcal{L}_v \, \Omega^2 = (\mathrm{d} \circ i_v + i_v \circ \mathrm{d}) \, \Omega^2$. Then,

$$\begin{aligned}
\mathcal{L}_v(\Omega^2) &= \mathrm{d} \circ i_v \Omega^2 + i_v \circ \mathrm{d} \, \Omega^2 \\
&= \mathrm{d}[(\boldsymbol{\omega} \times \boldsymbol{v}) \cdot \mathrm{d}\boldsymbol{x}] + \ i_v(\operatorname{div} \boldsymbol{\omega}) \mathrm{d}^3\boldsymbol{x} = [\nabla \times (\boldsymbol{\omega} \times \boldsymbol{v})] \cdot \boldsymbol{s}^2,
\end{aligned}$$

since $\operatorname{div} \boldsymbol{\omega} = 0$. Thus, we find the *vorticity equation* (7.195) from the equation above it.

It is remarkable that the vorticity equation (7.195) in the $\boldsymbol{x}$-space has been derived from the conservation law (7.188) associated with the rotation symmetry in the $\boldsymbol{a}$-space. In addition, using the vector identity,

$$\nabla \times (\boldsymbol{\omega} \times \boldsymbol{v}) = (\boldsymbol{v} \cdot \nabla)\boldsymbol{\omega} + (\nabla \cdot \boldsymbol{v})\boldsymbol{\omega} - (\boldsymbol{\omega} \cdot \nabla)\boldsymbol{v},$$

(since $\nabla \cdot \boldsymbol{\omega} = 0$) together with the continuity equation (7.102), Eq. (7.195) is transformed to the well-known form of the vorticity equation for a compressible fluid:

$$\frac{d}{dt}\left(\frac{\omega}{\rho}\right) = \left(\frac{\omega}{\rho} \cdot \nabla\right) \boldsymbol{v}. \tag{7.196}$$

7.16.4. *Kelvin's circulation theorem*

The local law (7.188) in the $\boldsymbol{a}$-space leads to the Kelvin's circulation theorem. Consider a closed loop C_a in $\boldsymbol{a}$-space and denote its line-element by $\mathrm{d}\boldsymbol{a}$. Using (7.180), we have

$$\boldsymbol{V}_a \cdot \mathrm{d}\boldsymbol{a} = \boldsymbol{v} \cdot \mathrm{d}\boldsymbol{x}_a, \tag{7.197}$$

where $\mathrm{d}\boldsymbol{x}_a$ is a corresponding line-element in the physical $\boldsymbol{x}$-space (Fig. 7.14). Integrating (7.197) along a loop C_a fixed in the $\boldsymbol{a}$-space, we obtain the following integration law:

$$\partial_\tau \oint_{C_a} \boldsymbol{V}_a \cdot \mathrm{d}\boldsymbol{a} = \partial_\tau \int_{S_a} (\nabla_a \times \boldsymbol{V}_a) \cdot \mathrm{d}\boldsymbol{S}_a = \int_{S_a} \partial_\tau (\nabla_a \times \boldsymbol{V}_a) \cdot \mathrm{d}\boldsymbol{S}_a = 0,$$

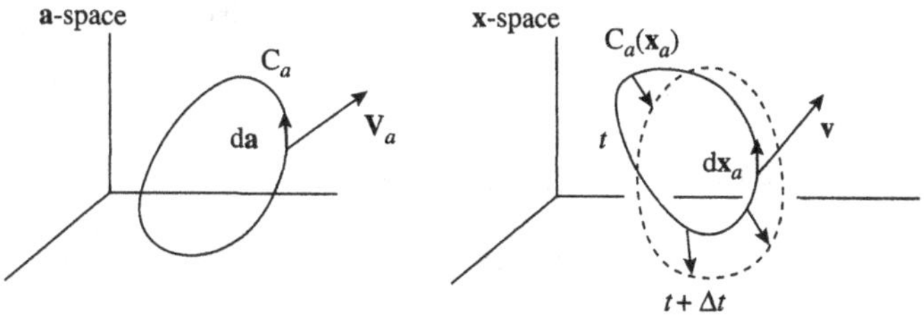

Fig. 7.14. Closed material loops in $\boldsymbol{a}$-space and $\boldsymbol{x}$-space.

by (7.188), where S_a and $\mathrm{d}\boldsymbol{S}_a$ are an open surface bounded by C_a and its surface element in $\boldsymbol{a}$-space, respectively. Thus, using (7.197), we obtain the *Kelvin's circulation theorem*:

$$\partial_\tau \oint_{C_a(\boldsymbol{x}_a)} \boldsymbol{v} \cdot \mathrm{d}\boldsymbol{x}_a = 0, \tag{7.198}$$

where $C_a(\boldsymbol{x}_a)$ is a closed material loop in $\boldsymbol{x}$-space corresponding to C_a.

Bretherton [Bre70] considered the invariance of the action integral under a reshuffling of indistinguishable particles which leaves the fields of velocity, density and entropy unaltered, and derived the Kelvin's circulation theorem directly. The present derivation is advantageous in the sense that the local form (7.188) is obtained first, and then the circulation theorem is derived.

7.16.5. *Lagrangian of the gauge field*

In view of the relation $\boldsymbol{\Omega}_a = \nabla_a \times \boldsymbol{V}_a$, we have the transformation,

$$\int_{M_a} \langle \boldsymbol{\Omega}_a, \boldsymbol{A}_a \rangle \mathrm{d}^3\boldsymbol{a} = \int_{M_a} \langle \boldsymbol{V}_a, \nabla_a \times \boldsymbol{A}_a \rangle \mathrm{d}^3\boldsymbol{a}, \tag{7.199}$$

by integration by parts (omitting the integrated terms). We seek a related expression in terms of $\boldsymbol{V}_a$ and $\boldsymbol{B}_a$. Using A^1, B^2, $\boldsymbol{B} = \nabla \times \boldsymbol{A}$ defined in §7.13 and recalling A^1 of (7.191) in terms of $\boldsymbol{A}_a$, we have a 2-form $B^2 = \mathrm{d}A^1$:

$$\begin{aligned} B^2 = \mathrm{d}A^1 &= B_a \mathrm{d}b \wedge \mathrm{d}c + B_b \mathrm{d}c \wedge \mathrm{d}a + B_c \mathrm{d}a \wedge \mathrm{d}b = \boldsymbol{B}_a \cdot \boldsymbol{S}^2 \\ &= B_1 \mathrm{d}x^2 \wedge \mathrm{d}x^3 + B_2 \mathrm{d}x^3 \wedge \mathrm{d}x^1 + B_3 \mathrm{d}x^1 \wedge \mathrm{d}x^2 = \boldsymbol{B} \cdot \boldsymbol{s}^2. \end{aligned}$$

This is analogous to $\Omega^2 = \mathrm{d}V^1$ of (7.189). Thus, an exterior product of B^2 and V^1 (defined by (7.181) and (7.182)) is

$$V^1 \wedge B^2 = \langle \boldsymbol{V}_a, \boldsymbol{B}_a \rangle \mathrm{d}^3\boldsymbol{a} = \langle \boldsymbol{V}_a, \nabla_a \times \boldsymbol{A}_a \rangle \mathrm{d}^3\boldsymbol{a}, \tag{7.200}$$

since $\boldsymbol{B}_a = (B_a, B_b, B_c) = \nabla_a \times \boldsymbol{A}_a$. This is also written as

$$V^1 \wedge B^2 = \langle \boldsymbol{V}, \boldsymbol{B} \rangle \, \mathrm{d}^3\boldsymbol{x} = \langle \boldsymbol{V}, \nabla \times \boldsymbol{A} \rangle \mathrm{d}^3\boldsymbol{x} = \langle \boldsymbol{V}, \rho \boldsymbol{V} \rangle \mathrm{d}^3\boldsymbol{x}, \quad (7.201)$$

where the gauge potential covector $\boldsymbol{A}$ is defined by

$$\nabla \times \boldsymbol{A} = \rho \boldsymbol{V}. \quad (7.202)$$

It is understood that $\rho \boldsymbol{V}$ is the divergence-free part of the mass flux $\rho \boldsymbol{v}$ (see (7.139)).

Thus, it is found that the Lagrangian L_B of (7.136) (which was given for constant ρ previously) can now be generalized to the case of variable density as

$$L_B = \tfrac{1}{2} \int \langle \boldsymbol{V}, \boldsymbol{B} \rangle \mathrm{d}^3\boldsymbol{x} = \tfrac{1}{2} \int \langle \boldsymbol{\Omega}_a, \boldsymbol{A}_a \rangle \mathrm{d}^3\boldsymbol{a}. \quad (7.203)$$

The last expression was shown to depend only on $\boldsymbol{a}$ in §7.16.3 and invariant with respect to change of time variable τ. Therefore, in the material variation with $\boldsymbol{a}$ fixed, the L_B is unchanged. Moreover, it is remarkable that the Lagrangian L_B has an internal symmetry, i.e. a local conservation law of vorticity in the $\boldsymbol{a}$-space, which is expressed by $\partial_\tau (\nabla_a \times \boldsymbol{V}_a) = 0$. This is analogous to the electromagnetic Lagrangian Λ_{em}, which has internal symmetries equivalent to the Faraday's law and the conservation of magnetic flux [LL75, §26].

7.17. Conclusions

Following the scenario of the gauge principle in the field theory of physics, it is found that the variational principle of fluid motions can be reformulated in terms of the covariant derivative and gauge fields for fluid flows. The gauge principle requires invariance of the Lagrangian and its variations with respect to both a translation group and a rotation group.

In complying with the local gauge invariance (with respect to both transformations), a gauge-covariant derivative is defined in terms of gauge fields. Local gauge invariance imply existence of a background *material* field.

Gauge invariance with respect to the SO(3) rotational transformations requires existence of a gauge potential which is found to be the vector potential of the mass flux field $\rho \boldsymbol{v}$. This results in the relation of Biot–Savart law between curl of a gauge field and the vorticity $\boldsymbol{\omega}$. This is in close analogy to the electromagnetic Biot–Savart law between curl of the magnetic field and current density. The invariance of Lagrangian with respect to the gauge transformation and Galilei invariance determine that the *vorticity* is the gauge field to the rotational gauge transformations. As a result, the covariant derivative of velocity is given by the *material derivative* of velocity (§7.14).

Using the gauge-covariant derivative, a variational principle is formulated by means of *isentropic material variations*, and the Euler's equation of motion is derived for isentropic flows from the Hamilton's principle in §7.15, where the Lagrangian consists of three terms: a Lagrangian of fluid flow, a Lagrangian of internal energy representing the background material, and a Lagrangian of rotational gauge field. This is also analogous to the electromagnetic case (see Appendix I and [LL75, Chap. 4]).

In addition, global gauge invariances of the Lagrangian with respect to two transfomations, translation and rotation, imply Noether's conservation laws which are interpreted from the point of view of classical field theory in [Sap76].[30] Those laws are the conservations of momentum and angular momentum, respectively. Furthermore, the Lagrangian has an internal symmetry with respect to particle permutation, which leads to a local law of vorticity conservation, i.e. the vorticity equation. The Kelvin's circulation theorem results from it.

The present gauge theory can provide a theoretical ground for physical analogy between the aeroacoustic phenomena associated with vortices [KM83; Kam86; KM87] and the electron and electromagnetic-field interactions.

[30]In [Sap76, Chap. 4], the particle coordinate labels are called basic fields and regarded as a sort of gauge fields.

Chapter 8

Volume-Preserving Flows of an Ideal Fluid

In the pervious chapter, we have considered a variational formulation of flows of an ideal fluid on the basis of the gauge principle. This leads to a picture that a fluid (massive) particle moves under interaction with a background material field (gauge field) having mass and internal energy and the material field is also moving rotationally and isentropically. From the invariance of the Lagrangian with respect to both translational and rotational gauge transformations as well as Galilei transformation, a covariant derivative ∇_t is defined, and the Euler's equation of motion has been derived. It is said that the fluid flows have *gauge groups* such as a group of translational transformations and a group of rotational transformations. The covariant derivative ∇_t is interpreted as the time derivative following the motion of a fluid particle, or a material derivative.

In this chapter, we are going to investigate volume-preserving flows of an ideal fluid. Corresponding gauge group of such flows is known to be the group of *volume-preserving diffeomorphisms*, for which mathematical machineries are well developed. Volume-preserving is another word for *constant* fluid density. Under the restricting conditions of constant density and constant entropy of fluid flows, the internal energy is kept constant during the motion. Therefore, the Lagrangian reduces to the kinetic energy integral only. In this respect, from a geometrical point of view it is said that the group of volume-preserving diffeomorphisms has the metric defined by the kinetic energy.

By the conventional approach, flows of an inviscid fluid are studied in great detail in the fluid dynamics. However, the present geometrical approach can reveal new aspects which are missing in the conventional fluid dynamics. Based on the Riemannian geometry and Lie group theory, developed first by [Arn66], it is found that Euler's equation of motion is a geodesic equation on a group of volume-preserving diffeomorphisms with the metric defined by the kinetic energy, and the behaviors of the geodesics are controlled by Riemannian (sectional) curvatures, which are quantitative characterizations of the flow field. In particular, the analysis shows that the curvatures are found to be mostly negative (with some exceptions), which can be related to mixing or ergodicity of the fluid particle motion in a bounded domain. Primary concern of the geometrical theory is the behaviors of particles and streamlines.

The present chapter is based on the works of [Arn66; EbMa70; Luk79; NHK92; Mis93; HK94; EM97]. It is known that the geodesic equation on a central extension of the group of volume-preserving diffeomorphisms is equivalent to the flow of a perfectly conducting fluid. Here, only the following references are noted: [Viz01] (and [Zei92]).

8.1. Fundamental Concepts

8.1.1. *Volume-preserving diffeomorphisms*

We consider flows of an inviscid incompressible fluid on a manifold M. The flow region M may be a bounded domain D, or T^3, or $\mathbb{R}^3$ (Fig. 8.1). In §7.8.1, a fluid flow was described by a map $\phi_\tau \boldsymbol{a} = \boldsymbol{x}(\tau, \boldsymbol{a})$ from the particle coordinate $\boldsymbol{a}$ to the spatial coordinate $\boldsymbol{x}$. Here, we are going to make the description more precise, and we represent the flow of an incompressible fluid on M by a continuous sequence of volume-preserving diffeomorphisms of M, because each volume element is conserved in flows of an *incompressible* fluid. Mathematically, the set of all *volume-preserving* diffeomorphisms of M composes a group $\mathcal{D}_\mu(M)$. An element $g \in \mathcal{D}_\mu(M)$ denotes a map, $g : M \to M$.

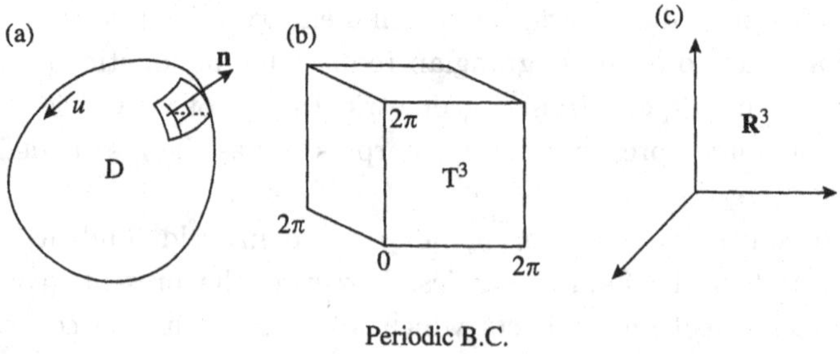

Fig. 8.1. Flow regions: (a) A bounded domain, (b) T^3, (c) $\mathbb{R}^3$.

Suppose that a flow is described by a curve, $t(\in \mathbb{R}) \to g_t(x)$ (with t a time parameter), where a particle initially ($t = 0$) located at a point x is mapped to the point $g_t(x)$ at a time t. All the particles at $^\forall x \in M$ (at $t = 0$) are mapped to $g_t(x)$ simultaneously. The product of two maps g_s and g_t is given by the composition law:

$$g_t \circ g_s(x) = g_t(g_s(x)).$$

The expression $g_t(x)$ is the *Lagrangian* description of flows. The Lagrangian description is alternatively written as

$$\boldsymbol{x} = \boldsymbol{x}(\tau, \boldsymbol{a}) = \boldsymbol{x}_a(\tau), \quad \boldsymbol{x}(0, \boldsymbol{a}) = \boldsymbol{a}, \tag{8.1}$$

(see §7.8.1), where $\tau = t$. In the expression of $g_t(x)$, the variable x assume the role of the particle coordinate $\boldsymbol{a}$. Hence, we may write (8.1) as $\boldsymbol{x} = g_t\boldsymbol{a}$. Its inverse is $\boldsymbol{a} = g_t^{-1}\boldsymbol{x}$.

The volume-preserving map g_t is characterized by unity of the Jacobian $J(g_t)$ of the map $\boldsymbol{a} \mapsto \boldsymbol{x}$ ((7.72), (7.73)):

$$J(g_t) := \frac{\partial(\boldsymbol{x})}{\partial(\boldsymbol{a})} = \frac{\partial(x^1, x^2, x^3)}{\partial(a^1, a^2, a^3)} = 1. \tag{8.2}$$

The group $\mathcal{D}_\mu(M)$, composed of all η satisfying $J(\eta) = 1$ for $\eta \in \mathcal{D}(M)$, is a closed submanifold of $\mathcal{D}(M)$ which is a group of all diffeomorphisms of M (see §8.5.2).

Displacement of a material particle x during a time Δt is denoted by $g_{t+\Delta t}(x) - g_t(x)$. Therefore, the particle velocity is given by

$$\begin{aligned} \dot{g}_t(x) &= \lim_{\Delta t \to 0} \frac{g_{t+\Delta t}(x) - g_t(x)}{\Delta t} = \lim_{\Delta t \to 0} \frac{g_{\Delta t} - e}{\Delta t} \circ g_t(x) \\ &= u_t \circ g_t(x) = u_t(g_t(x)), \end{aligned} \tag{8.3}$$

at a time t, where $g_{t+\Delta t} = g_{\Delta t} \circ g_t$ (Fig. 8.2). The velocity field

$$U_t(x) := \dot{g}_t(x) = u_t \circ g_t(x) \tag{8.4}$$

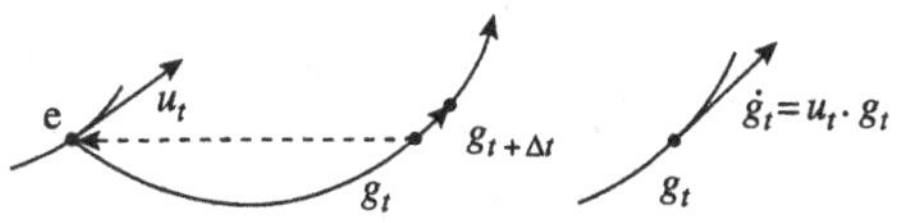

Fig. 8.2. Tangent vector $\dot{g}_t$.

is a *tangent* vector field at g_t for $^\forall x \in M$. The subscript t denotes that the tangent fields are time-dependent. In §7.15.1, $U_t(x)$ was denoted as $\boldsymbol{v}(\tau, \boldsymbol{a}) = \partial_\tau \boldsymbol{x}_a(\tau)$ in (7.143).

The expression $U_t = \dot{g}_t = u_t \circ g_t$ is a *right-invariant* representation by the definition (§1.7). Operating g_t^{-1} (a right-translation) on $U_t = u_t \circ g_t$, we obtain the velocity field at $g_0 = e(=\text{identity})$:

$$u_t = U_t \circ g_t^{-1} = \dot{g}_t \circ g_t^{-1}, \quad u_t \in T_e\mathcal{D}_\mu(M).$$

In §7.15.1, this was written as $\boldsymbol{v}(t, \boldsymbol{x})$, because the operation g_t^{-1} from the right of $\dot{g}_t$ (equivalent to $\boldsymbol{v}(\tau, \boldsymbol{a})$) means inverse transformation to express $\boldsymbol{a}$ in terms of $\boldsymbol{x}$. The $u_t(x)$ is the Eulerian representation of the velocity field, where

$$u_t(x) = \dot{g}_t \circ g_t^{-1} x, \quad = \dot{g}_t(\boldsymbol{a}),$$

hence $u_t(x) = u_t(\boldsymbol{x})$ with $x = \boldsymbol{x} = g_t \boldsymbol{a}$, whereas $U_t = \dot{g}_t(\boldsymbol{a})$ is the Lagrangian counterpart.

The volume-preserving condition (8.2), that is the *invariance* of the volume element $J_a = \mathrm{d}V/\mathrm{d}V_a = 1$, is equivalent to $\mathrm{D}_t(\mathrm{d}V) = 0$. From (7.76), this leads to the divergence-free condition, $\operatorname{div} u_t = 0$, for the Eulerian velocity field u_t.

In mathematical language, $u = u_t$ is an element of a Lie algebra $T_e\mathcal{D}_\mu(M)$, satisfying

$$\operatorname{div} u = 0. \tag{8.5}$$

The boundary condition (BC) for a bounded domain D is

$$BC : (u, \boldsymbol{n}) = 0 \quad \text{on} \quad \partial D, \tag{8.6}$$

where $\boldsymbol{n}$ is a unit outward normal to ∂D, i.e. the velocity vector u is tangent to the boundary ∂D. If $M = T^3$, periodic boundary condition should be imposed for the velocity u. If $M = \mathbb{R}^3$, the velocity field u is assumed to decay sufficiently rapidly at infinity so that volume integrals including u converge.

For the sake of mathematically rigorous analyses, it is useful to consider a manifold $\mathcal{D}^s_\mu(M)$ which is a subgroup of volume-preserving diffeomorphisms (of M) of Sobolev class H^s, where $s > n/2 + 1$ and $n = \dim M$ (see Appendix F). That is, $\mathcal{D}^s_\mu(M) = \{\eta \in \mathcal{D}^s : \eta^*(\mu) = \mu\}$, where μ is the volume form of M, and η is a bijective map (Appendix A.2): $M \to M$ such that η and η^{-1} are of Sobolev class H^s. The group $\mathcal{D}^s_\mu(M)$

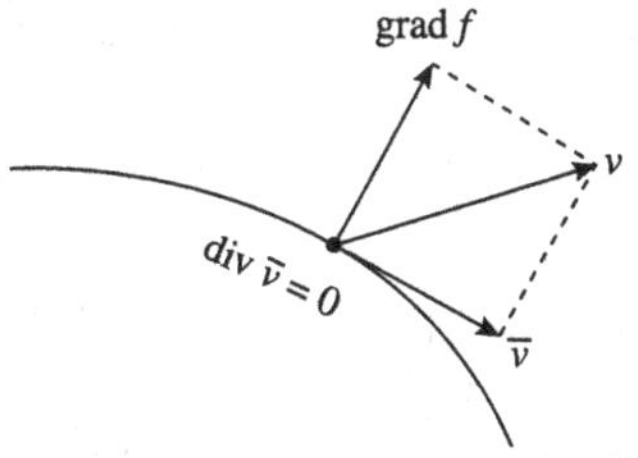

Fig. 8.3. Orthogonal projection.

is a weak Riemannian submanifold of the group $\mathcal{D}^s(M)$ of all Sobolev H^s-diffeomorphisms of M [Mis93].

An arbitrary tangent field $v \in T_\eta \mathcal{D}^s(M)$ can be decomposed into L^2-orthogonal components of divergence-free part $\bar{v}$ and gradient part (Fig. 8.3):

$$v = \bar{v} + \operatorname{grad} f, \quad f \in H^{s+1}(M), \tag{8.7}$$

(Appendix F), where $\operatorname{div} \bar{v} = 0$. Using the operator P to denote the projection to the divergence-free part, we have

$$\bar{v} = \mathsf{P}[v], \quad \mathsf{Q}[v] := \bar{v} - \mathsf{P}[v] = \operatorname{grad} f, \tag{8.8}$$

where Q is the projection operator orthogonal to the divergence-free part.

8.1.2. *Right-invariant fields*

For the geometrical theory of hydrodynamics, it is important to realize that the flow field can be represented in a *right-invariant* way. We saw already in (8.4) that the velocity field U_t had the form of a right-invariant field.

Consider a tangent field U_ξ at $\xi \in \mathcal{D}_\mu(M)$ (ξ: a volume-preserving diffeomorphism on M), and suppose that U_ξ is *right-invariant*:

$$U_\xi(x) := u \circ \xi(x), \quad \text{for} \quad {}^\forall \xi \in \mathcal{D}_\mu(M), \quad u \in T_e\mathcal{D}_\mu(M), \tag{8.9}$$

where $U_e = u$. Correspondingly, a right-invariant L^2-metric can be introduced on $\mathcal{D}_\mu(M)$. With any two tangent fields

$$u = u(x)\partial = u^k(x)\partial_k, \quad v = v(x)\partial = v^k(x)\partial_k$$

in the tangent space $T_e\mathcal{D}_\mu(M)$, an inner product is defined by

$$\langle u, v\rangle := \int_M (u(x), v(x))_x \mathrm{d}\mu(x), \tag{8.10}$$

where $(u,v)_x$ denotes the scalar product $u^k(x)v^k(x)$ and $\mathrm{d}\mu(x)$ the volume form, defined pointwisely at $x \in M$.

With two right-invariant fields $U_\xi = u\circ\xi$ and $V_\xi = v\circ\xi$ in the tangent space $T_\xi\mathcal{D}_\mu(M)$, the right-invariant metric is defined by

$$\langle U_\xi, V_\xi\rangle_e := \int_M (U_\xi\circ\xi^{-1}, V_\xi\circ\xi^{-1})\,(\xi^{-1})^*\mathrm{d}\mu_\xi = \int_M (u,v)_x \mathrm{d}\mu_e = \langle u, v\rangle_e, \tag{8.11}$$

where $(\xi^{-1})^*\mathrm{d}\mu_\xi$ is the pull-back of the volume form $\mathrm{d}\mu_\xi$ (Appendix B.6) which is equal to $\mathrm{d}\mu_e$ in the original space $M(x)$ by the volume-preserving property. Applying the right translation by ξ on the integrand of (8.11), we have

$$\int_{\xi(M)} (U_\xi\circ\xi^{-1}, V_\xi\circ\xi^{-1})_x\circ\xi\,\mathrm{d}\mu_\xi = \int_{\xi(M)} (U_\xi, V_\xi)_\xi\,\mathrm{d}\mu_\xi := \langle U_\xi, V_\xi\rangle_\xi. \tag{8.12}$$

The two integrals (8.11) and (8.12) are equal by the pull-back integration formula (B.50) of Appendix B.7.2. Thus the present L^2-metric is *isometric*, $\langle U_\xi, V_\xi\rangle_\xi = \langle u, v\rangle_e$, with respect to the right translation by any $\xi \in \mathcal{D}_\mu(M)$.

In the previous section, we saw that the right-invariant velocity $U_t = u_t\circ g_t$ is equivalent to the Lagrangian expression of the particle velocity $\boldsymbol{v}(\tau, \boldsymbol{a}) = \partial_\tau \boldsymbol{x}(\tau, \boldsymbol{a})$. Analogously, we may express the *particle acceleration* $\partial_\tau \boldsymbol{v}(\tau, \boldsymbol{a})$ in the right-invariant form. However, we must recall that the representation $\partial_\tau \boldsymbol{v}(\tau, \boldsymbol{a}) = \nabla_t \boldsymbol{v}$ of (7.141) has been obtained on the basis of the gauge principle. Correspondingly, we must regard the following connections of the right-invariant form as the *law of gauge principle*.

Perhaps, the most essential premise in the formulation of hydrodynamics given below is the right-invariance of the connections. For any right-invariant vector fields $U_\xi, V_\xi \in T_\xi\mathcal{D}^s_\mu(M)$, we define the following *right-invariant* connection (Fig. 8.4),

$$\hat{\nabla}_{U_\xi} V_\xi := (\nabla_u v)_e \circ \xi, \tag{8.13}$$

for $\xi \in \mathcal{D}_\mu(M)$ and $u, v \in T_e\mathcal{D}_\mu(M)$, where ∇ is the covariant derivative on M (manifold of Eulerian description). Similarly on the group

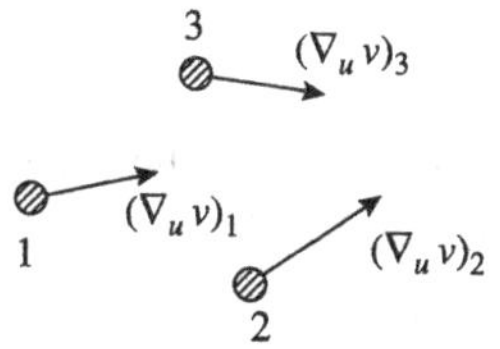

Fig. 8.4. Right-invariant connection.

$\mathcal{D}_\mu(M)$ of volume-preserving diffeomorphisms, we have the *right-invariant* connection $\bar{\nabla}$,

$$\bar{\nabla}_{U_\xi} V_\xi := \mathtt{P}[\nabla_u v] \circ \xi, \tag{8.14}$$

where the symbol $\mathtt{P}$ is the projection operator to the divergence-free part.

The difference between the two connections $\hat{\nabla}$ and $\bar{\nabla}$ is the second fundamental form S of $\mathcal{D}_\mu(M)$:

$$S(U_\xi, V_\xi) := \hat{\nabla}_{U_\xi} V_\xi - \bar{\nabla}_{U_\xi} V_\xi = \mathtt{Q}[\nabla_u v] \circ \xi, \tag{8.15}$$

(see (3.178)), where $\mathtt{Q}[\nabla_u v] = \nabla_u v - \mathtt{P}[\nabla_u v]$. This is right-invariant as well.

The curvature tensors are also defined in the right-invariant way. For tangent fields $U, V, W, Z \in T_\xi \mathcal{D}^s_\mu(M)$, the curvature tensor $\hat{R}$ is defined on $\mathcal{D}(M)$ by

$$(\hat{R}(U, V)W)_\xi := (R(U \circ \xi^{-1}, V \circ \xi^{-1})W \circ \xi^{-1}) \circ \xi, \tag{8.16}$$

where R is the curvature tensor on M:

$$R(u, v)w := \nabla_u(\nabla_v w) - \nabla_v(\nabla_u w) - \nabla_{[u,v]} w \tag{8.17}$$

for $u, v, w \in T_e\mathcal{D}_\mu(M)$ (see (3.97)). The curvature tensor $\bar{R}$ on the divergence-free fields $T\mathcal{D}_\mu(M)$ is given by replacing ∇ of (8.17) with $\bar{\nabla}$:

$$\bar{R}(u, v)w := \bar{\nabla}_u(\bar{\nabla}_v w) - \bar{\nabla}_v(\bar{\nabla}_u w) - \bar{\nabla}_{[u,v]} w. \tag{8.18}$$

Both curvature tensors R and $\bar{R}$ are related by the following Gauss–Codazzi equation (3.179) on $T_e\mathcal{D}_\mu(M)$:

$$\langle R(u, v)w, z\rangle_{L^2} = \langle \bar{R}(u, v)w, z\rangle + \langle S(u, w), S(v, z)\rangle - \langle S(u, z), S(v, w)\rangle. \tag{8.19}$$

8.2. Basic Tools

We are going to present basic tools for the geodesic formulation of hydrodynamics (of incompressible ideal fluids) on a group of volume-preserving diffeomorphisms of M, i.e. $\mathcal{D}_\mu(M)$. Let M be a compact (bounded and closed) flow domain of a Riemannian space. The Lie algebra corresponding to the group $\mathcal{D}_\mu(M)$ consists of all vector fields u on M, i.e. $u \in T_e\mathcal{D}_\mu$, such that

$$\operatorname{div} u(x) = 0 \quad \text{at} \quad x \in M, \qquad (u, \boldsymbol{n}) = 0 \quad \text{on} \quad \partial M, \tag{8.20}$$

where $\boldsymbol{n}$ is unit outward normal to the boundary ∂M (Fig. 8.5). The metric is defined by (8.10) or (8.11).

8.2.1. *Commutator*

Commutator of the present problem of volume-preserving diffeomorphisms is given by

$$[u, v] = \{u, v\} = u^k \partial_k v - v^k \partial_k u \tag{8.21}$$

(see (1.76) and (1.77)). The right-hand side can be shown as divergence-free if each one of the vector fields u and v is so, i.e. $u, v \in T_e\mathcal{D}_\mu$. In fact, we have the vector identity,

$$\nabla \times (u \times v) = (v \cdot \nabla)u + (\nabla \cdot v)u - (u \cdot \nabla)v - (\nabla \cdot u)v \tag{8.22}$$

$$= (v \cdot \nabla)u - (u \cdot \nabla)v = -\{u, v\}. \tag{8.23}$$

Obviously, the left-hand side is divergence-free.

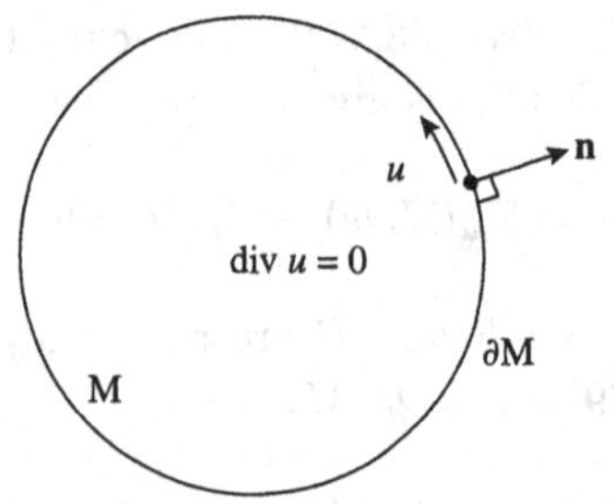

Fig. 8.5. Flow domain M and boundary ∂M.

By the torsion-free property,

$$[u, v] = \nabla_u v - \nabla_v u,$$

of the Riemannian connection (§3.3.1), the commutator $[U_\xi, V_\xi] = \nabla_{U_\xi} V_\xi - \nabla_{V_\xi} U_\xi$ is also represented in a right-invariant form, since each term on the right-hand side is so:

$$[U_\xi, V_\xi] = [u, v] \circ \xi. \tag{8.24}$$

8.2.2. *Divergence-free connection*

The Riemannian connection ∇ is defined by Eq. (3.30), which is reproduced here:

$$2\langle \nabla_u v, w \rangle = \langle [u, v], w \rangle - \langle [v, w], u \rangle + \langle [w, u], v \rangle, \tag{8.25}$$

for $u, v, w \in T_e\mathcal{D}_\mu(M)$, where $u = u^k \partial_k$, etc. This is assured by the right-invariance of the metric $\langle \cdot, \cdot \rangle$ defined in the previous section, and also by the right-invariance of the vector fields defined by (8.9) (see (3.87) and notes in §3.7.3 and 3.4.3).

The adjoint action $ad_v w = [v, w]$ is defined in §1.8, and the coadjoint action $ad_v^* u$ is defined by (3.64) in §3.6.2 as

$$\langle ad_v^* u, w \rangle = \langle u, ad_v w \rangle = \langle u, [v, w] \rangle. \tag{8.26}$$

From (8.25), we obtain

$$\nabla_u v = \frac{1}{2}([u, v] - ad_u^* v - ad_v^* u) + \operatorname{grad} f, \tag{8.27}$$

by the nondegeneracy of the metric (§3.1.2 and 1.5.2), where $f(x)$ is an arbitrary differentiable scalar function. The last term $\operatorname{grad} f$ can be added because

$$\langle w, \operatorname{grad} f \rangle = \int_M (w, \operatorname{grad} f) \mathrm{d}\mu = \int_M \operatorname{div}(f w) \mathrm{d}\mu = \int_{\partial M} f(w, \boldsymbol{n}) \mathrm{d}S = 0,$$

i.e. the $\operatorname{grad} f$ term does not provide any contribution to the inner product with a divergence-free vector w satisfying (8.20). This is also satisfied by periodic boundary condition when $M = T^3$.

Taking projection to the divergence-free part, we have

$$\bar{\nabla}_u v = \mathsf{P}\left[\frac{1}{2}([u,v] - ad_u^* v - ad_v^* u) + \operatorname{grad}\bar{f}\right], \tag{8.28}$$

where the function $\bar{f}(x)$ is to be determined so as to satisfy the condition, $\operatorname{div}\bar{\nabla}_u v = 0$.

8.2.3. *Coadjoint action ad^**

One can give an explicit expression to the coadjoint action $ad_u^* v$ defined by (8.26). In $M = \mathbb{R}^3$, this is represented as follows:

$$B \equiv ad_u^* v = -(\nabla \times v) \times u + \operatorname{grad} f. \tag{8.29}$$

This formula can be verified in two ways. First one is a direct derivation from (8.26). In fact, using the definition (8.10), and (8.21), (8.23), we obtain

$$\begin{aligned}
\langle ad_u^* v, w\rangle &= \langle v, [u,w]\rangle = \int_M (v,[u,w])_x \mathrm{d}x \\
&= -\int_M (v, \nabla\times(u\times w))_x \mathrm{d}x = -\int_M (\nabla\times v, u\times w)_x \mathrm{d}x \\
&= -\int_M ((\nabla\times v)\times u, w)_x \mathrm{d}x = -\langle (\nabla\times v)\times u, w\rangle.
\end{aligned}$$

This verifies the expression (8.29) for any triplet of $u, v, w \in T_e\mathcal{D}_\mu(M)$ satisfying (8.20), because $\langle \operatorname{grad} f, w\rangle = 0$.

There is a more general approach to verify (8.29), which relies on differential forms on a Riemannian manifold M^n (of dimension n). To every tangent vector $v = v^k\partial_k$, one can define a 1-form $\alpha_v^1 = v_i \mathrm{d}x^i$ where $v_i = g_{ik}v^k$ (§1.5.2). Then, we have $\alpha_v^1[w] = g_{ik}v^k w^i = \langle v, w\rangle$ for any tangent vector w. In addition, setting

$$B = B^i\partial_i \equiv ad_u^* v,$$

we can define a corresponding 1-form by $\alpha_B^1 = g_{ik}B^k \mathrm{d}x^i$.

Then, we have a theorem for the vector field $B = ad_u^* v \in T_e\mathcal{D}_\mu(M)$. That is, *the corresponding 1-form* α_B^1 *is given by the formula* [Arn66][1]:

$$\alpha_B^1 = -i_u \mathrm{d}\alpha_v^1 + \mathrm{d}f, \tag{8.30}$$

where i_u is the operator symbol of interior product (Appendix B.4), $u, v \in T_e\mathcal{D}_\mu(M)$, and $\mathrm{d}f = \partial_i f \mathrm{d}x^i$. Its proof is given in the last section §8.9 for a Riemannian manifold M^n generally.

If $M = \mathbb{R}^3$ and $g_{ij} = \delta_{ij}$, the machinery of vector analysis in the euclidean space is in order (Appendix B.5). First, since $\alpha_v^1 = v_x \mathrm{d}x + v_y \mathrm{d}y + v_z \mathrm{d}z$ (where $v_i = \delta_{ik} v^k = v^i$), we have

$$\mathrm{d}\alpha_v^1 = V_x \mathrm{d}y \wedge \mathrm{d}z + V_y \mathrm{d}z \wedge \mathrm{d}x + V_z \mathrm{d}x \wedge \mathrm{d}y,$$

where $V = (V_x, V_y, V_z) = \nabla \times v$. Then the first term of (8.30) is

$$\begin{aligned} i_u \mathrm{d}\alpha_v^1 &= (V_y u_z - V_z u_y)\mathrm{d}x + (V_z u_x - V_x u_z)\mathrm{d}y + (V_x u_y - V_y u_x)\mathrm{d}z \\ &= W_x \mathrm{d}x + W_y \mathrm{d}y + W_z \mathrm{d}z, \end{aligned}$$

where $W = (W_x, W_y, W_z) = V \times u = (\nabla \times v) \times u$. In view of $\alpha_B^1 = B_x \mathrm{d}x + B_y \mathrm{d}y + B_z \mathrm{d}z$, Eq. (8.30) implies $B \equiv ad_u^* v = -(\nabla \times v) \times u + \operatorname{grad} f$, which is (8.29) itself.

8.2.4. *Formulas in* $\mathbb{R}^3$ *space*

Following convention, we use bold face letters to denote vectors in $\mathbb{R}^3$ and a dot to denote the scalar product, in this chapter. Then, the expression (8.21) is written for $\boldsymbol{u}, \boldsymbol{v} \in T_e\mathcal{D}_\mu(\mathbb{R}^3)$ as

$$ad_{\boldsymbol{u}} \boldsymbol{v} \equiv [\boldsymbol{u}, \boldsymbol{v}] = (\boldsymbol{u} \cdot \nabla)\boldsymbol{v} - (\boldsymbol{v} \cdot \nabla)\boldsymbol{u}. \tag{8.31}$$

Equation (8.29) is

$$ad_{\boldsymbol{u}}^* \boldsymbol{v} = -(\nabla \times \boldsymbol{v}) \times \boldsymbol{u} + \nabla f_{uv}. \tag{8.32}$$

Noting the following vector identity,

$$(\nabla \times \boldsymbol{v}) \times \boldsymbol{u} = (\boldsymbol{u} \cdot \nabla)\boldsymbol{v} - u^k \nabla v^k, \tag{8.33}$$

the coadjoint action is also written as

$$ad_{\boldsymbol{u}}^* \boldsymbol{v} = -(\boldsymbol{u} \cdot \nabla)\boldsymbol{v} + u^k \nabla v^k + \nabla f_{uv}. \tag{8.34}$$

[1]Due to the difference of definition of $[v, w]$ by $\pm$, [Arn66] gives $\alpha_B^1 = i_u \mathrm{d}\alpha_v^1 + \mathrm{d}f$.

If $\boldsymbol{v} = \boldsymbol{u}$ and $u^k \nabla v^k = \nabla(u^2/2)$, we may write

$$ad^*_{\boldsymbol{u}}\boldsymbol{u} = -(\boldsymbol{u}\cdot\nabla)\boldsymbol{u} - \nabla p \tag{8.35}$$

$$= -(\nabla\times\boldsymbol{u})\times\boldsymbol{u} - \nabla\frac{1}{2}u^2 - \nabla p, \tag{8.36}$$

where $p = -f_{uu} - \frac{1}{2}u^2$. The scalar function p must satisfy

$$\nabla^2 p = -\mathrm{div}((\boldsymbol{u}\cdot\nabla)\boldsymbol{u}) = -\partial_i\partial_k(u^i u^k), \tag{8.37}$$

to ensure $\mathrm{div}(ad^*_{\boldsymbol{u}}\boldsymbol{u}) = 0$, where $\mathrm{div}\,\boldsymbol{u} = \partial_i u^i = 0$.

Using (8.31) and (8.34), the connection (8.27) is given by

$$\nabla_{\boldsymbol{u}}\boldsymbol{v} = \frac{1}{2}([\boldsymbol{u},\boldsymbol{v}] - ad^*_{\boldsymbol{u}}\boldsymbol{v} - ad^*_{\boldsymbol{v}}\boldsymbol{u}) + \mathrm{grad}\, f = (\boldsymbol{u}\cdot\nabla)\boldsymbol{v} + \nabla p', \tag{8.38}$$

where $p' = f - \frac{1}{2}(u^k v^k + f_{uv} + f_{vu})$. Furthermore, the divergence-free connection (8.28) is

$$\bar{\nabla}_{\boldsymbol{u}}\boldsymbol{v} = \mathsf{P}[(\boldsymbol{u}\cdot\nabla)\boldsymbol{v} + \nabla p'] = (\boldsymbol{u}\cdot\nabla)\boldsymbol{v} + \nabla p_*, \tag{8.39}$$

where the scalar function p_* must satisfy

$$\nabla^2 p_* = -\mathrm{div}((\boldsymbol{u}\cdot\nabla)\boldsymbol{v}) = -\partial_k\partial_i(u^i v^k), \tag{8.40}$$

since $\partial_i u^i = 0$. In particular, for $\boldsymbol{v} = \boldsymbol{u}$, we have

$$\bar{\nabla}_{\boldsymbol{u}}\boldsymbol{u} = (\boldsymbol{u}\cdot\nabla)\boldsymbol{u} + \nabla p. \tag{8.41}$$

The property $\mathrm{div}(\bar{\nabla}_{\boldsymbol{u}}\boldsymbol{u}) = 0$ is assured by p satisfying (8.37). Using the vector identity (8.33) with $\boldsymbol{v} = \boldsymbol{u}$, the divergence-free connection (8.41) can be written also as

$$\bar{\nabla}_{\boldsymbol{u}}\boldsymbol{u} = (\nabla\times\boldsymbol{u})\times\boldsymbol{u} + \nabla\left(\frac{1}{2}u^2\right) + \nabla p. \tag{8.42}$$

8.3. Geodesic Equation

In §3.6.2, we derived the geodesic equation of a time-dependent problem which is given by (3.67): $\partial_t X - ad^*_X X = 0$ for the tangent vector X (spatial part). Setting $X = \boldsymbol{u}$ and using (8.35), the geodesic equation becomes

$$\partial_t \boldsymbol{u} + (\boldsymbol{u} \cdot \nabla)\boldsymbol{u} + \nabla p = \nabla_t \boldsymbol{u} + \nabla p = 0, \tag{8.43}$$

where $\nabla_t = \partial_t + \boldsymbol{u} \cdot \nabla$, and p should satisfy (8.37). This is also written as

$$\partial_t \boldsymbol{u} + \bar{\nabla}_{\boldsymbol{u}} \boldsymbol{u} = 0. \tag{8.44}$$

Taking *divergence*, we obtain $\partial_t(\operatorname{div} \boldsymbol{u}) = 0$. Therefore, the divergence-free condition $\operatorname{div} \boldsymbol{u} = 0$ is satisfied at all times if it is satisfied initially at all points. This is nothing but the *Euler's equation of motion* for an incompressible fluid. Equivalently, the geodesic equation is also written as

$$\partial_t \boldsymbol{u} + (\nabla \times \boldsymbol{u}) \times \boldsymbol{u} + \nabla \frac{1}{2} u^2 = -\nabla p. \tag{8.45}$$

This is consisitent with (7.158) in Chapter 7.

The above is the Eulerian description for the velocity $\boldsymbol{u}(t, \boldsymbol{x})$. In order to obtain the right-invariant representation, consider a curve: $t \to g_t \equiv \xi$ and its tangent $\dot{g}_t = \dot{\xi}$.[2] Using (8.13) and (3.16) with $V_\xi = U_\xi$ and $\dot{\xi} = \boldsymbol{u} \circ \xi$, the right-invariant connection of a time-dependent problem is given by

$$\hat{\nabla}_{U_\xi} U_\xi = \frac{\partial}{\partial t}(\dot{\xi} \circ \xi^{-1}) \circ \xi + (\nabla_{(\dot{\xi} \circ \xi^{-1})} \dot{\xi} \circ \xi^{-1}) \circ \xi, \tag{8.46}$$

where $\dot{\xi} \circ \xi^{-1} = \boldsymbol{u}$ and $U_\xi = \dot{\xi}$. This right-invariant form must be regarded as representing the gauge principle, as noted in §8.1.2.

The geodesic equation on the group $\mathcal{D}_\mu(M)$ is given by vanishing of the projection to the divergence-free part as $\mathtt{P}[\hat{\nabla}_{\dot{g}_t} \dot{g}_t] = \mathtt{P}[\hat{\nabla}_{U_\xi} U_\xi] = 0$, i.e.

$$0 = \mathtt{P}[(\hat{\nabla}_{\dot{g}_t} \dot{g}_t)_{g_t}] = \mathtt{P}[\partial_t \boldsymbol{u} + \nabla_{\boldsymbol{u}} \boldsymbol{u}] \circ g_t. \tag{8.47}$$

The Euler's equation of motion is obtained by the right translation g_t^{-1} as $\mathtt{P}[\partial_t \boldsymbol{u} + \nabla_{\boldsymbol{u}} \boldsymbol{u}] = 0$, which is also written as

$$\partial_t \boldsymbol{u} + \nabla_{\boldsymbol{u}} \boldsymbol{u} = -\operatorname{grad} p, \quad \operatorname{div} \boldsymbol{u} = 0. \tag{8.48}$$

This form is valid in a general manifold M. In the euclidean manifold $\mathbb{R}^3$, we have

$$\nabla_{\boldsymbol{u}} \boldsymbol{u} = \nabla_{u^k \partial_k} \boldsymbol{u} = u^k \nabla_{\partial_k} \boldsymbol{u} = u^k \partial_k \boldsymbol{u} = (\boldsymbol{u} \cdot \nabla)\boldsymbol{u}, \tag{8.49}$$

by the definition of the connection (3.6) in the flat space where $\Gamma^k_{ij} = 0$. Using the projection operator $\mathtt{Q}$ orthogonal to the divergence-free part, the

[2] In order to simplify the notation, ξ is used instead of g_t.

above equation reduces to

$$\mathsf{Q}[\nabla_{\boldsymbol{u}}\boldsymbol{u}] = -\mathrm{grad}\, p. \tag{8.50}$$

This is another expression of (8.37).

8.4. Jacobi Equation and Frozen Field

Consider a family of geodesic curves $g = g(t, \alpha)$ with α as the variation parameter. A reference geodesic is given by $g_0(t) = g(t, 0)$. Behaviors of its nearby geodesics are described by the Jacobi field J, defined by $J = \partial g/\partial\alpha|_{\alpha=0}$. The equation governing J is given by (3.127), which is reproduced here,

$$\frac{\mathrm{d}^2}{\mathrm{d}t^2}\frac{\|J\|^2}{2} = \| \bar{\nabla}_T J\|^2 - K(T, J), \tag{8.51}$$

where $T = \partial g/\partial t|_{\alpha=0}$ is the tangent to the geodesic g_0, and J is the Jacobi vector where $\|J\|^2 = \langle J, J\rangle$. The $K(T, J)$ is the sectional curvature defined by

$$K(T, J) := \langle \bar{R}(J, T)T, J\rangle = R_{ijkl}J^i T^j J^k T^l.$$

The above Jacobi equation (8.51) has been derived from the definitions of the geodesic curve and Riemannian curvature tensor in §3.10. It is seen that the sectional curvature $K(T, J)$ controls the stability behavior of geodesics. Writing $J = \|J\|\boldsymbol{e}_J$ where $\|\boldsymbol{e}_J\| = 1$, Eq. (8.51) is rewritten as

$$\frac{\mathrm{d}^2}{\mathrm{d}s^2}\|J\| = (\|\nabla_T \boldsymbol{e}_J\|^2 - K(T, \boldsymbol{e}_J))\|J\|. \tag{8.52}$$

Let us consider the Jacobi field from a different point of view. According to the above definitions of T and J, Eq. (3.91) is written as $\bar{\nabla}_T J = \bar{\nabla}_J T$ by applying the definition (3.24) to the divergence-free connection $\bar{\nabla}$. Therefore, the vector fields T and J commute by the torsion-free property (3.18). Thus, it is found that the Lie derivative vanishes:

$$\mathcal{L}_T J = [T, J] = \bar{\nabla}_T J - \bar{\nabla}_J T = 0 \tag{8.53}$$

(see (1.81)). The argument given above the equation (3.174) asserts that the torsion-free is valid not only with the divergence-free connection $\bar{\nabla}$, but also with the general connection $\hat{\nabla}$ as well. Thus, we have $\hat{\nabla}_{\hat{T}}\hat{J} - \hat{\nabla}_{\hat{J}}\hat{T} = [\hat{T}, \hat{J}] = 0$.

In the time-dependent problem in $\mathbb{R}^3$, it is noted that the tangent vector (velocity vector) of a flow is written as $\hat{T} = (\partial_t, u^k\partial_k)$, whereas the Jacobi vector (non-velocity vector) is written as $\hat{J} = (0, J^k\partial_k)$. Then, the equation $[\hat{T}, \hat{J}] = 0$ is rewritten as

$$\partial_t \boldsymbol{J} + (\boldsymbol{u} \cdot \nabla)\boldsymbol{J} = (\boldsymbol{J} \cdot \nabla)\boldsymbol{u}, \tag{8.54}$$

by (8.21). This is equivalent to (1.81) under the divergence-free condition. Assuming $\nabla \cdot \boldsymbol{J} = 0$,[3] this equation is transformed to

$$\partial_t \boldsymbol{J} + \nabla \times (\boldsymbol{J} \times \boldsymbol{u}) = 0, \tag{8.55}$$

by using the vector identity (8.22), because $\nabla \cdot \boldsymbol{u} = 0$. This is usually called the equation of *frozen* field, (see Remark of §1.8) since it indicates that the divergence-free field $\boldsymbol{J}$ is carried along with the flow $\boldsymbol{u}$ and behaves as if $\boldsymbol{J}$ were frozen to the carrier fluid (Fig. 8.6).

The vorticity defined by $\boldsymbol{\omega} = \operatorname{curl} \boldsymbol{u}$ is regarded as an example of the Jacobi field. Taking the curl of Eq. (8.45), we obtain the *vorticity equation*,

$$\partial_t \boldsymbol{\omega} + \nabla \times (\boldsymbol{\omega} \times \boldsymbol{u}) = 0. \tag{8.56}$$

Obviously, this is an equation of frozen-field with $\boldsymbol{J}$ replaced by $\boldsymbol{\omega}$ in (8.55). Note that this equation can be rewritten as

$$\partial_t \boldsymbol{\omega} + (\boldsymbol{u} \cdot \nabla)\boldsymbol{\omega} - (\boldsymbol{\omega} \cdot \nabla)\boldsymbol{u} = [\hat{\boldsymbol{u}}, \hat{\boldsymbol{\omega}}] = 0. \tag{8.57}$$

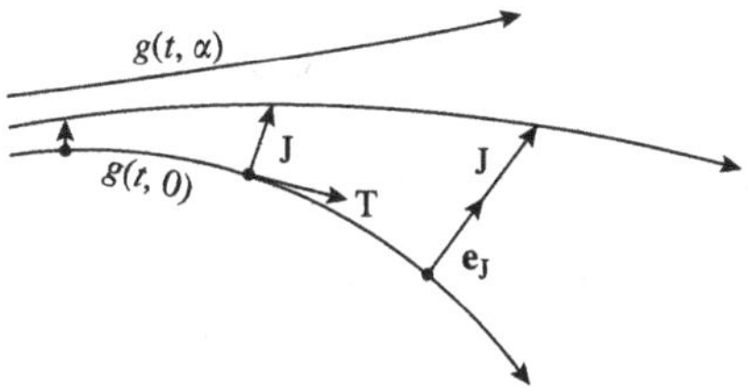

Fig. 8.6. Frozen-in field.

[3]Equation (8.54) is rewritten as $\partial_t \boldsymbol{J} + \nabla \times (\boldsymbol{J} \times \boldsymbol{u}) = (\nabla \cdot \boldsymbol{u})\boldsymbol{J} - (\nabla \cdot \boldsymbol{J})\boldsymbol{u}$. Taking div and using $\nabla \cdot \boldsymbol{u} = 0$, we obtain $\partial_t(\nabla \cdot \boldsymbol{J}) + (\boldsymbol{u} \cdot \nabla)(\nabla \cdot \boldsymbol{J}) = 0$. Hence, if $\nabla \cdot \boldsymbol{J} = 0$ initially, we have $\nabla \cdot \boldsymbol{J} = 0$ thereafter.

Comparing Eq. (8.54) with (1.82) in view of $\mathrm{D}/\mathrm{D}t = \partial_t + \boldsymbol{u}\cdot\nabla$, we obtain the Cauchy's solution (1.83) written as

$$J^k(t) = J^j(0)\frac{\partial x^k}{\partial a^j}, \tag{8.58}$$

where the Lagrangian description (8.1) of the map from the particle coordinate $\boldsymbol{a}$ to the space coordinate $\boldsymbol{x}$ is used.

Remark. Equation (8.55) can be extended to the *compressible* flow of $\mathrm{div}\,\boldsymbol{v} \neq 0$ but with keeping $\mathrm{div}\,\boldsymbol{J} = 0$ without change, where the velocity field is denoted by $\boldsymbol{v}$ instead of $\boldsymbol{u}$. This is the case of the vorticity equation (7.195) for the vorticity $\boldsymbol{\omega} = \nabla\times\boldsymbol{v}$ instead of $\boldsymbol{J}$. Introducing the fluid density ρ and using the continuity equation (7.160), the vorticity equation is transformed to

$$\frac{\mathrm{d}}{\mathrm{d}t}\left(\frac{\boldsymbol{\omega}}{\rho}\right) = \left(\frac{\boldsymbol{\omega}}{\rho}\cdot\nabla\right)\boldsymbol{v} \tag{8.59}$$

(see (7.196)). Comparing this equation with (1.82), we obtain the Cauchy's solution

$$\frac{\omega^k(t,\boldsymbol{x})}{\rho(t,\boldsymbol{x})} = \frac{\omega^j(0,\boldsymbol{a})}{\rho(0,\boldsymbol{a})}\frac{\partial x^k}{\partial a^j}, \tag{8.60}$$

since $\boldsymbol{\omega}/\rho$ takes the part of Y in (1.82).

It is well-known that the magnetic field $\boldsymbol{B}$ in the ideal magnetohydrodynamics is also governed by the same equation of frozen field as (8.59), where $\boldsymbol{\omega}$ is just replaced with $\boldsymbol{B}$ [Moff78].

8.5. Interpretation of Riemannian Curvature of Fluid Flows

8.5.1. *Flat connection*

It is instructive to consider a model system on a manifold $\mathcal{D}_*(M)$, governed by

$$\partial_t\boldsymbol{u} + (\boldsymbol{u}\cdot\nabla)\boldsymbol{u} = 0, \tag{8.61}$$

instead of the geodesic equation $\partial_t\boldsymbol{u} + \bar{\nabla}_{\boldsymbol{u}}\boldsymbol{u} = 0$ of (8.44) on $\mathcal{D}_\mu(M)$ (volume-preserving) for $M \subset \mathbb{R}^3$. In this model system, the connection is flat, i.e. the sectional curvatures vanish identically. This is verified as follows.

First, note the following equations,

$$\nabla_X Z = X \cdot \nabla Z = X^k \partial_k Z,$$
$$\nabla_X \nabla_Y Z = X^k \partial_k((Y_l \partial_l) Z) = (X^k \partial_k Y^l)\partial_l Z + X^k Y^l \partial_k \partial_l Z,$$
$$\nabla_{[X,Y]} Z = [X,Y] \cdot \nabla Z = (X^k \partial_k Y^l \partial_l - Y^l \partial_l X^k \partial_k) Z,$$

for three arbitrary tangent vectors $X = X^k\partial_k, Y = Y^l\partial_l, Z \in T_e\mathcal{D}_\mu(M)$. Then, the curvature tensor $R(X,Y)Z$ defined by (8.17) is given by

$$\begin{aligned} R(X,Y)Z &= \nabla_X(\nabla_Y Z) - \nabla_Y(\nabla_X Z) - \nabla_{[X,Y]} Z \\ &= X^k \partial_k((Y^l \partial_l) Z) - Y^l \partial_l((X^k \partial_k) Z) \\ &\quad - ((X^k \partial_k Y^l)\partial_l - (Y^l \partial_l X^k)\partial_k)) Z = 0, \end{aligned} \tag{8.62}$$

that is, the curvature tensor $R(X,Y)Z$ vanishes for any X, Y, Z. Therefore, all the sectional curvatures defined by $K(X,Y) = \langle R(X,Y)Y, X\rangle$ vanish. Such a connection is said to be *flat*.

Thus the motion governed by (8.61) is considered to be one on a flat manifold. However, the geodesic equation (8.43) on $\mathcal{D}_\mu(M)$ (volume-preserving) has an additional term grad p, and the fluid motion is characterized by non-vanishing Riemannian curvatures as investigated below. Therefore, it may be concluded that *the pressure term gives rise to curvatures of fluid motion.*

8.5.2. *Pressure gradient as an agent yielding curvature*

Let us consider how the fluid motion acquires a curvature and what the curvature of a divergence-free flow is. On the group $\mathcal{D}_\mu(M)$ of volume-preserving diffeomorphisms of M, the Jacobian $J(\eta(x))$ for $\eta \in \mathcal{D}(M)$ is always unity for any $x \in M$:

$$J(\eta(x)) := \frac{\partial(\eta)}{\partial(x)} = 1, \quad \forall x \in M.$$

From the implicit function theorem, the group $\mathcal{D}_\mu(M)$, composed of all η satisfying $J(\eta) = 1$, is a *closed* submanifold of $\mathcal{D}(M)$. According to the formulation of §3.13 and 8.1.2, the difference of the two connections, $\hat{\nabla}$ in $\mathcal{D}(M)$ and $\bar{\nabla}$ in $\mathcal{D}_\mu(M)$, is given by the second fundamental form S of (8.15) (see (3.178)).

The *curvature* of the closed submanifold $\mathcal{D}_\mu(M)$ is given by $\langle \bar{R}(U,V)W, Z\rangle$ in the Gauss–Codazzi equation (8.19). In particular, the sectional curvature of the section spanned by the tangent vectors $X, Y \in T_\eta \mathcal{D}_\mu(M)$ is given by

$$\begin{aligned}\bar{K}(X,Y)^{\mathcal{D}_\mu} &:= \langle \bar{R}(X,Y)Y, X\rangle_{L^2}^{\mathcal{D}_\mu} \\ &= \langle R(X,Y)Y, X\rangle^M + \langle S(X,X), S(Y,Y)\rangle - \|S(X,Y)\|^2,\end{aligned} \tag{8.63}$$

where the $\mathcal{D}_\mu$ raised to the superscript emphasizes that $\bar{K}(X,Y)$ denotes the sectional curvature on the group of volume-preserving diffeomorphisms. This describes a non-trivial fact that, even when the manifold $\mathcal{D}(M)$ is flat, i.e. $\langle R(X,Y)Y, X\rangle^M = 0$, the curvature $\bar{K}(X,Y)^{\mathcal{D}_\mu}$ of $\mathcal{D}_\mu(M)$ does not necessarily vanish due to the second and third terms associated with the second fundamental form S. Namely, the curvature of a *divergence-free* flow originates from the following part,

$$\bar{K}_S(X,Y) := \langle S(X,X), S(Y,Y)\rangle_{L^2} - \|S(X,Y)\|^2.$$

Thus it is found that *the restriction to the volume-preserving flows gives rise to the additional curvature* $\bar{K}_S$.

It is useful to see that the second fundamental form is related to the pressure gradient. In fact, we have

$$S(X,Y) := \hat{\nabla}_X Y - \bar{\nabla}_X Y = \mathsf{Q}[\nabla_X Y] \tag{8.64}$$

from (8.15). The decomposition theorem (Appendix F) says that an arbitrary vector field v can be decomposed orthogonally into a divergence-free part and a gradient part. Setting $v = \nabla_X Y$ in (F.1) of Appendix F, one obtains immediately

$$\mathsf{Q}[\nabla_X Y] = \operatorname{grad} G(\nabla_X Y), \quad G(v) = F_{\mathrm{D}}(v) + H_{\mathrm{N}}(v).$$

Thus it is found that the curvature is related to the "grad" part of the connection $\nabla_X Y$ which is orthogonal to $T_e \mathcal{D}_\mu(M)$.

In particular, for $\eta_t \in \mathcal{D}_\mu(M)$ and $\dot{\eta}_t(e) = X$, we have $S(X,X) = \mathsf{Q}[\nabla_X X]$ from (8.64). Using (8.50), we obtain

$$S(X,X) = \mathsf{Q}[\nabla_X X] = -\operatorname{grad} p_X, \tag{8.65}$$

where p_X is the pressure of the velocity field X. It is found that the second fundamental form $S(X,X)$ is given by the pressure gradient (Fig. 8.7).

The first term of $\bar{K}_S$ is represented as $\langle S(X,X), S(Y,Y)\rangle = \langle \operatorname{grad} p_X, \operatorname{grad} p_Y\rangle$, a correlation of two pressure gradients, and the second term is *non-positive*.

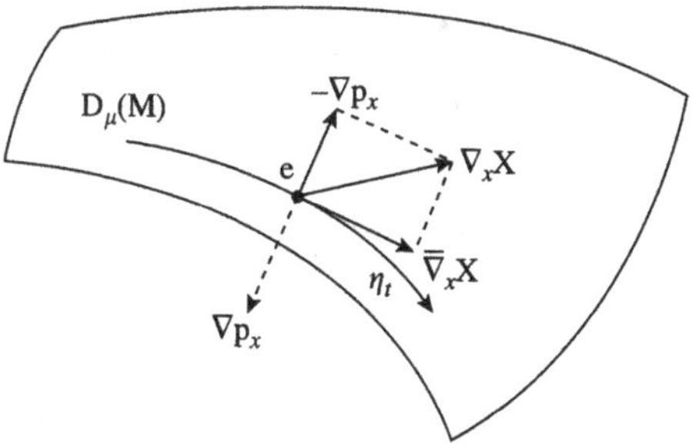

Fig. 8.7. Pressure gradient, $\operatorname{grad} p_X$.

Thus, the $\bar{K}_S$ part of the curvature is given by

$$\bar{K}_S(X,Y) = \langle \operatorname{grad} p_X, \operatorname{grad} p_Y \rangle - \|\operatorname{grad} G(\nabla_X Y)\|^2. \tag{8.66}$$

8.5.3. *Instability in Lagrangian particle sense*

Stability in Lagrangian particle sense is different from the stability of the velocity field in Eulerian sense. In the stability analysis of conventional fluid dynamics, growth or decay of velocity perturbations are of concern. A velocity field is said to be stable in the Eulerian sense if, at fixed points $\boldsymbol{x}$, small perturbations in the initial velocity field do not grow exponentially with time t.

Consider a steady *parallel shear flow* (Fig. 8.8), whose velocity field is given by

$$X = (U(y), 0, 0), \quad \text{and} \quad p = \text{const.} \tag{8.67}$$

This is an exact solution of the equation of motion (8.43). Since $\partial_t X = 0$ and $\nabla_X X = (X \cdot \nabla)X = U\partial_x X = 0$, Eq. (8.43) results in $\operatorname{grad} p = 0$ and hence we have $p = \text{const.}$

In [Mis93, Example 4.4], it is shown that the geodesic curve $g_t = (x + t\sin y, y, z)$ in $\mathcal{D}_\mu(M)$ is *stable in the Eulerian sense.* Noting $X = \dot{g}_t \circ g_t^{-1}$, the velocity X has the form (8.67), where

$$U(y) = \sin y \quad \text{for} \quad 0 \le y \le \pi, \tag{8.68}$$

(and $-\pi \le x, z \le \pi$, say). It is claimed that this flow is stable in the Eulerian sense because the velocity profile $U(y) = \sin y$ has no inflection point within the flow domain $0 < y < \pi$. The *Rayleigh's inflection point theorem* [DR81] requires existence of inflection points where $U'(y) = 0$ in

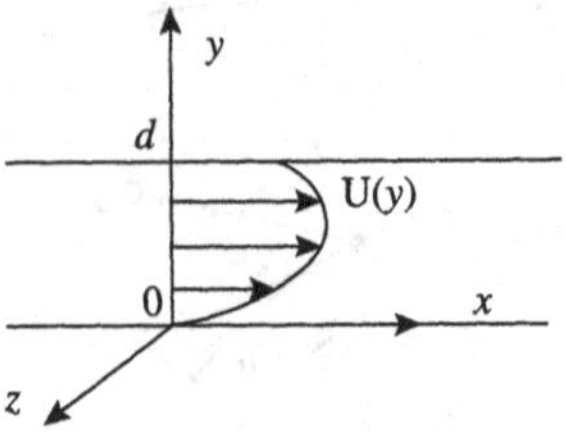

Fig. 8.8. Parallel shear flow.

the velocity profile as a necessary condition for instability of an inviscid fluid flow.[4]

However, in the Lagrangian particle sense, the above flow is regarded as unstable, since perturbations to the diffeomorphisms of particle configuration grow with time. This is seen by using Eq. (8.51) for the Jacobi field J and the sectional curvature (8.63), where $K(T,J)$ is replaced by $\bar{K}(X,J)^{\mathcal{D}_\mu}$. Suppose that the manifold M is the euclidean space $\mathbb{R}^3$ characterized with zero sectional curvatures, i.e. $\langle R(J,X)X,J\rangle^M = 0$ (shown in §8.5.1). Then, we have

$$\bar{K}(X,J)^{\mathcal{D}_\mu} = \langle S(X,X), S(J,J)\rangle - \|S(X,J)\|^2.$$

The above flow (8.68) is a constant-pressure flow, hence the first term on the right vanishes because $S(X,X) = 0$ due to (8.65) and (8.67). Therefore we obtain

$$\bar{K}(X,J)^{\mathcal{D}_\mu} = K(X,e_J)\|J\|^2 \le 0.$$

Then Eq. (8.52) is

$$\frac{1}{\|J\|}\frac{\mathrm{d}^2}{\mathrm{d}s^2}\|J\| = (\|\nabla_X e_J\|^2 - K(X,e_J)) \ge 0. \tag{8.69}$$

This equation implies that the Jacobi vector J grows exponentially at initial times, provided that the coefficient $\|\nabla_X e_J\|^2 - K(X,e_J)$ takes a positive value (i.e. not zero). Therefore, the flow (8.68) is *unstable in the Lagrangian particle sense.* This is because the variation of a nearby geodesic of the parameter α is given by αJ for an infinitesimal α.

[4]In the present geometrical theory, the fluid is regarded as *ideal*, i.e. the viscosity ν is zero. The ideal fluid is *inviscid.* The velocity profile $U = \sin y\ (0 \le y \le \pi)$ is chosen due to mathematical simplicity and is somewhat artificial.

The above analysis suggests that all the parallel shear flows represented by (8.67) are unstable in the Lagrangian particle sense, because they are constant-pressure flows. This is investigated again in §8.7 by calculating the sectional curvatures explicitly.

8.5.4. *Time evolution of Jacobi field*

Jacobi field $J(t)$ is uniquely determined by its value $J(0)$ and the value of $\nabla_T J$ at $t=0$ on the geodesic g_t in the neighborhood of $t=0$. Provided that $J(0)=0$ and

$$\|\nabla_T J\|_{t=0} = \|\partial_t J\|_{t=0} := a_0,$$

initial development of the magnitude of Jacobi field is given by Eq. (3.133) (with t replacing s) in §3.10.2,

$$\frac{\|J\|}{a_0 t} = 1 - \frac{t^2}{6}\kappa(0) + O(t^3), \quad \kappa(0) \equiv \left.\frac{\langle R(J,T)T, J\rangle}{|J|^2}\right|_{t=0}. \tag{8.70}$$

Therefore, if $\kappa(0) < 0$, then $\|J\|/a_0 t > 1$, and if $\kappa(0) > 0$, then $\|J\|/a_0 t < 1$ for sufficiently small t. Thus, the time development of the Jacobi vector is controlled by the curvature $\langle R(J,T)T, J\rangle$ and in particular by its sign.

8.5.5. *Stretching of line-elements*

Consider two nearby geodesic flows $g_t(x : v_1)$ and $g_t(x : v_2)$ emanating from the identity e with two different initial velocity fields v_1 and v_2 in a bounded domain D (hence $g_0(x : v_1) = g_0(x : v_2) = x$). An L^2-distance between the two flows at a time t may be defined (Fig. 8.9) by

$$d(v_1, v_2 : t) := \left(\int_D |g_t(x : v_1) - g_t(x : v_2)|^2 \mathrm{d}^3 x\right)^{1/2}.$$

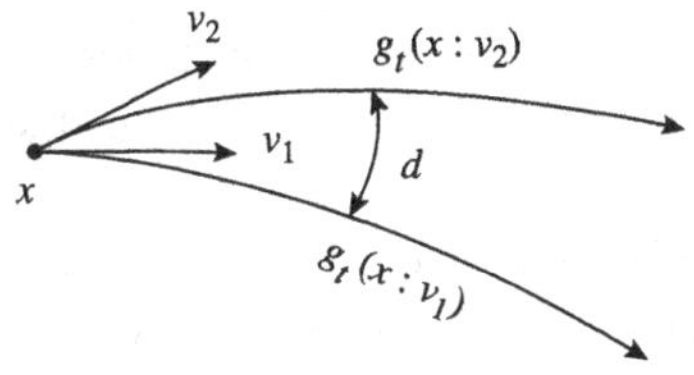

Fig. 8.9. Distance between two flows.

Evidently, one has $d(v_1, v_2 : 0) = 0$. By definition, we have a Taylor expansion with respect to the time t:

$$g_t(x : v) = x + tv(g_t(x), t) + \frac{1}{2}t^2\nabla_t v + \frac{t^3}{6}(\nabla_t)^2 v + O(t^4), \qquad (8.71)$$

where $\nabla_t = \partial_t + v \cdot \nabla$ (see (8.43)). The *distance* d is the mean L^2-distance between two particles starting at the same position x but evolving with different initial velocity fields v. Let us introduce $\bar{v}$ and v' by

$$\bar{v} = (v_1 + v_2)/2, \quad \varepsilon v' = (v_1 - v_2)/2,$$

with an infinitesimal constant ε, then the Jacobi field is defined by $J(t) = (\partial/\partial\varepsilon)g_t(x, : \bar{v} + \varepsilon v')|_{\varepsilon=0}$, and we have

$$d \approx 2\varepsilon\|J\|$$

for infinitesimally small ε, and $a_0 = \|v'\|$. Then, from the formula (8.69), we have

$$d(v_1, v_2 : t) = 2\varepsilon\|v'\| \left(t - \frac{t^3}{6}\|\bar{v}\|^2 \sin^2\theta \hat{K}(\bar{v}, v') \right) + O(\varepsilon^2 t, \varepsilon t^5),$$
$$\hat{K}(\bar{v}, v') = \langle R(v', \bar{v})\bar{v}, v'\rangle / (\|\bar{v}\|^2\|v'\|^2 - \langle \bar{v}, v'\rangle^2),$$

where $\cos\theta = \langle \bar{v}, v'\rangle / \|\bar{v}\|\|v'\|$. This formula was verified for two-dimensional flows on a flat two-torus T^2 by [HK94]. It is found that the sectional curvature appears as a factor to the t^3 term with a negative sign and determines the departure from the linear growth of the L^2-distance. This means, if the curvature is negative, the L^2-distance d grows faster than the linear behavior, and furthermore Eq. (8.52) implies an exponential growth of the distance $d \sim 2\varepsilon|v'|J$ for negative $\hat{K}(\bar{v}, v')$. The variable d is also interpreted as the mean distance between two neighboring particles in the same flow field [HK94].

In the case of flows in a *bounded* domain D without mean flow, the *negative* curvature implies *mixing* of particles. Consider a finite material segment δl connecting two neighboring particles. The segment δl will be *stretched* initially by the negative curvature. In due time, the segment δl will be *folded* by the boundedness of the domain. The stretching and folding of segments are two main factors for chaotic mixing of particles. Stretching of line segments was studied for a two-dimensional turbulence in [HK94], and it was found that average size of δl actually grows exponentially with t if the average value of sectional curvatures is negative initially.

8.6. Flows on a Cubic Space (Fourier Representation)

Explicit expressions can be given for space-periodic flows in a cube of $(2\pi)^3$ by Fourier representation, i.e. for flows on a flat 3-torus $M = T^3 = \mathbb{R}^3/(2\pi Z)^3$ [NHK92; HK94]. With $\boldsymbol{x} \in T^3$, we have $\boldsymbol{x} = \{(x^1, x^2, x^3); \bmod 2\pi\}$. The manifold T^3 is a bounded manifold without boundary (Fig. 8.10). The elements of the Lie algebra of the volume-preserving diffeomorphism group $\mathcal{D}_\mu(T^3)$ can be thought of as real periodic vector fields on T^3 with *divergence-free* property. Such periodic fields are represented by the real parts of corresponding *complex* Fourier forms. The Fourier bases are denoted by $e_k = e^{i\boldsymbol{k}\cdot\boldsymbol{x}}$ where $\boldsymbol{k} = (k_i)$ is a wave number covector with $k_i \in Z$ (integers) and $i = 1, 2, 3$.[5] Now the representations are complexified so that all the fields become linear (or multilinear) in the complex vector space of the complexified Lie algebra. The bases of this vector space are given by the functions e_k $(\boldsymbol{k} \in Z^3, \boldsymbol{k} \neq 0)$. The velocity field $\boldsymbol{u}(\boldsymbol{x}, t)$ is represented as

$$\boldsymbol{u}(\boldsymbol{x}, t) = \sum_{\boldsymbol{k}} \boldsymbol{u}_k(t) e_k,$$

where $\boldsymbol{u}_k(t)$ is the Fourier amplitude, also written as $u^i(\boldsymbol{k})$ $(i = 1, 2, 3)$. The amplitude must satisfy the two properties,

$$\boldsymbol{k} \cdot \boldsymbol{u}_k = 0, \quad \boldsymbol{u}_{-k} = \boldsymbol{u}_k^*, \tag{8.72}$$

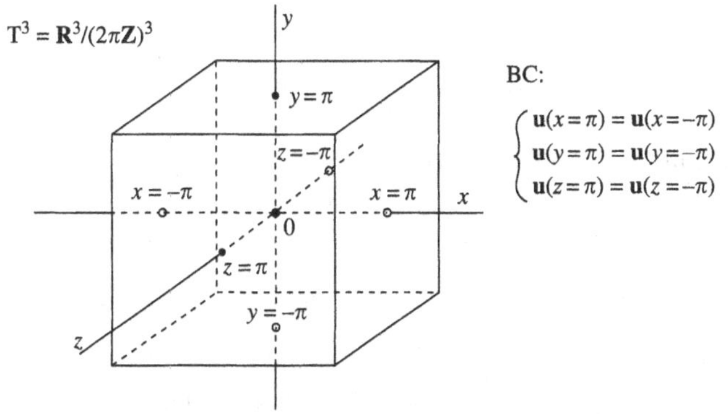

Fig. 8.10. 3-torus T^3 and BC.

[5]In this section, the subscript k and other roman letters in the subscript are understood to denote three-component vectors which are written with bold faces otherwise.

to describe the divergence-free condition and reality condition respectively, where the asterisk denotes the complex conjugate.[6] It should be noted that $\boldsymbol{u}_k$ has two independent polarization components. For example, if $\boldsymbol{k} = (k_x, 0, 0)$, then $\boldsymbol{u}_k = (0, u_k^y, u_k^z)$.

Let us take four tangent fields satisfying (8.72): $\boldsymbol{u}_k e_k, \boldsymbol{v}_l e_l, \boldsymbol{w}_m e_m, \boldsymbol{z}_n e_n$, and use the scalar product convention such as $(\boldsymbol{u} \cdot \boldsymbol{v}) = u^1 v^1 + u^2 v^2 + u^3 v^3$. Then we have the following metric, covariant derivative, commutator, etc.

The *metric* is defined by (8.10) with $M = T^3$, which results in

$$\langle \boldsymbol{u}_k e_k, \boldsymbol{v}_l e_l \rangle = (2\pi)^3 (\boldsymbol{u}_k \cdot \boldsymbol{v}_l) \delta_{0,k+l},$$

where $\delta_{0,k+l} = 1$ (if $\boldsymbol{k} + \boldsymbol{l} = 0$) and 0 (otherwise).

The *covariant derivative* is obtained from (8.39) with p_* satisfying (8.40). In fact, the simple connection is

$$\nabla_{(\boldsymbol{u}_k e_k)}(\boldsymbol{v}_l e_l) = e_k (\boldsymbol{u}_k \cdot \nabla) \boldsymbol{v}_l e_l = i(\boldsymbol{u}_k \cdot \boldsymbol{l}) \boldsymbol{v}_l e_{k+l}. \tag{8.73}$$

The function p_* must satisfy

$$\nabla^2 p_* = -\mathrm{div}[\nabla_{(\boldsymbol{u}_k e_k)}(\boldsymbol{v}_l e_l)] = -i^2 (\boldsymbol{u}_k \cdot \boldsymbol{l})(\boldsymbol{v}_l \cdot (\boldsymbol{k} + \boldsymbol{l})) e_{k+l}.$$

This is satisfied by

$$p_* = -\frac{1}{|\boldsymbol{k} + \boldsymbol{l}|^2} (\boldsymbol{u}_k \cdot \boldsymbol{l})(\boldsymbol{v}_l \cdot (\boldsymbol{k} + \boldsymbol{l})) e_{k+l}. \tag{8.74}$$

The divergence-free connection is defined by

$$\begin{aligned} \bar{\nabla}_{(\boldsymbol{u}_k e_k)}(\boldsymbol{v}_l e_l) &= \nabla_{(\boldsymbol{u}_k e_k)}(\boldsymbol{v}_l e_l) + \nabla p_* \\ &= i(\boldsymbol{u}_k \cdot \boldsymbol{l}) \frac{\boldsymbol{k} + \boldsymbol{l}}{|\boldsymbol{k} + \boldsymbol{l}|} \times \left(\boldsymbol{v}_l \times \frac{\boldsymbol{k} + \boldsymbol{l}}{|\boldsymbol{k} + \boldsymbol{l}|} \right) e_{k+l}. \end{aligned} \tag{8.75}$$

It is seen that the amplitude vector on the right is perpendicular to $\boldsymbol{k} + \boldsymbol{l}$. Therefore $\bar{\nabla}_{(\boldsymbol{u}_k e_k)}(\boldsymbol{v}_l e_l)$ is divergence-free.

The *commutator* is defined by (8.21), which results in

$$\begin{aligned} [\boldsymbol{u}_k e_k, \boldsymbol{v}_l e_l] &= \nabla_{(\boldsymbol{u}_k e_k)}(\boldsymbol{v}_l e_l) - \nabla_{(\boldsymbol{v}_l e_l)}(\boldsymbol{u}_k e_k) \\ &= i((\boldsymbol{u}_k \cdot \boldsymbol{l}) \boldsymbol{v}_l - (\boldsymbol{v}_l \cdot \boldsymbol{k}) \boldsymbol{u}_k) e_{k+l} \\ &= -i(\boldsymbol{k} + \boldsymbol{l}) \times (\boldsymbol{u}_k \times \boldsymbol{v}_l) e_{k+l}. \end{aligned} \tag{8.76}$$

The right side is also perpendicular to $\boldsymbol{k} + \boldsymbol{l}$. Hence $[\boldsymbol{u}_k e_k, \boldsymbol{v}_l e_l]$ is divergence-free as well.

---

[6] We obtain $\mathrm{div}\, \boldsymbol{u} = \sum_k i\boldsymbol{k} \cdot \boldsymbol{u}_k e_k = 0$, and $\boldsymbol{u}_k e_k + \boldsymbol{u}_{-k} e_{-k} = \boldsymbol{u}_k e_k + \boldsymbol{u}_k^* e_k^* = 2\Re[\boldsymbol{u}_k e_k]$.

The *geodesic equation* (8.44) reduces to

$$\frac{\partial}{\partial t}u^l(\boldsymbol{k}) + i\sum_{\boldsymbol{p}+\boldsymbol{q}=\boldsymbol{k}}\sum_{m,n}\left(\delta_{ln} - \frac{k_l k_n}{k^2}\right) k_m u^m(\boldsymbol{p})u^n(\boldsymbol{q}) = 0, \tag{8.77}$$

by using (8.75).

From the definition (8.17) with ∇ replaced by the divergence-free $\bar{\nabla}$ and by using (8.75), the *curvature tensor* is found to be nonzero only for $\boldsymbol{k}+\boldsymbol{l}+\boldsymbol{m}+\boldsymbol{n}=0$, and expressed as

$$\begin{aligned}\bar{R}_{klmn} &:= \langle \bar{R}(\boldsymbol{u}_k e_k, \boldsymbol{v}_l e_l)\boldsymbol{w}_m e_m, \boldsymbol{z}_n e_n\rangle \\ &= (2\pi)^3 \times \left(-\frac{(\boldsymbol{u}_k\cdot\boldsymbol{m})(\boldsymbol{w}_m\cdot\boldsymbol{k})}{|\boldsymbol{k}+\boldsymbol{m}|}\frac{(\boldsymbol{v}_l\cdot\boldsymbol{n})(\boldsymbol{z}_n\cdot\boldsymbol{l})}{|\boldsymbol{l}+\boldsymbol{n}|} \right. \\ &\qquad \left. +\frac{(\boldsymbol{v}_l\cdot\boldsymbol{m})(\boldsymbol{w}_m\cdot\boldsymbol{l})}{|\boldsymbol{l}+\boldsymbol{m}|}\frac{(\boldsymbol{u}_k\cdot\boldsymbol{n})(\boldsymbol{z}_n\cdot\boldsymbol{k})}{|\boldsymbol{k}+\boldsymbol{n}|}\right),\end{aligned} \tag{8.78}$$

and $\bar{R}_{klmn}=0$ if $\boldsymbol{k}+\boldsymbol{l}+\boldsymbol{m}+\boldsymbol{n}\neq 0$. In case when one of the denominators vanishes, then the term originally possessing it should be excluded. In other words, if $\boldsymbol{k}+\boldsymbol{m}=0$ (then $\boldsymbol{l}+\boldsymbol{n}=0$ too), the first term in the parenthesis of (8.78) should be annihilated, but the second term is retained as far as $\boldsymbol{k}+\boldsymbol{n}\neq 0$ and $\boldsymbol{l}+\boldsymbol{m}\neq 0$. When $\boldsymbol{k}+\boldsymbol{m}=0$, we obtain $\bar{\nabla}_{(\boldsymbol{u}_k e_k)}(\boldsymbol{w}_m e_m) = i(\boldsymbol{u}_k\cdot\boldsymbol{m})\boldsymbol{w}_m e_0 = \text{const}$ by using (8.39) and (8.40) with $p_* = \text{const}$. The vanishing of the first term is a consequence of this property.

The two-dimensional problem of flows on T^2 was first studied by Arnold [Arn66]. When two-dimensionality is imposed, the above formulas reduce to those of [Arn66]. Because of difference of the definitions, the signs of curvature tensors are reversed.

8.7. Lagrangian Instability of Parallel Shear Flows

8.7.1. *Negative sectional curvatures*

On the basis of the representations deduced in the previous section for the flows in a cubic space with periodic boundary conditions, we again consider sectional curvature of steady parallel shear flows of the form $Y = (U(y),0,0)$ (and $p=\text{const}$), where $U(y)=\sin ky$ or $\cos ky$ (Fig. 8.11). Introducing a constant wave vector $\boldsymbol{k}$ and Fourier amplitude $\boldsymbol{u}_k$ as

$$\boldsymbol{k}=(0,k,0), \quad \boldsymbol{u}_k = \left(\frac{1}{2i},0,0\right), \quad \text{or} \quad \boldsymbol{u}_k = \left(\frac{1}{2},0,0\right), \tag{8.79}$$

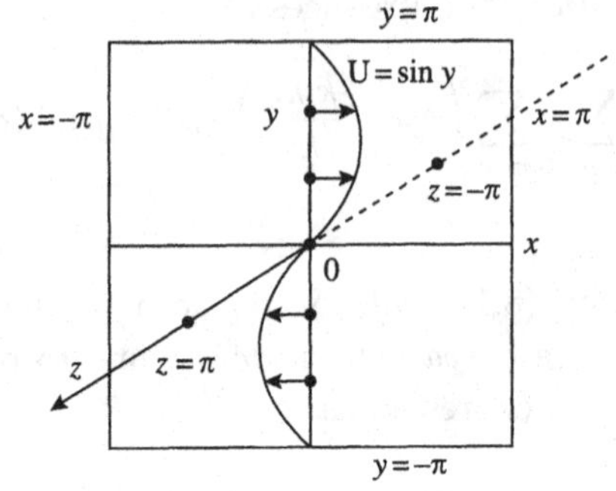

Fig. 8.11. Parallel shear flow $Y = (\sin y, 0, 0)$ for $k = 1$.

together with the cartesian coordinate $\boldsymbol{x} = (x, y, z)$, we have

$$Y = \boldsymbol{u}_k e_k + \boldsymbol{u}_{-k} e_{-k} = (\sin ky, 0, 0) \quad \text{or} \quad (\cos ky, 0, 0)$$

for $-\pi \le x, y, z \le \pi$.

Let $X = \sum \boldsymbol{v}_m e_m$ be any velocity field, satisfying the properties (8.72). Then, the sectional curvature $\langle \bar{R}(X,Y)Y, X\rangle$ can be shown to be non-positive for both $U(y) = \sin ky$ and $\cos ky$. Procedure of its verification is similar and the result is also the same for both flows, which is as follows.

Substituting the above Fourier representations of X and Y and using the property that nonzero $\bar{R}_{klmn}$ must satisfy $\boldsymbol{k}+\boldsymbol{l}+\boldsymbol{m}+\boldsymbol{n} = 0$, we obtain the following,

$$\begin{aligned}\langle \bar{R}(X,Y)Y, X\rangle = \sum_m \big(&\langle \bar{R}(\boldsymbol{v}_m e_m, \boldsymbol{u}_k e_k)\boldsymbol{u}_k e_k, \boldsymbol{v}_{-m-2k} e_{-m-2k}\rangle \\ &+ \langle \bar{R}(\boldsymbol{v}_m e_m, \boldsymbol{u}_k e_k)\boldsymbol{u}_{-k} e_{-k}, \boldsymbol{v}_{-m} e_{-m}\rangle \\ &+ \langle \bar{R}(\boldsymbol{v}_m e_m, \boldsymbol{u}_{-k} e_{-k})\boldsymbol{u}_k e_k, \boldsymbol{v}_{-m} e_{-m}\rangle \\ &+ \langle \bar{R}(\boldsymbol{v}_m e_m, \boldsymbol{u}_{-k} e_{-k})\boldsymbol{u}_{-k} e_{-k}, \boldsymbol{v}_{-m+2k} e_{-m+2k}\rangle\big).\end{aligned}$$

Substituting (8.78) and (8.79), we obtain

$$\begin{aligned}\langle \bar{R}(X,Y)Y, X\rangle = -(2\pi)^3 \frac{k^2}{4} \Bigg[&\sum_{\boldsymbol{m}(\neq -\boldsymbol{k})} \frac{m_x^2}{|\boldsymbol{m}+\boldsymbol{k}|^2} (v_m v_{-m} \pm v_m v_{-m-2k}) \\ &+ \sum_{\boldsymbol{m}(\neq \boldsymbol{k})} \frac{m_x^2}{|\boldsymbol{m}-\boldsymbol{k}|^2} (v_m v_{-m} \pm v_m v_{-m+2k}) \Bigg],\end{aligned}$$

where $\boldsymbol{u}_k \cdot \boldsymbol{m} = (\boldsymbol{u}_k)_x m_x$ and $\boldsymbol{v}_m \cdot \boldsymbol{k} = k(\boldsymbol{v}_m)_y = kv_m$, with the upper sign for $U = \sin ky$, the lower sign for $U = \cos ky$. Replacing $\boldsymbol{m}$ in the second term with $\boldsymbol{m} + 2\boldsymbol{k}$, the above is transformed to

$$\begin{aligned}\langle \bar{R}(X,Y)Y,X\rangle &= -(2\pi)^3 \frac{k^2}{4} \sum_{\boldsymbol{m}(\neq -\boldsymbol{k})} \frac{m_x^2}{|\boldsymbol{m}+\boldsymbol{k}|^2} \\ &\quad \times \left(v_m v_{-m} \pm v_m v_{-m-2k} \pm v_{m+2k} v_{-m} + v_{m+2k} v_{-m-2k}\right) \\ &= -(2\pi)^3 \frac{k^2}{4} \sum_{\boldsymbol{m}(\neq -\boldsymbol{k})} \frac{m_x^2}{|\boldsymbol{m}+\boldsymbol{k}|^2} \left|v_m \pm v_{m+2k}\right|^2, \end{aligned} \tag{8.80}$$

since $v_{-m} = v_m^*$ and $v_{-m-2k} = v_{m+2k}^*$. Thus, it is found

$$\langle \bar{R}(X,Y)Y,X\rangle \le 0. \tag{8.81}$$

It is remarkable that this result is valid for both $U = \sin ky$ and $U = \cos ky$. This *non-positive* sectional curvature is a three-dimensional counterpart of the Arnold's two-dimensional finding [Arn66].

8.7.2. *Stability of a plane Couette flow*

The sinusoidal parallel flows $U(y) = \sin ky$ or $\cos ky$ considered in the previous subsection are chosen by mathematical simplicity, i.e. they are Fourier normal modes. In the context of fluid mechanics, a simplest example is a flow with a linear velocity profile, i.e. a *plane Couette flow* $Y = (U(y), 0, 0)$, where

$$U(y) = y, \quad p = \text{const}, \tag{8.82}$$

for $-\pi \le x, y, z \le \pi$ (Fig. 8.12). The term *plane* means two-dimensional, but here, this is investigated as a flow in $M = T^3$. Obviously, this is a solution of the equation of motion (8.48) since $\partial_t Y = 0$ and $\nabla_Y Y = (Y \cdot \nabla)Y = U\partial_x Y = 0$. Therefore,

$$\bar{\nabla}_Y Y = Y \cdot \nabla Y + \operatorname{grad} p = 0. \tag{8.83}$$

In the linear stability theory with respect to the *velocity field*, it is well known that the plane Couette flow is (neutrally) stable.

In the hydrodynamic stability theory, the linear stability equation of the plane Couette flow takes an exceptional form owing to the linear velocity profile $U(y) = y$, i.e. the second derivative $U''(y)$ vanishes. For that reason, its linear stability is studied by a viscous theory taking account of fluid

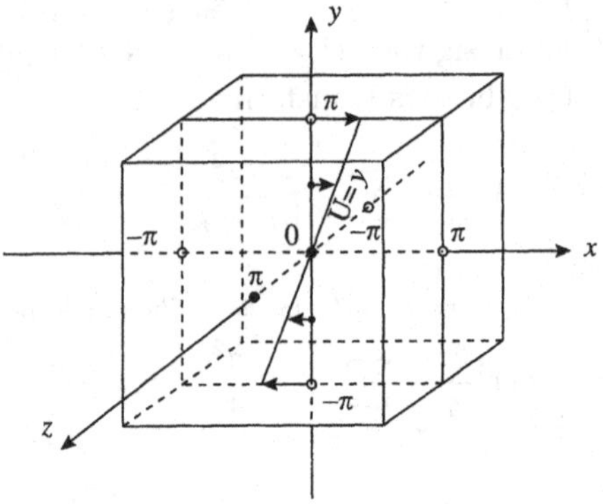

Fig. 8.12. Couette flow $(y, 0, 0)$.

viscosity ν. In the Couette flow of a visous fluid, all the disturbance modes (eigen-solutions) of velocity field are stable, i.e. decay exponentially [DR81]. In the limit $\nu \to +0$, all the modes become *neutrally stable*, i.e. do not decay nor grow exponentially with time. So that, the Couette flow of an ideal fluid is regarded as neutrally stable with respect to the velocity field.

Now we consider stability of the plane Couette flow in the Lagrangian particle sense. Denoting any velocity field as $X = \sum \boldsymbol{v}_l e_l$, let us calculate the sectional curvature $K(X,Y) = \langle \bar{R}(Y,X)X, Y\rangle$ for

$$X = \sum_l \boldsymbol{v}_l e_l, \quad Y = (U(y), 0, 0),$$

where $U(y) = y$ and $e_l = \exp[i\boldsymbol{l} \cdot \boldsymbol{x}] = \exp[i(l_x x + l_y y + l_z z)]$. It will be found that $K(X,Y)$ is *non-positive.*

Using the definition of the curvature tensor,

$$\bar{R}(X,Y)Z = \bar{\nabla}_X(\bar{\nabla}_Y Z) - \bar{\nabla}_Y(\bar{\nabla}_X Z) - \bar{\nabla}_{[X,Y]}Z,$$

the sectional curvature is given by

$$\begin{aligned} K(X,Y) &= \langle \bar{R}(X,Y)Y, X\rangle \\ &= -\langle \bar{\nabla}_Y Y, \bar{\nabla}_X X\rangle + \langle \bar{\nabla}_X Y, \bar{\nabla}_Y X\rangle - \langle \bar{\nabla}_{[X,Y]} Y, X\rangle, \end{aligned} \qquad (8.84)$$

where the following formula has been used: $Z\langle X, Y\rangle = \langle \nabla_Z X, Y\rangle + \langle X, \nabla_Z Y\rangle = 0$ for $X, Y, Z \in T\mathcal{D}_\mu(T^3)$ (see (3.88) of §3.7.3).[7] Note that $\bar{\nabla}_X Y, \bar{\nabla}_Y Y \in T\mathcal{D}_\mu(T^3)$.

[7]This is valid for the right-invariant metric $\langle X, Y\rangle$ of right-invariant fields X and Y along the flow generated by Z.

Because the steady tangent field Y is the solution of the geodesic equation (8.83), i.e. $\bar{\nabla}_Y Y = 0$, the first term vanishes:

$$\langle \bar{\nabla}_Y Y, \bar{\nabla}_X X \rangle = 0. \tag{8.85}$$

In order to obtain the second term, let us calculate $\bar{\nabla}_X Y$, which is given by (8.39) as

$$\bar{\nabla}_X Y = (X \cdot \nabla)Y + \operatorname{grad} p_{XY}. \tag{8.86}$$

The first simple covariant derivative is

$$(X \cdot \nabla)Y = e_l(\boldsymbol{v}_l \cdot \nabla)(U(y), 0, 0) = e_l(v_l^y U'(y), 0, 0), \tag{8.87}$$

where v_l^y is the y-component of $\boldsymbol{v}_l$, which is a function of the wave number $\boldsymbol{l} = (l_x, l_y, l_z)$ and independent of $\boldsymbol{x} = (x, y, z)$.[8] The scalar function p_{XY} must be determined so that $\bar{\nabla}_X Y$ is divergence-free. Hence,

$$\nabla^2 p_{XY} = -\operatorname{div}[(X \cdot \nabla)Y] = -il_x v_l^y U'(y) e_l \tag{8.88}$$

Since $U'(y) = 1$, the right side is a function of $e_l = \exp[i\boldsymbol{l} \cdot \boldsymbol{x}]$ with a coefficient depending on only the wave number $\boldsymbol{l}$. Therefore, we obtain

$$p_{XY} = i\frac{l_x}{l^2} v_l^y U'(y) \exp[i\boldsymbol{l} \cdot \boldsymbol{x}]. \tag{8.89}$$

Thus, substituting (8.87) and (8.89) into (8.86), we have

$$\bar{\nabla}_X Y = e_l(v_l^y U'(y), 0, 0) - \frac{l_x}{l^2} v_l^y U'(y) e_l (l_x, l_y, l_z). \tag{8.90}$$

Next, to obtain $\bar{\nabla}_Y X = Y \cdot \nabla X + \operatorname{grad} p_{YX}$, it is noted that

$$(Y \cdot \nabla)X = U(y)\partial_x(e_m \boldsymbol{v}_m) = im_x U(y) e_m \boldsymbol{v}_m. \tag{8.91}$$

Taking divergence, we obtain

$$\begin{aligned} \nabla^2 p_{YX} &= -\operatorname{div}[(Y \cdot \nabla)X] = -im_x v_m^y U'(y) e_m - im_x U(y) e_m (i\boldsymbol{m} \cdot \boldsymbol{v}_m) \\ &= -im_x v_m^y U'(y) e_m, \end{aligned}$$

since $\boldsymbol{m} \cdot \boldsymbol{v}_m = 0$. The right-hand side is seen to be the same as that of (8.88), hence we find that p_{YX} is given by (8.89) as well. Thus, we have found

$$\bar{\nabla}_Y X = im_x U(y) e_m \boldsymbol{v}_m - \frac{m_x}{m^2} v_m^y U'(y) e_m \boldsymbol{m}. \tag{8.92}$$

---

[8]In Eq. (8.87) (and below) and Eq. (8.91) (and below), the summation with respect to the wavenumber $\boldsymbol{l}$ or $\boldsymbol{m}$ is meant implicitly, although the symbol $\sum$ is omitted for simplicity. Only when considered to be necessary, it is written explicitly.

Then, the second term of (8.84) is

$$
\begin{aligned}
&\langle \bar{\nabla}_X Y, \bar{\nabla}_Y X \rangle \\
&= \sum_l \sum_m \Big[i m_x v_l^y v_m^x U' \langle e_l, U(y) e_m \rangle - i m_x \frac{l_x}{l^2} v_l^y (\boldsymbol{l} \cdot \boldsymbol{v}_m) U' \langle e_l, U(y) e_m \rangle \\
&\qquad - \frac{m_x^2}{m^2} v_l^y v_m^y (U')^2 \langle e_l, e_m \rangle + \frac{l_x m_x}{l^2 m^2} v_l^y v_m^y (\boldsymbol{l} \cdot \boldsymbol{m}) (U')^2 \langle e_l, e_m \rangle \Big] .
\end{aligned} \tag{8.93}
$$

In order to obtain the third term of (8.84), it is noted that

$$[X, Y] = X \cdot \nabla Y - Y \cdot \nabla X = e_l (v_l^y U'(y), 0, 0) - i l_x U(y) \boldsymbol{v}_l e_l .$$

Then, we have

$$
\begin{aligned}
\nabla_{[X,Y]} Y &= ([X, Y] \cdot \nabla) Y = (e_l v_l^y U'(y) \partial_x - i l_x U(y) e_l \boldsymbol{v}_l \cdot \nabla) Y \\
&= -i l_x v_l^y (U(y) U'(y) e_l, 0, 0),
\end{aligned} \tag{8.94}
$$

since $\partial_x Y = 0$. Writing the divergence-free connection $\bar{\nabla}$ as

$$\bar{\nabla}_{[X,Y]} Y = \nabla_{[X,Y]} Y + \operatorname{grad} p_{[*]}, \tag{8.95}$$

the scalar function $p_{[*]}$ must satisfy

$$\nabla^2 p_{[*]} = -\operatorname{div}[([X, Y] \cdot \nabla) Y] = -l_x^2 v_l^y U' U(y) e_l . \tag{8.96}$$

A difference from Eq. (8.88) should be remarked that there is a factor $U(y) = y$, a function of y, in addition to $e_l = \exp[i\boldsymbol{l} \cdot \boldsymbol{x}]$ $(U'(y) = 1)$. Due to this factor, the function $p_{[*]}$ satisfying the above equation is given by

$$p_{[*]} = \frac{l_x^2}{l^2} v_l^y U' U(y) e_l + 2i \frac{l_x^2 l_y}{l^2 l^2} v_l^y (U')^2 e_l . \tag{8.97}$$

It can be readily checked that operating ∇^2 on the above $p_{[*]}$ gives the right-hand side of (8.96). Thus, it is found that the expression (8.95) is written by

$$\bar{\nabla}_{[X,Y]}Y = -il_x v_l^y U'U(y)(e_l,0,0) + il_x \frac{l_x}{l^2} v_l^y U'U(y)e_l \boldsymbol{l}$$
$$+ \frac{l_x^2}{l^2} v_l^y (U')^2 (0,e_l,0) - 2\frac{l_x^2 l_y}{l^2 l^2} v_l^y (U')^2 e_l \boldsymbol{l}. \tag{8.98}$$

Then, the third term of (8.84) is

$$\langle \bar{\nabla}_{[X,Y]}Y, X\rangle$$
$$= \sum_l \sum_m \left[-il_x v_l^y v_m^x U' \langle U(y)e_l, e_m\rangle + il_x \frac{l_x}{l^2} v_l^y (\boldsymbol{l}\cdot \boldsymbol{v}_m) U' \langle U(y)e_l, e_m\rangle \right.$$
$$\left. + \frac{l_x^2}{l^2} v_l^y v_m^y (U')^2 \langle e_l, e_m\rangle - 2\frac{l_x^2 l_y}{l^2 l^2} v_l^y v_m^y (\boldsymbol{l}\cdot \boldsymbol{v}_m)(U')^2 \langle e_l, e_m\rangle \right]. \tag{8.99}$$

Thus, it is found from (8.85), (8.93) and (8.99) that the sectional curvature of (8.84) is given by

$$K(X,Y) = \langle \bar{\nabla}_X Y, \bar{\nabla}_Y X\rangle - \langle \bar{\nabla}_{[X,Y]}Y, X\rangle$$
$$= \sum_l \sum_m \left[i(m_x + l_x) v_l^y v_m^x U' \langle U(y)e_l, e_m\rangle \right.$$
$$- i(m_x + l_x)\frac{l_x}{l^2} v_l^y (\boldsymbol{l}\cdot \boldsymbol{v}_m) U' \langle U(y)e_l, e_m\rangle$$
$$+ \left(-\left(\frac{m_x^2}{m^2} + \frac{l_x^2}{l^2} \right) + \frac{l_x m_x}{l^2 m^2} (\boldsymbol{l}\cdot \boldsymbol{m}) \right) v_l^y v_m^y (U')^2 \langle e_l, e_m\rangle$$
$$\left. - 2\frac{l_x^2 l_y}{l^2 l^2} (\boldsymbol{l}\cdot \boldsymbol{v}_m) v_l^y v_m^y (U')^2 \langle e_l, e_m\rangle \right], \tag{8.100}$$

where

$$\langle e_l, e_m\rangle = \int_{T^3} \exp[i(\boldsymbol{l}+\boldsymbol{m})\cdot \boldsymbol{x}] \mathrm{d}^3\boldsymbol{x}$$
$$= (2\pi)^3 \delta(l_x + m_x)\delta(l_y + m_y)\delta(l_z + m_z), \tag{8.101}$$

$$\langle U(y)e_l, e_m\rangle = \int_{T^3} U(y) \exp[i(\boldsymbol{l}+\boldsymbol{m})\cdot \boldsymbol{x}]\, \mathrm{d}^3\boldsymbol{x}$$
$$= (2\pi)^2 \delta(l_x + m_x)\delta(l_z + m_z) \int_{\pi}^{\pi} U(y) \exp[i(l_y + m_y)y]\, \mathrm{d}y, \tag{8.102}$$

and the function $\delta(l_x + m_x)$ denotes 1 if $l_x + m_x = 0$, and 0 otherwise.

The first two terms of (8.100) vanish owing to (8.102) since $(m_x + l_x)\langle U(y)e_l, e_m\rangle = 0$. The fourth term vanishes as well since $(\boldsymbol{l}\cdot\boldsymbol{v}_m)\langle e_l, e_m\rangle = -(2\pi)^3(\boldsymbol{m}\cdot\boldsymbol{v}_m) = 0$ by (8.72). Thus, using (8.101), we obtain

$$K(X,Y) = -(2\pi)^3(U')^2\frac{m_x^2}{m^2}|v_m^y|^2. \qquad (8.103)$$

This states that the curvature $K(X,Y)$ is non-positive in the section spanned by the plane Couette flow $U = y$ and any tangent field X. It will be readily seen that this is consistent with (8.80) in §8.7.1 for $U = \sin ky$. Thus, *the plane Couette flow is unstable in the Lagrangian particle sense, but it is neutrally stable with respect to the velocity field,* as noted in the beginning.

8.7.3. *Other parallel shear flows*

Negative sectional curvature can be found for any parallel shear flow of the velocity field

$$X = (U(y), 0, 0),$$

as already remarked in §8.5.3. Since $\partial_t X = 0$ and $\nabla_X X = (X\cdot\nabla)X = U\partial_x X = 0$, we obtain that the pressure p is constant by the equation of motion (8.43). Hence, we have $S(X,X) = 0$. Therefore, the Jacobi equation is given by (8.69). Thus, any parallel shear flow of the form $X = (U(y), 0, 0)$ is unstable in the Lagrangian particle sense.

A realistic velocity field (more realistic than $U(y) = \sin y$) would be the *plane Poiseuille flow* (Fig. 8.13),

$$U(y) = 1 - y^2, \quad -1 \le y \le 1,$$

which is established between two parallel plate walls at $y = \pm 1$. This is an exact solution of a viscous fluid flow under *constant pressure gradient.*

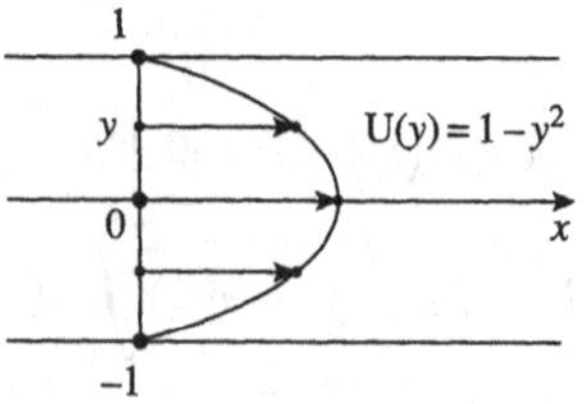

Fig. 8.13. Plane Poiseuille flow $U(y) = 1 - y^2$.

However, in the case of an inviscid flow (the viscosity ν being 0), this is an exact solution under *constant pressure* likewise as above. The plane Poiseuille flow is unstable as a viscous flow [DR81], but it tends to neutral stability (not unstable) in the limit as the viscosity ν tends to zero. In fact, the parabolic velocity profile does not have any inflection point. It is obvious that the plane Poiseuille flow is unstable in the Lagrangian particle sense by the reasoning given above.

The *axisymmetric Poiseuille flow* in a circular pipe of unit radius,

$$U(r) = 1 - r^2, \quad 0 \le r \le 1,$$

is also an exact solution of a viscous fluid flow under constant pressure gradient in the axisymmetric frame (x, r, θ) with x taken along the pipe axis and r the radial coordinate in the circular cross-section. In the case of an inviscid flow, this is a solution under constant pressure as well. The same is true for this axisymmetric Poiseuille flow as for the plane flow, except that this axisymmetric flow is linearly stable as a viscous fluid [DR81] with respect to velocity field. The axisymmteric Poiseuille flow is unstable in the Lagrangian particle sense.

8.8. Steady Flows and Beltrami Flows

In this section, we consider *steady* flows which do not depend on time.

8.8.1. *Steady flows*

From (8.43), the equation of a steady flow of an incompressible ideal fluid in a bounded domain $M \subset \mathbb{R}^3$ is given by

$$(\boldsymbol{u} \cdot \nabla)\boldsymbol{u} + \nabla p = 0, \tag{8.104}$$

where the velocity field $\boldsymbol{u}$ satisfies $\operatorname{div} \boldsymbol{u} = 0$,[9] or equivalently by

$$\boldsymbol{u} \times \operatorname{curl} \boldsymbol{u} = -\nabla B, \tag{8.105}$$

(see (8.45)), $B = p + \frac{1}{2}u^2$. The function $B : M \to \mathbb{R}$ is called the *Bernoulli function*.

[9]In arbitrary n-dimensional Riemannian manifold M^n with measure μ, the same equation takes the form: $\nabla_v v + \nabla p = 0$ for a velocity field v satisfying $\mathcal{L}_v \mu = 0$.

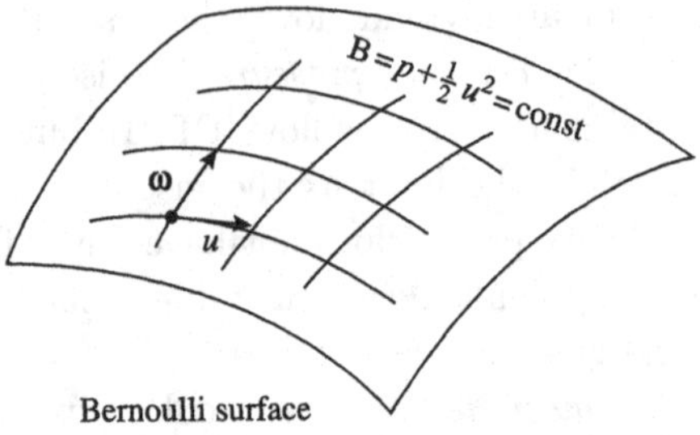

Fig. 8.14. Bernoulli surface.

Equation (8.105) describes that the velocity field $\boldsymbol{u}$ as well as the vorticity field $\boldsymbol{\omega} = \mathrm{curl}\,\boldsymbol{u}$ are perpendicular to the vector ∇B. If $\boldsymbol{u}$ and $\boldsymbol{\omega}$ are *not collinear*, i.e. if $\boldsymbol{u} \times \boldsymbol{\omega} \neq 0$, both of them are tangent to the level surface of the function B. This means that $B = p + \frac{1}{2}u^2$ is a first integral of motion in M. Indeed, because $(\boldsymbol{u} \cdot \nabla)B = 0$ and $(\boldsymbol{\omega} \cdot \nabla)B = 0$, B is constant over a surface composed of streamlines (generated by $\boldsymbol{u}$) and intersecting vortex-lines (generated by $\boldsymbol{\omega}$). Such an integral surface may be called a *Bernoulli surface* (Fig. 8.14). See Appendix J for the condition of integrability.

It is interesting to see that steady flows with such Bernoulli surfaces have *non-negative* (or positive) sectional curvatures. This is verified by noting that $\boldsymbol{u}$ satisfies the geodesic equation $\bar{\nabla}_{\boldsymbol{u}}\boldsymbol{u} = 0$, and that the steady vorticity equation is

$$0 = (\boldsymbol{u} \cdot \nabla)\boldsymbol{\omega} - (\boldsymbol{\omega} \cdot \nabla)\boldsymbol{u} = \bar{\nabla}_{\boldsymbol{u}}\boldsymbol{\omega} - \bar{\nabla}_{\boldsymbol{\omega}}\boldsymbol{u} = [\boldsymbol{u}, \boldsymbol{\omega}],$$

(see (8.53)), since the scalar function p_* is common for both $\bar{\nabla}_{\boldsymbol{u}}\boldsymbol{\omega}$ and $\bar{\nabla}_{\boldsymbol{\omega}}\boldsymbol{u}$ from (8.40). Thus we have $[\boldsymbol{u}, \boldsymbol{\omega}] = 0$, i.e. the two fields $\boldsymbol{u}$ and $\boldsymbol{\omega}$ commute with one another.

Noting that $\bar{\nabla}_{[\boldsymbol{u},\boldsymbol{\omega}]} = 0$ for $[\boldsymbol{u}, \boldsymbol{\omega}] = 0$, and that $\bar{\nabla}_{\boldsymbol{u}}\boldsymbol{\omega} = \bar{\nabla}_{\boldsymbol{\omega}}\boldsymbol{u}$, Eq. (8.84) of the sectional curvature reduces to

$$K(\boldsymbol{u}, \boldsymbol{\omega}) = \langle \bar{\nabla}_{\boldsymbol{u}}\boldsymbol{\omega}, \bar{\nabla}_{\boldsymbol{\omega}}\boldsymbol{u} \rangle = \|\bar{\nabla}_{\boldsymbol{u}}\boldsymbol{\omega}\|^2 \geq 0. \tag{8.106}$$

It is interesting to compare this non-negativity of the sectional curvature with the following theorem. From the property that steady velocity field $\boldsymbol{u}$ commutes with $\boldsymbol{\omega}$, one can deduce a topological structure of steady flows of ideal fluids in a bounded three-dimensional domain. The flow domain is partitioned by analytic submanifolds into a finite number of cells. Namely, the Bernoulli surface must be a torus (invariant under the flow), or an annnular surface (diffeomorphic to $S^1 \times \mathbb{R}$, invariant under

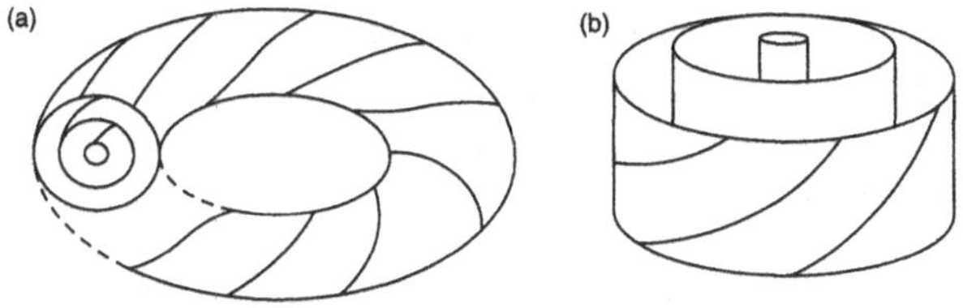

Fig. 8.15. Regions of a steady flow ($\boldsymbol{u} \times \boldsymbol{\omega} \neq 0$), fibered into (a) tori, and (b) annuli.

the flow) [Arn66; Arn78; AK98]. This is verified by using the Liouville's theorem [Arn78, §49]. Streamlines are either closed or dense on each torus, and closed on each annulus (Fig. 8.15).

In the case where $\boldsymbol{u}$ and $\operatorname{curl}\boldsymbol{u}$ are *collinear*, we may write that $\operatorname{curl}\boldsymbol{u} = \zeta(\boldsymbol{x})\boldsymbol{u}$ at each point $\boldsymbol{x}$. The function $\zeta(\boldsymbol{x})$ is a first integral of the field $\boldsymbol{u}$. In fact, we have

$$0 \equiv \operatorname{div}(\operatorname{curl}\boldsymbol{u}) = \operatorname{div}(\zeta\boldsymbol{u}) = (\boldsymbol{u}\cdot\nabla)\zeta,$$

since $\operatorname{div}\boldsymbol{u} = 0$. The function $\zeta(\boldsymbol{x})$ is constant along a streamline (as well as a overlapping vortexline). It may so happen that the streamline fills the entire space M. In this collinear case, we have $\boldsymbol{u} \times \operatorname{curl}\boldsymbol{u} = 0$, and $\nabla B = 0$. Therefore, $B = \text{const}$ at all points $\boldsymbol{x} \in M$. It is interesting to see that the condition of integrability (J.1) of Appendix J is violated in the collinear case. In the context of magnetohydrodynamics, the field satisfying $\boldsymbol{u} \times \operatorname{curl}\boldsymbol{u} = 0$ is called a *force-free* field [AK98].

8.8.2. *A Beltrami flow*

A *Beltrami field* is defined by a velocity field $\boldsymbol{u}(\boldsymbol{x})$ satisfying

$$\nabla \times \boldsymbol{u} = \lambda\boldsymbol{u}, \quad \lambda \in \mathbb{R}, \tag{8.107}$$

i.e. $\boldsymbol{u}$ is an eigenfield of the operator "curl" with a real eigenvalue λ. Such a force-free field can have a complicated structure, and flows with the Beltrami property are characterized by *negative* sectional curvatures.

Consider a velocity field in $M = T^3\{(x, y, z) \,|\, \operatorname{mod} 2\pi\}$, defined by

$$\boldsymbol{u}^B = \boldsymbol{u}_p e_p + \boldsymbol{u}_{-p} e_{-p}, \tag{8.108}$$

with the divergence-free property $\boldsymbol{p}\cdot\boldsymbol{u}_p = 0$ and the reality $\boldsymbol{u}_{-p} = \boldsymbol{u}_p^*$. If $\boldsymbol{p} = (0, p, 0)$, then $\boldsymbol{u}_p = (u_p^x, 0, u_p^z)$.

Suppose that the velocity field $\boldsymbol{u}(\boldsymbol{x})$ satisfies the Beltrami condition (8.107). Then, we have $\lambda^2 = |\boldsymbol{p}|^2$. In fact, the Beltrami condition is ($\nabla \times \boldsymbol{u}_p e_p =$) $i\boldsymbol{p} \times \boldsymbol{u}_p e_p = \lambda \boldsymbol{u}_p e_p$ and its complex conjugate. Taking cross product with $i\boldsymbol{p}$, we have

$$i^2 \boldsymbol{p} \times (\boldsymbol{p} \times \boldsymbol{u}_p) = p^2 \boldsymbol{u}_p = \lambda i \boldsymbol{p} \times \boldsymbol{u}_p = \lambda^2 \boldsymbol{u}_p.$$

Thus we obtain $p^2 = \lambda^2$. Hence, the flow field (8.108) has the Beltrami property $\nabla \times \boldsymbol{u}^B = |\boldsymbol{p}|\boldsymbol{u}^B$, if $\boldsymbol{u}_p = i|\boldsymbol{p}|^{-1}\boldsymbol{p} \times \boldsymbol{u}_p$ is satisfied. An example of such a velocity field is a *one-mode* Beltrami flow $\boldsymbol{u}^B = (u, v, w)$, with $\boldsymbol{p} = (0, p, 0)$ and $\boldsymbol{u}_p = \frac{1}{2}c(1, 0, -i)$, which is

$$u = c\cos py, \quad v = 0, \quad w = c\sin py. \tag{8.109}$$

It can be readily checked that this satisfies $\nabla \times \boldsymbol{u}^B = p\boldsymbol{u}^B$. As y (taken vertically upward) increases, the velocity vector $\boldsymbol{u}^B = c(\cos py, 0, \sin py)$, which is uniform in a horizontal (z, x)-plane, but rotates clockwise (seen from the y axis) in the (z, x)-plane together with the vorticity vector $\nabla \times \boldsymbol{u}^B$. This is a *directionally shearing* flow (Fig. 8.16).

According to Eq. (8.105), this velocity $\boldsymbol{u}$ is regarded as a steady solution to the equation where $p + \frac{1}{2}u^2$ is constant. Let $X = \sum \boldsymbol{v}_l e_l$ be any velocity field satisfying (8.72). Then one can show that $K(\boldsymbol{u}^B, X)$ is *non-positive* [NHK92]. This is another class of flows of negative sectional curvatures in addition to the parallel shear flows considered in the previous section. In this Beltrami flow (directional shearing), the vorticity vector $\nabla \times \boldsymbol{u}^B$ is parallel to the flow velocity, whereas in the parallel shear flows the vorticity is perpendicular to the flow velocity. The negative sectional curvature leads to exponential growth of the Jacobi vector $\|J\|$ according to (8.51), and means that infinitesimal line-elements are stretched on the average.

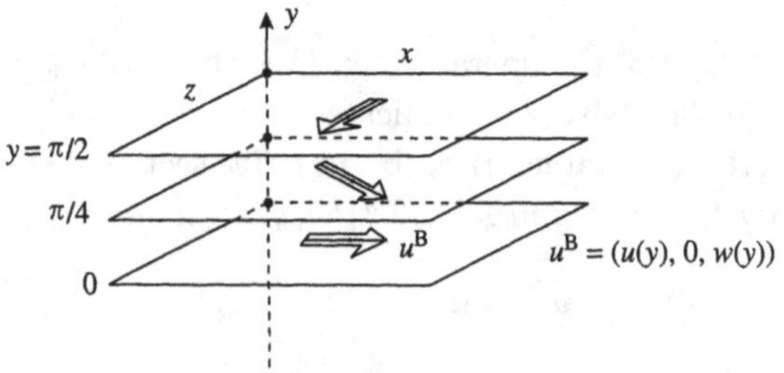

Fig. 8.16. One-mode Beltrami flow $\boldsymbol{u}^B$.

8.8.3. *ABC flow*

Another family of Beltrami field is given by the following *ABC flow* $\boldsymbol{v}^{ABC}$ in $T^3\{(x, y, z) \mid \mod 2\pi\}$:

$$\left.\begin{aligned} u &= \pm A \sin z + C \cos y, \\ v &= \pm B \sin x + A \cos z, \\ w &= \pm C \sin y + B \cos x. \end{aligned}\right\} \quad A, B, C \in \mathbb{R} \tag{8.110}$$

It is obvious that the velocity field $\boldsymbol{v}^{ABC} = (u, v, w)$ with the three parameters (A, B, C) is divergence-free, and it can be readily checked that this satisfies $\nabla \times \boldsymbol{v} = \pm \boldsymbol{v}$.[10] There is numerical evidence that certain trajectories are *chaotic* (Fig. 8.17), i.e. densely fill in a three-dimensional domain [DFGHMS86]. On the other hand, if one of the parameters (A, B, C) vanishes, the flow is integrable.

The quantity $(\boldsymbol{v}, \nabla \times \boldsymbol{v})_x / |\boldsymbol{v}|^2 = \pm 1$ is often called the *helicity*. It is obvious that the ABC flow of the helicity -1 is obtained by the transformation $(x, y, z) \to (-x, -y, -z)$ of the ABC flow of helicity $+1$.

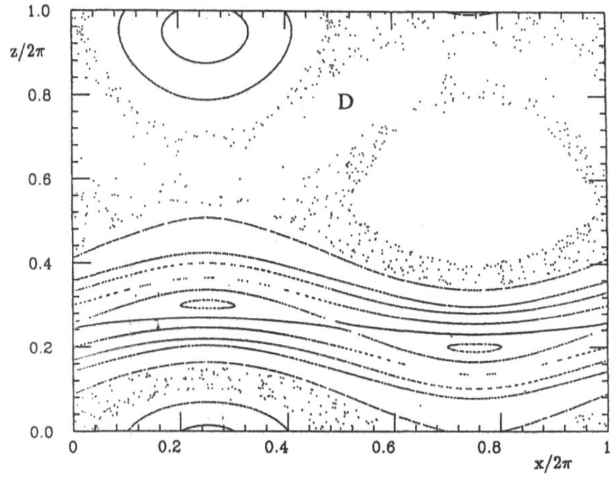

Fig. 8.17. A Poincare map at the section $y = 0$ for the streamlines of an ABC flow $(A = 1, B = C = 1/\sqrt{2})$. Each dot signifies crossing of a streamline through the section $y = 0$. Each dot uniquely determines a next dot. Irregular distribution of such dots in the part D of the diagram represents chaotic behavior of one streamline.

[10]The ABC flow was studied earlier by Gromeka (1881) and Beltrami (1889), and in the present context by Arnold (1965) and Childress (1967). [AK98].

The same ABC flow can be represented by a linear combination of three $\boldsymbol{u}^B$-type flows, i.e. a three-mode Beltrami flow:

$$\begin{aligned}\boldsymbol{v}^{ABC} &= A[(\mp i,1,0)e^{iz} + (\pm i,1,0)e^{-iz}] + B[(0,\mp i,1)e^{ix} + (0,\pm i,1)e^{-ix}] \\ &\quad + C[(1,0,\mp i)e^{iy} + (1,0,\pm i)e^{-iy}]. \qquad (8.111)\end{aligned}$$

With a single mode $\boldsymbol{u}^B$ flow considered in the previous subsection, the sectional curvature $K(\boldsymbol{u}^B, X)$ is already non-positive (for any velocity field X), it is highly likely that $K(\boldsymbol{v}^{ABC}, X)$ is non-positive. Even in an intgegrable case where one of (A, B, C) vanishes, $K(\boldsymbol{v}^{ABC}, X)$ will be negative.

Suppose that we have another ABC flow with $(A', B', C') \neq (A, B, C)$. It is straightforward to show [KNH92] that the normalized sectional curvature is a negative constant:

$$K_*(X,Y) = \frac{\langle \bar{R}(X,Y)Y, X\rangle}{|X|^2|Y|^2 - \langle X, Y\rangle^2} = -\frac{1}{64\pi^3},$$

where $X = \boldsymbol{v}^{ABC}$ and $Y = \boldsymbol{v}^{A'B'C'}$. This means that, even in the case that both (A, B, C) and (A', B', C') are close and the streamlines are not chaotic, the particle motion by $\boldsymbol{v}^{A'B'C'}$ will deviate from that of $\boldsymbol{v}^{ABC}$ and not be predicted from the particle motion of $\boldsymbol{v}^{ABC}$ in the course of time.

8.9. Theorem: $\boldsymbol{\alpha}_B^1 = -\boldsymbol{i}_u \mathrm{d}\boldsymbol{\alpha}_w^1 + \mathrm{d}\boldsymbol{f}$

In §8.2.3, associated with the vector field $B = ad_u^* w \in T_e\mathcal{D}_\mu(M)$, a corresponding 1-form α_B^1 was given by the formula (8.30),[11]

$$\alpha_B^1 = -i_u \mathrm{d}\alpha_w^1 + \mathrm{d}f, \qquad (8.112)$$

where i_u is the operator symbol of interior product (Appendix B.4), $u, w \in T_e\mathcal{D}_\mu(M)$, and $\mathrm{d}f = \partial_i f \mathrm{d}x^i$. Its proof for a general Riemannian manifold M^n is as follows [Arn66].

[11] To any tangent vector B, one can define a 1-form by $\alpha_B^1[\xi] = i_\xi \alpha_B^1 := (B, \xi)$, the scalar product with any tangent vector ξ, i.e. $\alpha_B^1 = g_{jk}B^k \mathrm{d}x^j$. By (8.26), B is defined by $(B, \xi) = (ad_u^* w, \xi) = (w, [u, \xi]) = \alpha_B^1[\xi]$.

Let τ be a Riemannian volume element on M (an n-form, $n = \dim M$). Then, for any tangent vector fields u, v, we have

$$\alpha_u^1 \wedge (i_v \tau) = (u, v)\tau,$$

by definition of exterior algebras (Appendix B, (B.43)), where (u, v) is the scalar product at a point in M. For any tangent field $u, v, w \in T_e\mathcal{D}_\mu(M)$, we have

$$\langle w, [u, v]\rangle = \int_M (w, \{u, v\})\tau = \int_M \alpha_w^1 \wedge i_{\{u,v\}}\tau \tag{8.113}$$

$$= \langle ad_u^* w, v\rangle = \langle B, v\rangle = \int_M i_v \alpha_B^1 \tau \tag{8.114}$$

(see (8.21) for $\{u, v\}$). Note that we have the following definition and identity:

$$i_{u\wedge v}\tau := i_v(i_u\tau), \tag{8.115}$$

$$\mathrm{d}(i_{u\wedge v}\tau) = -i_{\{u,v\}}\tau + i_u \mathrm{d}(i_v\tau) - i_v \mathrm{d}(i_u\tau). \tag{8.116}$$

Since $\mathrm{d}(i_v\tau) = (\mathrm{div}\, v)\tau = 0$ and $\mathrm{d}(i_u\tau) = (\mathrm{div}\, u)\tau = 0$, Eq. (8.116) leads to

$$i_{\{u,v\}}\tau = -\mathrm{d}(i_{u\wedge v}\tau).$$

Using this, the integration $\int_M \alpha_w^1 \wedge i_{\{u,v\}}\tau$ is transformed to

$$\int_M \alpha_w^1 \wedge i_{\{u,v\}}\tau = -\int_M \alpha_w^1 \wedge [\mathrm{d}(i_{u\wedge v}\tau)]$$

$$= -\int_M \mathrm{d}\alpha_w^1 \wedge (i_{u\wedge v}\tau) + \int_{\partial M} \alpha_w^1 \wedge (i_{u\wedge v}\tau), \tag{8.117}$$

by the Stokes theorem (Appendix B). The second integral over ∂M vanishes for u, v, w tangent to ∂M, satisfying (8.6).

From (8.115) and (B.10), we have $\mathrm{d}\alpha_w^1 \wedge (i_{u\wedge v}\tau) = i_v i_u \tau \wedge \mathrm{d}\alpha_w^1$. The factor $i_u\tau \wedge \mathrm{d}\alpha_w^1$ is equal to 0, because it is of degree $n+1$. Then we have

$$(i_v i_u \tau) \wedge \mathrm{d}\alpha_w^1 = (-1)^n (i_u\tau) \wedge (i_v \mathrm{d}\alpha_w^1).$$

The form $\tau \wedge (i_v \mathrm{d}\alpha_w^1)$ is also of degree $n+1$, hence vanishes. As a consequence, we have

$$(i_u\tau) \wedge (i_v \mathrm{d}\alpha_w^1) = -(-1)^n[\tau \wedge i_u i_v \mathrm{d}\alpha_w^1] = (-1)^n (i_v i_u \mathrm{d}\alpha_w^1)\tau. \tag{8.118}$$

Collecting the formulae (8.113), (8.114), (8.117) and (8.118), we obtain

$$\langle w, [u, v]\rangle = \int_M \alpha_w^1 \wedge i_{\{u,v\}}\tau = -\int_M i_v i_u \mathrm{d}\alpha_w^1 \tau = \int_M i_v \alpha_B^1 \tau,$$

by (8.114). Therefore, we find

$$-i_v i_u \mathrm{d}\alpha_w^1 = i_v(\alpha_B^1 - \mathrm{d}f) = (v, B - \mathrm{grad}\, f), \tag{8.119}$$

since the field $v \in T_e\mathcal{D}_\mu(M)$ is orthogonal to grad f: $\langle v, \mathrm{grad}\, f\rangle = 0$. Thus, it is found that $\alpha_B^1 = -i_u \mathrm{d}\alpha_w^1 + \mathrm{d}f$. This verifies the theorem given in the beginning.

Chapter 9

Motion of Vortex Filaments

The significance and importance of *vorticity* for description of fluid flows is already stated explicitly in Chapter 7 as being a gauge field and we saw that irrotational flow fields are integrable (§7.10.4). In the last part (§8.8) of Chapter 8, some consideration is given to the role of vorticity with respect to the Lagrangian instability, contributing to particle mixing. In this chapter, we consider another aspect, i.e. by contrast there are some vortex motions which are regarded as completely integrable.

Notion of a vortex tube is useful to describe a tube-like structure in flows where magnitudes of the vorticity $\boldsymbol{\omega}$ are much larger than its surrounding. Mostly, this tube is approximately parallel to the vector $\boldsymbol{\omega}$ at every section so as to satisfy $\operatorname{div} \boldsymbol{\omega} = 0$. A mathematical idealization (§9.1) is derived by supposing the vortex tube to contract on to a curve (a line-vortex) with the strength of the vortex tube remaining constant, equal to γ say, and assuming the vorticity being zero elsewhere. We then have a spatial curve, called a *vortex filament*.

The dynamics of a thin vortex filament, embedded in an ideal incompressible fluid, is known to be well approximated by the *local induction equation* (§9.2), when the filament curvature is sufficiently small. A vortex filament is assumed to be spatially periodic in the present geometrical analysis,[1] and given by a time-dependent C^∞-curve $\boldsymbol{x}(s,t)$ in $\mathbb{R}^3$ with the arc-length parameter $s \in S^1$ and the time parameter t.

As described in §9.3, this system is characterized with the rotation group $G = SO(3)$ pointwise on the S^1 manifold.[2] The group $G(S^1)$ of smooth

[1]The derivation here for the periodic case would be generalized to a non-periodic case by imposing appropriate conditions at infinity. See the footnote on the first page of Chapter 5 and the description of §8.1.1 with $M = \mathbb{R}^1$.
[2]Suzuki, Ono and Kambe [SOK96]; Kambe [Kam98].

mappings, $g : s(\in S^1) \mapsto G = SO(3)$, equipped with the pointwise composition law,

$$g''(s) = g'(s) \circ g(s) = g'(g(s)), \quad g, g', g'' \in G,$$

is an infinite-dimensional Lie group, i.e. a *loop group*. The corresponding loop algebra leads to the Landau–Lifshitz equation, which is derived as the geodesic equation (§9.4). Furthermore, the loop-group formulation admits a central extension[3] in §9.8.

Based on the Riemannian geometrical point of view, this chapter includes a new interpretation of the local induction equation and the equation of Fukumoto and Miyazaki [FM91] by applying the theory of loop group and its extension.

Two-dimensional analogue of a line-vortex is a point-vortex. It is well known that a system of finite number of point-vortices is described by a Hamiltonian function [Ons49; Bat67; Saf92]. A Riemannian geometrical derivation of the Hamiltonian system is given in [Kam03b], where the Finsler geometry [Run59; BCS00], rather than the Riemannian geometry, is applied to the Hamiltonian system of point vortices.

9.1. A Vortex Filament

We consider the motion of a vortex filament $\mathcal{F}$ embedded in an ideal incompressible fluid in $\mathbb{R}^3$. Given a velocity field $\boldsymbol{u}(\boldsymbol{x})$ (at $\boldsymbol{x} \in \mathbb{R}^3$) satisfying $\operatorname{div} \boldsymbol{u} = 0$, the vorticity is given by $\boldsymbol{\omega} = \operatorname{curl} \boldsymbol{u}$, which vanishes at all points except at points on $\mathcal{F}$. The vortex filament $\mathcal{F}$ of strength γ, expressed by a space curve, is assumed to move and change its shape (the time variable is omitted here for the time being). The Biot–Savart law[4] (Fig. 9.1) can represent the velocity $\boldsymbol{u}$ at a point $\boldsymbol{x}$ induced by an element of vorticity, $\boldsymbol{\omega} \mathrm{d}V = \gamma \boldsymbol{t} \mathrm{d}s$, at a point $\boldsymbol{y}(s)$ where s is an arc-length parameter along the filament $\mathcal{F}$, $\mathrm{d}s$ is an infinitesimal arc-length and $\boldsymbol{t}$ unit tangent vector to $\mathcal{F}$

[3] Azcárraga and Izquierdo [AzIz95].

[4] According to the Biot–Savart law in electromagnetism, an electric current element $\boldsymbol{I}\mathrm{d}s$ at a point $\boldsymbol{y}(s)$ induces magnetic field $\boldsymbol{H}$ at a point $\boldsymbol{x}$. The field $\boldsymbol{H}(\boldsymbol{x})$ is given by $-(1/4\pi)[(\boldsymbol{x}-\boldsymbol{y}(s)) \times \boldsymbol{I}\mathrm{d}s]/|\boldsymbol{x}-\boldsymbol{y}(s)|^3$. This is equivalent to the local differential relation, $\operatorname{curl} \boldsymbol{H} = \boldsymbol{J}$, where $\boldsymbol{J}\mathrm{d}V = \boldsymbol{I}\mathrm{d}s$ with $\mathrm{d}V$ as a volume element and $\mathrm{d}s$ a line element.

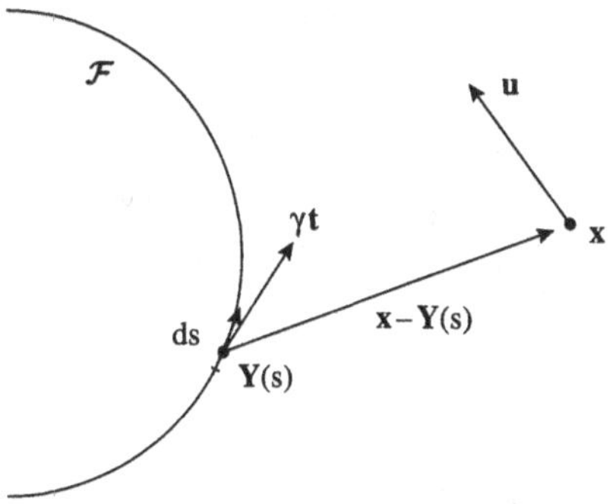

Fig. 9.1. Biot–Savart law.

at $\boldsymbol{y}(s)$. Namely, the velocity $\boldsymbol{u}(\boldsymbol{x})$ is expressed as

$$\boldsymbol{u}(\boldsymbol{x}) = -\frac{\gamma}{4\pi}\int_{\mathcal{F}} \frac{(\boldsymbol{x}-\boldsymbol{y}(s))\times \boldsymbol{t}(s)}{|\boldsymbol{x}-\boldsymbol{y}(s)|^3}\,\mathrm{d}s \tag{9.1}$$

$$= \frac{\gamma}{4\pi}\mathrm{curl}_x \int_{\mathcal{F}} \frac{\boldsymbol{t}(s)\mathrm{d}s}{|\boldsymbol{x}-\boldsymbol{y}(s)|}, \tag{9.2}$$

[Bat67, §7.1; Saf92, §2.3], where curl_x denotes taking curl with respect to the variable $\boldsymbol{x}$. In fact, using the property that $f = -1/4\pi|\boldsymbol{x}|$ is a fundamental solution of the Laplace equation, i.e. $\nabla^2 f = \delta(\boldsymbol{x})$ with $\delta(\boldsymbol{x})$ a 3D delta function, we have

$$\mathrm{curl}_x\ \mathrm{curl}_x \frac{1}{|\boldsymbol{x}-\boldsymbol{y}(s)|} = -\nabla^2 \frac{1}{|\boldsymbol{x}-\boldsymbol{y}(s)|} = 4\pi\delta(\boldsymbol{x}-\boldsymbol{y}(s)).$$

Recalling $\gamma \boldsymbol{t}\mathrm{d}s = \boldsymbol{\omega}\mathrm{d}V$, we obtain $\mathrm{curl}\,\boldsymbol{u} = \boldsymbol{\omega}$ from (9.2) as expected [Eq. (7.139)].

We consider the velocity induced in the neighborhood of a point P on the filament $\mathcal{F}$. We choose a local rectilinear frame K determined by three mutually-orthogonal vectors $(\boldsymbol{t}, \boldsymbol{n}, \boldsymbol{b})$, where $\boldsymbol{n}$ and $\boldsymbol{b}$ are unit vectors in the principal normal and binormal directions (and $\boldsymbol{t}$ unit tangent to $\mathcal{F}$), as indicated in Fig. D1 of Appendix D. With P taken as the origin of K, the position vector $\boldsymbol{x}$ of a point in the plane normal to the filament $\mathcal{F}$ (i.e. perpendicular to $\boldsymbol{t}$) at P can be written as (Fig. 9.2)

$$\boldsymbol{x} = y\boldsymbol{n} + z\boldsymbol{b}.$$

We aim at finding the form of velocity $\boldsymbol{u}(\boldsymbol{x})$ taken in the limit as $\boldsymbol{x}$ approaching the origin P, i.e. $r = (y^2 + z^2)^{1/2} \to 0$.

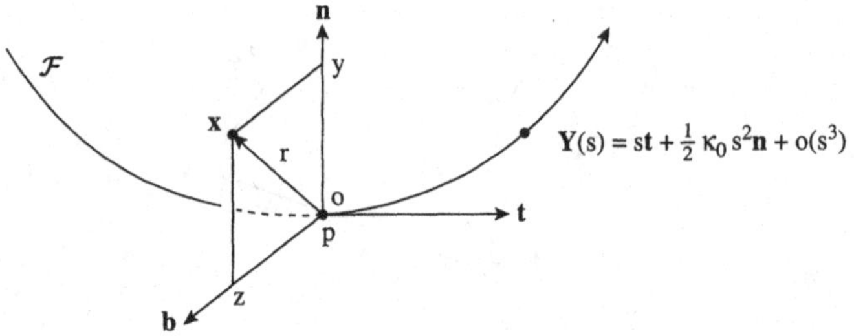

Fig. 9.2. Local representation.

Let us denote the origin P by $s = 0$. Then, the point $\boldsymbol{y}(s)$ on the curve $\mathcal{F}$ near P is expanded in the frame K as

$$\boldsymbol{y}(s) = \boldsymbol{y}'(0)s + \frac{1}{2}\boldsymbol{y}''(0)s^2 + O(s^3) = s\boldsymbol{t} + \frac{1}{2}\kappa_0 s^2 \boldsymbol{n} + O(s^3), \tag{9.3}$$

where $\boldsymbol{y}(0) = 0$, and κ_0 is the curvature of the curve $\mathcal{F}$ at $s = 0$ (see Appendix D.1). After some algebra, it is found that the integrand of (9.1) behaves like

$$\frac{z\boldsymbol{n} - y\boldsymbol{b}}{(r^2+s^2)^{3/2}}\left(1 + \frac{3}{2}\kappa_0 \frac{ys^2}{r^2+s^2}\right) - \kappa_0 \frac{\frac{1}{2}s^2\boldsymbol{b} + ys\boldsymbol{t}}{(r^2+s^2)^{3/2}} + O(\kappa_0^2). \tag{9.4}$$

To determine the behavior of $\boldsymbol{u}(\boldsymbol{x})$ as $r \to 0$, we substitute (9.4) into (9.1), and evaluate contribution to $\boldsymbol{u}$ from the above nearby portion of the filament $(-\lambda < s < \lambda)$. Changing the variable from s to $\sigma = s/r$ and taking limit $\lambda/r \to \infty$, it is found that the Biot–Savart integral (9.1) is expressed in the frame K as

$$\boldsymbol{u}(\boldsymbol{x}) = \frac{\gamma}{2\pi}\left(\frac{y}{r^2}\boldsymbol{b} - \frac{z}{r^2}\boldsymbol{n}\right) + \frac{\gamma}{4\pi}\kappa_0\left(\log\frac{\lambda}{r}\right)\boldsymbol{b} + (\text{b.t.}), \tag{9.5}$$

where (b.t.) denotes remaining bounded terms. The first term proportional to $\gamma/2\pi$ represents the circulatory motion about $\boldsymbol{t}$, i.e. about the vortex filament, anti-clockwise in the $(\boldsymbol{n}, \boldsymbol{b})$-plane. This is the right motion as it should be called a vortex. However, there is another term, i.e. the second term proportional to the curvature κ_0 which is not circulatory, but directed towards $\boldsymbol{b}$. These two terms are unbounded, whereas the remaining terms are bounded, as $r \to 0$.

The usual method to resolve the unboundedness is to use a cutoff. Namely, every vortex filament has a vortex core of finite size a, and r

should be bounded below at the order of a. If r is replaced by a, the second term is

$$\boldsymbol{u}_{\mathrm{LI}} = \frac{\gamma}{4\pi}\kappa_0 \left(\log\frac{\lambda}{a}\right)\boldsymbol{b} + (\text{b.t.}), \tag{9.6}$$

which is independent of y and z. This is interpreted such that the vortex core moves rectilinearly with the velocity $\boldsymbol{u}_{\mathrm{LI}}$ in the binormal direction $\boldsymbol{b}$. There is additional circulatory fluid motion about the filament axis. These are considered as main terms of $\boldsymbol{u}(\boldsymbol{x})$.

The magnitude of velocity $\boldsymbol{u}_{\mathrm{LI}}$ is proportional to the local curvature κ_0 of the filament at P, and called the *local induction*. This term vanishes with a rectilinear vortex because its curvature is zero. This is consistent with the known property that a rectilinear vortex has no self-induced velocity and its motion is determined solely by the velocities induced by other objects.

The same locally induced velocity $\boldsymbol{u}_{\mathrm{LI}}$ can be derived by the Biot–Savart integral for the velocity at a station $\boldsymbol{x} = \boldsymbol{x}(s_*)$ right on the filament. However, paradoxically, the portion $-ka < s - s_* < ka$ ($a \to 0$ and k: a constant of $O(1)$) must be excluded from the integral [Saf92, §11]. This is permitted because there is no contribution to the $\boldsymbol{u}_{\mathrm{LI}}$ from an infinitesimally small rectilinear portion.

9.2. Filament Equation

When interested only in the motion of a filament (without seeing circulatory motion about it), the velocity is given by $\boldsymbol{u}_{\mathrm{LI}}(s)$, whose dominant term can be expressed as

$$\boldsymbol{u}_{\mathrm{LI}}(s) = c\kappa(s)\boldsymbol{b}(s) = c\kappa(s)\boldsymbol{t}(s) \times \boldsymbol{n}(s), \tag{9.7}$$

where $c = (\gamma/4\pi)\log(\lambda/a)$ is a constant independent of s. Rates of change of the unit vectors $(\boldsymbol{t}, \boldsymbol{n}, \boldsymbol{b})$ (with respect to s) along the curve are described by the Frenet–Serret equation (D.4) in terms of the curvature $\kappa(s)$ and torsion $\tau(s)$ of the filament.

A vortex ring is a vortex in the form of a circle (of radius R, say), which translates with a constant speed in the direction of $\boldsymbol{b}$. The binormal vector $\boldsymbol{b}$ is independent of the position along the circle and perpendicular to the plane of circle (directed from the side where the vortex line looks clockwise to the side where it looks anti-clockwise). The direction of $\boldsymbol{b}$ is the same as that of the fluid flowing inside the circle. This is consistent with the expression (9.7) since $\kappa = 1/R = \text{const}$.

It has been just found from above that the vortex ring in the rectilinear translational motion (with a constant speed) depicts a cylindrical surface of circular cross-section in a three-dimensional euclidean space $\mathbb{R}^3$. The circular vortex filament coincides at every instant with a geodesic line of the surface which is to be depicted in the space $\mathbb{R}^3$. This is true for general vortex filaments in motion under the law given by Eq. (9.7), because the tangent plane to the surface to be generated by the vortex motion is formed by the two orthogonal tangent vectors $\boldsymbol{t}$ and $\boldsymbol{u}_{\text{LI}}\mathrm{d}t$. Therefore the normal $\boldsymbol{N}$ to the surface coincides with the normal $\boldsymbol{n}$ to the curve C of the vortex filament. This property is nothing but that C is a geodesic curve (§2.5.3, 2.6).

Suppose we have an *active* space curve C: $\boldsymbol{x}(s,t)$, which moves with the above velocity. Namely, the velocity $\partial_t \boldsymbol{x}$ at a station s is given by the local value $\boldsymbol{u}_* = c\kappa(s)\boldsymbol{b}(s)$, i.e. $\partial_t \boldsymbol{x} = c\kappa(s)\boldsymbol{b}(s)$ (the local induction velocity). It can be shown that the separation of two nearby particles on the curve, denoted by Δs, is unchanged by this motion. In fact,

$$\frac{\mathrm{d}}{\mathrm{d}t}\Delta s = (\Delta s\, \partial_s \boldsymbol{u}_*) \cdot \boldsymbol{t}, \tag{9.8}$$

where $\partial_s \boldsymbol{u}_* = c\kappa'(s)\boldsymbol{b} + c\kappa \boldsymbol{b}'(s)$. From the Frenet–Serret equation (D.4), it is readily seen that $\boldsymbol{b} \cdot \boldsymbol{t} = 0$ and $\boldsymbol{b}'(s) \cdot \boldsymbol{t} = 0$. Thus, it is found that

$$\frac{\mathrm{d}}{\mathrm{d}t}\Delta s = 0, \tag{9.9}$$

i.e. the length element Δs of the curve is *invariant* during the motion, and we can take s as representing the Lagrangian parameter.

Since $\boldsymbol{b} = \boldsymbol{t} \times \boldsymbol{n}$ and in addition $\partial_s \boldsymbol{x} = \boldsymbol{t}$ and $\partial_s^2 \boldsymbol{x} = \kappa \boldsymbol{n}$ (Appendix D.1), the local relation (9.7) for the curve $\boldsymbol{x}(s,t)$ is given by

$$\partial_t \boldsymbol{x} = \partial_s \boldsymbol{x} \times \partial_s^2 \boldsymbol{x}, \tag{9.10}$$

where the time is rescaled so that the previous ct is written as t here. This is termed the *filament equation* (Fig. 9.3). In fluid mechanics, the same equation is called the *local induction equation* (*approximation*).[5]

[5] This equation was given by Da Rios [DaR06] and has been rediscovered several times historically [Ric91; Ham88; Saf92].

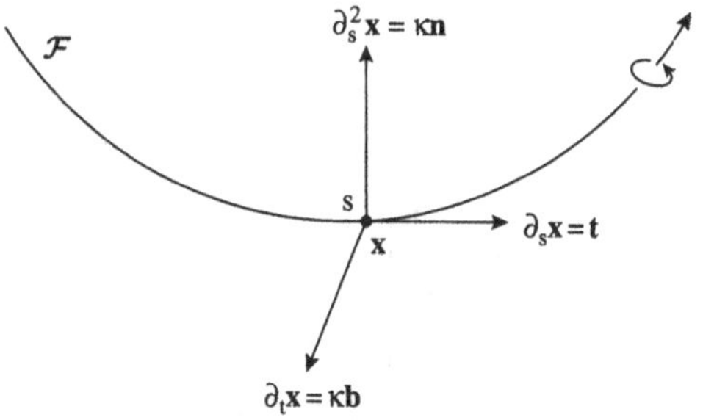

Fig. 9.3. Filament motion.

Some experimental evidence is shown in [KT71], where in addition, a circular vortex ring is shown to be neutrally stable with respect to small perturbations.

It is not difficult to see that there is a solution in the form of a rotating helical vortex $\boldsymbol{x}_h$ to Eq. (9.10). In fact, consider a *helix* $\boldsymbol{x}_h = (x, y, z)(s, t)$ and its tangent $\boldsymbol{t}_h$ defined by

$$\boldsymbol{x}_h = a(\cos\theta, \sin\theta, hks + \lambda\omega t), \tag{9.11}$$

$$\boldsymbol{t}_h = ak(-\sin\theta, \cos\theta, h), \tag{9.12}$$

(Fig. 9.4), where $\theta = ks - \omega t$, and a, k, h, ω, λ being constants. The vortex is directed towards $\boldsymbol{t}$, i.e. increasing s. The left-hand side of (9.10) is $a\omega(\sin\theta, -\cos\theta, \lambda)$, whereas the right-hand side is $a^2k^3h(\sin\theta, -\cos\theta, 1/h)$. Thus, Eq. (9.10) is satisfied if $\omega = ak^3h$ and $h\lambda = 1$. Requiring that s is an arc-length parameter, i.e. $\mathrm{d}s^2 = \mathrm{d}x^2 + \mathrm{d}y^2 + \mathrm{d}z^2$, we must have $a^2k^2(1 + h^2) = 1$. Hence, once the radius a and the wave number k of the helix are given, all the other constants h, ω, λ are determined. The helix translates towards positive z-axis and rotates clockwise (seen from above), whereas the circulatory fluid motion about the helical filament is anti-clockwise.

It is remarkable that the local induction equation can be transformed to the cubic-nonlinear Schrödinger equation. A complex function $\psi(s, t)$ is introduced by

$$\psi(s, t) = \kappa(s) \exp\left[i \int^s \tau(s')\mathrm{d}s'\right]$$

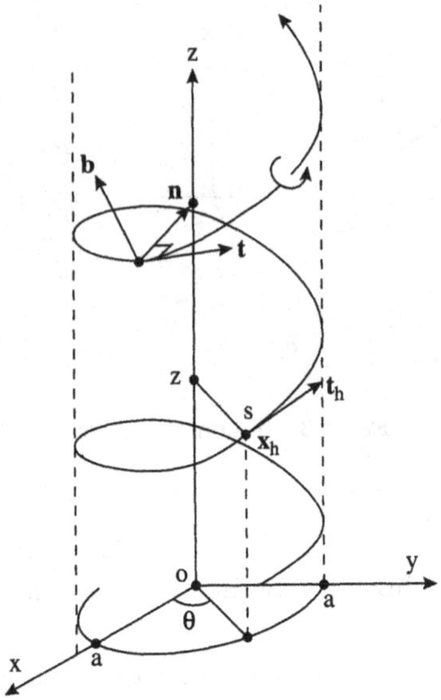

Fig. 9.4. Helical vortex.

(called Hasimoto transformation), where κ and τ are the curvature and torsion of the filament. The local induction equation (9.10) is transformed to

$$\partial_t \psi = i\left(\partial_s^2 \psi + \frac{1}{2}|\psi|^2\psi\right) \tag{9.13}$$

[Has72; LP91]. As is well known, this is one of the completely integrable systems, called the *nonlinear Schrödinger equation.* Naturally, this equation admits a soliton solution, which is constructed for an infinitely long vortex filament [Has72] as

$$\psi(s,t) = \kappa(\xi)\exp[i\tau_0 s], \quad \kappa = 2\tau_0 \text{ sech } \tau_0\xi, \tag{9.14}$$

where $\xi = s - ct$ and $\tau = \tau_0 = \frac{1}{2}c = \text{const}$.

In addition, there are planar solutions which are permanent in form in a rotating plane [Has71]. Their shapes are equivalent to those of the elastica [KH85; Lov27]. A thorough description of filaments of permanent form under (9.10) is given in [Kid81].

9.3. Basic Properties

9.3.1. *Left-invariance and right-invariance*

In conventional mechanics terms, the variable $\boldsymbol{x}(s,t) \in \mathbb{R}^3$ is a position vector of a point on the filament. Speaking mathematically, $\boldsymbol{x}(s,t)$ is an element of the C^∞-embeddings of S^1 into a three-dimensional euclidean space $\mathbb{R}^3$, i.e. $\boldsymbol{x} : S^1 \times \mathbb{R} \to \mathbb{R}^3$.

The motion of the curve $\boldsymbol{x}(s,t)$ is a map,

$$\begin{aligned} &\phi_t : x_0(s) \mapsto x_t(s) = \phi_t \circ x_0(s) := \Phi_t(s), \\ \text{or} \quad &\Phi_t : s \mapsto x_t, \end{aligned}$$

where $x_t = \Phi_t(s) = \boldsymbol{x}(s,t)$ is a position vector of the filament at a time t. Henceforth, the unit tangent vector is denoted by $T_t(s) = T(s,t) = \partial_s x_t$ instead of $\boldsymbol{t}$. Following the motion, the tangent $T_t(s)$ to the curve is left-translated:

$$T_t(s) = \partial x_t/\partial s = \phi_t \circ \partial x_0(s)/\partial s = (L_{\phi_t})_* T_0(s), \tag{9.15}$$

that is, $T_t(s)$ is a *left-invariant* vector field (Fig. 9.5). Likewise, its derivative is also left-invariant, i.e.

$$\partial_s T_t = (L_{\phi_t})_*(\partial_s T_0), \tag{9.16}$$

where $\partial_s T_t = \kappa_t(s)\boldsymbol{n}_t(s)$.

The tangent vector T_t can be expressed also as a function of the position vector $x_t = \Phi_t(s)$. Hence, the $T_t(s)$ is regarded as a *right-invariant* vector

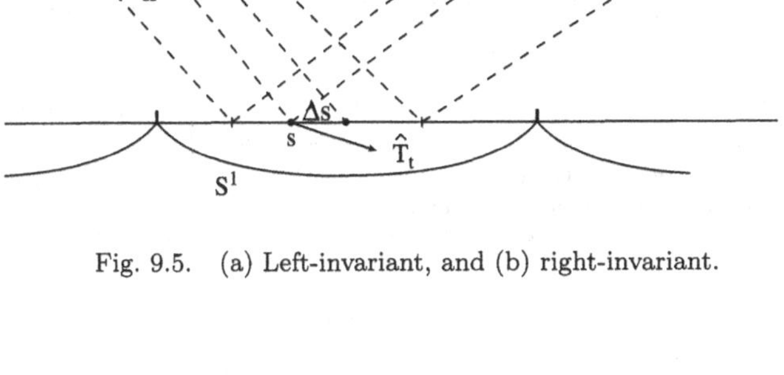

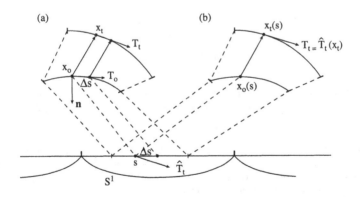

Fig. 9.5. (a) Left-invariant, and (b) right-invariant.

field. In fact, writing $T_t(s) = \hat{T}_t(x_t(s)) = \hat{T}_t \circ \Phi_t(s)$, we have

$$T_t = \hat{T}_t(x_t) = \hat{T}_t \circ \Phi_t = (R_{\Phi_t})_* \hat{T}_t \tag{9.17}$$

(Fig. 9.5). Likewise, the curvature is expressed as

$$\kappa_t = \hat{\kappa}_t(x_t) = (R_{\Phi_t})_* \hat{\kappa}_t. \tag{9.18}$$

Thus, the curvature $\kappa_t(s)$ is a *right-invariant* vector field as well, where $|\mathrm{d}x_t| = |\mathrm{d}s|$ by the property (9.9).

9.3.2. *Landau–Lifshitz equation*

Differentiating Eq. (9.10) with respect to s and setting $T = \partial_s \boldsymbol{x}$, one obtains

$$\partial_t T = T \times \partial_s^2 T = -T'' \times T, \tag{9.19}$$

since $\boldsymbol{x}'' \times \boldsymbol{x}'' = 0$, where a prime denotes the differentiation with respect to s and the subscript t of T_t being omitted henceforth. This is a particular case of *Landau–Lifshitz equation.*[6] In this regard, it would be useful to recall that the vector product plays the commutator of the Lie algebra **so**(3) (see §1.8.2 and 4.3).

9.3.3. *Lie–Poisson bracket and Hamilton's equation*

Equation (9.19) can be interpreted as a Hamilton's equation, in an analogous way to the case of the Euler's top in §4.1.3. To that end, we first verify that the following integral,

$$H = \frac{1}{2}\int_{S^1} \kappa^2(s)\mathrm{d}s = \frac{1}{2}\int_{S^1} (\partial_s T, \partial_s T)_{R^3}\mathrm{d}s \tag{9.20}$$

$$= -\frac{1}{2}\int_{S^1} T(\sigma)\cdot T''(\sigma)\mathrm{d}\sigma, \tag{9.21}$$

is an invariant under the Landau–Lifshitz equation (9.19), where $(A, B)_{R^3} = \delta_{kl}A^k B^l = A \cdot B$.[7] Noting $\partial_s T = T' = \kappa \boldsymbol{n}$ and $\kappa^2 = (T', T')$,

[6] *General* Landau–Lifshitz equation [AK98, Ch. VI] is given by replacing the vector product $A \times B$ by a Lie bracket $[A, B]$. The equation of the form $\partial_t X = \Omega \times X$ describes "rotation" of the vector X with the angular velocity Ω. Hence, the factor $-\partial_s^2 T = -T''(s)$ in (9.19) is interpreted as the angular velocity of $T(s)$ at s locally.

[7] Henceforth the subscript R^3 is omitted, and $A \cdot B$ will be often used in place of (A, B) in this chapter.

we obtain

$$\partial_t \kappa^2 = 2T' \cdot \partial_t T' = 2T' \cdot (T \times T''') = \partial_s(2\kappa^2 \tau), \tag{9.22}$$

from (9.19). The last equality can be shown by using the Frenet–Serret equation (D.4). In fact, we have

$$T \times T''' = (\kappa\tau' + 2\kappa'\tau)\boldsymbol{n} + K\boldsymbol{b},$$

where $K = \kappa'' - \kappa^3 - \kappa\tau^2$. Integrating (9.22) with respect to s, we obtain $\mathrm{d}H/\mathrm{d}t = 0$, since $\kappa^2\tau$ is a function on S^1.

Suppose that the functional H of (9.21) (or (9.20)) is the Hamiltoninan for the equation of motion:

$$\frac{\mathrm{d}}{\mathrm{d}t} T_\alpha = \{T_\alpha, H\}, \tag{9.23}$$

where $T_\alpha(s,t)$ is the αth component of the unit vector T in the fixed cartesian frame $\mathbb{R}^3$, and the bracket $\{\cdot,\cdot\}$ is defined by the following Poisson bracket (a kind of the Lie–Poisson bracket[8]):

$$\{T_\alpha, H\} := \int_{S^1} T(\sigma) \cdot \left(\frac{\delta T_\alpha}{\delta T} \times \frac{\delta H}{\delta T} \right) \mathrm{d}\sigma, \tag{9.24}$$

where the βth component of $\delta/\delta T(s)$ is given by $\delta/\delta T_\beta(s)$, the functional derivative with respect to $T_\beta(s)$ at a position s (where $\beta = 1, 2, 3$). Then, we have

$$\frac{\delta T_\alpha(s)}{\delta T(\sigma)} = \delta(\sigma - s)\boldsymbol{e}_\alpha, \qquad \frac{\delta H}{\delta T(\sigma)} = -T''(\sigma),$$

where $\boldsymbol{e}_\alpha$ is the unit vector in the direction of αth axis.[9] Then, Eq. (9.23) leads to

$$\begin{aligned} \frac{\mathrm{d}}{\mathrm{d}t} T_\alpha = \{T_\alpha, H\} &= \int_{S^1} T(\sigma) \cdot (\delta(\sigma - s)\boldsymbol{e}_\alpha \times (-T''(\sigma)))\mathrm{d}\sigma \\ &= -T(s) \cdot (\boldsymbol{e}_\alpha \times T''(s)) = \boldsymbol{e}_\alpha \cdot (T(s) \times T''(s)). \end{aligned} \tag{9.25}$$

This is nothing but the αth component of Eq. (9.19).

---

[8]See the footnote to §4.1.2 for its general definition [MR94; HMR98].

[9]$\delta H = -\int_{S^1} T''(\sigma) \cdot \delta T(\sigma)\mathrm{d}\sigma$, and $T(s) = T_\alpha(s)\boldsymbol{e}_\alpha$, and $T_\alpha(s) = \int T_\alpha(\sigma)\delta(\sigma - s)\mathrm{d}\sigma$.

9.3.4. *Metric and loop algebra*

According to the observation (given in the footnote in §9.3.2) that the term $-\partial_s^2 T = -T''(s)$ in (9.19) is interpreted as an angular velocity of rotation of $T(s)$, we may regard the vector $-T''(s)$ as an elelment Ω of the Lie algebra $\mathbf{so}(3)$, i.e. $\Omega = T''(s) \in \mathbf{so}(3)$ at each s.

Based on the Hamiltonian H of (9.21), the *metric* of the system of a vortex filament is defined as

$$\langle \Omega, \Omega \rangle := -\int_{S^1} (T, T'')\, ds = -\int_{S^1} (T, \Omega)\, ds = \int_{S^1} (\partial_s T, \partial_s T)\, ds = \int_{S^1} \kappa^2\, ds. \tag{9.26}$$

Integration by parts has been carried out on the second line. We may write $-T = A\Omega$ by introducing an operator $A = -\partial_s^{-2}$.[10] Then, the metric is rewritten as

$$\langle \Omega, \Omega \rangle = \int_{S^1} (A\Omega, \Omega)\, ds, \tag{9.27}$$

(see §4.3 for comparison). The operator $A = -\partial_s^{-2}$ is often called an *inertia operator* (or momentum map) of the system.

The invariance of H together with the left-invariance of $\partial_s T$ (shown in the previous sections) suggests that the metric $\langle \Omega, \Omega \rangle$ is *left-invariant.*

Thus, using the new symbol L instead of T, one may define

$$L'' \in \mathcal{L}\mathbf{g} := C^\infty(S^1, \mathbf{so}(3)),$$
$$L(= -AL'') \in \mathcal{L}\mathbf{g}^* := C^\infty(S^1, \mathbf{so}(3)^*),$$

where $C^\infty(S^1, \mathbf{so}(3))$ is the space of C^∞ functions of the fiber bundle on the base manifold S^1 with the fiber $\mathbf{so}(3)$, while $\mathcal{L}\mathbf{g} = \mathbf{so}(3)[S^1]$ denotes the *loop algebra* of the loop group, $\mathcal{L}G := SO(3)[S^1]$, and $C^\infty(S^1, \mathbf{so}(3)^*)$ is the dual space.[11]

9.4. Geometrical Formulation and Geodesic Equation

Now let us reformulate the above dynamical system in the following way. Let

$$X(s), Y(s), Z(s) \in \mathcal{L}\mathbf{g} = \mathbf{so}(3)[S^1] = C^\infty(s \in S^1, \mathbf{so}(3)) \tag{9.28}$$

[10] We have $AT''(s) = -T(s)$, under the conditions of periodicity and $|T| = 1$.

[11] In the sense of the pointwise locality, the group $\mathcal{L}G(S^1)$ is called a *local group*.

be the vector fields. Correspondingly, we define their respective dual fields by $AX, AY, AZ \in C^\infty(S^1, \mathbf{so}(3)^*)$ with $A = -\partial_s^{-2}$, together with $T_X = -AX$, $T_Y = -AY$ and $T_Z = -AZ$. The left-invariant *metric* is defined by

$$\langle X, Y\rangle := -\int_{S^1} (T_X, Y)\mathrm{d}s = -\int_{S^1} (T_Y, X)\mathrm{d}s \tag{9.29}$$

$$= \int_{S^1} (AX, Y)\mathrm{d}s = \int_{S^1} (AY, X)\mathrm{d}s, \tag{9.30}$$

where $T_X'' := X$ and $T_Y'' := Y$. The symmetry of the metric with respect to X and Y can be easily verified by integration by parts. The *commutator* is given by

$$[X, Y]^{(\mathrm{L})}(s) := X(s) \times Y(s) \tag{9.31}$$

(see (4.27)) at each s, i.e. pointwise.

In the case of the left-invariant metric (9.30), the connection satisfies Eq. (3.30), and in terms of the operators *ad* and *ad**, we have the expression (3.65), which is reproduced here:

$$\nabla_X Y = \frac{1}{2}(ad_X Y - ad_X^* Y - ad_Y^* X) \tag{9.32}$$

(see also [Fre88]). By using the definition $ad_X Y = [X, Y]^{(\mathrm{L})}$ and the definition $\langle ad_X^* Y, Z\rangle = \langle Y, ad_X Z\rangle$, we obtain

$$\langle ad_X^* Y, Z\rangle = \langle Y, [X, Z]^{(\mathrm{L})}\rangle = \int_{S^1} (AY, X \times Z)\mathrm{d}s$$

$$= \int_{S^1} (AA^{-1}(AY \times X), Z)\mathrm{d}s,$$

(see (4.35) for comparison), which leads to

$$ad_X^* Y = A^{-1}(AY \times X) = \partial_s^2 [T_Y, X]^{(\mathrm{L})}. \tag{9.33}$$

Using this, it is found from (3.65) that the *connection* ∇ of the filament motion is given in the following form [SOK96; Kam98]:

$$\nabla_X Y = \frac{1}{2}(X \times Y - \partial_s^2 (T_X \times Y) - \partial_s^2 (T_Y \times X)), \tag{9.34}$$

which leads to $\nabla_X X = -\partial_s^2 (T_X \times X)$.

In the time-dependent problem, a tangent vector is expressed in the form, $\tilde{X} = \partial_t + X^\alpha \partial_\alpha$, where $X = (X^\alpha) \in \mathbf{so}(3)$. Then the *geodesic equation* is

$$\nabla_{\tilde{X}} \tilde{X} = \partial_t X + \nabla_X X = 0, \tag{9.35}$$

on the loop group $\mathcal{L}G = SO(3)[S^1]$. This is also written as $\partial_t X - \partial_s^2 (T_X \times X) = 0$. Applying the operator $\partial_s^{-2} = -A$ and using $T_X = -AX$, we get an equation of motion in the dual space,

$$\partial_t T_X - (T_X \times X) = 0. \tag{9.36}$$

Replacing X by T_X'', we recover the Landau–Lifshitz equation:

$$\partial_t T_X = T_X \times T_X''. \tag{9.37}$$

Furthermore, integrating with respect to s, one gets back to Eq. (9.10).

9.5. Vortex Filaments as a Bi-Invariant System

9.5.1. *Circular vortex filaments*

In order to get an insight into the filament motions, let us consider a class of simple filaments for which $\partial_s^2 T = T'' = -c^2 T$ is satisfied with a constant c. From the Frenet–Serret equations (D.4), we have $T'' = -\kappa^2 T + \kappa' \boldsymbol{n} - \kappa\tau \boldsymbol{b}$. This requires $\tau = 0$ and $c = \kappa$, and we have a family of circular filaments (with c^{-1} denoting a radius). A family of circular vortex filaments (a sub-family of vortex filaments) has a particular symmetry, and is worthwhile investigating separately since its metric is analogous to that of a spherical top (§4.6.1).

Replacing $\Omega = T'' = -c^2 T$ by $-T$ in (9.26), the *metric* is defined as

$$\langle \Omega, \Omega \rangle := \int_{S^1} (T, T)\, \mathrm{d}s = \int_{S^1} \mathrm{d}s, \tag{9.38}$$

where $(T, T) = |\partial \boldsymbol{x}/\partial s|^2 = 1$ and $T = -A\Omega$. As described in §9.3.1, this metric is bi-invariant, i.e. both left- and right-invariant, because $(T_t, T_t) = (T_0, T_0) = 1$ in regard to (9.15), and $(T_t, T_t) = (\hat{T}_t, \hat{T}_t) = 1$ in regard to

(9.17) (see Fig. 9.5). The above metric induces the following inner product:

$$\langle X, Y \rangle := \int_{S^1} (X, Y) \mathrm{d}s. \tag{9.39}$$

Suppose that X, Y, Z are C^∞ functions taking values of the algebra $\mathbf{so}(3)[S^1]$. Then the commutator is given by (9.31) as before, and the connection is given by (9.32), where $ad_X Z = X \times Z$. In view of (9.39),

$$\langle ad_X^* Y, Z \rangle = \int_{S^1} (Y, X \times Z) \mathrm{d}s = \int_{S^1} ((Y \times X), Z) \mathrm{d}s.$$

Therefore, we obtain $ad_X^* Y = Y \times X$, which is consistent with (9.33) when A is set to $-I$. Then, the connection formula (9.32) reduces to

$$\nabla_X Y(s) = \frac{1}{2} X(s) \times Y(s) = \frac{1}{2} [X, Y](s), \tag{9.40}$$

(omitting the superscript (L)). Then the geodesic equation, $\partial_t X + \nabla_X X = 0$, reduces to

$$\partial_t X = 0, \quad \text{since} \quad \nabla_X X = \frac{1}{2} X \times X = 0. \tag{9.41}$$

Hence the covector T_X tangent to the space curve (defined by $T_X'' = -X$) is also time-independent, where $T_X \in C^\infty(S^1)$ and $|T_X| = 1$. This is interpreted as *steady* translational motion of a *vortex ring*, described at the beginning of §9.2.

The Hamilton's equation (9.23) results in $(\mathrm{d}/\mathrm{d}t)T_\alpha = 0$, for the Hamiltonian $H = \frac{1}{2} \int_{S^1} (T, T) \mathrm{d}s = \frac{1}{2} \int_{S^1} \mathrm{d}s$, since $\delta H / \delta T = T$. Starting from the Hamilton's equation with the Hamiltonian $H = \frac{1}{2} \int \mathrm{d}s$ (without resorting to the fact that it is derived under the assumption of a circular filament), the property $T_\alpha = \text{const}$ is understood such that the filament form is invariant (unchanged) with the Hamiltonian of the filament length.

In §4.6.1, a finite-dimensional system (i.e. free rotation of a rigid body) with a bi-invariant metric was considered. The metric (9.39) of a circular filament is analogous to the metric (4.53) of a spherical top. The local form of the connection (9.40) at s is equivalent in form to (4.55). Note that $\nabla_X(\nabla_Y Z) = \frac{1}{4}[X, [Y, Z]]$ by (9.40), for example. Then, we obtain the curvature tensor,

$$R(X, Y)Z = -\frac{1}{4} [[X, Y], Z],$$

by the Jacobi identity (1.60). This is equivalent to (4.56).

The sectional curvature is found to be non-negative as before:

$$K(X,Y) := \langle R(X,Y)Y, X\rangle = \frac{1}{4}\int_{S^1} |X \times Y|^2 \, ds \geq 0.$$

In this steady problem, the Jacobi equation is given by (3.127):

$$\frac{d^2}{dt^2}\frac{\|J\|^2}{2} = \|\nabla_T J\|^2 - K(T, J).$$

The right-hand side vanishes because $\nabla_T J = \frac{1}{2}T \times J$ and $K(T, J) = \frac{1}{4}\|T \times J\|^2$, where $\|J\|^2 = \langle J, J\rangle$. Thus, we obtain $\|J\| = \text{const}$, as it should be. This implies that a vortex ring in steady translational motion is neutrally stable, and that the circular vortex is a Killing field, which is considered in §9.6.2.

9.5.2. *General vortex filaments*

Above formulae regarding a vortex ring are analogous to those of a spherical top in §4.6.1 and the note given pertaining to (4.62) in §4.6.2. A general rigid body is characterized by a general symmetric inertia tensor J, which is regarded as a *Riemannian* metric tensor, and its rotational motion is a bi-invariant metric system (§4.5.2). Formulations in §4.4 and 4.6.2 describe the *extension* to such general metric tensor J. It would be interesting to recall the property of complete-integrablility of the free rotation of a rigid body (see §4.1).

The general case of vortex filaments under the metric (9.30) is analogous to the asymmetrical top of the metric (4.58). The filament motions of general inertia operator A are governed by the filament equation (9.37) (or (9.10)) and regarded as a system of bi-invariant metric. In this regard, it is to be noted that this system is known to be *completely integrable*, i.e. there are inifinite numbers of integral invariants [LP91]. Two invariants of lowest orders are found as follows.

9.5.3. *Integral invariants*

According to the Poisson bracket (9.24), we define the Lie–Poisson bracket as

$$\{I, H\} = \int_{S^1} T(\sigma) \cdot \left(\frac{\delta I}{\delta T} \times \frac{\delta H}{\delta T}\right) d\sigma, \tag{9.42}$$

for the Hamiltonian H of (9.20) and an integral I:

$$H = \frac{1}{2}\int_{S^1} \kappa^2(s)\mathrm{d}s, \quad I = \int_{S^1} f(s)\mathrm{d}s.$$

The Hamilton's equation (§9.3.3) for I is written as

$$\frac{\mathrm{d}}{\mathrm{d}t} I = \{I, H\}. \tag{9.43}$$

If the bracket $\{I, H\}$ vanishes, then the integral I is an invariant of motion.

A first simplest integral invariant is given by

$$I_1 = \int_{S^1} (T, T)\mathrm{d}s = \int_{S^1} \mathrm{d}s. \tag{9.44}$$

In fact, since $\delta I_1/\delta T(\sigma) = 2T(\sigma)$, the bracket vanishes:

$$\{I_1, H\} = \int_{S^1} T(\sigma) \cdot \left(2T(\sigma) \times \frac{\delta H}{\delta T}\right) \mathrm{d}\sigma = 0.$$

A second integral invariant I_2 is given by

$$I_2 = \int_{S^1} T \cdot (T' \times T'')\mathrm{d}s = -\int_{S^1} \kappa^2(s)\tau(s)\mathrm{d}s, \tag{9.45}$$

where $T' \times T'' = -\kappa^2\tau T + \kappa^3 \boldsymbol{b}$ by the Frenet–Serret equations (D.4). Its invariance is verified as follows. Taking functional derivatives of H and I_2, we obtain $\delta H/\delta T(\sigma) = -T''(\sigma)$, and

$$\frac{\delta I_2}{\delta T} = T' \times T'' + (T \times T'')' + (T \times T')'' = 3T' \times T'' + 2T \times T'''.$$

Therefore,

$$\{H, I_2\} = -\int_{S^1} T \cdot (T'' \times [3T' \times T'' + 2T \times T'''])\mathrm{d}\sigma.$$

Recalling the formula of vector triple product, $A \times (B \times C) = (A \cdot C)B - (A \cdot B)C$ for three vectors $A, B, C \in \mathbb{R}^3$, we obtain

$$T'' \times (T' \times T'') = |T''|^2 T' - (T' \cdot T'')T'',$$
$$T'' \times (T \times T''') = (T'' \cdot T''')T - (T \cdot T'')T'''.$$

Substituting these and using $|T|^2 = 1$, $T \cdot T' = 0$, we obtain

$$\{H, I_2\} = -\int_{S^1} [3(T \cdot T'')(T' \cdot T'') - 2(T'' \cdot T''') + 2(T \cdot T'')(T \cdot T''')]\mathrm{d}s.$$

The second term can be written in the form of a derivative, $2T'' \cdot T''' = (\mathrm{d}/\mathrm{d}s)(T'' \cdot T'')$. Likewise, the first and third terms are written in the form of derivatives with respect to s as follows:

$$\begin{aligned}
&3(T \cdot T'')(T' \cdot T'') + 2(T \cdot T'')(T \cdot T''') \\
&= 2(T \cdot T'')[(T' \cdot T'') + (T \cdot T''')] + \frac{1}{2}(T' \cdot T')'(T \cdot T'') \\
&= 2(T \cdot T'')\frac{\mathrm{d}}{\mathrm{d}s}(T \cdot T'') + \frac{1}{2}\frac{\mathrm{d}}{\mathrm{d}s}(T' \cdot T')(-(T' \cdot T') + (T \cdot T')') \\
&= \frac{\mathrm{d}}{\mathrm{d}s}(T \cdot T'')^2 - \frac{1}{4}\frac{\mathrm{d}}{\mathrm{d}s}(T' \cdot T')^2
\end{aligned}$$

since $T \cdot T' = 0$. Therefore, we have

$$\{H, I_2\} = \int_{S^1} \frac{\mathrm{d}}{\mathrm{d}s}\left((T'' \cdot T'') - (T \cdot T'')^2 + \frac{1}{4}(T' \cdot T')^2\right) \mathrm{d}s = 0.$$

Thus, it is found that the integral I_2 is an integral invariant.

9.6. Killing Fields on Vortex Filaments

Motions of the following three vortex filaments are characterized by translation and rigid-body rotation, without change of their forms.

9.6.1. *A rectilinear vortex*

A rectilinear vortex filament $\boldsymbol{x}_l$ in the direction of z-axis (say) is defined by the unit tangent field $\partial_s \boldsymbol{x}_l = T_{X_l} = (0, 0, 1)$ in the cartesian (x, y, z)-frame with the algebra element $X_l = T''_X = (0, 0, 0)$. One can immediately show that this field is regarded as a Killing field satisfying (3.161).

In fact, from the definition (9.34) of connection, we have $\nabla_Y X_l = -\frac{1}{2}\partial_s^2(T_{X_l} \times Y)$. Then, using $A = -\partial_s^{-2}$,

$$\begin{aligned}
\langle \nabla_Y X_l, Z \rangle &= \int_{S^1} (A\nabla_Y X_l, Z)\mathrm{d}s = \frac{1}{2}\int_{S^1} (T_{X_l} \times Y, Z)\mathrm{d}s \\
&= \frac{1}{2}\int_{S^1} T_{X_l} \cdot (Y \times Z)\mathrm{d}s.
\end{aligned}$$

Likewise, we have $\langle \nabla_Z X_l, Y \rangle = \frac{1}{2}\int T_{X_l} \cdot (Z \times Y)\mathrm{d}s$. Therefore, the Killing equation (3.161),

$$\langle \nabla_Y X_l, Z \rangle + \langle Y, \nabla_Z X_l \rangle = 0, \tag{9.46}$$

is satisfied, since $Y \times Z + Z \times Y = 0$, for any $Y, Z \in \mathcal{L}\mathbf{g} = \mathbf{so}(3)[S^1]$. Thus it is found that T_{X_l} is a *Killing field* (§3.12.1).

The corresponding conserved quantity is given by

$$\langle X_l, Y \rangle = \int_{S^1} (T_{X_l}, Y) \mathrm{d}s = \int_{S^1} Y_z \mathrm{d}s,$$

where Y_z is the component of Y in the direction of $T_{X_l} = (0, 0, 1)$.

9.6.2. *A circular vortex*

A circular vortex is also considered to generate a Killing field X_c. The position vector of a circular vortex of radius a is denoted by $\boldsymbol{x}_c = a(\cos s, \sin s, 0)$ for $s \in [0, 2\pi] = S^1$, where the origin of a cartesian frame is taken at the center of the vortex and the (x, y)-plane coincides with the plane of the circular vortex. According to Eq. (9.10), the ring radius a can be made unity by rescaling the time t to at. Then, the variable s becomes a length parameter along the circumference of the unit circle. The tangent to the circle is given by

$$T_{X_c} = \partial_s \boldsymbol{x}_c = (-\sin s, \cos s, 0) := \boldsymbol{t}_c(s), \tag{9.47}$$

and $X_c = T''_{X_c} = \boldsymbol{t}''_c = -\boldsymbol{t}_c$ (Fig. 9.6). The Landau–Lifshitz equation (9.37) reduces to

$$\partial_t T_{X_c} = 0.$$

From the definition (9.34) of connection, we have

$$\nabla_Y X_c = \frac{1}{2}(Y \times X_c - \partial_s^2 [T_Y \times X_c + T_{X_c} \times Y]), \tag{9.48}$$

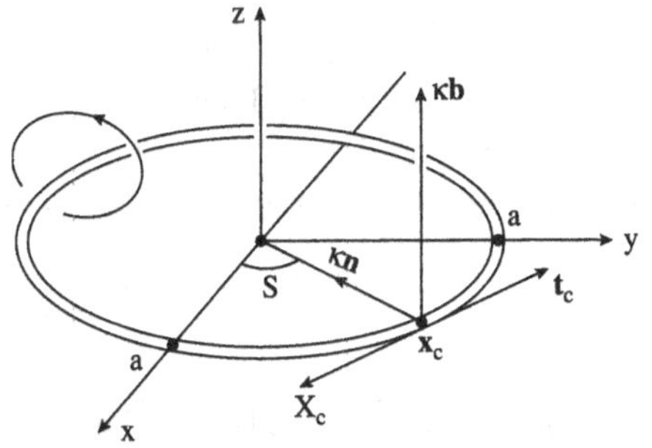

Fig. 9.6. Circular vortex.

since $\partial_t T_{X_c} = 0$. Then, we have

$$\begin{aligned}\langle \nabla_Y X_c, Z \rangle &= -\frac{1}{2}\int_{S^1} T_Z \cdot (Y \times X_c - \partial_s^2 [T_Y \times X_c + T_{X_c} \times Y])\mathrm{d}s \\ &= \frac{1}{2}\int_{S^1} X_c \cdot (Y \times T_Z + Z \times T_Y)\mathrm{d}s + \frac{1}{2}\int_{S^1} T_{X_c} \cdot (Y \times Z)\mathrm{d}s,\end{aligned} \tag{9.49}$$

for $Y, Z \in \mathcal{L}\mathbf{g} = \mathbf{so}(3)[S^1]$, where integration by parts is carried out two times except for the first term, and then T_Z'' is replaced with Z. Another term $\langle Y, \nabla_Z X_c \rangle$ is obtained by interchanging Y and Z. Then, we have

$$\langle \nabla_Y X_c, Z \rangle + \langle Y, \nabla_Z X_c \rangle = \int_{S^1} X_c \cdot (Y \times T_Z + Z \times T_Y)\mathrm{d}s, \tag{9.50}$$

where the last term of (9.49) cancels out with its counterpart.

It is required that all of Y, Z and $Y + Z$ satisfy the Landau–Lifshitz equation:

$$\begin{aligned}&\partial_t T_Y - T_Y \times Y = 0, \quad \partial_t T_Z - T_Z \times Z = 0, \\ &\partial_t (T_Y + T_Z) - (T_Y + T_Z) \times (Y + Z) = 0.\end{aligned} \tag{9.51}$$

Hence, we have

$$T_Z \times Y + T_Y \times Z = 0.$$

Thus, it is found the Killing equation is satisfied:

$$\langle \nabla_Y X_c, Z \rangle + \langle Y, \nabla_Z X_c \rangle = 0,$$

for any $Y, Z \in \mathcal{L}\mathbf{g} = \mathbf{so}(3)[S^1]$ under the condition that the conservation integral of the form (9.52) makes sense. This verifies that a circular vortex (called a vortex ring) represented by $T_{X_c} = \boldsymbol{t}_c = (-\sin s, \cos s, 0)$ is a Killing field. The corresponding conserved quantity is given by

$$\langle X_c, Y \rangle = -\int_{S^1} (T_{X_c}, Y)\mathrm{d}s = -\int_{S^1} \boldsymbol{t}_c \cdot Y \mathrm{d}s, \tag{9.52}$$

where $\boldsymbol{t}_c \cdot Y$ is the component of Y in the tangential direction of the circular vortex. This implies that the circumferential length projected onto the circle is invariant.

9.6.3. *A helical vortex*

A similar analysis applies to a helical vortex (9.11) (with $a = 1$), represented by

$$T_{X_h} = \partial_s \boldsymbol{x}_h = \boldsymbol{t}_h = k(-\sin\theta, \cos\theta, h), \tag{9.53}$$

and $X_h = -T''_{X_h} = -\boldsymbol{t}''_h = k^3\boldsymbol{t}_c = k^3(-\sin\theta, \cos\theta, 0)$, where $\theta = k(s - ct)$ and $c = \omega/k = hk^2$. Here, some modification is necessary, because the helical vortex rotates with respect to the z-axis and translates along it without change of form.

Introducing a pair of new variables (τ, σ) by $\sigma = s - ct$ and $\tau = t$ where $\theta = k\sigma$, the T_{X_h} and X_h are functions of σ only, hence $\partial_\tau T_h = 0$, $\partial_\tau X_h = 0$. The derivatives are transformed as

$$\partial_s = \partial_\sigma, \quad \partial_t = \partial_\tau - c\partial_\sigma. \tag{9.54}$$

Then, Eq. (9.19) is transformed to

$$\partial_\tau T_h - c\partial_\sigma T_h = T_h \times \partial_\sigma^2 T_h.$$

We may call *slide-Killing* (according to [LP91]) if the Killing equation,

$$\langle \nabla_Y X_h, Z\rangle + \langle Y, \nabla_Z X_h\rangle = 0, \tag{9.55}$$

is satisfied for the variable σ. Following the derivation of formulae in the previous section, it is found that the left-hand side of (9.55) is given by

$$\int_{S^1} X_h \cdot (Y \times T_Z + Z \times T_Y)\mathrm{d}\sigma, \tag{9.56}$$

which is obtained just by replacing X_c and s with X_h and σ in (9.50). Equations (9.51) are replaced by

$$\partial_t T_Y - c\partial_\sigma T_Y - T_Y \times Y = 0, \quad \partial_t T_Z - c\partial_\sigma T_Z - T_Z \times Z = 0,$$
$$\partial_t (T_Y + T_Z) - c\partial_\sigma (T_Y + T_Z) - (T_Y + T_Z) \times (Y + Z) = 0.$$

From these, we immediately obtain

$$T_Z \times Y + T_Y \times Z = 0.$$

Thus, it is found that Eq. (9.55) of slide-Killing is satisfied, and that the helical vortex of X_h with T_h is a Killing field. The corresponding conserved quantity is given in the form (9.52) if T_{X_c} and $\boldsymbol{t}_c$ are replaced by T_{X_h} and $\boldsymbol{t}_h$.

9.7. Sectional Curvature and Geodesic Stability

Stability of a vortex filament is described by the Jacobi equation. In other words, an infinitesimal variation field εJ between two neighboring geodesics (for an infinitesimal parameter ε and the Jacobi field J) is governed by the Jacobi equation (3.127):

$$\frac{\mathrm{d}^2}{\mathrm{d}s^2}\frac{1}{2}\|J\|^2 = \|\nabla_X J\|^2 - K(X,J), \tag{9.57}$$

where X is the tangent vector to the reference geodesic.

The sectional curvature $K(X,J)$ is defined by (3.166) in §3.12.3, which is reproduced here:

$$\begin{aligned} K(X,J) &= \langle R(X,J)J, X\rangle \\ &= -\langle \nabla_X X, \nabla_J J\rangle + \langle \nabla_J X, \nabla_X J\rangle + \langle \nabla_{[X,J]} X, J\rangle. \end{aligned} \tag{9.58}$$

9.7.1. *Killing fields*

In §3.12.3, it was verified that the sectional curvature $K(X,J)$ is positive if the tangent vector X is a Killing field and that the right-hand side of the Jacobi equation (9.57) vanishes.

First example is the section spanned by a straight-line vortex and an arbitrary variation field J. The straight-line vortex is characterized by the (cotangent) vector $T_{X_l} = (0,0,1)$ and the vector $X_l = (0,0,0)$. From the connection formula (9.34), we have

$$\nabla_J X_l = -\frac{1}{2}\partial_s^2 (T_{X_l} \times J).$$

Hence, the curvature formula (3.167) for a Killing field X_l gives

$$K(X_l, J) = \|\nabla_J X_l\|^2 = \|\nabla_{X_l} J\|^2 = \frac{1}{4}\int_{S^1} (T_{X_l} \times J')^2 \mathrm{d}s, \tag{9.59}$$

i.e. the curvature is positive except for J' parallel to T_{X_l}. It is readily seen that the right-hand side of (9.57) vanishes.

Second example is the section between a circular filament and an arbitrary variation field J. A circular vortex is characterized by the tangent vector field $X_c = -\boldsymbol{t}_c$ of (9.47). From the curvature formula (3.167) for a Killing field X_c, we find

$$K_c(X_c, J) = \|\nabla_J X_c\|^2 = \|\nabla_{X_c} J\|^2, \tag{9.60}$$

where $\nabla_J X_c$ is given by (9.48) with $Y = J$. Thus, it is found that *the sectional curvature of a vortex ring with an arbitrary field J is positive.* Again, the right-hand side of (9.57) vanishes. This implies that a vortex ring in steady translational motion is neutrally stable, which is consistent with the perturbation analysis of [KT71].

Regarding the helical vortex X_h, we have the same properties.

9.7.2. *General tangent field X*

Using the connection of (9.34) and the curvature formula (9.58), it is found that the *sectional curvature* is given by

$$K(X, J) = \langle R(X, J)J, X\rangle = \int_{S^1} f(s)\mathrm{d}s, \tag{9.61}$$

for $X, J \in C^\infty(S^1, \mathbf{so}(3))$, where[12]

$$\begin{aligned} f(s) &= (AX \times X) \cdot (AJ \times J)'' - (3/4)[\partial_s^{-1}(X \times J)]^2 \\ &\quad + \frac{1}{4}[\partial_s(AX \times J + AJ \times X)]^2 + \frac{1}{2}[|X|^2(J \cdot AJ) + |J|^2(X \cdot AX)] \\ &\quad - \frac{1}{2}(X \cdot J)[(AX \cdot J) + (X \cdot AJ)] \end{aligned} \tag{9.62}$$

[SOK96; Kam98].

Regarding a helical vortex, it is assumed to be stable since it is a Killing field as far as the perturbations belong to those on S^1. However, fluid-dynamically speaking, a physical aspect must be taken into account. If the wavelength of the perturbation is sufficiently large and the condition of S^1-periodic field is not satisfied, then the stability of a helical vortex is not guaranteed [SOK96]. In addition, the local induction equation is an approximate equation in the sense that the filament must be very thin and in addition, its curvature must be relatively small.

9.8. Central Extension of the Algebra of Filament Motion

The loop algebra $\mathcal{L}\mathbf{g} = \mathbf{so}(3)[S^1]$ yielded the Landau–Lifshitz equation (9.19). It is possible to formulate its central extension analogously according to the KdV problem (§5.3, 5.4 and Appendix H). The result is the Kac–Moody algebra known in the gauge theory [AzIz95]. However, its

[12] The present X and J correspond to X'' and J'' in [Kam98].

application to the motion of a vortex filament is not seen in any existing textbook. It is remarkable to find that the resulting geodesic equation is that obtained in [FM91] in the context of fluid dynamics.

Let us introduce extended *algebra* elements defined as

$$\hat{X}, \hat{Y}, \hat{Z} \in \mathbf{so}(3)[S^1] \oplus \mathbb{R},$$

where $\hat{X} := (X, a), \hat{Y} := (Y, b), \hat{Z} := (Z, c)$, for $a, b, c \in \mathbb{R}$. The extended *metric* is defined as

$$\langle \hat{X}, \hat{Y} \rangle := \int_{S^1} (AX, Y) \mathrm{d}s + ab$$

where $A = -\partial_s^{-2}$. The *extended algebra* is defined as

$$[\hat{X}, \hat{Y}] := ([X, Y]^{(\mathrm{L})}(s), c(X, Y)), \tag{9.63}$$

where

$$c(X, Y) := \int_{S^1} (X(s), Y'(s)) \mathrm{d}s = -c(Y, X),$$

and the Jacobi identity is satisfied by the new commutator:

$$[[\hat{X}, \hat{Y}], \hat{Z}] + [[\hat{Y}, \hat{Z}], \hat{X}] + [[\hat{Z}, \hat{X}], \hat{Y}] = 0.$$

It is not difficult to show that the commutator (9.63) is equivalent to that of the Kac–Moody algebra [AzIz95]. The extended *connection* is found to be given by

$$\nabla_{\hat{X}} \hat{Y} = \left(\nabla_X Y, \frac{1}{2} \int_{S^1} (X, \partial_s Y) \mathrm{d}s \right),$$
$$\nabla_X Y := \frac{1}{2} ([X, Y]^{(\mathrm{L})} + \partial_s^2 [AX, Y]^{(\mathrm{L})} + \partial_s^2 [AY, X]^{(\mathrm{L})} - \partial_s^2 (a \partial_s Y + b \partial_s X)).$$

Then the *geodesic equation* $(\partial_t \hat{X} + \nabla_{\hat{X}} \hat{X} = 0)$ for the extended system is obtained as

$$\partial_t X + \partial_s^2 (AX \times X) - a \partial_s^3 X = 0,$$
$$\partial_t a = 0.$$

The second equation follows from $\int_{S^1} (X, \partial_s X) \mathrm{d}s = 0$. Applying the operator A, we obtain the equation for $\boldsymbol{x}_s = L = -AX$ $(X = L'')$:

$$\partial_t L - (L \times L'') - a \partial_s^3 L = 0.$$

Integrating this with respect to s, we return to the equation for the *space curve* $\boldsymbol{x}(s,t)$:

$$\boldsymbol{x}_t = \boldsymbol{x}_s \times \boldsymbol{x}_{ss} + a\boldsymbol{x}_{sss},$$

in $\mathbb{R}^3$. From the Frenet–Serret equation (D.4), we have

$$\boldsymbol{x}_{sss} = -\kappa^2 \boldsymbol{t} + \kappa' \boldsymbol{n} - \kappa\tau \boldsymbol{b}.$$

Denoting $\boldsymbol{v}_3 = a\boldsymbol{x}_{sss}$, we obtain

$$\partial_s \boldsymbol{v}_3 = a(-3\kappa\kappa' \boldsymbol{t} + \beta \boldsymbol{n} + \gamma \boldsymbol{b}),$$

where β and γ are certain scalar functions of s. The rate of change of an arc-length Δs between two nearby points along the curve is given by (9.8):

$$(\Delta s)^{-1} \frac{\mathrm{d}}{\mathrm{d}t} \Delta s = (\partial_s \boldsymbol{v}_3) \cdot \boldsymbol{t} = -3\kappa\kappa' a,$$

where $\kappa(s)$ is the curvature of the filament at a point s. Therefore the new term $a\boldsymbol{x}_{sss}$ induces change of Δs. However this local change can be annihilated by adding a tangential velocity $\boldsymbol{v}_{3*} = (3/2)a\kappa^2 \boldsymbol{x}_s$ without affecting the velocity component perpendicular to $\boldsymbol{t}$. In fact, we have $(\partial_s \boldsymbol{v}_{3*}) \cdot \boldsymbol{t} = 3\kappa\kappa' a$. The shape of the filament is not changed by the additional term.

Thus, we have found a new equation of motion conserving the arc-length parameter s:

$$\boldsymbol{x}_t = \boldsymbol{x}_s \times \boldsymbol{x}_{ss} + a(\boldsymbol{x}_{sss} + (3/2)\kappa^2 \boldsymbol{x}_s). \tag{9.64}$$

This is equivalent to the equation obtained by Fukumoto and Miyazaki [FM91] (FM equation). This was originally derived for the motion of a thin vortex tube with an axial flow along it (Fig. 9.7). The two equations, (9.10) and (9.64), are known to be the first two members of the hierarchy of completely integrable equations for the filament motion [LP91].

In Chapter 5, the KdV equation is derived as a geodesic equation on the diffeomorphism group of a circle S^1 with a central extension. Here,

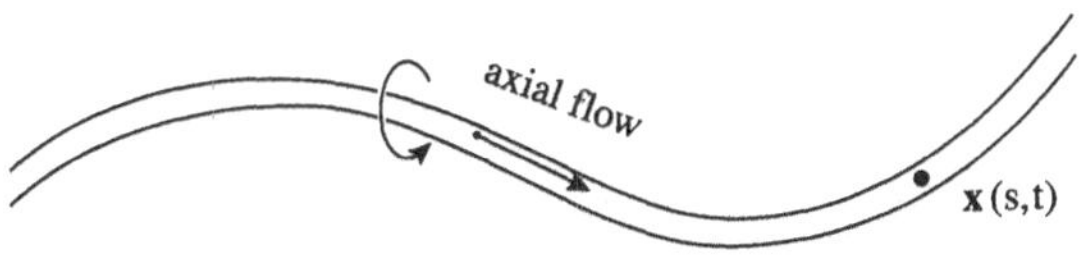

Fig. 9.7. Thin vortex tube with an axial flow.

it is verified that the motion of a vortex filament governed by Eq. (9.10) is a geodesic on the loop group $\mathcal{L}G = SO(3)[S^1]$, with $SO(3)$-valued and pointwise multiplication. Furthermore, the infinite-dimensional loop algebra $\mathcal{L}\mathbf{g}$ has a non-trivial central extension equivalent to the Kac–Moody algebra. This is a new formulation verifying that the extended system leads to another geodesic equation with an additional third derivative term, which was derived earlier [FM91] and shown to be a completely integrable system. It is remarkable that there is a similarity in the forms between the KdV equation and FM equation. These are two integrable systems defined over the S^1 manifold: one is a geodesic equation over the extended diffeomorphism group $\hat{D}(S^1)$ and the other is over the extended loop group $\hat{SO}(3)[S^1]$.

Part IV

Geometry of Integrable Systems

It is well known that some soliton equations admit certain geometric interpretation. An oldest example is the sine–Gordon equation on a pseudo-spherical surface in $\mathbb{R}^3$ [Eis47]. The Gauss and Mainardi–Codazzi equations (§2.4) of the differential geometry of surfaces in $\mathbb{R}^3$ yield the sine–Gordon equation when the Gaussian curvature is constant with a negative value. On the basis of the surface geometry, the Bäcklund transformation can be explained as a transformation from one surface to another in $\mathbb{R}^3$. Namely, the relation between the new and old surfaces is nothing else than the so-called Bäcklund transformation. Both were already known before modern soliton theory. This will be described in Chapter 10.

In order to understand a certain background of the Lax representation in the soliton theory, modern approaches of group-theoretic and differential-geometric theories were developed on the integrable systems [LR76; Herm76; Cra78; Lun78], which later led to theories of soliton surfaces. Among them, two kinds of approaches have been recognized. One is based on the structure equations which express integrability conditions for surfaces, which was proposed by [Sas79] and developed by [CheT86]. Second is an immersion problem of an integration surface in an envelop space, which was initiated by [Sym82], and later systematically developed by [Bob94; FG96; CFG00]. This is an approach by defining surfaces on Lie groups and Lie algebras. Both are described in Chapter 11. Firstly, we start to consider a historical geometrical problem to derive the sine–Gordon (SG) equation on a surface of constant Gaussian curvature.

Chapter 10

Geometric Interpretations of Sine–Gordon Equation

This Chapter 10, §11.3.4 and §11.5 are regarded as some applications of the formulation of Chapter 2 for surfaces in $\mathbb{R}^3$.

10.1. Pseudosphere: A Geometric Derivation of SG

We consider surfaces of *constant* Gaussian curvature K. As K is positive or negative, the surface is called *spherical* or *pseudospherical.* On the basis of the theory of surfaces in $\mathbb{R}^3$ (Chapter 2, §3.5.2, Appendix K), we consider the coordinate curves that are defined by the *lines of curvature*[1] on the surfaces of constant K in $\mathbb{R}^3$, and hence, the coordinate curves constitute an orthogonal coordinate net (u^1, u^2), and a line-element $\mathrm{d}s$ is defined by $\mathrm{d}s^2 = g_{11}(\mathrm{d}u^1)^2 + g_{22}(\mathrm{d}u^2)^2$. We assume that K is equal to ε/a^2, where ε is $+1$ or -1 according as K is positive or negative with a being a positive constant. Such a surface can be covered by a conjugate net. According to the representation (K.7), the second fundamental tensors (Appendix K and §2.2) are given by

$$b_{11} = \frac{\sqrt{g}}{a}, \qquad b_{12} = 0, \qquad b_{22} = \varepsilon\frac{\sqrt{g}}{a}, \tag{10.1}$$

where $g = \det g_{\alpha\beta} = g_{11}g_{22}$.

[1]A line of curvature is defined by the property that its tangent coincides at each point with one of two principal directions which are orthogonal to each other. The lines of curvature intersecting orthogonally satisfy the condition of conjugate directions, and are hence called *self-conjugate* [Appendix K].

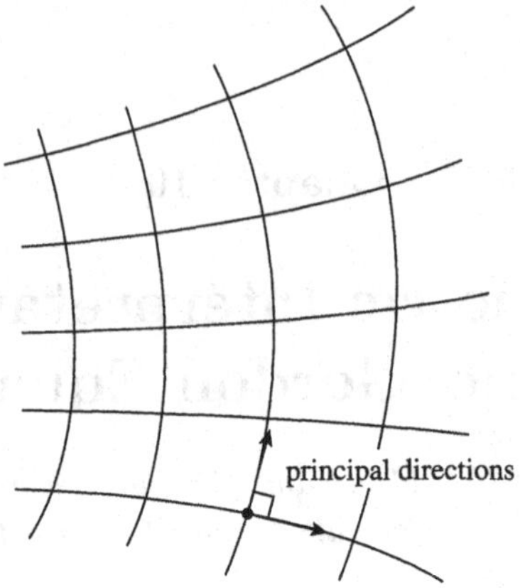

Fig. 10.1. An orthogonal coordinate net.

By the properties $g_{12} = 0$ and $b_{12} = 0$, the coordinate curves are orthogonal and coincide with the lines of curvature (see (K.5)). On such a coordinate system as given by the lines of curvature, the tangent to the coordinate curve at each point coincides with one of the principal directions (Fig. 10.1).

Now, the Mainardi–Codazzi equation derived in §2.4 is useful to determine the surface. Taking $\gamma = \alpha$ in Eq. (2.49), we have

$$\partial_\beta b_{\alpha\alpha} - \partial_\alpha b_{\alpha\beta} - \Gamma^\nu_{\alpha\beta} b_{\nu\alpha} + \Gamma^\nu_{\alpha\alpha} b_{\nu\beta} = 0,$$

where $\alpha \neq \beta$. Setting $(\alpha, \beta) = (1, 2)$, one obtains $\partial_2 b_{11} - \Gamma^1_{12} b_{11} + \Gamma^2_{11} b_{22} = 0$, since $b_{12} = 0$. Substituting the above expressions of $b_{\alpha\beta}$ and using the definition (2.40) of $\Gamma_{\beta\nu,\alpha}$ with $g_{12} = 0$, this reduces to

$$\frac{g_{11}}{2a\sqrt{g}} \partial_2 (g_{22} - \varepsilon g_{11}) = 0,$$

where $g = g_{11} g_{22}$. Another pair $(\alpha, \beta) = (2, 1)$ gives the equation: $(g_{22}/2a\sqrt{g})\ \partial_1 (g_{22} - \varepsilon g_{11}) = 0$. Thus, we obtain

$$\frac{\partial}{\partial u^1}(g_{22} - \varepsilon g_{11}) = 0, \qquad \frac{\partial}{\partial u^2}(g_{22} - \varepsilon g_{11}) = 0,$$

since $g_{11} \neq 0$ and $g_{22} \neq 0$. These lead to the solution,

$$g_{22} - \varepsilon g_{11} = \text{const.}$$

For the case of the pseudospherical surface ($\varepsilon = -1$), this equation is satisfied by

$$g_{11} = a^2 \cos^2 \phi, \qquad g_{22} = a^2 \sin^2 \phi, \qquad (g_{12} = 0), \tag{10.2}$$

with an appropriate scaling of u^1 and u^2 where a is a positive constant. Then, the line-element length $\mathrm{d}s$ is given by

$$\mathrm{d}s^2(\phi) = a^2 \cos^2 \phi \, (\mathrm{d}u^1)^2 + a^2 \sin^2 \phi \, (\mathrm{d}u^2)^2.$$

Therefore, we have $\mathrm{d}s(0) = a|\mathrm{d}u^1|$ and $\mathrm{d}s(\pi/2) = a|\mathrm{d}u^2|$. Hence, the coordinate curves divide the surfaces into small infinitesimal squares. In this sense, the coordinate net is called an *isometric orthogonal net.* From (10.1), we have $b_{11} = -b_{22} = a \sin\phi \cos\phi$. (Likewise, the case of the spherical surface $\varepsilon = 1$ can be solved.)

Substituting the above expressions (10.2) of $g_{\alpha\beta}$ in the representation of the Gaussian curvature (2.63), and using $K = -1/a^2$, one obtains finally

$$\left(\frac{\partial}{\partial u^1}\right)^2 \phi - \left(\frac{\partial}{\partial u^2}\right)^2 \phi = \sin\phi \cos\phi \tag{10.3}$$

[Eis47, §49]. For each solution $\phi(u^1, u^2)$ of (10.3), the metric tensors (10.2) together with the values of $b_{\alpha\beta}$ determine a pseudospherical surface with the coordinate net of lines of curvtures. Introducing the variable $\Phi = 2\phi$, the above equation becomes

$$\left(\frac{\partial}{\partial u^1}\right)^2 \Phi - \left(\frac{\partial}{\partial u^2}\right)^2 \Phi = \sin\Phi. \tag{10.4}$$

The first and second fundamental forms are

$$\mathrm{I} = a^2 \cos^2(\Phi/2)(\mathrm{d}u^1)^2 + a^2 \sin^2(\Phi/2)(\mathrm{d}u^2)^2, \tag{10.5}$$

$$\mathrm{II} = \frac{a}{2} \sin\Phi(\mathrm{d}u^1)^2 - \frac{a}{2} \sin\Phi(\mathrm{d}u^2)^2. \tag{10.6}$$

Performing the coordinate transformation defined by $x = (u^1 + u^2)/2$ and $y = (u^1 - u^2)/2$, Eq. (10.4) is also written as

$$\frac{\partial}{\partial x}\frac{\partial}{\partial y}\Phi = \sin\Phi. \tag{10.7}$$

Equation (10.7), or (10.4), is called the *sine–Gordon equation* (SG) in the soliton theory.

Each solution $\Phi(u^1, u^2)$ gives an explicit representation of the tensor fields $g_{\alpha\beta}(u^1, u^2)$ and $b_{\alpha\beta}(u^1, u^2)$ of the first and second fundamental forms, which can determine a pseudospherical surface in $\mathbb{R}^3$, *uniquely* within a rigid motion (§2.11).

10.2. Bianchi–Lie Transformation

Geometric interpretation of the Bäcklund transformation is illustrated by the Bianchi–Lie transformation [AnIb79]. The outline of the Bianchi's geometrical construction [Bia1879] is as follows. Consider a surface Σ of *constant negative curvature* $-1/a^2$ in the euclidean space $\mathbb{R}^3$. Another surface Σ' is related to Σ in the following way. To each point $p \in \Sigma$, there corresponds a point $p' \in \Sigma'$, such that

$$\text{(i)} \;:\; |pp'| = a, \qquad \text{(ii)} \;:\; \overline{pp'} \in T_p\Sigma,$$
$$\text{(iii)} \;:\; \overline{pp'} \in T_{p'}\Sigma', \qquad \text{(iv)} \;:\; T_p\Sigma \perp T_{p'}\Sigma',$$

where $|pp'|$ is the length of the line segment $\overline{pp'}$, and $T_p\Sigma, T_{p'}\Sigma'$ are tangent planes to Σ, Σ' at p, p' respectively (Fig. 10.2). It was shown by Bianchi [Bia1879] that the surface Σ' thus constructed is also a surface of the same constant curvature $-1/a^2$.

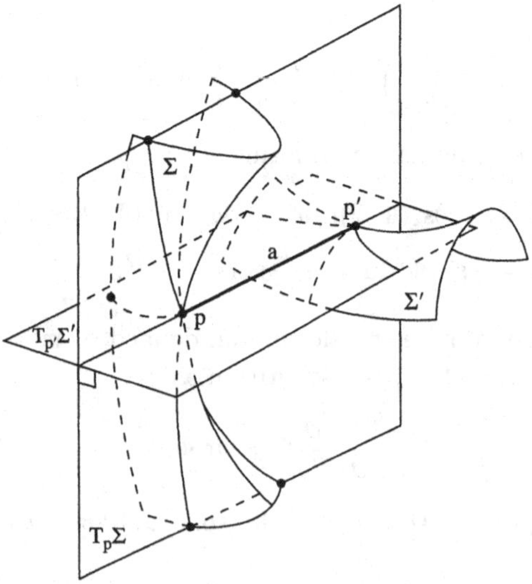

Fig. 10.2. Sketch of Bianchi's transformation.

An analytical interpretation equivalent to the above transformation was given by Lie [Lie1880]. Any surface $z = f(x, y)$ in $\mathbb{R}^3(x, y, z)$ of constant curvature $K = -1/a^2$ satisfies the following second order partial differential equation:

$$f_{xx}f_{yy} - (f_{xy})^2 = -\frac{1}{a^2}\left(1 + (f_x)^2 + (f_y)^2\right)^2 \tag{10.8}$$

(see (D.10)). At a point $k = (x, y, z)$ on such a surface Σ, the normal vector N to the tangent plane $(T_p\Sigma)$ is given by $N = (p, q, -1)$ where $p = f_x, q = f_y$, since $p\mathrm{d}x + q\mathrm{d}y - \mathrm{d}z = 0$. Therefore, a surface element of Σ is defined by the set (x, y, z, p, q), satisfying the following consistency condition:

$$\frac{\partial p}{\partial y} = \frac{\partial^2 f}{\partial y \partial x} = \frac{\partial q}{\partial x}$$

(Fig. 10.3). The corresponding surface element of the transformed surface Σ' (represented by $Z = F(X, Y), P = F_X, Q = F_Y$) is denoted by (X, Y, Z, P, Q). Then the above Bianchi's conditions (i) $\sim$ (iv) are expressed in the following way:

$$\left.\begin{array}{ll} \text{(i)} : & (x - X)^2 + (y - Y)^2 + (z - Z)^2 = a^2, \\ \text{(ii)} : & p(x - X) + q(y - Y) - (z - Z) = 0, \\ \text{(iii)} : & P(x - X) + Q(y - Y) - (z - Z) = 0, \\ \text{(iv)} : & pP + qQ + 1 = 0. \end{array}\right\} \tag{LT}$$

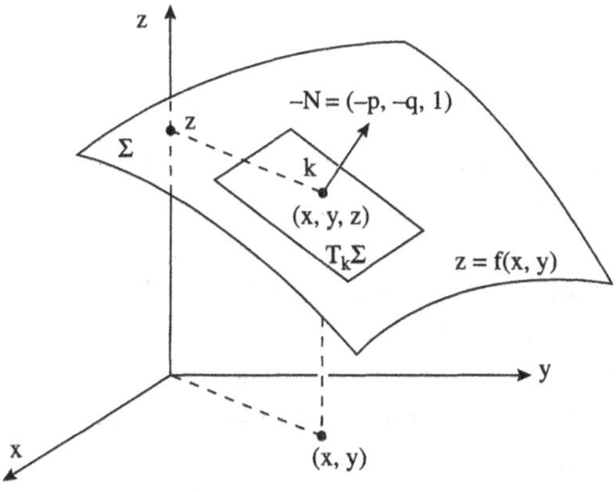

Fig. 10.3. Surface element (x, y, z, p, q).

It is observed that, given any surface element (x, y, z, p, q), the equations (LT) give four relationships between the five quantities X, Y, Z, P, Q. Therefore there is a one-fold infinity of surface elements (X, Y, Z, P, Q) satisfying (LT), that is, not unique but multi-valued. Next, we quote Lie's lemma and theorem without proof [AnIb79; Lie1880].

Lemma. *Suppose that a surface element* (x, y, z, p, q) *of a surface* Σ *is given together with the relations* (LT). *If* (X, Y, Z, P, Q) *is a surface element, that is, if the integrability condition* $\partial P/\partial Y = \partial Q/\partial X$ *is satisfied on* Σ, *then* Σ *is a surface of constant negative curvature, i.e.* Σ *satisfies Eq.* (10.8).

A statement dual to the above Lemma holds for the symmetry between the surface elements of Σ and Σ'. As a consequence,

Theorem. *The partial differential equation* (10.8) *is invariant under the transformation (LT) in the following sense. Suppose that* Σ *is a surface of constant curvature* $-1/a^2$ *and* Σ' *is an image of* Σ *under the action of (LT), then* Σ' *is also a surface of constant curvature* $-1/a^2$.

Accordingly, one can construct a family of surfaces of constant negative curvature starting from a given initial surface. This is recognized as a possibility that the transformation (LT) converts a solution of Eq. (10.8) into a family of solutions of the same equation. It also gives us a geometrical hint to the Bäcklund transformation to be considered in the next section. It is well known that the sine–Gordon equation equivalent to Eq. (10.8) is one of the soliton equations.

10.3. Bäcklund Transformation of SG Equation

Bäcklund [Bac1880] generalized the Bianchi's construction of surfaces Σ and Σ', by replacing the orthogonality condition of the two tangent planes $\mathcal{T} = T_p\Sigma$ and $\mathcal{T}' = T_{p'}\Sigma'$ with the condition that the angle between $\mathcal{T}$ and $\mathcal{T}'$ is fixed and not necessarily at right angles. Namely, the Bianchi's condition (iv) is replaced by (iv') $\angle(\mathcal{T}, \mathcal{T}') = \text{const}$. Although Bäcklund's construction generalized Bianchi's construction geometrically, it turned out that it is analytically only a simple extension up to a one-parameter group of dilatations [AnIb79].

Heuristically, it is as follows. Suppose that a surface $z = z(x, y)$ in (x, y, z)-space is specified pointwise by the values of z and its derivatives

$p, q, r, s, t, \ldots$ at (x, y). The *Bäcklund transformation* is represented by the four conditions:

$$\left.\begin{aligned} X &= x, \\ Y &= y, \\ P &= f(x, y, z, p, q, Z), \\ Q &= g(x, y, z, p, q, Z). \end{aligned}\right\} \tag{10.9}$$

Suppose that a surface $z = h(x, y)$ is given, and that this is substituted into the third and fourth equation of (10.9), the two relationships represent an overdetermined system of two first order partial differential equations in one unknown function $Z(X, Y)$. A consistency condition $\partial P/\partial Y = \partial Q/\partial X$ must be satisfied. If $z(x, y)$ satisfies this condition, then the system (10.9) is regarded as a transformation from a surface $z = z(x, y)$ into a surface $Z = Z(X, Y)$, and considered as an integrable system.

Suppose that the function $z(x, y)$ satisfies the sine–Gordon equation (10.7). In addition, consider a particular transformation of four relationships represented by $X = x, Y = y$ and,

$$p - P = 2a \sin \frac{1}{2}(z + Z), \tag{10.10}$$

$$q + Q = \frac{2}{a} \sin \frac{1}{2}(z - Z), \tag{10.11}$$

where $P = Z_X, Q = Z_Y$. Recall that $p = z_x, q = z_y$ and $p_y = q_x = z_{xy} = \sin z$ by (10.7). Differentiation of (10.10) with respect to y leads to

$$\begin{aligned} p_y - P_Y &= 2a \cos \frac{1}{2}(z + Z) \frac{1}{2}(q + Q) \\ &= 2 \cos \frac{1}{2}(z + Z) \sin \frac{1}{2}(z - Z) = \sin z - \sin Z. \end{aligned}$$

Likewise, differentiation of (10.11) with respect to x results in $q_x + Q_X = \sin z + \sin Z$. Sine $p_y = q_x = \sin z$, we obtain

$$\frac{\partial P}{\partial Y} = \sin Z, \quad \text{and} \quad \frac{\partial Q}{\partial X} = \sin Z,$$

respectively. Thus the consistency condition $\partial P/\partial Y = \partial Q/\partial X = Z_{XY}$ is satisfied, and we find that $Z(X, Y)$ satisfies the sine–Gordon equation:

$$Z_{XY} = \sin Z. \tag{10.12}$$

The pair of transformations (10.10) and (10.11) is usually called the (self-)*Bäcklund transformation* for the sine–Gordon equation. A systematic derivation of the Bäcklund transformation will be considered later in §11.4.

Chapter 11

Integrable Surfaces: Riemannian Geometry and Group Theory

11.1. Basic Ideas

Soliton theory concerns solvable systems of nonlinear partial differential equations such as the sine–Gordon equation, nonlinear Schrödinger equation, KdV equation, modified KdV (mKdV) equation and so on. The inverse scattering transform is one of the methods to solve them [AS81; AKNS73]. In this framework, a pair of linear systems are introduced:

$$\psi_x = X\psi, \qquad \psi_t = T\psi, \tag{11.1}$$

where ψ is an n-dimensional vector wave function of variables x and t, and X, T are traceless $n \times n$ matrices, called the *Lax pair*, including a certain spectral parameter ζ and functions $u(x,t), \cdots$ ($n = 2$ in the following examples). A solvability condition is obtained by the cross-differentiation of (11.1) with respect to x and t, and equating ψ_{xt} with ψ_{tx}:

$$X_t - T_x + [X, T] = 0, \tag{11.2}$$

where $[X, T] = XT - TX$ (the commutator of X and T). According to the scenario of the method, given a matrix operator X, there is a simple deductive procedure to find T such that the system (11.2) yields nonlinear evolution equations for $u(x,t), \ldots$. In order for Eq. (11.2) to be useful, the operator X should have an eigenvalue parameter ζ which is time-independent, $\partial_t \zeta = 0$. This problem is solved by the inverse scattering transform, called the AKNS method coined after the authors of the seminal work [AKNS73].

A geometric aspect, pointed out by Lund and Regge [LR76], was why a particular linear problem like (11.1) is helpful in solving a certain nonlinear equation. Later, geometrical interpretations for the inverse scattering problem were developed: some integrable equations describe pseudospherical surfaces [Sas79; CheT86], and some other integrable equations describe spherical surfaces [AN84; Kak91]. In addition, a theory of immersion of a two-dimensional surface described by integrable equations into a three-dimensional euclidean space has been developed by [Sym82; Bob90; Bob94; FG96]. Recently, it has been shown that integrable systems are mapped to the surface of a sphere [CFG00]. We will consider these problems one by one below.

11.2. Pseudospherical Surfaces: SG, KdV, mKdV, ShG

Let M^2 be a two-dimensional differentiable manifold with coordinates (x, t). The two equations in (11.1) are combined into the following form,

$$\mathrm{d}\psi = \Omega\psi, \qquad \psi = \begin{pmatrix} \psi_1 \\ \psi_2 \end{pmatrix}, \tag{11.3}$$

where $\mathrm{d}\psi = \psi_x \mathrm{d}x + \psi_t \mathrm{d}t$ is a vector-valued 1-form, and $\Omega = X\mathrm{d}x + T\mathrm{d}t$ is a traceless real 2×2 matrix with 1-form entry,[1] and expressed as

$$\Omega = \frac{1}{2}\begin{pmatrix} -\sigma^2 & \sigma^1 - \varpi \\ \sigma^1 + \varpi & \sigma^2 \end{pmatrix}, \tag{11.4}$$

where $\{\sigma^1, \sigma^2, \varpi\}$ are three 1-forms on $\mathbb{R}^2 = (x, t)$, depending on the function $u(x, t)$ and its partial derivatives, including a certain spectral parameter ζ.

Equation (11.3), $\mathrm{d}\psi - \Omega\psi = 0$, is read as vanishing of the covariant derivative of a vector ψ (see Eqs. (3.41) and (3.27)), describing *parallel* transport of v, and the matrix Ω is the connection 1-form (§3.5.1). This observation is the motivation for the following formulation. The key step is to find appropriate 1-forms $\{\sigma^1, \sigma^2, \varpi\}$ for a nonlinear partial differential equation which is completely integrable.

Integrability condition for the Pfaffian system (11.3) is described by $\mathrm{d}(\mathrm{d}\psi) = (\psi_{xt} - \psi_{tx})\mathrm{d}t \wedge \mathrm{d}x = 0$ (footnote to §1.5, §2.7, §3.5):

$$\mathrm{d}(\mathrm{d}\psi) = \mathrm{d}\Omega\psi - \Omega \wedge \mathrm{d}\psi = (\mathrm{d}\Omega - \Omega \wedge \Omega)\psi = 0.$$

[1] A system of 1-form equations like (11.3) is often called as a *Pfaffian* system.

This requires vanishing of the 2-form,

$$\mathrm{d}\Omega - \Omega \wedge \Omega = 0, \tag{11.5}$$

which is equivalent to the solvability condition (11.2).[2] Writing with components, Eq. (11.5) reduces to

$$\mathrm{d}\sigma^1 = \sigma^2 \wedge \varpi, \tag{11.6}$$

$$\mathrm{d}\sigma^2 = \varpi \wedge \sigma^1, \tag{11.7}$$

$$\mathrm{d}\varpi = -\sigma^1 \wedge \sigma^2. \tag{11.8}$$

Comparing this[3] with (3.52) of §3.5.2, the first two equations correspond to the structure equations describing the first integrability condition, and the third equation, written as

$$\mathrm{d}\varpi = K\sigma^1 \wedge \sigma^2 \tag{11.9}$$

in (3.52), is the second integrability condition. This requires the Gaussian curvature K to be -1. In the formulation of §3.5, the two-dimensional manifold M^2 is structured with 1-forms σ^1 and σ^2 in the orthonomal directions $\boldsymbol{e}_1$ and $\boldsymbol{e}_2$ respectively, and the first fundamental form on M^2 is given by $\mathrm{I} = \sigma^1\sigma^1 + \sigma^2\sigma^2$. It is said that *the manifold M^2 is a pseudospherical surface, if $K = -1$ (or a negative constant)*.

Given the three 1-forms $\{\sigma^1, \sigma^2, \varpi\}$ appropriately dependent on the function $u(x,t)$ and its partial derivatives (including a certain spectral parameter ζ), the integrability conditions (11.6)–(11.8) require that a certain evolution equation must be satisfied. This determines a differential equation which describes a pseudospherical surface.

Explicit representations of Ω of (11.4) are now given for four soliton equations. According to [Sas79; CheT86], the connection 1-form matrix Ω is for

(a) sine–Gordon (SG) equation:

$$\Omega_{\mathrm{SG}} = \frac{1}{2}\begin{pmatrix} \zeta\mathrm{d}x + \zeta^{-1}(\cos u)\mathrm{d}t & -u_x\mathrm{d}x + \zeta^{-1}(\sin u)\mathrm{d}t \\ u_x\mathrm{d}x + \zeta^{-1}(\sin u)\mathrm{d}t & -\zeta\mathrm{d}x - \zeta^{-1}(\cos u)\mathrm{d}t \end{pmatrix}, \tag{11.10}$$

where the real parameter ζ plays the role of the eigenvalue in the scattering problem of (11.3).

---

[2] This is known as the *Maurer–Cartan equation.*

[3] Our ϖ corresponds to $-\omega$ of [Sas79]. Our $(\sigma^1, \sigma^2, \varpi)$ correspond to $(\omega_1, -\omega_2, \omega_3)$ of [CheT86].

Comparing with (11.4), we have

$$\left.\begin{aligned} &\sigma^1 = \zeta^{-1}(\sin u)\mathrm{d}t, \quad \sigma^2 = -\zeta \mathrm{d}x - \zeta^{-1}(\cos u)\mathrm{d}t, \\ &\varpi = u_x \mathrm{d}x. \end{aligned}\right\} \tag{11.11}$$

Using these, both Eqs. (11.6) and (11.7) are found to be *identity* equations, hence yielding no new relation, whereas the structure equation (11.9) reduces to $u_{xt} = -K \sin u$. When $K = -1$, this becomes the sine–Gordon equation (10.12):

$$u_{xt} = \sin u. \tag{11.12}$$

Thus, the sine–Gordon equation describes a pseudospherical surface. Note that this is obtained from the integrability condition $\mathrm{d}^2\psi = 0$, equivalent to $\psi_{xt} = \psi_{tx}$. See §11.3.2 for the case $K = 1$.

(b) KdV equation is described by

$$\Omega_{\mathrm{KdV}} = \begin{pmatrix} \zeta \mathrm{d}x - (4\zeta^3 + 2\zeta u + u_x)\mathrm{d}t & u\mathrm{d}x - (u_{xx} + 2\zeta u_x + 4\zeta^2 u + 2u^2)\mathrm{d}t \\ -\mathrm{d}x + (4\zeta^2 + 2u)\mathrm{d}t & -\zeta \mathrm{d}x + (4\zeta^3 + 2\zeta u + u_x)\mathrm{d}t \end{pmatrix}. \tag{11.13}$$

Comparing with (11.4), we have

$$\begin{aligned} \sigma^1 &= (u-1)\mathrm{d}x + (-u_{xx} - 2\zeta u_x - 4\zeta^2 u - 2u^2 + 2u + 4\zeta^2)\mathrm{d}t, \\ \sigma^2 &= -2\zeta \mathrm{d}x + 2(u_x + 2\zeta u + 4\zeta^3)\mathrm{d}t, \\ \varpi &= -(u+1)\mathrm{d}x + (u_{xx} + 2\zeta u_x + 4\zeta^2 u + 2u^2 + 2u + 4\zeta^2)\mathrm{d}t. \end{aligned}$$

Using these, both of Eqs. (11.6) and (11.8) reduce to

$$[u_t + 6uu_x + u_{xxx}]\mathrm{d}x \wedge \mathrm{d}t = 0,$$

whereas (11.7) reduces to an identity such as $[0]\mathrm{d}x \wedge \mathrm{d}t = 0$. Thus, we find that the only nontrivial equation to be satisfied is the KdV equation:

$$u_t + 6uu_x + u_{xxx} = 0.$$

This is a differential equation which defines M^2 to be a pseudospherical surface. Likewise, we have

(c) Modified KdV (mKdV) equation,

$$\Omega_{\mathrm{mKdV}} = \begin{pmatrix} \zeta \mathrm{d}x - (4\zeta^3 + 2\zeta u^2)\mathrm{d}t & u\mathrm{d}x - (u_{xx} + 2\zeta u_x + 4\zeta^2 u + 2u^3)\mathrm{d}t \\ -u\mathrm{d}x + (u_{xx} - 2\zeta u_x + 4\zeta^2 u + 2u^3)\mathrm{d}t & -\zeta \mathrm{d}x + (4\zeta^3 + 2\zeta u^2)\mathrm{d}t \end{pmatrix};$$

(d) sinh–Gordon (ShG) equation,

$$\Omega_{\text{ShG}} = \frac{1}{2}\begin{pmatrix} \zeta \mathrm{d}x + \frac{1}{\zeta}(\cosh u)\mathrm{d}t & \frac{1}{2}u_x \mathrm{d}x - \frac{1}{\zeta}(\sinh u)\mathrm{d}t \\ u_x \mathrm{d}x + \frac{1}{\zeta}(\sinh u)\mathrm{d}t & -\zeta \mathrm{d}x - \frac{1}{\zeta}(\cosh u)\mathrm{d}t \end{pmatrix}. \tag{11.14}$$

From (c) and (d), we obtain corresponding nonlinear differential equations:

$$\text{mKdV equation}: u_t + 6u^2 u_x + u_{xxx} = 0, \tag{11.15}$$

$$\text{ShG equation}: u_{xt} - \sinh u = 0, \tag{11.16}$$

respectively for $K = -1$. If these equations are satisfied, we obtain a *pseudospherical* surface.

11.3. Spherical Surfaces: NLS, SG, NSM

Integrable equations are mapped to the surface of a sphere [CFG00]. This may sound puzzling after learning that some integable equations describe pseudospherical surfaces. However, it can be shown that there exists another integrable system whose underlying surfaces have Gaussian curvature equal to +1. Here, we consider such systems. The immersion problem of integrable surfaces will be considered in §11.5 and 6.

11.3.1. *Nonlinear Schrödinger equation*

According to the formulation in the previous section, the AKNS linear problem of the inverse scattering method for the nonlinear Schrödinger (NLS) equation (9.13) of §9.2 may be written as $\mathrm{d}\phi = \Omega_{\text{NLS}}\,\phi$ for a two-component wave function $\phi = (\phi_1, \phi_2)^T$, where the connection 1-form Ω_{NLS} is defined as

$$\Omega_{\text{NLS}} = \begin{pmatrix} -i\zeta \mathrm{d}x - i\Lambda \mathrm{d}t & q(x,t)\mathrm{d}x + B\mathrm{d}t \\ -q^*(x,t)\mathrm{d}x - B^*\mathrm{d}t & i\zeta \mathrm{d}x + i\Lambda \mathrm{d}t \end{pmatrix}. \tag{11.17}$$

Here, a complex function $q(x,t) = q^{(r)} + iq^{(i)}$ (equivalent to ψ of (9.13)) is defined with two real functions $q^{(r)}$ and $q^{(i)}$ independent of a real spectral parameter ζ, and

$$\Lambda = 2\zeta^2 - |q|^2, \quad B = 2\zeta q + i q_x, \quad (B^* = 2\zeta q^* - i q_x^*), \tag{11.18}$$

(Λ is a real function).[4]

The complex matrix $\Omega_{\rm NLS}$ is found to be an element of $su(2)$, i.e. skew hermitian matrices (with trace 0). The Lie algebra $su(2)$ is regarded as a three-dimensional vector space over *real* coefficients with the orthonomal basis (e_1, e_2, e_3) defined by

$$e_1 = \frac{1}{2}\begin{pmatrix} 0 & -i \\ -i & 0 \end{pmatrix}, \quad e_2 = \frac{1}{2}\begin{pmatrix} 0 & -1 \\ 1 & 0 \end{pmatrix}, \quad e_3 = \frac{1}{2}\begin{pmatrix} -i & 0 \\ 0 & i \end{pmatrix}, \tag{11.19}$$

which are related to the Pauli matrices (7.23) by $\sigma_k = 2ie_k$, and satisfy the commutation relations,

$$[e_j, e_k] = \epsilon_{jkl} e_l, \tag{11.20}$$

which is consistent with (7.24).

Let us define three *real* 1-forms on $\mathbb{R}^2 = (x, t)$ by

$$\left.\begin{aligned} \sigma^1 &= 2(q^{(i)}\mathrm{d}x + B^{(i)}\mathrm{d}t), \\ \sigma^2 &= 2(\zeta\mathrm{d}x + \Lambda\mathrm{d}t), \\ \varpi &= -2(q^{(r)}\mathrm{d}x + B^{(r)}\mathrm{d}t), \end{aligned}\right\} \tag{11.21}$$

where $B^{(r)}$ and $B^{(i)}$ are the real and imaginary parts of B. Then, we have

$$\Omega_{\rm NLS} = \frac{1}{2}\begin{pmatrix} -i\sigma^2 & i\sigma^1 - \varpi \\ i\sigma^1 + \varpi & i\sigma^2 \end{pmatrix} = -\sigma^1 e_1 + \varpi e_2 + \sigma^2 e_3. \tag{11.22}$$

Suppose that the three 1-forms satisfy

$$\left.\begin{aligned} \mathrm{d}\sigma^1 &= \sigma^2 \wedge \varpi, \\ \mathrm{d}\sigma^2 &= \varpi \wedge \sigma^1, \\ \mathrm{d}\varpi &= \sigma^1 \wedge \sigma^2. \end{aligned}\right\} \tag{11.23}$$

Comparing with (11.9), this describes the underlying surface M^2 that has the Gaussian curvature $+1$. The manifold M^2 (structured with 1-forms

[4]The formulation of [AS81, §1.2] includes the case where q and Λ are replaced by $\pm q$ and $2\zeta^2 \mp |q|^2$, respectively. Soliton solutions are found with the upper signs.

σ^1 and σ^2 in the orthonormal directions $\boldsymbol{e}_1$ and $\boldsymbol{e}_2$, §3.5.2) might be said to be *spherical*. The first fundamental form of M^2 is given by

$$\mathrm{I} = \langle \mathrm{d}\boldsymbol{x}, \mathrm{d}\boldsymbol{x} \rangle = \sigma^1\sigma^1 + \sigma^2\sigma^2, \tag{11.24}$$

where $\mathrm{d}\boldsymbol{x} = \sigma^1 \boldsymbol{e}_1 + \sigma^2 \boldsymbol{e}_2$.

Using (11.21), the first and third equations of (11.23) reduce to

$$q_t^{(i)} = q_{xx}^{(r)} + 2|q|^2 q^{(r)}, \tag{11.25}$$

$$q_t^{(r)} = -q_{xx}^{(i)} - 2|q|^2 q^{(i)}. \tag{11.26}$$

The second of (11.23) reduces to an identity such as $[0]\,\mathrm{d}x \wedge \mathrm{d}t = 0$. Multiplying (11.25) by i and summing up with (11.26) result in

$$q_t = i(q_{xx} + 2q|q|^2), \tag{11.27}$$

i.e. the nonlinear Schrödinger equation equivalent to (9.13) with $q = \frac{1}{2}\psi$. This is a differential equation which defines M^2 to be a spherical surface, with the metric defined by (11.24).

Remark. If the problem under investigation is periodic, then there may be no problem. However, some problems of motion of vortex filaments may include an unbounded domain $x \in \mathbb{R}^1$. The vortex soliton (9.14) of §9.2 is such an example. In this regard, it is to be noted that the 1-form σ^1 and the connection form ϖ include linearly the real functions $q^{(r)}$ and $q^{(i)}$ (since $B = 2\zeta q + iq_x$), without nonhomogeneous terms. In a problem for $x \in \mathbb{R}^1$, if $|q^{(r)}|$ and $|q^{(i)}|$ decay more rapidly than x^{-1} as $|x| \to \pm\infty$, then the distance in the e_1 direction will be bounded as $|x| \to \pm\infty$. In fact, the function q of (9.14) decays as $e^{-|\tau_0 x|}$. In addition, the inverse scattering problem with the potential q of a vortex soliton (9.14) will be characterized with a negative value of the spectral parameter ζ, say $2\zeta = -c$ $(c > 0)$. Then as $|x| \to \pm\infty, \sigma^2 \to 2\zeta(\mathrm{d}x - c\mathrm{d}t)$ since $|q| \to 0$. It might be reasonable that the solution q is regarded as a limiting form of a periodic function of $\xi = x - ct$ with its periodicity length tending to infinity.

11.3.2. *Sine–Gordon equation revisited*

It may be puzzling to find that the sine–Gordon equation has a underlying surface of Gaussian curvature +1 as well. This can be verified according to the formulation in the previous section. Instead of (11.21), we introduce the three *real* 1-forms on $\mathbb{R}^2 = (u, v)$ by

$$\left.\begin{aligned} \sigma^1 &= \Phi_u du, \\ \sigma^2 &= -\zeta du + \zeta^{-1}\cos\Phi dv, \\ \varpi &= \zeta^{-1}\sin\Phi dv. \end{aligned}\right\} \tag{11.28}$$

In this case, the connection 1-form matrix $\Omega^*_{\rm SG}$ is defined by the same form as (11.22):

$$\Omega^*_{\rm SG} = \frac{1}{2}\begin{pmatrix} -i\sigma^2 & i\sigma^1 - \varpi \\ i\sigma^1 + \varpi & i\sigma^2 \end{pmatrix} = -\sigma^1 e_1 + \varpi e_2 + \sigma^2 e_3. \tag{11.29}$$

The first of the structure equation (11.23) leads to the sine–Gordon equation:

$$\Phi_{uv} = \sin\Phi,$$

whereas the remaining two equations results in identities, yielding no new relation. Thus, it is found that the sine–Gordon equation defines a spherical surface (just like the nonlinear Schrödinger equation does).

The problem of spherical and pseudospherical surfaces M^2 will be considered in the next subsection again by presenting particular solutions explicitly and also by viewing from the enveloping space $\mathbb{R}^3$, i.e. transforming the surfaces M^2 to surfaces of revolution in R^3.

11.3.3. *Nonlinear sigma model and SG equation*

A nonlinear sigma model (NSM) is obtained from a relativistic conformal field theory in one time and one space dimension, which is described by an $O(3)$-invariant Lagrangian for a real field variable $\boldsymbol{n}$ [Poh76; AN84]. The variable $\boldsymbol{n}(u, v)$ is represented as a three-dimensional vector $\boldsymbol{n}(u, v) = (n^1, n^2, n^3)$ depending on two real parameters (u, v) and its magnitude is constrained to be unity:

$$\langle \boldsymbol{n}, \boldsymbol{n} \rangle := (n^1)^2 + (n^2)^2 + (n^3)^2 = 1.$$

The vector $\boldsymbol{n}$ describes a unit sphere S^2 in $\mathbb{R}^3$, and (u, v) are regarded as parameters on it. The tangent vectors $\boldsymbol{n}_u$ and $\boldsymbol{n}_v$ are normalized such that

$$\langle \boldsymbol{n}_u, \boldsymbol{n}_u \rangle = 1, \qquad \langle \boldsymbol{n}_v, \boldsymbol{n}_v \rangle = 1,$$

and satisfy the following conditions:

$$\langle \boldsymbol{n}_u, \boldsymbol{n} \rangle = 0, \qquad \langle \boldsymbol{n}_v, \boldsymbol{n} \rangle = 0, \qquad -1 \le \langle \boldsymbol{n}_u, \boldsymbol{n}_v \rangle \le 1.$$

The equation of motion of the sigma model is described by

$$\boldsymbol{n}_{uv} + \langle \boldsymbol{n}_u, \boldsymbol{n}_v \rangle \boldsymbol{n} = 0. \tag{11.30}$$

It is useful to introduce the angle variable Φ between the two tangents by

$$\cos\Phi = \langle \boldsymbol{n}_u, \boldsymbol{n}_v \rangle. \tag{11.31}$$

In order to investigate the integrability of the sigma model, we introduce a pair of (original *physical*) variables (t, x) by

$$u = \frac{1}{2}(x+t), \qquad v = \frac{1}{2}(x-t), \qquad \partial_u\partial_v = -\partial_t\partial_t + \partial_x\partial_x,$$

and define two *basis* 1-forms σ^1, σ^2 and connection 1-form ϖ on $(t, x) \in R^2$ by

$$\left.\begin{aligned} \sigma^1 &= \cos(\Phi/2)\mathrm{d}t, \\ \sigma^2 &= \sin(\Phi/2)\mathrm{d}x, \\ \varpi &= -\frac{1}{2}(\partial_x\Phi\mathrm{d}t + \partial_t\Phi\mathrm{d}x) \end{aligned}\right\}. \tag{11.32}$$

Using (11.32), the structure equations (11.6) and (11.7) are satisfied identically, yielding no new relation, whereas Eq. (11.9) becomes $\Phi_{xx} - \Phi_{tt} = K\sin\Phi$. When $K = 1$, this is written as

$$\Phi_{xx} - \Phi_{tt} = \sin\Phi, \qquad (\Phi_{uv} = \sin\Phi), \tag{11.33}$$

the sine–Gordon equation (11.12) [AN84; Kak91].

It should be noted that the equation of the case $K = -1$,

$$\Phi_{xx} - \Phi_{tt} = -\sin\Phi, \qquad (\Phi_{uv} = -\sin\Phi), \tag{11.34}$$

is another sine–Gordon equation. Obviously, the difference is only the interchange of roles of x and t. In other words, the variable v is replaced with $-v$. This is illustrated by considering the following two particular solutions to (11.33):

$$\tan(\Phi_1/4) = \exp[\zeta u + \zeta^{-1} v], \tag{11.35}$$

$$\tan(\Phi_2/4) = \sqrt{(1-\omega^2)/\omega^2}\,\sin\omega t\ \mathrm{sech}\sqrt{1-\omega^2}x, \tag{11.36}$$

where ζ and ω are real constants. The first one $\Phi_1(u, v)$ is called the *kink* solution, while the second $\Phi_2(t, x)$ is called the *breather* solution [AS81, §1.4]. It is straightforward to see that $\Phi_1(u, v)$ satisfies (11.33), and that Eq. (11.34) is satisfied by $\Phi_1(u, v)$ as well if v is replaced with $-v$. Similarly,

$\Phi_2(t, x)$ satisfies (11.33), and if t and x are interchanged, it satisfies (11.34). Kakuhata [Kak91] considered a dual transformation to the spherical surface of the sigma model, and found that the transformed system is characterized with negative curvatures.

11.3.4. *Spherical and pseudospherical surfaces*

On a surface $\Sigma^2(u^1, u^2)$ in $\mathbb{R}^3$, geodesic curves are described by (2.65), which is reproduced here for $\gamma = 2$:

$$\frac{\mathrm{d}^2 u^2}{\mathrm{d}s^2} + \Gamma^2_{\alpha\beta} \frac{\mathrm{d}u^\alpha}{\mathrm{d}s} \frac{\mathrm{d}u^\beta}{\mathrm{d}s} = 0. \tag{11.37}$$

Suppose that the coordinate curves $u^2 = \text{const}$ are geodesics, then we must have $\Gamma^2_{11} = 0$. When the coordinate curves form an orthogonal net (Appendix K), this reduces to

$$\frac{\partial}{\partial u^2} g_{11} = 0, \quad \text{hence} \quad g_{11} = g_{11}(u^1), \tag{11.38}$$

by the definition of the Christoffel symbols (2.39) and (2.40). Using this property, let us rescale $\mathrm{d}u^1$ as $g_{11}(u^1)\,(\mathrm{d}u^1)^2 \to (\mathrm{d}u^1)^2$.

Therefore, when the surface $\Sigma^2(u^1, u^2)$ is referred to a family of geodesics $u^2 = \text{const}$ and their orthogonal coordinate u^1, the line-element can be written in the form,

$$\mathrm{d}s^2 = (\mathrm{d}u^1)^2 + g_{22}(\mathrm{d}u^2)^2. \tag{11.39}$$

In view of $g_{11} = 1$ and $g_{12} = 0$, the formula (2.63) for the Gaussian curvature K reduces to

$$\left(\frac{\partial}{\partial u^1}\right)^2 \sqrt{g_{22}} = -K\sqrt{g_{22}}. \tag{11.40}$$

In the case of sperical surfaces of $K = 1/a^2$, this gives

$$\sqrt{g_{22}} = \varphi(u^2)\cos(u^1/a) + \psi(u^2)\sin(u^1/a).$$

Choosing $\psi(u^2) = 0$, and rescaling u^2 suitably this time, we may write the line-element

$$\mathrm{d}s^2 = (\mathrm{d}u^1)^2 + c^2\cos^2(u^1/a)(\mathrm{d}u^2)^2. \tag{11.41}$$

Although all spherical surfaces have the same intrinsic properties, there is a distinction among them, when viewed from the envelope space $\mathbb{R}^3$. The same applies to pseudospherical surfaces of $K = -1/a^2$ as well.

This is seen when we consider surfaces of revolution in $\mathbb{R}^3$ corresponding to spherical or pseudospherical surfaces. Taking the x^3-axis for the axis of revolution, the transformation is defined by

$$x^1 = u\cos v, \qquad x^2 = u\sin v, \qquad x^3 = \varphi(u), \tag{11.42}$$

[Eis47, §49], where the line-element in this case is given by

$$\mathrm{d}s^2 = (1 + \varphi'(u)^2)(\mathrm{d}u)^2 + u^2(\mathrm{d}v)^2,$$

and the function $\varphi(u)$ is determined by equating this metric with (11.41). By Eq. (11.38), the curve $v = \text{const}$ is a geodesic. Hence the contour $x^3 = \varphi(x^1)$ is a geodesic since it is given by $v = 0$. (Appendix K)

In this transformation, generic cases are represented by periodic sequence of a *zonal* surface of revolution along the x^3-axis in both cases of a spherical (Fig.11.1(a), an elliptic type) and a pseudospherical (Fig.11.2(a), a hyperbolic type) surface [Eis47, §49], where the contours $x^3 = \varphi(x^1)$ represent geodesic curves.

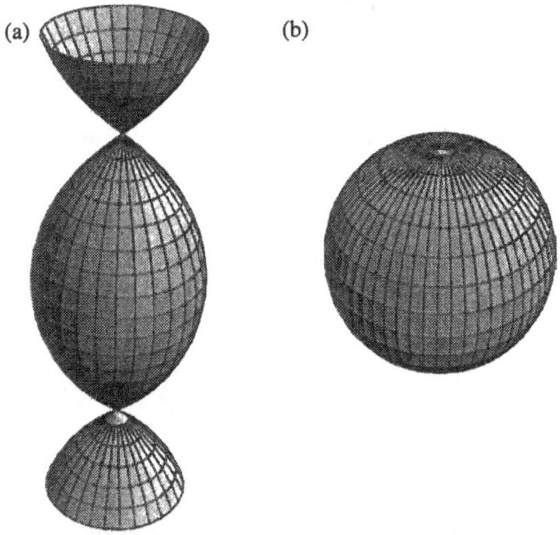

Fig. 11.1. Spherical surfaces of revolution (drawn by *Mathematica*): (a) An elliptic type, where $u = c\cos(u^1/a)$ and $\varphi(u) = \int[1 - (c/a)^2 \sin^2(u^1/a)]^{1/2}\mathrm{d}u^1$; (b) a sphere.

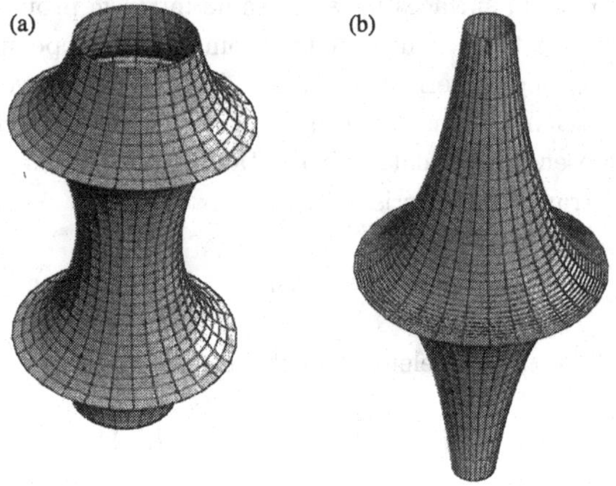

Fig. 11.2. Pseudospherical surfaces of revolution (drawn by *Mathematica*): (a) A hyperbolic type where $u = c\cos(u^1/a)$ and $\varphi(u) = \int [1-(c/a)^2\sinh^2(u^1/a)]^{1/2}\mathrm{d}u^1$; (b) a parabolic type, where $u = c\sin\theta$ and $\varphi(\theta) = a[\cos\theta - \log(\sin^{-1}\theta + \cot\theta)]$ with $a = c = 1$.

A particular case in the spherical surface is a sphere (Fig.11.1(b)), while another particular case in the pseudospherical surface is a surface of *parabolic* type extending to infinity along x_3-axis (Fig.11.2(b)).

11.4. Bäcklund Transformations Revisited

11.4.1. *A Bäcklund transformation*

The geometrical properties of a *pseudospherical* surface provide a systematic method to obtain Bäcklund transformations [CheT86]. Let M^2 be a surface endowed with a Riemannian metric. Consider a local frame field (e_1, e_2) and its dual co-frame (σ^1, σ^2) with ϖ as the connection form. Then the structure equations are given by (11.6), (11.7) and (11.9) with $K = -1$.

Theorem. *Suppose that we have Eqs.* (11.6), (11.7) *and* (11.9) *with* $K = -1$, *i.e. the structure equations describe a pseudospherical surface. Then, the following 1-form equation,*

$$\mathrm{d}w = \varpi + \sigma^1 \sin w - \sigma^2 \cos w, \tag{11.43}$$

is integrable, i.e. $\mathrm{d}^2 w = 0$. [Propositions 4.2, 4.3 of [CheT86]].

Proof. Taking the external differential, we have

$$\begin{aligned}\mathrm{d}^2 w = \mathrm{d}\varpi + \sin w\, \mathrm{d}\sigma^1 - \cos w\, \mathrm{d}\sigma^2 \\ + \cos w\, \mathrm{d}w \wedge \sigma^1 + \sin w\, \mathrm{d}w \wedge \sigma^2.\end{aligned}$$

Substituting (11.6), (11.7), (11.9) and (11.43) to eliminate $\mathrm{d}\sigma^1, \mathrm{d}\sigma^2, \mathrm{d}\varpi$ and $\mathrm{d}w$, we obtain

$$\mathrm{d}^2 w = (K + \cos^2 w + \sin^2 w)\sigma^1 \wedge \sigma^2 = 0, \quad \text{if} \quad K = -1. \tag{11.44}$$

□

Corollary. *Suppose that we have Eqs.* (11.6), (11.7) *and* (11.9) *with* $K = 1$, *i.e. the structure equations describe a spherical surface. Then, the following 1-form equation,*

$$\mathrm{d}w = \varpi + i\sigma^1 \sin w - i\sigma^2 \cos w, \tag{11.45}$$

is integrable, i.e. $\mathrm{d}^2 w = 0$.

In this case, Eq. (11.44) is replaced by

$$\begin{aligned}\mathrm{d}^2 w = (K - \cos^2 w - \sin^2 w)\sigma^1 \wedge \sigma^2 = 0, \\ \text{if} \quad K = 1.\end{aligned}$$

□

Example. *Sine–Gordon equation* (11.12) with $K = -1$. □

The corresponding 1-forms are given by (11.11). Then, Eq. (11.43) reduces to

$$\begin{aligned}\mathrm{d}w &= w_x \mathrm{d}x + w_t \mathrm{d}t \\ &= [u_x + \zeta \cos w]\mathrm{d}x + \zeta^{-1} \cos(u - w)\mathrm{d}t.\end{aligned} \tag{11.46}$$

This is equivalent to

$$\left.\begin{aligned} w_x - u_x &= \zeta \cos w, \\ w_t &= \zeta^{-1} \cos(u - w).\end{aligned}\right\} \tag{11.47}$$

Taking differential of (11.46) and using (11.47), we obtain

$$\mathrm{d}^2 w = [u_{xt} - \sin u]\mathrm{d}t \wedge \mathrm{d}x.$$

Thus, if the sine–Gordon equation,

$$u_{xt} = \sin u, \tag{11.48}$$

is satisfied, then dw is integrable. In fact, from (11.47), we obtain

$$w_{xt} = \cos w \sqrt{1 - (\zeta w_t)^2}, \tag{11.49}$$

$$u = w + \cos^{-1}(\zeta w_t). \tag{11.50}$$

The system of equations (11.47) is interpreted as a Bäcklund transformation beween a function $u(x,t)$ satisfying the sine–Gordon equation (11.48) and a function $w(x,t)$ satisfying its associated equation (11.49). *Given a solution u of the sine–Gordon equation* (11.48), *then the system of equations* (11.47) *is integrable and w is a solution of* (11.49) *for each constant ζ*. Conversely, if w is a solution of (11.49) for a constant ζ, then u of (11.50) satisfies (11.48).

11.4.2. *Self-Bäcklund transformation*

A Bäcklund transformation which relates solutions of the same equation is called a *self-Bäcklund transformation.* For the sine–Gordon equation, we observe that Eq. (11.49) is invariant under the transformation: $(w,\zeta) \to (\pi - w, -\zeta)$. If u is a solution of (11.48) and (w,ζ) satisfies (11.47), then (11.50) holds and w is a solution of (11.49). From the preceding considerations together with transformation invariance of (11.49), it follows that U defined by

$$U = \pi - w + \cos^{-1}(\zeta w_t) \tag{11.51}$$

is another solution of (11.48). Eliminating $\cos^{-1}(\zeta w_t)$ between (11.50) and (11.51), we find

$$w = \frac{1}{2}(u - U + \pi).$$

Substituting this into (11.47), we obtain

$$\left.\begin{aligned} (u+U)_x &= 2\zeta \sin\frac{1}{2}(u-U), \\ (u-U)_t &= 2\zeta^{-1} \sin\frac{1}{2}(u+U), \end{aligned}\right\} \tag{11.52}$$

where U satisfies $U_{xt} = \sin U$. This is the self-Bäcklund transformation for the sine–Gordon equation (11.48), and equivalent to (10.10) and (10.11) by the replacement, $(\zeta, u, U) \to (a, z, -Z)$.

This formulation of Bäcklund transformation can be applied to other integrable equations describing pseudospherical surfaces such as the KdV equation, mKdV equation, or sinh–Gordon equation considered in §11.2 [CheT86].

11.5. Immersion of Integrable Surfaces on Lie Groups

This section and the next section are concerned with another geometrical problem of integrable systems where integration surfaces are to be constructed in an envelope space, i.e. an *immersion problem.*

11.5.1. *A surface Σ^2 in $\mathbb{R}^3$*

As an introduction to the present problem, we consider a smooth surface Σ^2 in $\mathbb{R}^3$ according to Chapter 2, i.e. immersion of a domain $D^2 \subset \mathbb{R}^2$ into a three-dimensional euclidean space,

$$F(u,v) = (F_1, F_2, F_3) : D^2 \to \mathbb{R}^3,$$

for $(u,v) \in D^2$. The euclidean metric of $\mathbb{R}^3$ *induces* a certain metric $g_{\alpha\beta}$ on Σ^2. Noting that $\mathrm{d}F = F_u \mathrm{d}u + F_v \mathrm{d}v$, the first fundamental form defined by $\mathrm{I} = \langle \mathrm{d}F, \mathrm{d}F \rangle$ is written as

$$\mathrm{I} = g_{uu}(\mathrm{d}u)^2 + 2g_{uv}\mathrm{d}u\mathrm{d}v + g_{vv}(\mathrm{d}v)^2.$$

Correspondingly, the second fundamental form is denoted by

$$\mathrm{II} = b_{uu}(\mathrm{d}u)^2 + 2b_{uv}\mathrm{d}u\mathrm{d}v + b_{vv}(\mathrm{d}v)^2.$$

The surface Σ^2 is uniquely defined within rigid motions by the first and second fundamental forms (§2.11).

Let $N(u,v)$ be the normal vector field defined at each point on Σ^2. Then the triplet (F_u, F_v, N) defines a basis of a moving frame on Σ^2. According to Chapter 2, the motion of this basis on Σ^2 is characterized by the Gauss–Weingaten equations. The compatibility of these equations are the Gauss–Mainardi–Codazzi equations (2.48) and (2.49), which are coupled nonlinear differential equations for $g_{\alpha\beta}$ and $b_{\alpha\beta}$. In terms of differential forms, this was formulated by structure equations in §2.7 and 3.5, and some relation to integrable equations has been investigated in this chapter.

11.5.2. *Surfaces on Lie groups and Lie algebras*

Regarding integrable equations, an immersion problem of a two-dimensional surface into an envelope space was investigated by Sym [Sym82] first, and later developed systematically by [Bob94; FG96; CFG00]. This has been accomplished by defining surfaces on Lie groups and surfaces on Lie algebras of finite dimensions in general.

As an example of the general formulation [FG96; CFG00], we consider the case that the group is $SU(2)$. We define an $SU(2)$-valued function $\Psi(u, v, \zeta)$ satisfying the Lax pair equations:

$$\Psi_u = U\Psi, \qquad \Psi_v = V\Psi, \tag{11.53}$$

where $U(u,v), V(u,v) \in su(2)$, and ζ is a spectral parameter. In addition, we introduce an $su(2)$-valued function $F(u, v, \zeta)$,

$$F_u = \Psi^{-1} A\Psi, \qquad F_v = \Psi^{-1} B\Psi, \tag{11.54}$$

where $A(u,v), B(u,v) \in su(2)$. The functions U, V, A, B are all differentiable functions of u, v in some neighborhood of $\mathbb{R}^2$, and $\zeta \in \mathbb{C}$. This may be interpreted as follows. The Ψ_u and Ψ_v are tangent vectors at a point $\Psi \in SU(2)$, while F_u and F_v are vectors in the tangent space at the identity of $SU(2)$, i.e. the Lie algebra. The tangents F_u and F_v are vectors pulled back from the tangents $A\Psi$ and $B\Psi$ (at a point Ψ) respectively.

The compatibility condition for (11.53), i.e. $\partial_v \Psi_u = \Psi_{uv} = \partial_u \Psi_v$, results in

$$U_v - V_u + [U, V] = 0. \tag{11.55}$$

where $[U, V] = UV - VU$. Equations (11.53) and (11.55) are equivalent to the Lax pair and its solvability (11.1) and (11.2) respectively, and define a two-dimensional surface $\Psi(u, v) \in SU(2)$. The compatibility condition of (11.54) reduces to

$$A_v - B_u + [A, V] + [U, B] = 0, \tag{11.56}$$

since $\Psi\partial(\Psi^{-1}) = -(\partial\Psi)\Psi^{-1}$ from $\Psi\Psi^{-1} = I$.

Then, for each ζ, the $su(2)$-valued function $F(u,v,\zeta)$ defines a two-dimensional surface $\boldsymbol{x}(u,v)$ in $\mathbb{R}^3$:

$$\boldsymbol{x}(u,v) = (F^1, F^2, F^3), \tag{11.57}$$
$$F = F^1 e_1 + F^2 e_2 + F^3 e_3, \tag{11.58}$$

where (e_1, e_2, e_3) is an orthonormal basis defined by (11.19), which are related to the Pauli matrices (7.23) by $\sigma_k = 2ie_k$, and satisfy the commutation relations,

$$[e_j, e_k] = \epsilon_{jkl} e_l. \tag{11.59}$$

We define the inner product by

$$\langle A, B\rangle = -2\ \mathrm{Tr}(AB), \qquad A, B \in su(2). \tag{11.60}$$

Then we have the orthonomal property, $\langle e_k, e_l\rangle = \delta_{kl}$.

The first and second fundamental forms of the surface $\boldsymbol{x}(u,v)$ are given by

$$\mathrm{I} = \langle A, A\rangle(\mathrm{d}u)^2 + 2\langle A, B\rangle \mathrm{d}u\mathrm{d}v + \langle B, B\rangle(\mathrm{d}v)^2, \tag{11.61}$$
$$\begin{aligned}\mathrm{II} = {}& \langle A_u + [A,U], C\rangle(\mathrm{d}u)^2 + 2\langle A_v + [A,V], C\rangle \mathrm{d}u\mathrm{d}v \\ & + \langle B_v + [B,V], C\rangle(\mathrm{d}v)^2,\end{aligned} \tag{11.62}$$
$$C = [A,B]/\|[A,B]\|, \qquad \|A\| = \langle A, A\rangle^{1/2} \tag{11.63}$$

[FG96]. In fact, the first fundamental form (11.61) is obtained by noting that

$$\mathrm{I} = \langle \mathrm{d}\boldsymbol{x}, \mathrm{d}\boldsymbol{x}\rangle, \qquad \mathrm{d}\boldsymbol{x} = F_u \mathrm{d}u + F_v \mathrm{d}v$$

(§2.1), and using (11.54), (11.57), (11.58) and (11.60). According to §2.2, the coefficient $b_{\alpha\beta}$ of the second fundamental form is given by (2.23): $b_{\alpha\beta} = \langle \boldsymbol{x}_{\alpha\beta}, \boldsymbol{N}\rangle$, where, e.g. for $\alpha = \beta = u$,

$$\begin{aligned}\boldsymbol{x}_{uu} &= \partial_u(\Psi^{-1}A\Psi) = \Psi^{-1}(A_u + [A,U])\Psi, \\ \boldsymbol{N} &= \frac{[F_u, F_v]}{\|[F_u, F_v]\|} = \frac{\Psi^{-1}[A,B]\Psi}{\|[A,B]\|}.\end{aligned} \tag{11.64}$$

A moving frame on this surface $\boldsymbol{x}(u,v)$ is

$$\Psi^{-1}A\Psi(= F_u), \qquad \Psi^{-1}B\Psi(= F_v), \qquad \Psi^{-1}C\Psi(= \boldsymbol{N}).$$

The Gauss curvature K is given by

$$K = \frac{\det(b_{\alpha\beta})}{\det(g_{\alpha\beta})}. \tag{11.65}$$

In terms of U and V satisfying (11.55), the functions A, B and the immersion function F are given explicitly [FG96] as:

$$A = \mu\partial_\zeta U + [R, U] + \partial_u(fU) + g_u V + g\partial_v U, \tag{11.66}$$

$$B = \mu\partial_\zeta V + [R, V] + \partial_v(gV) + f_v U + f\partial_u V, \tag{11.67}$$

$$F = \Psi^{-1}[\mu\partial_\zeta + R + fU + gV]\Psi, \tag{11.68}$$

where $\mu, f(u,v), g(u,v)$ are scalar functions depending on ζ, and R is a constant $su(2)$-valued matrix. A, B, F are augmented with an additional term associated with a symmetry of the integrable system [CFG00]. The first two terms are illustrated below in §11.5.3 and 11.6, term by term.

11.5.3. *Nonlinear Schrödinger surfaces*

We consider one example of such surfaces described by the nonlinear Schrödinger equation (NLS). The connection 1-form of NLS is given by Ω_{NLS} of (11.17). The $su(2)$ functions U and V of (11.53) are defined by the relation, $\Omega_{\mathrm{NLS}} = U\,dx + V\,dt$, where $u = x$ and $t = v$. Thus, we have

$$U = \begin{pmatrix} -i\zeta & q(x,t) \\ -q^*(x,t) & i\zeta \end{pmatrix} = -2q^{(i)}e_1 - 2q^{(r)}e_2 + 2\zeta e_3,$$

$$V = \begin{pmatrix} -i\Lambda & B \\ -B^* & i\Lambda \end{pmatrix} = -2B^{(i)}e_1 - 2B^{(r)}e_2 + 2\Lambda e_3$$

where $q(x,t)$ is a complex function to be determined, and $\Lambda(x,t), B(x,t)$ are defined by (11.18), with ζ as a real parameter. The integrability condition (11.55) results in the NLS equation:

$$q_t = i(q_{xx} + 2q|q|^2). \tag{11.69}$$

Next, let A and B be defined by the first terms of (11.66) and (11.67):

$$A = \frac{1}{2}\mu\frac{\partial U}{\partial\zeta} = \frac{1}{2}\mu\begin{pmatrix} -i & 0 \\ 0 & i \end{pmatrix} = \mu e_3, \tag{11.70}$$

$$\begin{aligned} B &= \frac{1}{2}\mu\frac{\partial V}{\partial\zeta} = \frac{1}{2}\mu\begin{pmatrix} -4i\zeta & 2q \\ -2q^* & 4i\zeta \end{pmatrix} \\ &= -2\mu q^{(i)}e_1 - 2\mu q^{(r)}e_2 + 4\mu\zeta e_3 = \mu U + 2\mu\zeta e_3. \end{aligned} \tag{11.71}$$

The integrability condition (11.56) reduces to a form obtained by differentiation of (11.55) with respect to ζ. Therefore, (11.56) is satisfied. Thus, we have a integration surface $\boldsymbol{x}(x,t)$.

The first and second fundamental forms of the surface $\boldsymbol{x}(x,t)$ are given by

$$\begin{aligned} \mathrm{I} &= \mu^2\left[(\mathrm{d}x + 4\zeta\,\mathrm{d}t)^2 + 4|q|^2(\mathrm{d}t)^2\right], \\ \mathrm{II} &= 2\mu|q|[\mathrm{d}x + (\phi_x - 2\zeta)\mathrm{d}t]^2 - 2\mu|q|_{xx}\mathrm{d}t^2 \end{aligned} \tag{11.72}$$

where, setting $q = |q|\exp[i\phi]$, $|q|$ and ϕ must satisfy

$$|q|\phi_t = |q|_{xx} - |q|\phi_x^2 + 2|q|^3, \qquad |q|_t = -|q|\phi_{xx} - 2|q|_x\phi_x, \tag{11.73}$$

by Eq. (11.69).

The definition (11.65) of the Gaussian curvature K leads to

$$K = \frac{\det(b_{\alpha\beta})}{\det(g_{\alpha\beta})} = -\frac{|q|_{xx}}{\mu^2|q|} = \frac{1}{\mu^2}(2|q|^2 - \phi_x^2 - \phi_t) \tag{11.74}$$

from the above expressions [CFG00]. It is seen that K can take both positive and negative values.

The immersion function F is given by

$$F = \frac{1}{2}\mu\Psi^{-1}(\partial\Psi/\partial\zeta).$$

This is obtained since

$$\begin{aligned} F_x &= \frac{1}{2}\mu[(\Psi^{-1})_x\Psi_\zeta + \Psi^{-1}(\partial_x\Psi)_\zeta] \\ &= \frac{1}{2}\mu[-\Psi^{-1}U\Psi_\zeta + \Psi^{-1}(U\Psi)_\zeta] = \frac{1}{2}\mu\Psi^{-1}U_\zeta\Psi = \Psi^{-1}A\Psi, \end{aligned}$$

where $(\Psi^{-1})_x = -\Psi^{-1}\Psi_x\Psi^{-1} = -\Psi^{-1}U$, and similarly $(\Psi^{-1})_t = -\Psi^{-1}V$.

Let us consider a particular case of the surface $\boldsymbol{x}(x,t)$. When the parameter $\zeta = 0$, we have

$$\mathrm{I} = \mu^2\left[(\mathrm{d}x)^2 + 4|q|^2(\mathrm{d}t)^2\right],$$

from the first fundamental form (11.72) of the surface $\boldsymbol{x}(x,t)$. In the case when $\phi = -ct$ (c: a positive constant), the amplitude $|q|$ is a function of x only: $|q| = f(x)$, and satisfies $f''(x) = -2f^3 - cf$ by (11.73). According to §11.3.4, it is therefore seen that the curves of t = const are *geodesic*, and we obtain

$$K = (2f^2 + c)/\mu^2 \ (> 0)$$

from (11.74), i.e. positive Gaussian curvature.

This case is analogous to the surface of revolution considered in §11.3.4, and the variable $t = -\phi/c$ in the present problem corresponds to v there. The corresponding surface $\boldsymbol{x}(x,t)$ (obtained by [CFG00, Fig.1]) looks similar to the spherical surface of Fig.11.1(a). This is based on the similarity between the transformation (11.42) and the definitions of tangent vectors A and B by (11.70) and (11.71), the latter being

$$B = -2\mu|q|(\sin\phi)e_1 - 2\mu|q|(\cos\phi)e_2, \qquad A = \mu e_3.$$

By Eq. (11.38), the curve $\phi = \text{const}$ (equivalently $t = \text{const}$) is a geodesic. Hence the contours in the plane (e_1, e_3) defined by $\phi = \pi/2$ and $3\pi/2$ are geodesics.

11.6. Mapping of Integrable Systems to Spherical Surfaces

Remarkably, every integrable equation having a Lax pair induces a map to the surface of a sphere [CFG00]. We consider this problem according to the formulation of the previous section. An $SU(2)$-valued function $\Psi(u,v)$ and an $su(2)$-valued function $F(u,v)$ are defined by (11.53) and (11.54):

$$\Psi_u = U\Psi, \quad \Psi_v = V\Psi; \qquad F_u = \Psi^{-1}A\Psi, \quad F_v = \Psi^{-1}B\Psi,$$

where U and V correspond to the Lax pair in the integrable system. We assume that the integrability condition (11.55) is satisfied:

$$U_v - V_u + [U,V] = 0. \tag{11.75}$$

This determines a differential equation which is claimed to be as above, according to the general theory studied in this chapter.

Now, let us take a *constant* $su(2)$ matrix R, and assume that the $su(2)$-valued functions U and A, and V and B, are connected by

$$A = [R,U], \qquad B = [R,V], \tag{11.76}$$

which correspond to the second terms of (11.66) and (11.67). This is understood as an $su(2)$-rotation by R. These transformations satisfy the compatibility condition (11.56). In fact, one can immediately show the following equalities:

$$A_v - B_u = [R, U_v] - [R, V_u] = -[R,[U,V]]$$
$$[A,V] + [U,B] = [[R,U],V] + [U,[R,V]] = [R,[U,V]]$$

where (11.75) is used in the first equation, wheras in the second equation the Jacobi identity of the triplet $\{R, U, V\}$ is used as the second equality. Thus, it is seen that the condition (11.56) is satisfied. So that, one can expect an integration surface $\boldsymbol{x}(u, v)$ in the three-dimensional $su(2)$-space.

Let us represent the matrix functions R, U, V with reference to the orthonormal basis (e_1, e_2, e_3) defined by (11.19) as

$$R = R^j e_j, \quad U = U^j e_j, \quad V = V^j e_j,$$

where R^j are constants, and U^j, V^j are scalar functions $(j = 1, 2, 3)$. Corresponding 3-vectors in $\mathbb{R}^3$ are written as $\hat{R} = (R^j)$, $\hat{U} = (U^j)$ and $\hat{V} = (V^j)$. Furthermore, writing $A = A^j e_j$ and $B = B^j e_j$, Eqs. (11.76) are rewritten as

$$\hat{A} = \hat{R} \times \hat{U}, \qquad \hat{B} = \hat{R} \times \hat{V},$$

on account of the commutation relation (11.59).[5] This means that the 3-vectors $\hat{A} = (A^j)$ and $\hat{B} = (B^j)$ are determined from $\hat{U}$ and $\hat{V}$ by a rotation with the constant vector $\hat{R}$.

A two-dimensional surface $\boldsymbol{x}(u, v)$ is defined in the three-dimensional $su(2)$-manifold. The tangent space spanned by the tangents F_u and F_v (associated with A and B) is characterized by the first and second fundamental tensors defined by (11.61) and (11.62). Its Gaussian curvature K is given by

$$K = \frac{1}{\|R\|^2}, \tag{11.77}$$

where $\|R\|^2 = \langle R, R \rangle = \hat{R} \cdot \hat{R}$ is a constant. Namely, the surface is regarded as a *spherical surface.* This is verified as follows.

Using the above relations, it is readily shown that

$$[A, B] = kR, \qquad k = \hat{R} \cdot (\hat{U} \times \hat{V}). \tag{11.78}$$

The $su(2)$ matrix C defined by (11.63) is given by

$$C = \frac{[A, B]}{\|[A, B]\|} = \varepsilon \frac{R}{\|R\|}, \qquad \varepsilon = \frac{k}{|k|} = \pm 1.$$

It can be shown that A_u, A_v and B_v are orthogonal to R. Hence,

$$\langle A_u, C \rangle = 0, \qquad \langle A_v, C \rangle = 0, \qquad \langle B_v, C \rangle = 0.$$

[5]The symbols $\times$ and $\cdot$ are the external and inner products of 3-vectors, respectively.

Using these relations, the second fundamental form of (11.62) is

$$\begin{aligned}\mathrm{II} &= \langle [A,U],C\rangle (\mathrm{d}u)^2 + 2\langle [A,V],C\rangle \mathrm{d}u\mathrm{d}v + \langle [B,V],C\rangle (\mathrm{d}v)^2 \\ &= \frac{\varepsilon}{\|R\|}\left[\langle [A,U],R\rangle (\mathrm{d}u)^2 + 2\langle [A,V],R\rangle \mathrm{d}u\mathrm{d}v + \langle [B,V],R\rangle (\mathrm{d}v)^2\right] \\ &= b_{uu}(\mathrm{d}u)^2 + 2b_{uv}\mathrm{d}u\mathrm{d}v + b_{vv}(\mathrm{d}v)^2\end{aligned}$$

where $b_{\alpha\beta} = -(\varepsilon/\|R\|)g_{\alpha\beta}$, and the first fundamental tensors are

$$\begin{aligned}g_{uu} &= \langle A,[R,U]\rangle = (\hat{U}\cdot\hat{U})\hat{R}^2 - (\hat{U}\cdot\hat{R})(\hat{U}\cdot\hat{R}), \\ g_{uv} &= \langle A,[R,V]\rangle = (\hat{U}\cdot\hat{V})\hat{R}^2 - (\hat{U}\cdot\hat{R})(\hat{V}\cdot\hat{R}), \\ g_{vv} &= \langle B,[R,V]\rangle = (\hat{V}\cdot\hat{V})\hat{R}^2 - (\hat{V}\cdot\hat{R})(\hat{V}\cdot\hat{R})\end{aligned}$$

from (11.76) and (11.61).

Thus, the definition (11.65) of the Gauss curvature K leads to

$$K = \frac{\det(b_{\alpha\beta})}{\det(g_{\alpha\beta})} = \left(\frac{\varepsilon}{\|R\|}\right)^2 \frac{\det(g_{\alpha\beta})}{\det(g_{\alpha\beta})} = \frac{1}{\|R\|^2}.$$

This verifies (11.77).

Furthermore, the immersion surface is given by

$$F = \Psi^{-1}R\Psi. \tag{11.79}$$

In fact, we have

$$F_u = \Psi^{-1}A\Psi = \Psi^{-1}(RU - UR)\Psi = \Psi^{-1}R\Psi_u + (\Psi^{-1})_u R\Psi = (\Psi^{-1}R\Psi)_u,$$

since $(\Psi^{-1})_u = -\Psi^{-1}\Psi_u\Psi^{-1} = -\Psi^{-1}U$, and $R_u = 0$. Similarly, $F_v = (\Psi^{-1}R\Psi)_v$. Thus, we have (11.79).

The above result implies the following. Variations of the parameters u and v are associated with the evolution of an integrable system. The point Ψ in the $SU(2)$ space translates according to the variations. The tangent space $T_\Psi\, SU(2)$ at Ψ is pulled back to the tangent space $T_{\mathrm{id}}SU(2)$ at the identity, i.e. the Lie algebra $su(2)$ which is three-dimensional. According to the motion of u and v, the function $F(u,v)$ describes a surface in the space $su(2)$. If the map is characterized by a constant $su(2)$-rotation R, the surface $F(u,v)$ is a *spherical surface* of Gaussian curvature $1/\|R\|^2$.

Thus, it has been found that *the evolution of an integrable system describes a spherical surface.* This is a most *impressive* characterization of an integrable system.

Appendix A

Topological Space and Mappings

Some basic mathematical notions and definitions of topology and mappings are presented to help the main text.
(Ref. §1.2)

A.1. Topology

A manifold is a topological space. A *topological space* is a set M with a collection of subsets called *open* sets. An example of open sets is a *ball* in the euclidean space $\mathbb{R}^n$ defined by

$$B_a(\epsilon) = \{x \in \mathbb{R}^n |\ \|x - a\| < \epsilon, a \in \mathbb{R}^n,\ \epsilon\ (> 0) \in \mathbb{R}\},$$

where $\|\cdot\|$ is the euclidean norm. As a generalization of balls, the open sets are defined to satisfy the following:

(i) If U and V are open, so is their intersection $U \cap V$.
(ii) The union of any collection of open sets (possibly infinite in number) is open.
(iii) The empty set is open.
(iv) The topological space M is open, that is a generalization of the entire $\mathbb{R}^n$ which is open.

Just as the topology of $\mathbb{R}^n$ is said to be induced by the euclidean norm $\|\cdot\|$, the topology of M is defined by the open subsets. A subset of M is said to be *closed* if its complement is open.

A.2. Mappings

A *map* F from a space U to a space V, $F : U \to V$, is a rule by which, for every element x of U, a *unique* element y of V is associated with $y = F(x)$, that is, $F : x \mapsto y$. For example, a real-valued function f of n real variables is represented as $f : \mathbb{R}^n \to \mathbb{R}$. Note that, for every $y = F(x) \in V$, x is not necessarily unique, and such a map is called *many-to-one*. For any subset U_* in U, the elements $F(x)$ in V mapped from $x \in U_*$ form a set V_* called the *image* of U_* under F. The image is denoted by $F(U_*)$. Conversely, the set U_* is called the *inverse image* of V_*, denoted by $F^{-1}(V_*)$. If the map is many-to-one in the sense given above, the correspondence F^{-1} from V to U is not called a map since every *map* is defined to have a unique image.

If every point in $F(U_*)$ has a unique inverse image in U_*, then the map F is said to be one-to-one, or simply 1-1. In this case, two different preimages $x_1, x_2 \in U$ have two different images $F(x_1) \neq F(x_2)$ in V. Then the operation F^{-1} is another 1-1 map, called the *inverse map* of F. The one-to-one map is also called an *injection*. If a map $F : U \to V$ has the property $V = F(U)$, then F is said to be an *onto-mapping* or a *surjection*. A map which is both 1-1 and onto is called a *bijection*.

Let us define two maps as $F : U \to V$, and $G : V \to W$. The result of the two successive maps is a *composition map*, denoted by $G \circ F$, that is, $G \circ F : U \to W$. This is understood as follows. For a point $x \in U$, we obtain $F(x) \in V$, from which we obtain a point $G(F(x)) = G \circ F(x)$ in W.

Suppose that we have a map $F : U \to V$ for two topological spaces U and V. The function $F : x \mapsto F(x)$ is said to be *continuous* at x if any open set of V containing $F(x)$ contains the image of an open set of U containing x.

A *homeomorphism* F takes an open set M into an open set N in the following sense. Namely, $F : M \to N$ is one-to-one and onto (thus the inverse map $F^{-1} : N \to M$ exists). In addition, both F and F^{-1} are continuous.

Appendix B

Exterior Forms, Products and Differentials

In physics and engineering, we always encounter integrals along a line, over an area, or over a volume. Usually, the integrands are represented in exterior differential forms (see (B.28), (B.30), (B.33)). Here is a brief account of the exterior algebra. See [Arn78; Fra97] for more details.

(Ref. §1.5, 1.6, 2.2, 2.7.1, 7.6.3, 7.11.3, 8.2.3)

B.1. Exterior Forms

Another name of a covector ω^1 is a 1-form. The 1-form ω^1 is a linear function $\omega^1(v)$ of a vector $v \in E = \mathbb{R}^n$, i.e. $\omega^1 : \mathbb{R}^n \to \mathbb{R}$, such that

$$\omega^1(c_1 v_1 + c_2 v_2) = c_1 \omega^1(v_1) + c_2 \omega^1(v_2), \tag{B.1}$$

where $c_1, c_2 \in \mathbb{R}$ and $v_1, v_2 \in \mathbb{R}^n$. The collection of all 1-forms on $E = \mathbb{R}^n$ constitutes an n-dimensional vector space dual to the vector space E and called the *dual* space E^*.

Similarly, a 2-form ω^2 is defined as a function on pairs of vectors $\omega^2(v_1, v_2) : E \times E \to \mathbb{R}$, which is bilinear and skew-symmetric with respect to two vectors v_1 and v_2:

$$\omega^2(c_1 v_1' + c_2 v_1'', v_2) = c_1 \omega^2(v_1', v_2) + c_2 \omega^2(v_1'', v_2),$$
$$\omega^2(v_1, v_2) = -\omega^2(v_2, v_1),$$

where $v_1', v_1'' \in \mathbb{R}^n$. From the second property, $\omega^2(v, v) = 0$.

Consider a uniform fluid flow of a constant velocity, $\boldsymbol{U} = (U^1, U^2, U^3) = U^1 \boldsymbol{e}_1 + U^2 \boldsymbol{e}_2 + U^3 \boldsymbol{e}_3$, in a three-dimensional euclidean space with the cartesian bases $\boldsymbol{e}_1, \boldsymbol{e}_2, \boldsymbol{e}_3$. An example of 2-form is the flux F^2 of the fluid through an area $S^2(v,w) = v \times w$ of a parallelogram spanned by $v = (v^1, v^2, v^3)$ and $w = (w^1, w^2, w^3)$. In fact, using the vector analysis in the euclidean space, the flux F^2 is given by

$$F^2 = \boldsymbol{U} \cdot S^2 = U^1 S^2_{23} + U^2 S^2_{31} + U^3 S^2_{12}, \tag{B.2}$$

where $S^2(v,w) = (S^2_{23}, S^2_{31}, S^2_{12})$ with S^2_{ij} defined by (B.8) below, and S^2 is an area 2-form. The flux $F^2(v,w)$ is bilinear and skew-symmetric with respect to v and w, as easily confirmed.

The collection of all 2-forms on $E \times E$ becomes a vector space if laws of addition and multiplication (by a scalar) are introduced appropriately. The space is denoted by $E^*(2) = E^* \wedge E^*$, whose dimension is $\dim(E^*(2)) = \binom{n}{2} = \frac{1}{2}n(n-1)$ (see (B.6)). For $n = 3$, the dimension is 3.

A 0-form ω^0, is specially defined as a scalar. The set of all 0-forms is real numbers $\mathbb{R}$, whose dimension is one. A 0-form field on a manifold M^n, $[f : x \in M^n \to \mathbb{R}]$ is a differentiable function $f(x)$.

An *exterior form* of degree k, i.e. k-form, is a function of k vectors $\omega^k(v_1, \ldots, v_k) : E \times \cdots \times E$ (k-product) $\to \mathbb{R}$. The k-form ω^k is k-linear and skew-symmetric:

$$\omega^k(c_1 v_1' + c_2 v_1'', v_2, \ldots, v_k) = c_1 \omega^k(v_1', v_2, \ldots, v_k) + c_2 \omega^k(v_1'', v_2, \ldots, v_k), \tag{B.3}$$

$$\omega^k(v_{a_1}, \ldots, v_{a_k}) = (-1)^\sigma \omega^k(v_1, \ldots, v_k), \tag{B.4}$$

where $\sigma = 0$ if the permutation $(a_1, \ldots, a_k)$ with respect to $(1, \ldots, k)$ is even, and $\sigma = 1$ if it is odd, and

$$v_a = v_a^j \partial_j, \quad j = [1, \ldots, n]; \;\; a = 1, \ldots, k. \tag{B.5}$$

If the same vector appears in two different entries, the value of ω^k is zero. Therefore, $\omega^k = 0$ if $k > n$.

The set of all k-forms becomes a vector space if addition and multiplication (by a scalar) are defined as

$$(\omega_1^k + \omega_2^k)(\boldsymbol{v}) = \omega_1^k(\boldsymbol{v}) + \omega_2^k(\boldsymbol{v}),$$
$$(\lambda \omega^k)(\boldsymbol{v}) = \lambda \omega^k(\boldsymbol{v}),$$

where $\boldsymbol{v} = \{v_1, \ldots, v_k\}$. Using (B.5) and the two laws (B.3) and (B.4), we have

$$\omega^k(\boldsymbol{v}) = v_{a_1}^{j_1} \cdots v_{a_k}^{j_k} \omega^k(\partial_{j_1}, \ldots, \partial_{j_k}),$$

where $j_1 < \cdots < \partial_{j_k}$. The number of distinct k-combinations such as $(j_1, \cdots, j_k)$ from $(1, \cdots, n)$ gives the dimension of the vector space of all k-forms, $E^*(k) = E^* \wedge \cdots \wedge E^*$ (k-product), which is

$$\dim(E^*(k)) = \binom{n}{k} = \frac{n!}{k!(n-k)!}. \tag{B.6}$$

B.2. Exterior Products (Multiplications)

We now introduce an exterior product of two 1-forms, which associates to every pair $(\omega_\alpha^1, \omega_\beta^1)$ of 1-forms on E a 2-form on $E \times E$. The exterior multiplication $\omega_\alpha^1 \wedge \omega_\beta^1$ is defined by

$$\omega_\alpha^1 \wedge \omega_\beta^1(v_a, v_b) = \omega_\alpha^1(v_a)\omega_\beta^1(v_b) - \omega_\beta^1(v_a)\omega_\alpha^1(v_b) \tag{B.7}$$

where $\omega_\alpha^1(v_a)$ is a linear function of v_a, etc. The right-hand side is obviously bilinear with respect to v_a and v_b and skew-symmetric. For example, if ω_α^1 and ω_β^1 are *differential 1-forms*, defined by $\omega_\alpha^1 = \mathrm{d}x^i$ and $\omega_\beta^1 = \mathrm{d}x^j$, then we have

$$\begin{aligned} \mathrm{d}x^i \wedge \mathrm{d}x^j(v, w) &= \mathrm{d}x^i(v)\mathrm{d}x^j(w) - \mathrm{d}x^j(v)\mathrm{d}x^i(w) \\ &= v^i w^j - v^j w^i = \begin{vmatrix} v^i & w^i \\ v^j & w^j \end{vmatrix} := S_{ij}^2, \end{aligned} \tag{B.8}$$

since $\mathrm{d}x^i(v) = v^i$, etc. (see (1.23)). This S_{ij}^2 is a projected area of the parallelogram spanned by two vectors v and w in the space $\mathbb{R}^n$ onto the (x^i, x^j)-plane.

A general *differential* k-form on $E \times \cdots \times E$ (k-product) can be written in the form,

$$\omega^k = \sum_{i_1 < \cdots < i_k} a_{i_1 \cdots i_k} \, \mathrm{d}x^{i_1} \wedge \cdots \wedge \mathrm{d}x^{i_k}, \quad i_1, \ldots, i_k \in [1, \ldots, n].$$

If the set $v_1, \ldots, v_k$ is a k-tuple vector, then

$$\mathrm{d}x^1 \wedge \cdots \wedge \mathrm{d}x^k(v_1, \ldots, v_k) = \det[\mathrm{d}x^i(v_j)] = \det[v_j^i]. \tag{B.9}$$

Definition of exterior multiplication: The exterior multiplication of an arbitrary k-form ω^k by an arbitrary l-form ω^l is a $(k+l)$-form, and satisfies the following properties:

$$\textit{skew-commutative:}\quad \omega^k \wedge \omega^l = (-1)^{kl}\omega^l \wedge \omega^k, \tag{B.10}$$

$$\textit{associative:}\quad (\omega^k \wedge \omega^l) \wedge \omega^m = \omega^k \wedge (\omega^l \wedge \omega^m), \tag{B.11}$$

$$\textit{distributive:}\quad (c_1\omega_1^k + c_2\omega_2^k) \wedge \omega^l = c_1\omega_1^k \wedge \omega^l + c_2\omega_2^k \wedge \omega^l. \tag{B.12}$$

Example (i): An exterior product of two 1-forms α^1 and β^1,

$$\alpha^1 = a_1 \mathrm{d}x + a_2 \mathrm{d}y + a_3 \mathrm{d}z, \quad \beta^1 = b_1 \mathrm{d}x + b_2 \mathrm{d}y + b_3 \mathrm{d}z,$$

(in the space (x, y, z)), is a 2-form:

$$\begin{aligned}\alpha^1 \wedge \beta^1 = (a_2b_3 - a_3b_2)\mathrm{d}y \wedge \mathrm{d}z + (a_3b_1 - a_1b_3)\mathrm{d}z \wedge \mathrm{d}x \\ + (a_1b_2 - a_2b_1)\mathrm{d}x \wedge \mathrm{d}y,\end{aligned} \tag{B.13}$$

which represents the cross product. From (B.10), we obtain

$$\mathrm{d}x \wedge \mathrm{d}x = 0, \quad \mathrm{d}x \wedge \mathrm{d}y = -\mathrm{d}y \wedge \mathrm{d}x = 0, \quad \text{etc.}$$

Example (ii): An exterior product of three 1-forms α^1, β^1 and γ^1 (where $\gamma^1 = c_1 \mathrm{d}x + c_2 \mathrm{d}y + c_3 \mathrm{d}z$) is a 3-form:

$$\begin{aligned}\alpha^1 \wedge \beta^1 \wedge \gamma^1 &= (a_2b_3 - a_3b_2)c_1 \mathrm{d}y \wedge \mathrm{d}z \wedge \mathrm{d}x + (a_3b_1 - a_1b_3)c_2 \mathrm{d}z \wedge \mathrm{d}x \wedge \mathrm{d}y \\ &\quad + (a_1b_2 - a_2b_1)c_3 \mathrm{d}x \wedge \mathrm{d}y \wedge \mathrm{d}z \\ &= \det[\boldsymbol{a}, \boldsymbol{b}, \mathbf{c}]\mathrm{d}x \wedge \mathrm{d}y \wedge \mathrm{d}z\end{aligned} \tag{B.14}$$

where $\boldsymbol{a} = (a_1, a_2, a_3)^T$, $\boldsymbol{b} = (b_1, b_2, b_3)^T$, and $\mathbf{c} = (c_1, c_2, c_3)^T$.

Example (iii): An exterior product of n 1-forms $\alpha_1^1, \ldots, \alpha_n^1$ is

$$\alpha_1^1 \wedge \cdots \wedge \alpha_n^1 = \det[\boldsymbol{a}_1, \ldots, \boldsymbol{a}_n]\mathrm{d}x_1 \wedge \cdots \wedge \mathrm{d}x_n \tag{B.15}$$

where $\alpha_k^1 = a_{kl}\mathrm{d}x_l$ and $\boldsymbol{a}_k = (a_{k1}, \ldots, a_{kn})^T$. This can be verified by the inductive method.

Example (iv): As an application of (iii), suppose that local transformation of coordinates from $\boldsymbol{x} = (x_1, \ldots, x_n)$ to $\boldsymbol{a} = (a_1, \ldots, a_n)$ in n-dimensional space is represented as

$$\mathrm{d}a_k = \frac{\partial a_k}{\partial x_l}\,\mathrm{d}x_l.$$

Then the n-form $\alpha^n = F(\boldsymbol{a})\mathrm{d}a_1 \wedge \cdots \wedge \mathrm{d}a_n$ is transformed as

$$F(\boldsymbol{a})\mathrm{d}a_1 \wedge \cdots \wedge \mathrm{d}a_n = \frac{\partial(\boldsymbol{a})}{\partial(\boldsymbol{x})} F(\boldsymbol{a}(\boldsymbol{x}))\mathrm{d}x_1 \wedge \cdots \wedge \mathrm{d}x_n, \tag{B.16}$$

where $\partial(\boldsymbol{a})/\partial(\boldsymbol{x})$ is the Jacobian of the transformation, and $F(\boldsymbol{a})$ is a 0-form.

B.3. Exterior Differentiations

Here, we define a differential operator d that takes exterior k-form fields into exterior $(k+1)$-form fields. A scalar function f is a 0-form, then its differential $\mathrm{d}f = (\partial_i f)\mathrm{d}x^i$ is a 1-form (see (1.27)). The $\omega^1 = a_i(x)\mathrm{d}x^i$ is a 1-form field, then its differential $\mathrm{d}\omega^1$ is a 2-form. The operator d of the *exterior differentiation* is defined to have the following properties:

(i) $\mathrm{d}\alpha^0 = \partial_i \alpha^0 \mathrm{d}x^i$,
(ii) $\mathrm{d}(\alpha + \beta) = \mathrm{d}\alpha + \mathrm{d}\beta$,
(iii) $\mathrm{d}(\alpha^k \wedge \beta^l) = \mathrm{d}\alpha^k \wedge \beta^l + (-1)^k \alpha^k \wedge \mathrm{d}\beta^l$,
(iv) $\mathrm{d}^2\alpha = \mathrm{d}(\mathrm{d}\alpha) := 0$, *for all forms*,

where a form α without upper index denotes any degree. The property (i) is defined in Sec. 1.5.1 of the main text. Properties (ii) and (iii) are taken as definitions.

To see (iv), consider a scalar function $f(x)$. Then, $\mathrm{d}^2 f$ is defined as

$$\mathrm{d}^2 f = \mathrm{d}((\partial_i f)\mathrm{d}x^i) := \mathrm{d}(\partial_i f) \wedge \mathrm{d}x^i.$$

The first factor is $\mathrm{d}(\partial_i f) = (\partial_j \partial_i f)\mathrm{d}x^j$ by (i). Substituting this,

$$\mathrm{d}^2 f = \sum_i \sum_j \frac{\partial^2 f}{\partial x^j \partial x^i}\,\mathrm{d}x^j \wedge \mathrm{d}x^i = 0,$$

since $\partial_i \partial_j f = \partial_j \partial_i f$ and $\mathrm{d}x^i \wedge \mathrm{d}x^j = -\mathrm{d}x^j \wedge \mathrm{d}x^i$, where (B.10) and (B.12) are used. Next, for any two scalar functions f and g, we obtain $\mathrm{d}(\mathrm{d}f \wedge \mathrm{d}g) = 0$ by (iii) and the above equation. By induction, one can verify that $\mathrm{d}^2\alpha = 0$ for any form α.

B.4. Interior Products and Cartan's Formula

If X is a vector and ω^p is a p-form, their **interior product** (a $p-1$ form) is *defined* by

$$i_X\omega^0 = 0 \qquad \text{for 0-form,} \tag{B.17}$$

$$i_X\omega^1 = \omega^1(X) \qquad \text{for 1-form,} \tag{B.18}$$

$$(i_X\omega^p)(X_2,\ldots,X_p) = \omega^p(X,X_2,\ldots,X_p) \qquad \text{for } p\text{-form.} \tag{B.19}$$

The Lie derivative $\mathcal{L}_X$ is defined by the **Cartan's formula,**

$$\mathcal{L}_X = i_X \circ \mathrm{d} + \mathrm{d} \circ i_X, \tag{B.20}$$

where $i_X\omega^p$ (for example) is an interior product acting on a p-form ω^p, yielding a $(p-1)$-form. This formula (B.20) can be verified by the method of *induction.* Operating $\mathcal{L}_X$ on a function f, we obtain

$$\mathcal{L}_X f = i_X \mathrm{d}f = \mathrm{d}f(X) = Xf,$$

by using (B.17) and (B.18). This is equivalent to Eq. (1.68). Operating on a differential $\mathrm{d}f$, we obtain

$$\mathcal{L}_X \mathrm{d}f = [i_X\mathrm{d} + \mathrm{d}i_X]\mathrm{d}f = \mathrm{d}i_X(\mathrm{d}f) = \mathrm{d}[i_X(\mathrm{d}f)] = \mathrm{d}[Xf] = \mathrm{d}\mathcal{L}_X f,$$

since $ddf = 0$. The commutability

$$\mathcal{L}_X \mathrm{d} = \mathrm{d}\mathcal{L}_X$$

can be shown from the definition of the Lie derivative [Fra97, §4.2; AM78, §2.4]. Furthermore, assuming that the equality (B.20) holds for p-forms, the formula can be verified for $p+1$ forms [Fra97; AM78].

B.5. Vector Analysis in $\mathbb{R}^3$

Let (x, y, z) be a cartesian coordinate frame in $\mathbb{R}^3$, and let 3-vectors in $\mathbb{R}^3$ be given by

$$\boldsymbol{a} = (a_x, a_y, a_z), \quad \boldsymbol{b} = (b_x, b_y, b_z), \quad \boldsymbol{c} = (c_x, c_y, c_z).$$

The inner product of $\boldsymbol{a}$ and $\boldsymbol{b}$ is defined by $\langle \boldsymbol{a}, \boldsymbol{b} \rangle$ (§1.4.2). Using the euclidean metric (1.30) of $\mathbb{R}^3$, the *inner product* is expressed as

$$\langle \boldsymbol{a}, \boldsymbol{b} \rangle = (\boldsymbol{a}, \boldsymbol{b})_{R^3} := a_x b_x + a_y b_y + a_z b_z. \tag{B.21}$$

The right-hand side is also written simply as

$$a_x b_x + a_y b_y + a_z b_z := \boldsymbol{a} \cdot \boldsymbol{b}. \tag{B.22}$$

The magnitude of the vector $\boldsymbol{a}$ is defined by $\|\boldsymbol{a}\| = \langle \boldsymbol{a}, \boldsymbol{b} \rangle^{1/2}$. The angle θ between two vectors $\boldsymbol{a}$ and $\boldsymbol{b}$ is defined by

$$\cos\theta = \frac{\langle \boldsymbol{a}, \boldsymbol{b} \rangle}{\|\boldsymbol{a}\| \|\boldsymbol{b}\|}. \tag{B.23}$$

If $\langle \boldsymbol{a}, \boldsymbol{b} \rangle = \boldsymbol{a} \cdot \boldsymbol{b} = 0$, the vector $\boldsymbol{a}$ is perpendicular to $\boldsymbol{b}$.

The *vector product* (cross product) $\boldsymbol{a} \times \boldsymbol{b}$ in $\mathbb{R}^3$ is defined such that

$$\langle \boldsymbol{a} \times \boldsymbol{b}, \boldsymbol{c} \rangle := \mathcal{V}^3[\boldsymbol{a}, \boldsymbol{b}, \boldsymbol{c}] = (\boldsymbol{a} \times \boldsymbol{b}) \cdot \boldsymbol{c}, \tag{B.24}$$

[Fra97], where $\mathcal{V}^3$ is the volume form in $\mathbb{R}^3$ defined by (B.35) below and we have

$$\mathcal{V}^3[\boldsymbol{a}, \boldsymbol{b}, \boldsymbol{c}] = (a_y b_z - a_z b_y) c_x + (a_z b_x - a_x b_z) c_y + (a_x b_y - a_y b_x) c_z.$$

It can be readily shown that $\mathcal{V}^3[\boldsymbol{a}, \boldsymbol{b}, \boldsymbol{c}] = \mathcal{V}^3[\boldsymbol{b}, \boldsymbol{c}, \boldsymbol{a}] = \mathcal{V}^3[\boldsymbol{c}, \boldsymbol{a}, \boldsymbol{b}]$. In components, we have

$$\boldsymbol{a} \times \boldsymbol{b} = (a_y b_z - a_z b_y, a_z b_x - a_x b_z, a_x b_y - a_y b_x). \tag{B.25}$$

From the definition (B.24) and the definition below, it is readily verified that $\boldsymbol{a} \times \boldsymbol{b}$ is perpendicular to both $\boldsymbol{a}$ and $\boldsymbol{b}$, i.e. $(\boldsymbol{a} \times \boldsymbol{b}) \cdot \boldsymbol{a} = 0$ and $(\boldsymbol{a} \times \boldsymbol{b}) \cdot \boldsymbol{b} = 0$.

In order to represent the cross product in component form, it is useful to introduce a third order skew-symmetric tensor $\varepsilon_{\mathbf{ijk}}$, defined by

$$\varepsilon_{ijk} = \begin{cases} 1, & \text{for } (1,2,3) \to (i,j,k) : \text{even permutation} \\ -1, & \text{for } (1,2,3) \to (i,j,k) : \text{odd one} \\ 0, & \text{otherwise: (for repeated indices)} \end{cases}. \tag{B.26}$$

Using the notation of vectors as $\boldsymbol{a} = (a_1, a_2, a_3)$ and $\boldsymbol{b} = (b_1, b_2, b_3)$, the vector product is written compactly as

$$(\boldsymbol{a} \times \boldsymbol{b})_i = \varepsilon_{ijk} a_j b_k, \quad (i = 1, 2, 3). \tag{B.27}$$

Let us introduce a position vector $\boldsymbol{x} = (x, y, z)$ and its infinitesimal variation $\mathrm{d}\boldsymbol{x} = (\mathrm{d}x, \mathrm{d}y, \mathrm{d}z)$. Owing to the eucledian metric structure and the inner product, differential forms are rephrased with inner products of vectors in $\mathbb{R}^3$ as follows.

(**a**) Exterior differential of a 0-form f is a 1-form $\mathrm{d}f$:

$$\mathrm{d}f = (\partial_x f)\mathrm{d}x + (\partial_y f)\mathrm{d}y + (\partial_z f)\mathrm{d}z = \nabla f \cdot \mathrm{d}\boldsymbol{x}. \tag{B.28}$$

(**b**) Definition of "curl": Let α^1 be a 1-form given by

$$\alpha^1 = a_x(\boldsymbol{x})\mathrm{d}x + a_y(\boldsymbol{x})\mathrm{d}y + a_z(\boldsymbol{x})\mathrm{d}z = \boldsymbol{a} \cdot \mathrm{d}\boldsymbol{x}.$$

Exterior differential of α^1 is a 2-form $\mathrm{d}\alpha^1$:

$$\begin{aligned}\mathrm{d}\alpha^1 &= \mathrm{d}a_x \wedge \mathrm{d}x + \mathrm{d}a_y \wedge \mathrm{d}y + \mathrm{d}a_z \wedge \mathrm{d}z \\ &= (\partial_y a_z - \partial_z a_y)\mathrm{d}y \wedge \mathrm{d}z + (\partial_z a_x - \partial_x a_z)\mathrm{d}z \wedge \mathrm{d}x \\ &\quad + (\partial_x a_y - \partial_y a_x)\mathrm{d}x \wedge \mathrm{d}y. \end{aligned} \tag{B.29}$$

This is rewritten in a vectorial form as

$$\mathrm{d}\alpha^1 = \mathrm{d}(\boldsymbol{a} \cdot \mathrm{d}\boldsymbol{x}) = (\mathrm{curl}\, \boldsymbol{a}) \cdot \boldsymbol{s}^2, \quad \boldsymbol{s}^2 = (\mathrm{d}y \wedge \mathrm{d}z, \mathrm{d}z \wedge \mathrm{d}x, \mathrm{d}x \wedge \mathrm{d}y) \tag{B.30}$$

where $\boldsymbol{s}^2$ is an *oriented* surface 2-form. Equations (B.29) and (B.30) define the curl $\boldsymbol{a}$:

$$\mathrm{curl}\, \boldsymbol{a} = (\partial_y a_z - \partial_z a_y, \partial_z a_x - \partial_x a_z, \partial_x a_y - \partial_y a_x). \tag{B.31}$$

(**c**) Definition of "div": Let β^2 be a 2-form given by

$$\beta^2 = b_x(\boldsymbol{x})\mathrm{d}y \wedge \mathrm{d}z + b_y(\boldsymbol{x})\mathrm{d}z \wedge \mathrm{d}x + b_z(\boldsymbol{x})\mathrm{d}x \wedge \mathrm{d}y = \boldsymbol{b} \cdot \boldsymbol{s}^2.$$

The exterior differential of β^2 is

$$\begin{aligned}\mathrm{d}\beta^2 &= \mathrm{d}b_x \wedge \mathrm{d}y \wedge \mathrm{d}z + \mathrm{d}b_y \wedge \mathrm{d}z \wedge \mathrm{d}x + b_z \wedge \mathrm{d}x \wedge \mathrm{d}y \\ &= (\partial_x b_x + \partial_y b_y + \partial_z b_z)\mathrm{d}x \wedge \mathrm{d}y \wedge \mathrm{d}z. \end{aligned} \tag{B.32}$$

This is rewritten in a vectorial form as

$$\mathrm{d}(\boldsymbol{b} \cdot \boldsymbol{s}^2) = (\mathrm{div}\, \boldsymbol{b})\mathcal{V}^3, \tag{B.33}$$

where $\mathcal{V}^3 = \mathrm{d}x \wedge \mathrm{d}y \wedge \mathrm{d}z$ is a volume 3-form. Equations (B.32) and (B.33) define div $\boldsymbol{b}$:

$$\mathrm{div}\, \boldsymbol{b} = \partial_x b_x + \partial_y b_y + \partial_z b_z. \tag{B.34}$$

B.6. Volume Form and Its Lie Derivative

Let us consider a volume form. Let (x^1, x^2, x^3) be a local cartesian coordinate of the three-dimensional space $M = \mathbb{R}^3$. Then the volume form $\mathcal{V}^3$ is a 3-form:

$$\mathcal{V}^3(x) := \mathrm{d}x^1 \wedge \mathrm{d}x^2 \wedge \mathrm{d}x^3. \tag{B.35}$$

Let $y = f(x) = y(x)$ be a coordinate transformation and $x = F(y) = x(y)$ be its inverse transformation between $x = (x^1, x^2, x^3)$ and $y = (y^1, y^2, y^3)$ in a neighborhood of the point $x \in M$. Transformation of a differential 1-form is represented by

$$\mathrm{d}x^i = \frac{\partial x^i}{\partial y^k}\, \mathrm{d}y^k, \tag{B.36}$$

which is regarded as the pull-back transformation F from x to y (see (1.50) with x and y interchanged). Next, the first fundamental form is defined as

$$\mathrm{I} := \mathrm{d}x^i \mathrm{d}x^i = \frac{\partial x^i}{\partial y^j}\frac{\partial x^i}{\partial y^k} \mathrm{d}y^j\, \mathrm{d}y^k := g_{jk} \mathrm{d}y^j\, \mathrm{d}y^k, \tag{B.37}$$

where the metric tensor g_{jk} and its determinant $g(y)$ are

$$g_{jk} = \frac{\partial x^i}{\partial y^j}\frac{\partial x^i}{\partial y^k}, \quad g(y) = \det(g_{jk}) = \left(\left[\frac{\partial(x)}{\partial(y)}\right]\right)^2, \tag{B.38}$$

where $[\partial(x)/\partial(y)]$ is the Jacobian determinant of the coordinate transformation, which is now represented as

$$\left[\frac{\partial(x)}{\partial(y)}\right] = \pm\sqrt{g(y)}, \quad (\sqrt{g(y)} > 0). \tag{B.39}$$

Using (B.36), the volume form (B.35) is transformed (pull-back to the y-frame) to

$$\mathcal{V}^3(y) = F^*[\mathcal{V}^3(x)] \tag{B.40}$$

$$= \mathrm{sgn}(y)\sqrt{g(y)}\, \mathrm{d}y^1 \wedge \mathrm{d}y^2 \wedge \mathrm{d}y^3, \tag{B.41}$$

where $\mathrm{sgn}(y) = \pm 1$ is an *orientation* factor of the local frame $y = (y^1, y^2, y^3)$ and chosen according to the sign of the Jacobian determinant $[\partial(x)/\partial(y)]$, namely, $\mathrm{sgn}(y) = 1$ if the (y^1, y^2, y^3)-frame has the same orientation as the (x^1, x^2, x^3)-frame (assumed to be right-handed usually), and $\mathrm{sgn}(y) = -1$ otherwise.

Next, we consider the **Lie derivative** $\mathcal{L}$ of a volume form $\mathcal{V}^3$. If $X = (X^1, X^2, X^3)$ is a vector field on M^3, the *divergence* of X, a scalar denoted by $\operatorname{div} X$, is defined by the formula,

$$(\operatorname{div} X)\mathcal{V}^3 := \mathcal{L}_X \mathcal{V}^3 = \frac{\mathrm{d}}{\mathrm{d}t}\mathcal{V}^3(Y_1, Y_2, Y_3). \tag{B.42}$$

The Lie derivative $\mathcal{L}_X\mathcal{V}^3$ is defined by the time derivative of $\mathcal{V}^3$ as one moves along the flow ϕ_t generated by X, while $\mathcal{V}^3$ is given by the value on three vector fields Y_1, Y_2, Y_3 that are invariant under the flow ϕ_t, i.e. $Y_i(\phi_t x) = \phi_t^* Y_i(x)$. Equation (B.42) defines the $\operatorname{div} X$ as the *relative* change of the volume element $\mathcal{V}^3$ along the flow generated by X.

When operating $\mathcal{L}_X$ on the volume form $\mathcal{V}^3 = \mathrm{d}x^1 \wedge \mathrm{d}x^2 \wedge \mathrm{d}x^3$, the *Cartan's formula* (B.20) is useful. Applying the formula (B.20) to $\mathcal{V}^3$, and noting that $\mathrm{d}\mathcal{V}^3 = 0$ and

$$\begin{aligned} i_X\mathcal{V}^3 &= i_X[\mathrm{d}x^1 \wedge \mathrm{d}x^2 \wedge \mathrm{d}x^3] = (i_X \mathrm{d}x^1) \wedge \mathrm{d}x^2 \wedge \mathrm{d}x^3 \\ &\quad - \mathrm{d}x^1 \wedge (i_X \mathrm{d}x^2) \wedge \mathrm{d}x^3 + \mathrm{d}x^1 \wedge \mathrm{d}x^2 \wedge (i_X\, \mathrm{d}x^3) \\ &= X^1 \mathrm{d}x^2 \wedge \mathrm{d}x^3 - X^2 \mathrm{d}x^1 \wedge \mathrm{d}x^3 + X^3 \mathrm{d}x^1 \wedge \mathrm{d}x^2, \end{aligned} \tag{B.43}$$

we obtain

$$\begin{aligned} \mathcal{L}_X\mathcal{V}^3 &= \mathrm{d} \circ i_X \mathcal{V}^3 = \mathrm{d}[X^1 \mathrm{d}x^2 \wedge \mathrm{d}x^3 - X^2 \mathrm{d}x^1 \wedge \mathrm{d}x^3 + X^3 \mathrm{d}x^1 \wedge \mathrm{d}x^2] \\ &= \left(\sum_i \frac{\partial X^i}{\partial x^i}\right) \mathrm{d}x^1 \wedge \mathrm{d}x^2 \wedge \mathrm{d}x^3 = (\operatorname{div} X)\mathcal{V}^3, \end{aligned} \tag{B.44}$$

which is consistent with (B.34) and (B.42), and

$$\operatorname{div} X = \frac{\partial X^1}{\partial x^1} + \frac{\partial X^2}{\partial x^2} + \frac{\partial X^3}{\partial x^3}. \tag{B.45}$$

B.7. Integration of Forms

B.7.1. *Stokes's theorem*

For a continuously differentiable k-form ω^k on a manifold M^n, the Stokes's Theorem reads

$$\int_V \mathrm{d}\omega^k = \int_{\partial V} \omega^k, \tag{B.46}$$

where $V = V^{k+1} \subset M^n$ is a compact oriented submanifold with boundary $(\partial V)^k$ in M^n. This formula suggests that (integral of) the exterior derivative

$d\omega^k$ on a manifold V is defined by an integral of the k-form ω^k over a boundary of V.

Examples

(a) $k = 0$: Integration of (B.28) along a smooth curve l $(= V^1)$ from a point $\boldsymbol{x}_1$ to an end point $\boldsymbol{x}_2$ $(= (\partial V)^0)$:

$$\int_l \nabla f \cdot \mathrm{d}\boldsymbol{x} = f(\boldsymbol{x}_2) - f(\boldsymbol{x}_1). \tag{B.47}$$

(b) $k = 1$: Integration of (B.30) over an oriented compact smooth surface $S^2 = V^2$ with a circumferential curve $l^1 = (\partial V)^1$:

$$\int_{S^2} (\mathrm{curl}\, \boldsymbol{a}) \cdot \boldsymbol{s}^2 = \oint_{l^1} \boldsymbol{a} \cdot \mathrm{d}\boldsymbol{x}. \tag{B.48}$$

(c) $k = 2$: Integration of (B.33) over an oriented compact 3-volume $V = V^3$ with a smooth closed 2-surface $S^2 = (\partial V)^2$:

$$\int_{V^3} (\mathrm{div}\, \boldsymbol{b}) \mathcal{V}^3 = \oint_{S^2} \boldsymbol{b} \cdot \boldsymbol{s}^2. \tag{B.49}$$

□

B.7.2. *Integral and pull-back*

Let f be a differentiable map of an orientation-preserving diffeomorphism, $f : M^n \to V^r$, from an interior subset σ of M^n onto an interior subset $f(\sigma)$ of V^r. Then, for any differential k-form ω^k on V^r, the following general formula of pull-back integration holds:

$$\int_{f(\sigma)} \omega^k = \int_\sigma f^* \omega^k, \tag{B.50}$$

which is a generalization of the integral formula (1.51) for 1-form. The integral of a k-form ω^k over the image $f(\sigma)$ is equal to the integral of the pull-back $f^*\omega^k$ over the original subset σ.

Appendix C

Lie Groups and Rotation Groups

Most dynamical systems are characterized by some invariance property with respect to a certain group of transformations, i.e. a Lie group. Rotation groups are typical symmetry groups with which some familiar dynamical systems are represented. This appendix is a brief summary of Lie groups, one-parameter subgroup and a particular Lie group $SO(n)$, complementing §1.7, 1.8, 3.8, 7.7 and 11.3, 11.4.

C.1. Various Lie Groups

The set of all rotations of a rigid body in the three-dimensional space $\mathbb{R}^3$ is a differentiable manifold, since it is parametrized continuously and differentiably by the three *Eulerian angles* [LL76]. It is a Lie group, $SO(3)$, its dimension being accidentally 3.

Let M be one of n-dimensional manifolds including $\mathbb{R}^n$. The structure group of TM is the set of all real $n \times n$ matrices with nonzero determinant, which is a Lie group $GL(n, \mathbb{R})$ called *general linear group*. Topologically, $GL(n, \mathbb{R})$ is an open subset of euclidean space of n^2-dimensions.

The *special linear group* $SL(n, \mathbb{R})$ is a set of all real $n \times n$ matrices $g \in SL(n)$ with $\det g = 1$, a subgroup of $GL(n, \mathbb{R})$, and has dimension $n^2 - 1$.

The *orthogonal group* $O(n)$ is a set of all real $n \times n$ matrices $\omega \in O(n)$ satisfying $\omega\omega^T = I$ (the orthogonality condition), where T denotes transpose, i.e. $(\omega^T)_{ij} = \omega_{ji}$. The matrix satisfying $\omega\omega^T = I$ is said to be orthogonal. The $O(n)$ is a subgroup of $GL(n, \mathbb{R})$ of dimension $\frac{1}{2}n(n-1)$. Since the matrix product $\omega\omega^T$, is a symmetric $n \times n$ matrix, the matrix equation $\omega\omega^T = I$ gives $(1 + \cdots + n) = \frac{1}{2}n(n+1)$ restricting conditions. Therefore the dimension of the group $O(n)$ is $n^2 - \frac{1}{2}n(n+1) = \frac{1}{2}n(n-1)$. From $\omega\omega^T = I$, we obtain $(\det\omega)^2 = 1$.

Thus the orthogonal group $O(n)$ consists of a subgroup $SO(n)$, the *rotation group* (the *special orthogonal group*), where $\det \omega = 1$, and a disjoint submanifold where $\det \omega = -1$. The latter does not include the identity matrix I. The dimensions of $SO(2)$, $SO(3)$ and $SO(n)$ are 1, 3 and $\frac{1}{2}n(n-1)$ respectively.

The simplest rotation group is $SO(2)$, which describes rotations of a plane. Matrix representation of its element $R \in SO(2)$ is

$$R(\theta) = \begin{bmatrix} \cos\theta & -\sin\theta \\ \sin\theta & \cos\theta \end{bmatrix}, \quad \theta \in [0, 2\pi].$$

To a rotation of a plane through an angle θ, we associate a point on the unit circle S^1 at angle θ. This is also represented by the point of $e^{i\theta}$ in the complex plane. Two successive rotations are represented by the multiplication $e^{i\theta}e^{i\phi} = e^{i(\theta+\phi)}$, i.e. given by addition of angles. Thus, $SO(2)$ is abelian (commutable), whereas other rotation groups $SO(n)$ for $n \geq 3$ are non-commutable (non-abelian). The rotation $e^{i\theta}$ is also an element of the unitary group $U(1)$.

General linear group $GL(n, \mathbb{C})$ is composed of all $n \times n$ matrices of complex numbers ($\in \mathbb{C}$) with nonzero determinant, whose dimension is $2n^2$.

Unitary group $U(n)$ consists of all complex $n \times n$ matrices $z \in U(n)$ satisfying $zz^\dagger = I$, where $z^\dagger = \bar{z}^T = z^{-1}$ (the overbar denotes complex conjugate, the dagger † denotes hermitian adjoint and ^{T} the transpose). $U(n)$ is a submanifold of complex n^2-space or real $2n^2$-space. Since the matrix product $zz^\dagger$ is a hermitian $n \times n$ complex matrix, the matrix equation $zz^\dagger = I$ gives $(1 + \cdots + (n-1)) = \frac{1}{2}(n-1)n$ complex conditions and n real (diagonal) conditions. Therefore the dimension of the group $U(n)$ is $2n^2 - (2\frac{1}{2}(n-1)n + n) = n^2$. From $zz^\dagger = I$, we obtain $|\det z|^2 = 1$.

$SU(n)$ is the *special unitary group* with $\det z = 1$ for $^\forall z \in SU(n)$, and has dimension $n^2 - 1$ because of the extra condition $\det z = 1$, since in general, $\det z$ is of the form $e^{i\varphi}$ for $\varphi \in \mathbb{R}$. See [Fra97; Sch80] for more details.

C.2. One-Parameter Subgroup and Lie Algebra

One-parameter subgroup of a group G is defined by a trajectory, $t \in \mathbb{R} \to g(t) \in G$ (with $g(0) = e$), satisfying the rule,

$$g(s+t) = g(s)g(t) = g(t)g(s),$$

i.e. a homomorphism (preserving products) of the additive group of $\mathbb{R}$ (real numbers) to the multiplicative group of G. Differentiating $g(s+t) = g(t)g(s)$ with respect to s and putting $s = 0$,

$$g'(t) = g(t)X_e (= g_* X_e), \quad X_e = g'(0). \tag{C.1}$$

This indicates that the tangent vector X_e at the identity $e = g(0)$ is left-translated along $g(t)$. In other words, the one-parameter subgroup $g(t)$ is an integral curve through e resulting from left-translation of the tangent vector X_e at e over G. The vector X_e is called the *infinitesimal generator* of the one-parameter subgroup $g(t)$.

Consider a matrix group G_m with $A = g'_m(0)$ a constant matrix. Then Eq. (C.1), $g'_m(t) = g_m(t)A$, can be integrated to give

$$g_m(t) = g_m(0)\exp[tA] = \exp[tA] \tag{C.2}$$

$$\exp[tA] := e + tA + \frac{(tA)^2}{2!} + \cdots = e + \sum_{k=1}^{\infty} \frac{(tA)^k}{k!}, \tag{C.3}$$

where $e = I$ (unit matrix).

Analogously, for any Lie group G, a one-parameter subgroup with the tangent vector X at e is denoted by

$$g(t) := \exp[tX] = e^{tX} = e + tX + O(t^2), \quad X \in T_e G \tag{C.4}$$

The tangent space $T_e G = \mathbf{g}$ at e of a Lie group G is called the *Lie algebra*. The algebra $\mathbf{g}$ is equipped with the Lie bracket $[X, Y,]$ for $^\forall X, Y \in T_e G$ (§1.7).

C.3. Rotation Group $SO(n)$

Let us consider the rotation group $SO(n)$. An element $g \in SO(n)$ is represented by an $n \times n$ orthogonal matrix ($gg^T = I$) satisfying $\det g = 1$. Let $\xi(t)$ be a curve (on $SO(n)$) issuing from the identity $e (= I)$ with a tangent vector $\mathbf{a}$ at e. Then one has $\xi(t) = e + t\mathbf{a} + O(t^2)$ for an infinitesimal parameter t. The vector $\mathbf{a} = \xi'(0)$ is an element of the tangent space $T_e SO(n)$, i.e. the Lie algebra $\mathbf{so}(n)$ (using lower case letters of bold face).

The orthogonality condition $\xi(t)\xi^T(t) = e$ requires $\mathbf{a} = (a_{ij}) \in \mathbf{so}(n)$ to be skew-symmetric. In fact, we obtain

$$(e + t\mathbf{a} + \cdots)(e + t\mathbf{a}^T + \cdots) = e, \quad \therefore \ \mathbf{a} = -\mathbf{a}^T, \quad \text{i.e. } a_{ij} = -a_{ji}. \tag{C.5}$$

C.4. so(3)

The dimension of the vector space of Lie algebra $\mathbf{so}(3)$ is 3, represented by the following (skew-symmetric) basis (E_1, E_2, E_3):

$$E_1 = \begin{bmatrix} 0 & 0 & 0 \\ 0 & 0 & -1 \\ 0 & 1 & 0 \end{bmatrix}, \quad E_2 = \begin{bmatrix} 0 & 0 & 1 \\ 0 & 0 & 0 \\ -1 & 0 & 0 \end{bmatrix}, \quad E_3 = \begin{bmatrix} 0 & -1 & 0 \\ 1 & 0 & 0 \\ 0 & 0 & 0 \end{bmatrix}. \tag{C.6}$$

Their commutation relations are given by

$$[E_1, E_2] = E_3, \quad [E_2, E_3] = E_1, \quad [E_3, E_1] = E_2, \tag{C.7}$$

where $[E_j, E_k] = E_j E_k - E_k E_j$. In addition, we have the following properties, $-\frac{1}{2}\mathrm{tr}(E_k E_l) = \delta_{kl}$, since

$$\mathrm{tr}(E_1 E_1) = \mathrm{tr}(E_2 E_2) = \mathrm{tr}(E_3 E_3) = -2, \tag{C.8}$$

$$\mathrm{tr}(E_1 E_2) = \mathrm{tr}(E_2 E_3) = \mathrm{tr}(E_3 E_1) = 0, \tag{C.9}$$

where tr denotes taking *trace* of the matrix that follows.

For $\boldsymbol{a}, \boldsymbol{b} \in \mathbf{so}(3)$, we have the following representations,

$$\boldsymbol{a} = a_1 E_1 + a_2 E_2 + a_3 E_3, \quad a_1, a_2, a_3 \in \mathbb{R}, \tag{C.10}$$

and a similar expression for $\boldsymbol{b}$, which are obviously skew-symmetric. Their scalar product is defined by

$$\langle \boldsymbol{a}, \boldsymbol{b} \rangle_{so(3)} := -\frac{1}{2}\mathrm{tr}(\boldsymbol{ab}) = -\frac{1}{2} a_k b_l \ \mathrm{tr}(E_k E_l) = a_k b_k. \tag{C.11}$$

Their commutation is given by

$$\mathbf{ab} - \mathbf{ba} = (a_2 b_3 - a_3 b_2)E_1 + (a_3 b_1 - a_1 b_3)E_2 + (a_1 b_2 - a_2 b_1)E_3. \tag{C.12}$$

Let us define a skew-symmetric matrix $\mathbf{c}$ and an associated column vector $\hat{\mathbf{c}}$ by

$$\mathbf{c} = c_1 E_1 + c_2 E_2 + c_3 E_3 := \mathbf{ab} - \mathbf{ba}, \quad \hat{\mathbf{c}} = (c^1, c^2, c^3)^T. \tag{C.13}$$

Similarly, we define the vectors $\hat{\mathbf{a}}$ and $\hat{\mathbf{b}}$ associated with the matrices $\mathbf{a}$ and $\mathbf{b}$. The representation of $\mathbf{ab} - \mathbf{ba}$ in the form of (C.12) is equivalent to $\boldsymbol{a} \times \boldsymbol{b}$ in the form of (B.25). Thus, we can represent the Eq. (C.13) by the following cross-product,

$$\hat{\mathbf{c}} = \hat{\mathbf{a}} \times \hat{\mathbf{b}}. \tag{C.14}$$

Given a point $\boldsymbol{r} = (r^1, r^2, r^3) \in M^3$, an infinitesimal transformation of $\boldsymbol{r}$ by $\xi(t) = e + t\mathbf{a} + O(t^2)$ is (up to the $O(t)$ term)

$$\boldsymbol{r} \mapsto \boldsymbol{r}' = \xi(t)\boldsymbol{r} = \boldsymbol{r} + t\mathbf{a}\boldsymbol{r} = \boldsymbol{r} + t(a_1 E_1 + a_2 E_2 + a_3 E_3)\boldsymbol{r} = \boldsymbol{r} + t\hat{\mathbf{a}} \times \boldsymbol{r}. \tag{C.15}$$

Thus, the infinitesimal transformation $\xi(t)$ represents a rotational transformation of *angular velocity* $\hat{\mathbf{a}}$.

Appendix D

A Curve and a Surface in $\mathbb{R}^3$

(Ref. §2.1, 2.5.3, 2.5.4, 2.7.1, 2.8, 9.1)

D.1. Frenet–Serret Formulas for a Space Curve

Let a space curve C be defined by $\boldsymbol{x}(s)$ in $\mathbb{R}^3$ with s as the arc-length parameter. Then, the unit tangent vector is given by

$$\boldsymbol{t} = \frac{\mathrm{d}\boldsymbol{x}}{\mathrm{d}s}, \quad \|\boldsymbol{t}\| = \langle \boldsymbol{t}, \boldsymbol{t}\rangle^{1/2} = 1.$$

Differentiating the equation $\langle \boldsymbol{t}, \boldsymbol{t}\rangle = 1$ with respect to s, we have $\langle \boldsymbol{t}, (\mathrm{d}\boldsymbol{t}/\mathrm{d}s)\rangle = 0$. Hence, $\mathrm{d}\boldsymbol{t}/\mathrm{d}s$ is orthogonal to $\boldsymbol{t}$ and is so to the curve C, and the vector $\mathrm{d}\boldsymbol{t}/\mathrm{d}s$ defines a unique direction (if $\mathrm{d}\boldsymbol{t}/\mathrm{d}s \neq 0$) in a plane normal to C at $\boldsymbol{x}$ called the direction of principal normal, represented by

$$\frac{\mathrm{d}\boldsymbol{t}}{\mathrm{d}s} = \frac{\mathrm{d}^2\boldsymbol{x}}{\mathrm{d}s^2} = \kappa(s)\boldsymbol{n}(s), \quad \|\boldsymbol{n}\| = \langle \boldsymbol{n}, \boldsymbol{n}\rangle^{1/2} = 1, \tag{D.1}$$

where $\boldsymbol{n}$ is the vector of unit *principal normal* and $\kappa(s)$ is the curvature. Then, we can define the unit *binormal* vector $\boldsymbol{b}$ by the equation, $\boldsymbol{b} = \boldsymbol{t} \times \boldsymbol{n}$, which is normal to the osculating plane spanned by $\boldsymbol{t}$ and $\boldsymbol{n}$. Thus we have a local right-handed orthonormal frame $(\boldsymbol{t}, \boldsymbol{n}, \boldsymbol{b})$ at each point $\boldsymbol{x}$ (Fig. D.1).

Analogously to $\mathrm{d}\boldsymbol{t}/\mathrm{d}s \perp \boldsymbol{t}$, the vector $\mathrm{d}\boldsymbol{n}/\mathrm{d}s$ is orthogonal to $\boldsymbol{n}$. Hence, it may be written as $\boldsymbol{n}' = \alpha\boldsymbol{t} - \tau\boldsymbol{b}$, where the prime denotes $\mathrm{d}/\mathrm{d}s$, and $\alpha, \tau \in \mathbb{R}$. Differentiating $\boldsymbol{b}$, we obtain

$$\frac{\mathrm{d}\boldsymbol{b}}{\mathrm{d}s} = \boldsymbol{t} \times \boldsymbol{n}' + \boldsymbol{t}' \times \boldsymbol{n} = -\tau\boldsymbol{t} \times \boldsymbol{b} = \tau(s)\boldsymbol{n}(s), \tag{D.2}$$

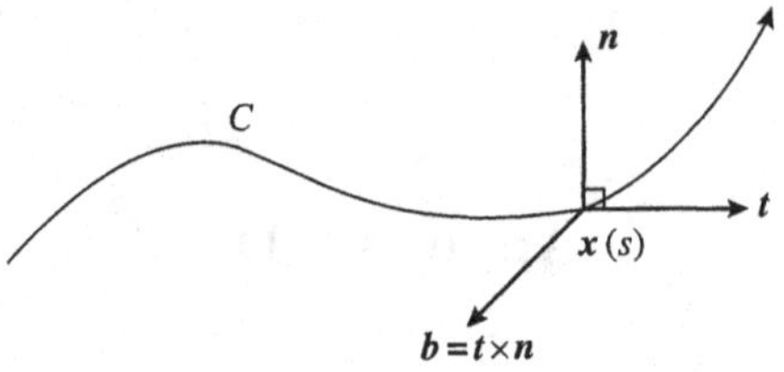

Fig. D.1. A space curve.

where (D.1) is used, and τ is the *torsion* of the curve C at s. Since $\boldsymbol{n} = \boldsymbol{b}\times\boldsymbol{t}$, we have

$$\frac{\mathrm{d}\boldsymbol{n}}{\mathrm{d}s} = \boldsymbol{b}\times\boldsymbol{t}' + \boldsymbol{b}'\times\boldsymbol{t} = -\kappa\boldsymbol{t} - \tau\boldsymbol{b}, \tag{D.3}$$

where (D.1) and (D.2) are used.

Collecting the three equations (D.1)–(D.3), we obtain a set of differential equations for $(\boldsymbol{t}(s), \boldsymbol{n}(s), \boldsymbol{b}(s))$:

$$\frac{\mathrm{d}}{\mathrm{d}s}\begin{pmatrix}\boldsymbol{t}\\ \boldsymbol{n}\\ \boldsymbol{b}\end{pmatrix} = \begin{pmatrix}0 & \kappa & 0\\ -\kappa & 0 & -\tau\\ 0 & \tau & 0\end{pmatrix}\begin{pmatrix}\boldsymbol{t}\\ \boldsymbol{n}\\ \boldsymbol{b}\end{pmatrix}. \tag{D.4}$$

This is called the *Frenet–Serret equations* for a space curve.

D.2. A Plane Curve in $\mathbb{R}^2$ and Gauss Map

A plane curve $\mathrm{C_p}$: $\boldsymbol{p}(s)$ in the plane $\mathbb{R}^2$ is a particular case of the space curve considered in the previous section D.1. The unit tangent $\boldsymbol{t}(s)$ and unit principal normal $\boldsymbol{n}(s)$ are defined in the same way. However, the binormal $\boldsymbol{b}$ defined by $\boldsymbol{t}\times\boldsymbol{n}$ is always perpendicular to the plane $\mathbb{R}^2$. Hence $\boldsymbol{b}$ is a constant unit vector, and $\mathrm{d}\boldsymbol{b}/\mathrm{d}s = 0$, resulting in the vanishing torsion, $\tau = 0$. Thus, Eq. (D.4) reduces to

$$\frac{\mathrm{d}}{\mathrm{d}s}\begin{pmatrix}\boldsymbol{t}\\ \boldsymbol{n}\end{pmatrix} = \begin{pmatrix}0 & \kappa\\ -\kappa & 0\end{pmatrix}\begin{pmatrix}\boldsymbol{t}\\ \boldsymbol{n}\end{pmatrix}. \tag{D.5}$$

As an interpretation of the curvature κ (which can be generalized to the surface case), let us consider the *Gauss map* G. The map G is defined by $G : \boldsymbol{p}(s) \mapsto \boldsymbol{n}(s)$, where the unit vector $\boldsymbol{n}(s)$ is plotted as a vector extending from the common origin O of a plane. Therefore, the end point of $\boldsymbol{n}(s)$ will draw an arc over a unit circle S^1_n (*Gauss circle*) as the point

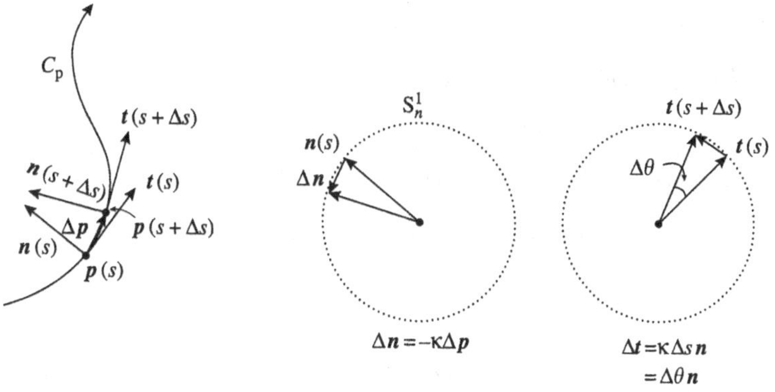

Fig. D.2. Gauss map.

$\boldsymbol{p}(s)$ moves along the curve C_p (Fig. D.2). Corresponding to an infinitesimal translation of the point $\boldsymbol{p}(s)$ along C_p by the length Δs, where the vectorial displacement is $\Delta\boldsymbol{p} = (\Delta s)\boldsymbol{t}$, the displacement of $\boldsymbol{n}(s)$ along the unit circle is given by $\Delta\boldsymbol{n} = -\kappa\Delta s\boldsymbol{t}$. Thus the curvature κ is given by the ratio of the two lengths $\|\Delta\boldsymbol{n}\|/\|\Delta\boldsymbol{p}\|$. More precisely, we have $\Delta\boldsymbol{n} = -\kappa\Delta\boldsymbol{p}$.

In order to define the basis $(\boldsymbol{t}, \boldsymbol{n})$ in the same sense of the right-handed (x, y)-frame, it is convenient to introduce both positive and negative values for the curvature κ. Then the first equation $\Delta\boldsymbol{t} = \kappa\boldsymbol{n}\Delta s$ of (D.5) is understood as $\Delta\boldsymbol{t}$ being clockwise or anti-clockwise with respect to $\boldsymbol{t}$ according as the κ is negative or positive. Since $\boldsymbol{t}$ is a unit vector, the infinitesimal change $\Delta\boldsymbol{t}$ is given by $(\Delta\theta)\boldsymbol{n}$, where $\Delta\theta$ is the rotation angle of the vector $\boldsymbol{t}$ (Fig. D.2). Thus the equation $\Delta\boldsymbol{t} = \kappa\boldsymbol{n}\Delta s$ reduces to another interpretation of the curvature:

$$\kappa = \frac{\mathrm{d}\theta}{\mathrm{d}s}. \tag{D.6}$$

D.3. A Surface Defined by $z = f(x, y)$ in $\mathbb{R}^3$

Suppose that a surface is defined by $z = f(x, y)$ in the three-dimensional cartesian space (x, y, z). One may take $u^1 = x, u^2 = y$ and $\boldsymbol{x} = (u^1, u^2, f(u^1, u^2))$ in the formulation in §2.1. The line-element is $\mathrm{d}\boldsymbol{x} = (\mathrm{d}x, \mathrm{d}y, \mathrm{d}f)$, where $\mathrm{d}f = p\mathrm{d}x + q\mathrm{d}y$ and

$$p = f_x = \frac{\partial f}{\partial x}, \quad q = f_y = \frac{\partial f}{\partial y}. \tag{D.7}$$

Then we have

$$\begin{aligned} ds^2 = \langle d\boldsymbol{x}, d\boldsymbol{x}\rangle &= (dx)^2 + (dy)^2 + (pdx + qdy)^2 \\ &= (1+p^2)(du^1)^2 + 2pq du^1 du^2 + (1+q^2)(du^2)^2. \end{aligned}$$

Hence the metric tensors are given by

$$g_{11} = 1 + p^2, \quad g_{12} = pq, \quad g_{22} = 1 + q^2.$$

The angle ϕ of the coordinate curves is given by

$$\cos\phi = \frac{pq}{\sqrt{(1+p^2)(1+q^2)}},$$

according to (2.16) in §2.1. The second fundamental form and its tensors $b_{\alpha\beta}$ are defined by (2.20) and (2.21), where the unit normal $\boldsymbol{N}$ is given by

$$\boldsymbol{N} = \left(-\frac{p}{\sqrt{W}}, -\frac{q}{\sqrt{W}}, \frac{1}{\sqrt{W}}\right) \perp d\boldsymbol{x}, \tag{D.8}$$

where $W = 1 + p^2 + q^2$. Then we have

$$\begin{aligned} \boldsymbol{N}_1 &= \partial_x \boldsymbol{N}, & \boldsymbol{N}_2 &= \partial_y \boldsymbol{N}, \\ \boldsymbol{x}_1 &= \partial_x \boldsymbol{x} = (1, 0, p), & \boldsymbol{x}_2 &= \partial_y \boldsymbol{x} = (0, 1, q). \end{aligned}$$

Using the notations of the second derivatives defined by

$$r = f_{xx} = \frac{\partial^2 f}{\partial x^2}, \quad s = f_{xy} = \frac{\partial^2 f}{\partial x \partial y}, \quad t = f_{yy} = \frac{\partial^2 f}{\partial y^2}, \tag{D.9}$$

we obtain the tensor $b_{\alpha\beta}$ as

$$\begin{aligned} b_{11} &= -\langle \boldsymbol{x}_1, \boldsymbol{N}_1\rangle = \frac{r}{\sqrt{W}} \\ b_{12} &= -\langle \boldsymbol{x}_1, \boldsymbol{N}_2\rangle = \frac{s}{\sqrt{W}} \\ b_{22} &= -\langle \boldsymbol{x}_2, \boldsymbol{N}_2\rangle = \frac{t}{\sqrt{W}}. \end{aligned}$$

Finally, we obtain the Gaussian curvature,

$$K = \frac{\det(b_{\alpha\beta})}{\det(g_{\alpha\beta})} = \frac{rt - s^2}{W^2}. \tag{D.10}$$

Appendix E

Curvature Transformation

(Ref. §3.8.1)

Suppose that we have a vector field $Z \in TM$. We consider parallel translation of a tangent vector $Z \in T_pM$ at p along a small curvilinear (deformed) parallelogram Π_ε. The sides are constructed by using two arcs $\xi_\varepsilon = e^{\varepsilon X}$ and $\eta_\varepsilon = e^{\varepsilon Y}$ emanating from p, where $X, Y \in T_pM$ and ε is an infinitesimal parameter. As considered in §1.7.3, there is a gap between $\eta_\varepsilon \circ \xi_\varepsilon$ and $\xi_\varepsilon \circ \eta_\varepsilon$, and the gap is given by $\eta_\varepsilon \xi_\varepsilon - \xi_\varepsilon \eta_\varepsilon = \varepsilon^2[X, Y]$ in the leading order (see (1.75)). Hence the circuit is actually five-sided.

The parallel translation of Z is carried out as follows. We make a circuit Π_ε in the sense ↻ from p along the side η_ε first and back to p along the side ξ_ε (Fig. E.1). The starting point p is denoted by 4 (where $Z_p = Z_4$) and the end of η_ε is 3 (where $Z = Z_3$), and then the end of $\xi_\varepsilon \circ \eta_\varepsilon$ is denoted as $2'$ (where $Z_{2'}$). The gap is denoted by an arc from $2'$ to 2 (where Z_2). After passing through the point 1 (where Z_1), we trace back to p denoted by 0 (where $Z_0 = Z_4$).

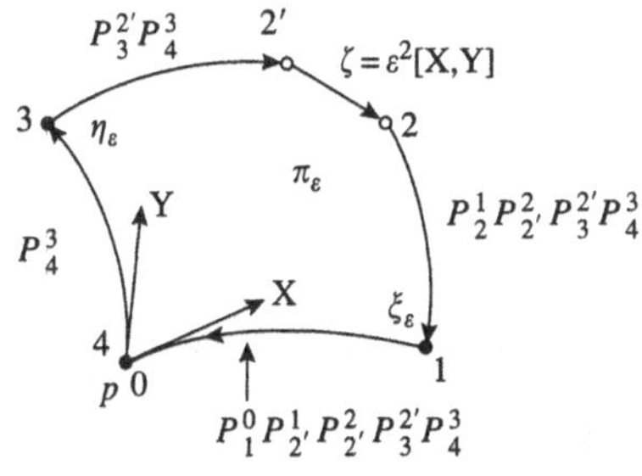

Fig. E.1. Parallel translation.

By the definition of covariant derivative (3.84), the parallel translation of X_t at γ_t back to γ_0 is represented by

$$P_t^0 X_t = X_0 + t(\nabla_T X)_0 + \frac{1}{2}t^2(\nabla_T \nabla_T X)_0 + O(t^3) \tag{E.1}$$

where P_0^t is the operator of a parallel translation from γ_0 to γ_t. In the present case, the parallel translation of Z_0 from 0 to 1 is

$$P_0^1 Z_0 = Z_1 - \varepsilon(\nabla_X Z)_1 + \frac{1}{2}\varepsilon^2(\nabla_X \nabla_X Z)_1 + O(\varepsilon^3). \tag{E.2}$$

Similarly, for the parallel translation of Z_4 from 4 to 3, we have

$$P_4^3 Z_4 = Z_3 - \varepsilon(\nabla_Y Z)_3 + \frac{1}{2}\varepsilon^2(\nabla_Y \nabla_Y Z)_3 + O(\varepsilon^3). \tag{E.3}$$

Subsequent translation from 3 to $2'$ is

$$\begin{aligned}(P_3^{2'} P_4^3) Z_4 = &\left(Z_{2'} - \varepsilon(\nabla_X Z)_{2'} + \frac{1}{2}\varepsilon^2(\nabla_X \nabla_X Z)_{2'}\right)\\ &- \varepsilon((\nabla_Y Z)_{2'} - \varepsilon(\nabla_X(\nabla_Y Z))_{2'}) + \frac{1}{2}\varepsilon^2(\nabla_Y \nabla_Y Z)_{2'},\end{aligned} \tag{E.4}$$

up to $O(\varepsilon^2)$. Performing the next translation from $2'$ to 2 which is of the order ε^2 distance, we obtain

$$\begin{aligned}(P_{2'}^2 P_3^{2'} P_4^3) Z_4 = &\, Z_2 - \varepsilon((\nabla_X Z)_2 + (\nabla_Y Z)_2) - \varepsilon^2(\nabla_{[X,Y]} Z)_2\\ &+ \varepsilon^2\left((\nabla_X(\nabla_Y Z))_2 + \frac{1}{2}(\nabla_X \nabla_X Z)_2 + \frac{1}{2}(\nabla_Y \nabla_Y Z)_2\right),\end{aligned} \tag{E.5}$$

up to $O(\varepsilon^2)$. It is instructive to obtain the consecutive parallel translation of Z_0 from 0 to 1 (given by (E.2)) and from 1 to 2, which is given by

$$\begin{aligned}(P_1^2 P_0^1) Z_0 = &\, Z_2 - \varepsilon((\nabla_X Z)_2 + (\nabla_Y Z)_2)\\ &+ \varepsilon^2\left((\nabla_Y(\nabla_X Z))_2 + \frac{1}{2}(\nabla_X \nabla_X Z)_2 + \frac{1}{2}(\nabla_Y \nabla_Y Z)_2\right).\end{aligned} \tag{E.6}$$

In view of this form, it is finally found that the consecutive parallel translation of $(P_{2'}^2 P_3^{2'} P_4^3) Z_4$ from 2 to 1 and from 1 to 0 is given by

$$\begin{aligned}(P_1^0 P_2^1 P_{2'}^2 P_3^{2'} P_4^3) Z_4 = &\, Z_0 - \varepsilon^2(\nabla_{[X,Y]} Z)_0\\ &+ \varepsilon^2((\nabla_X(\nabla_Y Z))_0 - (\nabla_Y(\nabla_X Z))_0).\end{aligned} \tag{E.7}$$

Now, the transformation operator $g_\varepsilon(X,Y)$ of the parallel translation (3.94) is given by $(P_1^0 P_2^1 P_{2'}^2 P_3^{2'} P_4^3)$. Thus it is found that

$$g_\varepsilon(X,Y) = e + \varepsilon^2 R(X,Y)$$

where $R(X,Y)$ is the operator of the *curvature transformation*:

$$R(X,Y) = \nabla_X \nabla_Y - \nabla_Y \nabla_X - \nabla_{[X,Y]}. \tag{E.8}$$

Appendix F

Function Spaces L_p, H^s and Orthogonal Decomposition

(Ref. §3.12, 8.1, footnote to Chapter 5)

The totality of functions, which are differentiable up to the qth order with all the derivatives being continuous over the manifold M^n, is denoted by $C^q(M)$. A special case is C^0, a class of continuous functions, and another is C^∞ which is a class of infinitely differentiable functions, i.e. all the derivatives exist and are continuous.

A function $f(x)$ is said to belong to the function space $L_p(M)$ if the integral $\int_M |f(x)|^p \, \mathrm{d}\mu(x)$ exists ($\mathrm{d}\mu$ is a volume form). The $L_2(M)$ denotes the functions which are square-integrable over the manifold M^n.

The Sobolev space $W_p^s(M)$ denotes the totality of the functions $f(x) \in L_p(M)$ which have the property $D^s f(x) \in L_p(M)$, where $x = (x_1, \ldots, x_n)$ and $D^s f$ denotes a generalized sth derivative (in the sense of the theory of generalized functions) including the ordinary sth derivative defined by $\partial_1^{s_1} \cdots \partial_n^{s_n} f$ with $s = s_1 + \cdots + s_n$.

The space $W_2^s(M)$ is written as $H^s(M)$. If $s > n/2$, then $H^s \subset C^q(M)$ by the *Sobolev's imbedding theorem*, where $q \equiv [s - n/2]$ is the maximum integer not larger than $s - n/2$. Therefore, if $s > n/2 + 1$, then $q \geq 1$, and a function $g \in H^s$ is continuously differentiable at least once.

An arbitrary vector field v on M can be decomposed orthogonally into divergence-free and gradient parts. In fact, an H^s vector field $v \in T_e M$ is written as

$$v = P_e(v) + Q_e(v), \tag{F.1}$$

where

$$Q_e(v) = \operatorname{grad} F_D + \operatorname{grad} H_N := \operatorname{grad} f,$$
$$P_e(v) = v - Q_e(v)$$

($f \in H^{s+1}$). The scalar functions F_D and H_N are the solutions of the following Dirichlet problem and Neumann problem, respectively,

$$\Delta F_D(v) = \operatorname{div} v, \quad \text{where} \quad \operatorname{supp} F_D \subset M,$$
$$\Delta H_N(v) = 0, \quad \text{and} \quad \langle \nabla H_N, \boldsymbol{n} \rangle = \langle v - \nabla F_D, \boldsymbol{n} \rangle,$$

(hence $\langle P_e(v), \boldsymbol{n} \rangle = 0$) where $\boldsymbol{n}$ is the unit normal on the boundary ∂M. There is orthogonality, $\langle \operatorname{grad} F_D, \operatorname{grad} H_N \rangle = 0$. Then, it can be shown that

$$\operatorname{div} P_e(v) = 0,$$
$$\langle P_e(v), Q_e(v) \rangle = 0.$$

This orthogonal decomposition into the divergence-free part $P_e(v)$ and gradient part $Q_e(v)$ is called the *Helmholtz decomposition* or *Hodge decomposition* or Weyl decomposition [Mis93].

Appendix G

Derivation of KdV Equation for a Shallow Water Wave

(Ref. §5.1)

G.1. Basic Equations and Boundary Conditions

We consider a surface wave of water of depth h. Suppose that water is incompressible and inviscid, and waves are excited on water otherwise at rest. Then the water motion is irrotational, and the velocity field $\boldsymbol{v}$ is represented by a velocity potential Φ: $\boldsymbol{v} = \mathrm{grad}\,\Phi$. We consider the two-dimensional problem of waves in the (x, y)-plane with the horizontal coordinate x and the vertical coordinate y, and denote the velocity as

$$\boldsymbol{v} = (u, v) = (\Phi_x, \Phi_y),$$

where $\Phi_x = \partial_x \Phi$, etc. The undisturbed horizontal free surface is specified by $y = 0$. Let the water surface be described by

$$y = z(x, t). \tag{G.1}$$

Incompressible irrotational motion in the (x, y)-plane must satisfy $u_x + v_y = 0$, which reduces to the following Laplace equation:

$$\Delta\Phi = \Phi_{xx} + \Phi_{yy} = 0, \quad \text{for} \quad -h < y < z(x, t). \tag{G.2}$$

The boundary condition at the horizontal bottom at $y = -h$ is

$$v = \Phi_y = 0, \quad \text{at} \quad y = -h. \tag{G.3}$$

The surface deforms freely subject to the following two boundary conditions. The first is the *pressure condition*, that is, the pressure p must be equal to the atmospheric pressure p_0 over the surface. This is represented by

$$\Phi_t + \frac{1}{2}|v|^2 + gz = 0, \quad \text{at} \quad y = z(x,t) \tag{G.4}$$

[Ach90, §3.2], where g is the acceleration of gravity.

The second is the *kinematic condition*, that is, the fluid particle on the free surface $y = z(x,t)$ must move with the surface and remain on the surface. This is represented by

$$z_t + uz_x = v, \quad \text{at} \quad y = z(x,t). \tag{G.5}$$

G.2. Long Waves in Shallow Water

There exist three length scales in the problem of long waves in a shallow water channel: water depth h, wave amplitude a and a horizontal scale of the wave λ. In order to derive an equation expressing the balance of the finiteness of wave amplitude and wave dispersion, it is supposed that the following two dimensionless parameters are small,

$$\alpha = \frac{a}{h}, \quad \beta = \left(\frac{h}{\lambda}\right)^2, \tag{G.6}$$

and, in addition, their orders of magnitude are as follows:

$$\alpha \approx \beta \quad (\ll 1). \tag{G.7}$$

The governing equation and the boundary conditions given in the previous section can be normalized by using the following dimensionless variables:

$$\xi = \left(\frac{\alpha}{\beta}\right)^{1/2} \frac{x - c_* t}{\lambda}, \quad \tau = \left(\frac{\alpha^3}{\beta}\right)^{1/2} \frac{c_* t}{\lambda},$$

$$\phi = \left(\frac{\alpha}{\beta}\right)^{1/2} \frac{c_* \Phi}{g a \lambda}, \quad \zeta = \frac{z}{a}, \quad \eta = \frac{y}{h},$$

where $c_* = \sqrt{gh}$. Using these dimensionless variables, the set of equations (G.2)–(G.5) are transformed to

$$\begin{aligned}
&\text{(i) } \phi_{\eta\eta} + \alpha\phi_{\xi\xi} = 0 && (-1 < \eta < \alpha\zeta) \\
&\text{(ii) } \phi_\eta = 0 && (\eta = -1) \\
&\text{(iii) } \zeta - \phi_\xi + \alpha\phi_\tau + \frac{1}{2}(\phi_\eta^2 + \alpha\phi_\xi^2) = 0 && (\eta = \alpha\zeta) \\
&\text{(iv) } \phi_\eta + \alpha(\zeta_\xi - \alpha\zeta_\tau - \alpha\phi_\xi\zeta_\xi) = 0 && (\eta = \alpha\zeta).
\end{aligned}$$

It is not difficult to see that the following function $\phi(\xi, \eta, \tau)$ satisfies both (i) and (ii):

$$\phi(\xi, \eta, \tau) = \sum_{m=0}^{\infty} \frac{(-1)^m}{(2m)!}\alpha^m(\eta+1)^{2m}\left(\frac{\partial}{\partial\xi}\right)^{2m} f(\xi, \tau). \tag{G.8}$$

Let us expand $f(\xi, \tau)$ and $\zeta(\xi, \tau)$ in the power series of α as

$$f(\xi, \tau) = f_0(\xi, \tau) + \alpha f_1(\xi, \tau) + \alpha^2 f_2(\xi, \tau) + \cdots, \tag{G.9}$$

$$\zeta(\xi, \tau) = \zeta_0(\xi, \tau) + \alpha\zeta_1(\xi, \tau) + \alpha^2\zeta_2(\xi, \tau) + \cdots. \tag{G.10}$$

Substituting (G.9) in (G.8), we have

$$\phi = f_0 + \alpha\left(f_1 - \frac{1}{2}Y^2\partial_\xi^2 f_0\right) + \alpha^2\left(f_2 - \frac{1}{2}Y^2\partial_\xi^2 f_1 + \frac{1}{24}Y^4\partial_\xi^4 f_0\right) + O(\alpha^3),$$

where $Y = 1 + \eta$. Using this and (G.10) and setting $Y = 1 + \alpha\zeta$, the boundary conditions (iii) and (iv) are expanded with respect to α as

$$\text{(iii)}' \ \zeta_0 - \partial_\xi f_0 + \alpha\left(\zeta_1 - \partial_\xi f_1 + \frac{1}{2}\partial_\xi^3 f_0 + \partial_\tau f_0 + \frac{1}{2}(\partial_\xi f_0)^2\right) + O(\alpha^2) = 0,$$

$$\begin{aligned}
\text{(iv)}' \ -\partial_\xi^2 f_0 + \partial_\xi\zeta_0 + \alpha\Big(&-\zeta_0\partial_\xi^2 f_0 - \partial_\xi^2 f_1 + \frac{1}{6}\partial_\xi^4 f_0 \\
&+ \partial_\xi\zeta_1 - \partial_\tau\zeta_0 - \partial_\xi\zeta_0\partial_\xi f_0\Big) + O(\alpha^2) = 0.
\end{aligned}$$

Vanishing of $O(\alpha^0)$ terms of (iii)′ and (iv)′ leads to $\zeta_0 - \partial_\xi f_0 = 0$ and $\partial_\xi(\zeta_0 - \partial_\xi f_0) = 0$, respectively. Thus, we obtain the first compatibility relation,

$$\zeta_0 = \partial_\xi f_0, \tag{G.11}$$

stating that the zeroth order elevation ζ_0 is equal to the zeroth order horizontal velocity $\partial_\xi f_0$.

Next, vanishing of $O(\alpha^1)$ terms of (iii)′ and (iv)′ leads to

$$\zeta_1 - \partial_\xi f_1 = -\partial_\tau f_0 - \frac{1}{2}(\zeta_0)^2 - \frac{1}{2}\partial_\xi^2 \zeta_0, \tag{G.12}$$

$$\partial_\xi(\zeta_1 - \partial_\xi f_1) = \partial_\tau \zeta_0 + 2\zeta_0 \partial_\xi \zeta_0 - \frac{1}{6}\partial_\xi^3 \zeta_0, \tag{G.13}$$

respectively, where $\partial_\xi f_0 = \zeta_0$ is used. Compatibility of both equations requires that the right-hand side of (G.13) should be equal to ∂_ξ of (G.12). Thus, we finally obtain the following KdV equation for $u = \zeta_0$:

$$2\partial_\tau u + 3u\partial_\xi u + \frac{1}{3}\partial_\xi^3 u = 0. \tag{G.14}$$

Appendix H

Two-Cocycle, Central Extension and Bott Cocycle

(Ref. §5.3, 5.4 and 9.8)

H.1. Two-Cocycle and Central Extension

The elements of the group $D(S^1)$ describe diffeomorphisms of a circle S^1, $g : z \in S^1 \mapsto g(z) \in S^1$. We may write $z = e^{ix}$ and consider the map, $x \mapsto g(x)$ such that $g(x+2\pi) = g(x)+2\pi$ (in the main text the variable x is written as ϕ here), with the composition law, $g''(x) = (g' \circ g)(x) = g'(g(x))$, where $g, g', g'' \in D(S^1)$. Writing $e^{ix} =: F_e(x)$, we can consider the following transformation by the mapping $x' = g(x)$:

$$\begin{aligned} F_g(x') := \exp[i\eta(g)] \exp[ig(x)] &= \exp[i\Delta(g,x)]F_e(x) \\ &= \exp[i(\Delta(g,x) + x)], \end{aligned} \tag{H.1}$$

i.e. there is a phase shift $\eta(g) : D(S^1) \to \mathbb{R}$ in the transformed function F_g, where

$$\Delta(g,x) = g(x) - x + \eta(g). \tag{H.2}$$

We are going to show that F_g is a function on $\hat{D}(S^1)$, whereas $e^{ig(x)}$ is a function on $D(S^1)$. The above transformation allows us to define the composition law for two successive transformations as follows. For $x'' = g'x' = g'g(x)$, we may write

$$F_{g'g}(x'') = \exp[i\Delta(g'g, x)]F_e(x). \tag{H.3}$$

The composition is written as

$$F_{g'}F_g(x'') = \exp[i\Delta(g', x')]F_g(x') = \exp[i\Delta(g', x') + i\Delta(g, x)]F_e(x). \tag{H.4}$$

Thus, eliminating $F_e(x)$ between (H.3) and (H.4), we obtain

$$F_{g'}F_g = \omega(g', g)F_{g'g}, \quad \omega(g', g) = \exp[i\Theta(g', g)], \tag{H.5}$$

$$\Theta(g', g) := \Delta(g', x') + \Delta(g, x) - \Delta(g'g, x), \tag{H.6}$$

where $\Theta(g', g)$ is called the *local* exponent. The transformation (H.5) is called the *projective* representation. Requiring that associativity (associative property) holds for F_g, we obtain the following *two-cocycle condition*:

$$\omega(g'', g')\omega(g''g', g) = \omega(g'', g'g)\omega(g', g). \tag{H.7}$$

In fact, we have $[F_{g''}F_{g'}]F_g = \omega(g'', g')\omega(g''g', g)F_{g''g'g}$ and $F_{g''}[F_{g'}F_g] = \omega(g'', g'g)\omega(g', g)F_{g''g'g}$. The $\omega(g', g)$ satisfying (H.7) is called the *two-cocycle*. In terms of the exponents $\Theta(g', g)$, the two-cocycle condition reads

$$\Theta(g'', g') + \Theta(g''g', g) = \Theta(g'', g'g) + \Theta(g', g). \tag{H.8}$$

Substituting (H.2) into (H.6), we find

$$\Theta(g', g) = \eta(g') + \eta(g) - \eta(g'g) : D \times D \to \mathbb{R},$$

$$\omega(g', g) = \gamma_{g'}\gamma_g\gamma_{g'g}^{-1} := \omega_{cob}(g', g), \quad \text{where} \quad \gamma_g = \exp[i\eta(g)].$$

With this form, the cocycle condition (H.7) is identically satisfied. In general, two *two-cocycles* ω and Ω are said to be equivalent, if there exists a factor ω_{cob} such that $\omega(g', g) = \Omega(g', g)\omega_{cob}(g', g)$.[1] The $\omega_{cob}(g', g)$ itself is a two-cocycle corresponding to $\Omega = 1$ and is called a *trivial* two-cocycle, or two-coboundary. The present problem is such a case.

Let us now consider briefly the problem of projective (or *ray*) representations in order to define the central extension. The ray operators $\bar{F}$ satisfy the relation:

$$\bar{F}_{g'}\bar{F}_g = \bar{F}_{g'g}, \quad g, g' \in D, \tag{H.9}$$

where the bar indicates the class of equivalent operators which differ in a phase $\theta \in \mathbb{R}$, i.e. $F, F' \in \bar{F} \Leftrightarrow F' = \gamma F$, where $\gamma = e^{i\theta}$ (said to be an element of the group $U(1)$). If a representative class F_g is selected in the

[1] The classes of inequivalent two-cocycles define the *second cohomology group* $H^2(G, U(1))$ for a group G.

class $\bar{F}_g$, (H.9) will be written as $F_{g'}F_g = \omega(g',g)F_{g'g}$ like (H.5). Next, inside the class $\bar{F}_g$, let us take other operators in the form, $e^{i\theta}F_g$ with a new variable θ. Then

$$e^{i\theta'}F_{g'}e^{i\theta}F_g = e^{i(\theta'+\theta)}e^{i\Theta(g',g)}F_{g'g} := e^{i\theta''}F_{g''}.$$

Suppose that, associated with a group D, we are given a local exponent $\Theta : D \times D \to \mathbb{R}$. Then, we can define a new group $\hat{D}$ consisting of elements $\hat{g} = (g, \theta)$ together with the group operation:

$$\hat{g'} \circ \hat{g} = (g', \theta') \circ (g, \theta) = (g' \circ g, \theta' + \theta + \Theta(g', g)). \tag{H.10}$$

In summary, the extended group $\hat{D}$ is such that: (1) it contains $U(1)$ as an invariant subgroup and $\hat{D}/U(1) = D$. In fact, the invariant subgroup $U(1)$ is a center i.e. the element (id, θ) commutes with any element $(g, \theta') \in \hat{D}$. Namely, $\hat{D}$ is a *central extension* of D by $U(1)$; (2) D is not a subgroup of $\hat{D}$.

H.2. Bott Cocycle

It can be verified that the following [Bott77] satisfies the two-cocycle condition (H.8) for the local exponent $B(g', g)$ in place of $\Theta(g', g)$:

$$B(g', g) = \frac{1}{2}\int_{S^1} \ln \partial_x(g' \circ g)\mathrm{d} \ln \partial_x g. \tag{H.11}$$

In fact, noting that $\partial_x(g' \circ g)(x) = \partial_x(g'(g(x)) = g'_x g_x$, its right-hand side is

$$\begin{aligned} B(g', g) + B(g'', g'g) &= \frac{1}{2}\int_{S^1} [\ln(g'_x g_x)\mathrm{d}\ln(g_x) + \ln(g''_x g'_x g_x)\mathrm{d}\ln(g'_x g_x)] \\ &= \frac{1}{2}\int_{S^1} [\ln(g'_x)\mathrm{d}\ln(g_x) + \ln(g''_x)\mathrm{d}[\ln(g'_x) + \ln(g_x)]], \end{aligned}$$

since $\ln(g'_x g_x) = \ln(g'_x) + \ln(g_x)$ and $\int_{S^1} \ln(g_x)\mathrm{d}\ln(g_x) = \int_{S^1} \mathrm{d}(\ln(g_x))^2/2 = 0$. Similarly, the left-hand side is

$$B(g'', g') + B(g''g', g) = \frac{1}{2}\int_{S^1} [\ln(g''_x)\mathrm{d}\ln(g'_x) + \ln(g''_x g'_x)\mathrm{d}\ln(g_x)].$$

Thus, it is found that the two-cocycle condition $B(g', g) + B(g'', g'g) = B(g'', g') + B(g''g', g)$ is satisfied.

H.3. Gelfand–Fuchs Cocycle: An Extended Algebra

Here is a note given about the relation between the group cocycle $B(g, f)$ and the algebra cocycle $c(u, v)$ defined in §5.4. On the extended space $\hat{D}(S^1)$, we consider two flows $\hat{\xi}_t$ and $\hat{\eta}_s$ generated by $\hat{u} = (u\partial_x, \alpha)$ and $\hat{v} = (v\partial_x, \beta)$ respectively, defined by

$$t \mapsto \hat{\xi}_t \quad \text{where} \quad \hat{\xi}_0 = (e, 0), \quad \left.\frac{\mathrm{d}}{\mathrm{d}t}\right|_{t=0} \hat{\xi}_t = \hat{u},$$
$$s \mapsto \hat{\eta}_s \quad \text{where} \quad \hat{\eta}_0 = (e, 0), \quad \left.\frac{\mathrm{d}}{\mathrm{d}s}\right|_{s=0} \hat{\eta}_s = \hat{v}$$

(see §1.7.3). Lie bracket of the two tangent vectors $\hat{u}$ and $\hat{v}$ is defined by

$$[\hat{u}, \hat{v}](f) = \hat{u}(\hat{v}(f))|_{(e,0)} - \hat{v}(\hat{u}(f))|_{(e,0)}.$$

According to Eq. (1.71), we have

$$[\hat{u}, \hat{v}] = (\partial_t\partial_s\hat{\eta}_s \circ \hat{\xi}_t - \partial_s\partial_t\hat{\xi}_t \circ \hat{\eta}_s)|_{(e,0)}.$$

Denoting only the extended component of the product $\hat{\eta} \circ \hat{\xi}$ of (5.19) (or (H.10)) as $\mathrm{Ext}\{\hat{\eta}_s \circ \hat{\xi}_t\}$, we have

$$\mathrm{Ext}\{\hat{\eta}_s \circ \hat{\xi}_t\} = a_t + b_s + B(\eta_s, \xi_t),$$

where suffices s and t are parameters. Therefore,

$$\partial_t\partial_s \,\mathrm{Ext}\{\hat{\eta}_s \circ \hat{\xi}_t\} = \partial_t\partial_s B(\eta_s, \xi_t),$$

and

$$\mathrm{Ext}\{[\hat{u}, \hat{v}]\} = \partial_t\partial_s B(\eta_s, \xi_t) - \partial_s\partial_t B(\xi_t, \eta_s)|_{(e,0)}.$$

Carrying out the calculation, we have

$$\partial_s B(\eta_s, \xi_t) = \partial_s \int_{S^1} \ln \partial_x(\eta_s \circ \xi_t) \mathrm{d} \ln \partial_x \xi_t = \int_{S^1} \frac{\partial_x(v \circ \eta_s \circ \xi_t)}{\partial_x(\eta_s \circ \xi_t)} \mathrm{d} \ln \partial_x \xi_t.$$

Since the Taylor expansion of ξ_t is $\xi_t = e + tu(x) + O(t^2)$, its x-derivative is $\partial_x\xi_t = 1 + t\partial_x u + O(t^2)$. Hence, we obtain

$$\ln \partial_x\xi_t = \ln(1 + t\partial_x u + O(t^2)) = t\partial_x u + O(t^2)$$
$$\mathrm{d} \ln \partial_x \xi_t = (tu_{xx})\mathrm{d}x$$

and furthermore,

$$\begin{aligned}
\eta_s \circ \xi_t &= (x + sv(x) + \cdots) \circ (x + tu(x) + \cdots) \\
&= x + tu(x) + sv(x + tu(x) + \cdots) + \cdots \\
&= x + tu(x) + sv(x) + O(s^2, st, t^2) \\
\partial_x(\eta_s \circ \xi_t) &= 1 + tu_x + sv_x + O(s^2, st, t^2).
\end{aligned}$$

Therefore we obtain

$$\partial_s B(\eta_s, \xi_t) = \int_{S^1} \frac{\partial_x v(x + O(s,t))}{1 + tu_x + sv_x + O(s^2, st, t^2)} (tu_{xx}) \mathrm{d}x.$$

Differentiating with respect to t and setting $t = 0$ and $s = 0$,

$$\partial_t \partial_s B(\eta_s, \xi_t)|_{s=0,t=0} = \int_{S^1} v_x u_{xx} \mathrm{d}x.$$

Thus finally we find

$$\begin{aligned}
c(u,v) = \mathrm{Ext}\{[\hat{u}, \hat{v}]\} &= \int_{S^1} v_x u_{xx} \mathrm{d}x - \int_{S^1} u_x v_{xx} \mathrm{d}x \\
&= 2 \int_{S^1} u_{xx} v_x \mathrm{d}x = -c(v,u).
\end{aligned}$$

This verifies the expression (5.22) of §5.4. This form of $c(u,v)$ is called the Gelfand–Fuchs cocycle [GF68]. The anti-symmetry of $c(u,v)$ can be shown by performing integration by parts and using the periodicity.

Appendix I

Additional Comment on the Gauge Theory of §7.3

By the replacement of ∂_μ with $\nabla_\mu = \partial_\mu - iqA_\mu(x)$ defined by (7.20) in the quantum electrodynamics of §7.3 (i), the Lagrangian density Λ_{free} of (7.16) is transformed to $\Lambda(\psi, A_\mu)$. A new term thus introduced is an interaction term $-A_\mu J^\mu$ between the gauge field A_μ and the electromagnetic current density (matter field) $J^\mu = -q\bar{\psi}\gamma^\mu\psi$.

To arrive at the complete Lagrangian, it remains to add an electromagnetic field term $\Lambda_F = -\frac{1}{16\pi}F_{\mu\nu}F^{\mu\nu}$ to the Lagrangian, where $F_{\mu\nu} = \partial_\mu A_\nu - \partial_\nu A_\mu$, and $F^{\mu\nu} = g^{\mu\alpha}F_{\alpha\beta}g^{\beta\nu}$ with the metric tensor $g^{\alpha\beta} = \text{diag}(-1, 1, 1, 1)$. Assembling all the pieces, we have the total Lagrangian for quantum electrodynamics,

$$\Lambda_{\text{qed}} = \Lambda(\psi, A_\mu) + \Lambda_F. \tag{I.1}$$

Thus, variation with respect to A_μ yields the equations for the gauge field A_μ, i.e. the Maxwell's equations in electromagnetism, whereas variation of Λ with respect to ψ yields the equation of quantum electrodynamics, i.e. the Dirac equation with electromagnetic field.

Using the notations $\boldsymbol{B} = \nabla \times \boldsymbol{A}$ and $\boldsymbol{E} = -\nabla\phi - c^{-1}\partial_t \boldsymbol{A}$ of the magnetic 3-vector $\boldsymbol{B}$ and electric 3-vector $\boldsymbol{E}$, we obtain

$$F_{\mu\nu} = \partial_\mu A_\nu - \partial_\nu A_\mu = \begin{pmatrix} 0 & -E_1 & -E_2 & -E_3 \\ E_1 & 0 & B_3 & -B_2 \\ E_2 & -B_3 & 0 & B_1 \\ E_3 & B_2 & -B_1 & 0 \end{pmatrix}.$$

In the Yang–Mills's system of §7.3(ii), the gauge-covariant derivative is defined by $\nabla_\mu = \partial_\mu - iq\boldsymbol{\sigma} \cdot \boldsymbol{A}_\mu(x)$, where $\boldsymbol{\sigma} = (\sigma_1, \sigma_2, \sigma_3)$ are the three-components Pauri matrices given in the main text. In addition, the following three gauge fields (colors) are defined: $\boldsymbol{A}^k = (A_0^k, A_1^k, A_2^k, A_3^k)$ with

$k = 1, 2, 3$. The new fields $\boldsymbol{A}^1, \boldsymbol{A}^2, \boldsymbol{A}^3$ are the Yang–Mills gauge fields. The connection $iq\boldsymbol{\sigma} \cdot \boldsymbol{A}_\mu$ leads to the interaction term, that couples the gauge field with quark current, corresponding to the middle term of (I.1). Finally, a gauge field term (called a kinetic term, corresponding to the third term of (I.1)) should be added to complete the Yang–Mills action functional. [Fra97, Chap. 20]

Appendix J

Frobenius Integration Theorem and Pfaffian System

(Ref. §1.5.1, 8.8.1 and 11.2)

We make a local consideration in a neighborhood U of the origin 0 in $\mathbb{R}^n$. Let $x = (x^1, \ldots, x^n) \in U$ and let $\omega(x) = a_1 \mathrm{d}x^1 + \cdots + a_n \mathrm{d}x^n = a_i \mathrm{d}x^i$ be a 1-form which does not vanish at 0. We look for an integrating factor for the first order differential equation $\omega(x) = 0$, called a Pfaffian equation. In other words, we find the conditions for which functions f and g satisfy $\omega = f \mathrm{d}g$. If $\omega = f \mathrm{d}g$, then f does not vanish in a neighborhood of 0, hence

$$\mathrm{d}\omega = \mathrm{d}f \wedge \mathrm{d}g = \mathrm{d}f \wedge f^{-1}\omega = \theta \wedge \omega,$$

where $\theta = f^{-1}\mathrm{d}f = \mathrm{d}\ln|f|$. So that,

$$\omega \wedge \mathrm{d}\omega = \omega \wedge \theta \wedge \omega = 0.$$

For a 1-form $\omega = P\mathrm{d}x + Q\mathrm{d}y + R\mathrm{d}z$ in $\mathbb{R}^3$, this condition leads to

$$\boldsymbol{X} \cdot \mathrm{curl}\, \boldsymbol{X} = P(R_y - Q_z) + Q(P_z - R_x) + (R(Q_x - P_y) = 0, \qquad \text{(J.1)}$$

where $\boldsymbol{X} = (P, Q, R)$. Thus, if $\omega = f \mathrm{d}g$, then the equation $\omega = 0$ and $\mathrm{d}g = 0$ are the same, and hence the solutions or the integral surfaces are given by the hypersurfaces $g = $ constant. This corresponds to the existence of the Bernoulli surface (§8.8.1).

For example, let $\omega = yz\mathrm{d}x + xz\mathrm{d}y + \mathrm{d}z$, so that $\mathrm{d}\omega = y\mathrm{d}z \wedge \mathrm{d}x - x\mathrm{d}y \wedge \mathrm{d}z$ and $\boldsymbol{X} = (yz, xz, 1)$, and let us consider the integral surfaces obtained from the Pfaffian equation $\omega = 0$. Since we have $\boldsymbol{X} \cdot \mathrm{curl}\, \boldsymbol{X} = (yz, xz, 1) \cdot (-x, y, 0) = 0$, we can expect to obtain the expression, $\mathrm{d}\omega = \theta \wedge \omega$. In fact, $\theta = -y\mathrm{d}x - x\mathrm{d}y$ will do. This suggests the representation, $\omega = f\mathrm{d}g$. In fact, it can be easily checked that the two functions, $f = e^{-xy}, g = ze^{xy}$, yield

$f\mathrm{d}g = e^{-xy}(z(y\mathrm{d}x + x\mathrm{d}y) + \mathrm{d}z)e^{xy} = \omega$. Thus, the integral surfaces are found to be $ze^{xy} =$ constant.

Theorem (Euler's integrability condition). *Let $\omega = f_i \mathrm{d}x^i$ be a 1-form which does not vanish at 0. Suppose there is a 1-form θ satisfying $\mathrm{d}\omega = \theta \wedge \omega$. Then there are functions f and g in a sufficiently small neighborhood of 0 which satisfy $\omega = f\mathrm{d}g$.*

Here, we only refer to textbooks (e.g. [Fla63; Fra97; AM78]) for its proof.

We now pass to the general problem. Let $\omega^1, \ldots, \omega^k$ be 1-forms in $n(= k + m)$-dimensional space, linearly independent at 0. Set $\Omega = \omega^1 \wedge \cdots \wedge \omega^k$. The system $\omega^1 = 0, \ldots, \omega^k = 0$ is called the *Pfaffian system*. The system is called *completely integrable* if it satisfies any of the conditions of the following lemma.

Lemma. *The following three conditions are equivalent:*

(i) *There exist 1-forms θ^i_j satisfying*

$$\mathrm{d}\omega^i = \sum_{j=1}^{n} \theta^i_j \wedge \omega^j \quad (i = 1, \ldots, k).$$

(ii) $\mathrm{d}\omega^i \wedge \Omega = 0$ $(i = 1, \ldots, k)$.

(iii) *There exists a 1-form λ satisfying*

$$\mathrm{d}\Omega = \lambda \wedge \Omega.$$

Theorem (Frobenius Integration Theorem). *Let $\omega^1, \ldots, \omega^k$ be 1-forms in $\mathbb{R}^n (n = k+m)$ linearly independent at 0. Suppose there are 1-forms θ^i_j satisfying*

$$\mathrm{d}\omega^i = \sum_{j=1}^{k} \theta^i_j \wedge \omega^j \quad (i = 1, \ldots, k).$$

Then there are functions f^i_j and g^j satisfying

$$\omega^i = \sum_{j=1}^{k} f^i_j \, \mathrm{d}g^j \quad (i = 1, \ldots, k).$$

See Flanders [Fla63] for the proof of the Lemma and Theorem.

Appendix K

Orthogonal Coordinate Net and Lines of Curvature

(Ref. §10.1)

The principal directions T^β corresponding to the extremum of the normal curvature κ_N of a surface Σ^2 are determined by (2.61) of §2.5.4:

$$(b_{\alpha\beta} - \lambda g_{\alpha\beta})T^\beta = 0, \quad (\alpha, \beta = 1, 2). \tag{K.1}$$

The equation $|b_{\alpha\beta} - \lambda g_{\alpha\beta}| = 0$ is necessary and sufficient for the non-trivial solution. This is a quadratic equation for the eigenvalue (principal value) λ. The discriminant D of the quadratic equation reduces to

$$D = (b_{11}g_{22} - b_{22}g_{11})^2 + 4(b_{12}g_{11} - b_{11}g_{12})(b_{12}g_{22} - b_{22}g_{12}).$$

Suppose that the coordinate curves, u^1 = const and u^2 = const, form an orthogonal net, then we have $g_{12} = 0$ from (2.16). For such a system, we have $D = (b_{11}g_{22} - b_{22}g_{11})^2 + 4b_{12}^2 g_{11}g_{22}$, which is non-negative, since g_{11} and g_{22} are positive (§2.1). Hence the principal values are real.[1]

In the case of two distinct real eigenvalues, we denote the larger one (the maximum κ_N) by κ_1 and the smaller one (the minimum κ_N) by κ_2, and the corresponding vectors by T_1^α and T_2^α, respectively. Thus we have

$$(b_{\alpha\beta} - \kappa_1 g_{\alpha\beta})T_1^\beta = 0, \quad (b_{\alpha\beta} - \kappa_2 g_{\alpha\beta})T_2^\beta = 0.$$

We multiply these equations by T_2^α and T_1^α respectively, sum-up with respect to α in each equation, and subtract the resulting equations. Then,

[1] They are distinct unless $b_{12} = 0$ and $b_{\alpha\beta} = cg_{\alpha\beta}$. Satisfying these two conditions results in $\lambda = c$ and the principal direction is not determined (c is a real constant). When this holds at every point, the surface is either a plane or sphere.

we get

$$(\kappa_2 - \kappa_1) g_{\alpha\beta} T_1^\alpha T_2^\beta = 0. \tag{K.2}$$

Since $\kappa_1 \neq \kappa_2$, we find $g_{\alpha\beta} T_1^\alpha T_2^\beta = \langle T_1, T_2 \rangle = 0$. It follows that the two principal directions T_1 and T_2 are orthogonal.

Thus, the two vectors T_1 and T_2 are the tangents to the curves of orthogonal net on the surface. In order to obtain the differential equation for the curves, we replace T^β by $\mathrm{d}u^\beta$ in (K.1) for $\alpha = 1, 2$. Eliminating λ, one obtains the following equation,

$$(b_{1\alpha} \mathrm{d}u^\alpha)(g_{2\alpha} \mathrm{d}u^\alpha) - (b_{2\alpha} \mathrm{d}u^\alpha)(g_{1\alpha} \mathrm{d}u^\alpha) = 0 \tag{K.3}$$

which defines the *lines of curvature.*

Suppose that the *coordinate curves* are the lines of curvature. Setting $(\mathrm{d}u^1, \mathrm{d}u^2) = (\Delta_1 u^1, 0)$ or $(0, \Delta_2 u^2)$, we obtain the following equations:

$$b_{11} g_{12} - b_{12} g_{11} = 0, \quad b_{12} g_{22} - b_{22} g_{12} = 0. \tag{K.4}$$

This states that, unless $g_{12} = b_{12} = 0$, the two fundamental forms $g_{\alpha\beta}$ and $b_{\alpha\beta}$ are proportional, which is the case only when the surface is a plane or sphere.

Since we are considering general surfaces, we have the following. A necessary and sufficient condition that the coordinate curves are orthogonal and coincide with the lines of curvature is,

$$g_{12} = 0 \quad \text{and} \quad b_{12} = 0. \tag{K.5}$$

However, on a plane or a sphere, these conditions are satisfied by any orthogonal net.

Thus, the lines of curvature form an *orthogonal coordinate net* when $g_{12} = 0$ and $b_{12} = 0$. In addition, we ask what condition is necessary if the coordinate curves are geodesics. If the curve $u^1 = \text{const}$ is a geodesic, we have $\mathrm{d}u^1/\mathrm{d}s = 0$ along the curve. Then the geodesic equation (2.65) of Chapter 2 gives $\Gamma^1_{22} = 0$. Likewise, if the curve $u^2 = \text{const}$ is a geodesic, we have $\Gamma^2_{11} = 0$. From the definition (2.40), we obtain

$$\Gamma^1_{22} = -\frac{g_{22}}{2g} \frac{\partial g_{22}}{\partial u^1} = -\frac{1}{2g_{11}} \frac{\partial g_{22}}{\partial u^1},$$

since $g = g_{11} g_{22}$. Therefore, if the coordinate curves $u^1 = \text{const}$ are geodesic, we must have $\partial g_{22}/\partial u^1 = 0$. Likewise, the coordinate curves $u^2 = \text{const}$ are geodesic, if $\partial g_{11}/\partial u^2 = 0$. These are also adequate. Thus, the necessary and sufficient condition that the curves $u^\alpha = \text{const}$ be geodesics is that $g_{\beta\beta}$ (where $\beta \neq \alpha$) be a function of u^β alone.

Suppose that the curves $u^2 = \text{const}$ are geodesics. Then we have $g_{11} = g_{11}(u^1)$, and a coordinate u can be chosen so that $du = g_{11}(u^1)du^1$, and the line-element is

$$\mathrm{I} = ds^2 = (du)^2 + g_{22}(v)(dv)^2, \tag{K.6}$$

where v is used instead of u^2.

Suppose that we have an orthogonal *conjugate* coordinate net.[2] In mathematical terms, two conjugate directions denoted by two tangents T_1 and T_2 must satisfy the following: $B \equiv b_{\alpha\beta}T_1^\alpha T_2^\beta = 0$. In fact, writing the tangents as $T_1 = (T_1^1, 0)$ and $T_2 = (0, T_2^2)$ for the two lines of curvature intersecting orthogonally at a point, it is satisfied, since

$$B = b_{11}T_1^1T_2^1 + b_{12}T_1^1T_2^2 + b_{21}T_1^2T_2^1 + b_{22}T_1^2T_2^2 = b_{12}T_1^1T_2^2 = 0$$

by (K.5). Obviously, the conjugacy of directions is reciprocal, and the principal directions T_1 and T_2 are self-conjugate.

For a surface of nonzero Gaussian curvature ($K \neq 0$), we have $b = \det(b_{\alpha\beta}) \neq 0$ by the definition (2.62). In such a case, the second fundamental form can be expressed as

$$\mathrm{II} = A((du^1)^2 + \varepsilon(du^2)^2),$$

by an appropriate transformation [Eis47, §42], where ε is $+1$ or -1 according as K is positive or negative. Such coordinate curves are said to form an *(isometric-)conjugate net.* In this coordinate system, $b_{22} = \varepsilon b_{11}$. For a surface of constant $K \neq 0$, one may write $K = \varepsilon/a^2$. Then the definition (2.62) leads to the equation, $(b_{11})^2 = g/a^2$, where $g = \det(g_{\alpha\beta}) > 0$. Thus, we have

$$b_{11} = \frac{\sqrt{g}}{a}, \quad b_{22} = \varepsilon\frac{\sqrt{g}}{a}, \quad b_{12} = 0. \tag{K.7}$$

[2]Through each point of a curve C on a surface, there passes a generator of a developable surface which is formed by the envelope of a one-parameter family of tangent planes to the surface at points of the curve. The directions of the generator and the tangent are said to have *conjugate directions*. Suppose that two families of curves form a net, such that a curve of each family passes through a point of the surface and at the point their directions are conjugate. Such a net is called a *conjugate net.*

References

[AA68]: Arnold, V.I. and Avez, A., 1968. *Ergodic Problems of Classical Mechanics.* W.A. Benjamin, Inc.

[Ach90]: Acheson, D.J., 1990. *Elementary Fluid Dynamics.* Oxford University Press.

[ACPRZ02]: Angelani, L., Casetti, L., Pettini, M., Ruocco, G. and Zamponi, F., 2002. Topological signature of first order phase transitions. *cond-mat/* 0205483.

[AH82]: Aitchison, I.J.R. and Hey, A.J.G., 1982. *Gauge Theories in Particle Physics.* Adam Hilger Ltd. (Bristol).

[AK98]: Arnold, V.I. and Khesin, B.A., 1998. *Topological Methods in Hydrodynamics.* Springer.

[AKNS73]: Ablowitz, M.A., Kaup, D.J., Newell, A.C. and Segur, H., 1973. Nonlinear evolution equations of physical significance. *Phys. Rev. Lett.* **31**, 125–127; 1974. The inverse scattering transform — Fourier analysis for nonlinear problems. *Stud. Appl. Math.* **53**, 249–315.

[AM78]: Abraham, R. and Marsden, J.E., 1978. *Foundations of Mechanics.* Benjamin/Cunnings.

[AN84]: Arik, M. and Neyzi, F., 1984. Geometrical properties and invariance of the generalized sigma modcel. *J. Math. Phys.* **25**, 2009–2012.

[AnIb79]: Anderson, R.L., Ibragimov, N.H., 1979. *Lie-Bäcklund Transformations in Applications*, SIAM (Philadelphia).

[Ano67]: Anosov, V.I., 1967. Geodesic flows on compact manifolds with negative curvature. *Proc. Steklov Math. Inst.* **90**, 1.

[Arn66]: Arnold, V.I., 1966. Sur la géométrie différentielle des groupes de Lie de dimension infinie et ses applications a l'hydrodynamique des fluides parfaits. *Ann. Inst. Fourier* (Grenoble) **16**, 319–361.

[Arn78]: Arnold, V.I., 1978. *Mathematical Methods of Classical Physics.* Springer.

[AS81]: Ablowitz, M.A. and Segur, H., 1981. *Solitons and the Inverse Scattering Transform*, SIAM (Philadelphia).

[AzIz95]: de Azcárraga, J.A. and Izquierdo, J.M., 1995. *Lie Groups, Lie Algebras, Cohomology and Some Applications in Physics.* Cambridge University Press.

[Bac1880]: Bäcklund, A.V., 1880. Zur Theorie der partiellen Differentialgleichungen erster ordnung. *Math. Ann.* XVII, 285–328.

[Bat67]: Batchelor, G.K., 1967. *An Introduction to Fluid Dynamics.* Cambridge University Press, Cambridge.

[BCS00]: Bao, D., Chern, S.S. and Shen, Z., 2000. *An Introduction to Riemann-Finsler Geometry.* Springer.

[Bia1879]: Bianchi, L., 1879. Richerce sulle superficie a curvatura costante e sulle elicoidi. *Ann. Secuola Norm. Sup. Pisa*, II, 285.

[Birk27]: Birkhoff, G.D., 1927. *Dynamical Systems* (AMS Colloquium Publs. Vol. 9). American Mathematical Society, providence, Rhode Island.

[BJPV98]: Bohr, T., Jensen, M.H., Paladin, G. and Vulpiani, A., 1998. *Dynamical Systems Approach to Turbulence.* Cambridge University Press, Cambridge.

[Bob90]: Bobenko, A.I., 1990. Integrable surfaces. *Funkts. Anal. Prilozh.* **24**, 68–69.

[Bob94]: Bobenko, A.I., 1994. Surfaces in terms of 2 by 2 matrices. Old and new integrable cases. In *Harmonic maps and integrqable systems* (eds., Fordy and Wood), Aspects of Mathematics, Vieweg.

[Bott77]: Bott, R., 1977. On the characteristic classes of groups of diffeomorphisms. *Enseign. Math.* **23**, 209–220.

[Bre70]: Bretherton, F.P., 1970. A note on Hamilton's principle for perfect fluids. *J. Fluid Mech.* **44**, 19–31.

[Cau1816]: Cauchy, A.L., 1816. Théolie de la propagation des ondes á la surface d'un fluide pesant d'une profonder indéfinie (1815). Mém Divers Savants (2)1; Oevres (Cauchy), 1 Serie Tome I, 2nd Part, Sect 1, 33–47. (Paris, Gauthier-Villar, 1908).

[CCCP97]: Caiani, L., Casetti, L., Clementi, C. and Pettini, M., 1997. Geometry of dynamics, Lyapunov exponents, and phase transitions. *Phys. Rev. Lett.* **79**, 4361–4364.

[CCP96]: Casetti, L., Clementi, C. and Pettini, M., 1996. Riemannian theory of Hamiltonian chaos and Lyapunov exponents. *Phys. Rev. E* **54**, 5969–5984.

[CCP02]: Casetti, L., Cohen, E.G.D. and Pettini, M., 2002. Exact result on topology and phase transition at any finite N. *Phys. Rev. E* **65**, 036112.

[CeP95]: Cerruti-Sola, M. and Pettini, M., 1995. Geometrical description of chaos in self-gravitating systems. *Phys. Rev. E* **51**, 53–64.

[CeP96]: Cerruti-Sola, M. and Pettini, M., 1996. Geometrical description of chaos in two-degrees-of-freedom Hamiltonian systems. *Phys. Rev. E* **53**, 179–188.

[CFG00]: Ceyhan, Ö., Fokas, A.S. and Gürses, M., 2000. Deformations of surfaces associated with integrable Gauss-Mainardi-Codazzi equations. *J. Math. Phys.* **41**, 2251–2270.

[CheT86]: Chern, S.S. and Tenenblat, K., 1986. Pseudospherical surfaces and evolution equations. *Stud. Appl. Math.* **74**, 55–83.

[CiP02]: Ciraolo, G. and Pettini, M., 2002. Geometry of chaos in models of stellar dynamics. *Cele. Mech. Dyn. Astron.* **83**, 171–190.

[CLU99]: Coste, C., Lund, F. and Umeki, M., 1999. Scattering of dislocated wave fronts by vertical vorticity and the Aharonov-Bohm effect. I. Shallow water. *Phys. Rev. E* **60**, 4908–4916.

[ClP02]: Clementi, C. and Pettini, M., 2002. A geometric interpretation of integrable motions. *Cele. Mech. Dyn. Astron.* **84**, 263–281.

[Cod1869]: Codazzi, D., 1869. Sulle coordinate curvilinee d'una superficie e dello spazio. *Annali di matem.* **2** (ser.2), 269–287.

[Cont86]: Contopoulos, G., 1986. Qualitative changes in 3-dimensional dynamical systems. *Astron. and Astrophys.* **161**, 244–256.

[CP93]: Casetti, L. and Pettini, M., 1993. *Phys. Rev. E* **48**, 4320.

[CPC00]: Casetti, L., Pettini, M. and Cohen, E.G.D., 2000. Geometric approach to Hamiltonian dynamics and statistical mechanics. *Phys. Rep.* **337**, 237–341.

[Cra78]: Crampin, M., 1978. Solitons and $SL(2, \mathbb{R})$. *Phys. Lett.* **66**A, 170–172.

[CTW82]: Chang, Y.F., Tabor, M. and Weiss, J., 1982. *J. Math. Phys.* **23**A, 531.

[DaR06]: Da Rios, L.S., 1906. Sul moto d'un liquido indefinito con un filetto vorticoso di forma qualunque. *Rend. Circ. Mat. Palermo* **22**, 117–135.

[DFGHMS86]: Dombre, T., Frisch, U., Grenne, J.M., Henon, M., Mehr, A. and Soward, A.M., 1986. Chaotic streamlines in the ABC flows. *J. Fluid Mech.* **167**, 353–391.

[DR81]: Drazin, P.G. and Reid, W.H., 1981. *Hydrodynamic Stability.* Cambridge University Press.

[Eck60]: Eckart, C., 1960. Variation principles of hydrodynamics. *Phys. Fluids* **3**, 421–427.

[EbMa70]: Ebin, D.G. and Marsden, J., 1970. Groups of diffeomorphisms and the motion of an incompressible fluid. *Ann. Math.* **90**, 102–163.

[EM97]: Ebin, D.G. and Misiołek, G., 1997. The exponential map on D^s_μ. The Arnoldfest (Toronto, 1997), 153–163; 1999. (the same), *Fields Inst. Commun.* **24**; 1999 *Amer. Math. Soc.* (Providence, RI).

[Eis29]: Eisenhart, L.P., 1929. Dynamical trajectories and geodesics. *Ann. Math.* (Princeton) **30**, 591–606.

[Eis47]: Eisenhart, L.P., 1947. *An Introduction to Differential Geometry.* Princeton University Press.

[FG96]: Fokas, A.S. and Gelfand, I.M., 1996. Surfaces on Lie groups, on Lie algebras, and their integrability. *Commun. Math. Phys.* **177**, 203–220.

[Fla63]: Flanders, H., 1963. *Differential Forms (with Applications to the Physical Sciences).* Academic Press.

[FM91]: Fukumoto, Y. and Miyazaki, T., 1991. Three-dimensional distortions of a vortex filament with axial velocity. *J. Fluid Mech.* **222**, 369–416.

[Fra97]: Frankel, T., 1997. *The Geometry of Physics (An Introduction).* Cambridge University Press, Cambridge.

[Fre88]: Freed, D.S., 1988. The geometry of loop groups. *J. Differential Geometry* **28**, 223–276.

[FU86]: Fujikawa, K. and Ui, H., 1986. Nuclear rotation, Nambu-Goldstone mode and Higgs mechanism. *Prog.Theor. Phys.* **75**, 997–1013.

[GF68]: Gelfand, I.M., Fuchs, D.B., 1968. The cohomology of the Lie algebra of vector fields on a circle. *Functs. Anal. Prirozhen.* **2**.

[Ham88]: Hama, F.R., 1988. Genesis of of the LIA. *Fluid Dyn. Res.* **3**, 149–150.

[Has71]: Hasimoto, H., 1971. Motion of a vortex filament and its relation to elastica. *J. Phys. Soc. Jpn.* **31**, 293–294.

[Has72]: Hasimoto, H., 1972. A soliton on a vortex filament. *J. Fluid Mech.* **51**, 477–485.

[Heri55]: Herivel, J.W., 1955. The derivation of the equations of motion of an ideal fluid by Hamilton's principle. *Proc. Cambridge Phil. Soc.* **51**, 344–349.

[Herm76]: Hermann, R., 1976. Pseudopotential of Estabrook and Wahlquist, the geometry of solitons, and the theory of connections. *Phys. Rev. Lett.* **36**, 835–836.

[HH64]: Hénon, M. and Heiles, C., 1964. *Astron. J.* **69**, 73.

[Hic65]: Hicks, N.J., 1965. *Notes on Differential Geometry* (D. van Nostrand Co., Inc.).

[HK94]: Hattori, Y. and Kambe, T., 1994. Kinematical instability and line-stretching in relation to the geodesics of fluid motion. *Fluid Dyn. Res.* **13**, 97–117.

[HM01]: Himonas, A.A. and Misiołek, G., 2001. The Cauchy problem for an integrable shallow water equation. *Differential Integral Equations* **14**.

[HMR98]: Holm, D.D., Marsden, J.E. and Ratiu, T.S., 1998. The Euler–Poincaré equations and semidirect products with applications to continuum theories. *Adv. in Math.* **137**, 1–81.

[Kak91]: Kakuhata, H., 1991. The dual transformation and the nonlinear sigma model. *J. Phys. Soc. Jpn.* **60**, 3664–3668.

[Kam82]: Kambe, T., 1982. Scattering of sound wave by a vortex system. *Nagare* (J. Japan Soc. of Fluid Mech.) **1**, 149–165 [in Japanese].

[Kam86]: Kambe, T., 1986. Acoustic emissions by vortex motions. *J. Fluid Mech.* **173**, 643–666.

[Kam98]: Kambe, T. 1998. Geometrical aspects in hydrodynamics and integrable systems. *Theor. and Comput. Fluid Dynamics* **10**, 249–261.

[Kam01a]: Kambe, T., 2001. Geometrical Theory of Dynamical Systems and Fluid Flows. *Isaac Newton Institute Preprint Series* NI01041-GTF, 1–53.

[Kam01b]: Kambe T. 2001. Physics of flow-acoustics. *Nagare (J. Japan Soc. of Fluid Mech.)* **20**, 174–186 [in Japanese].

[Kam02]: Kambe, T., 2002. Geometrical theory of fluid flows and dynamical systems. *Fluid Dyn. Res.* **30**, 331–378.

[Kam03a]: Kambe, T., 2003. Gauge principle for flows of a perfect fluid. *Fluid Dyn. Res.* **32**, 193–199.

[Kam03b]: Kambe, T., 2003. Geometrical theory of two-dimensional hydrodynamics with special reference to a system of point vortices, *Fluid Dyn. Res.* **33**, 223–249.

[Kam03c]: Kambe, T., 2003. Gauge principle and variational formulation for flows of an ideal fluid. *Acta Mech. Sinica* **19**(5), 437–452.

[Kat83]: Kato, T., 1983. On the Cauchy problem for the (generalized) Korteweg–de Vries equation. *Studies in Applied Mathematics, Adv. Math. Suppl. Stud.* **8**, 93–128; Himonas, A.A. and Misiołek, G., 2001. The Cauchy problem for an integrable shallow water equation. *Differential and Integral Equations* **14**.

[KdV1895]: Korteweg, D.J. and de Vries, G., 1895. On the change of form of long waves advancing in a rectangular canal, and on a new type of long stationary waves. *Phil. Mag.* Ser.5 **39**, 422–443.

[KH85]: Kambe, T. and Hasimoto, H., 1985. Simulation of invariant shapes of a vortex filament with an elastic rod. *J. Phys. Soc. Jpn.* **54**, 5–7.

[KI90]: Kambe T. and Iwatsu, R., 1990. Chaotic behaviour of streamlines. Parity (Physical science magazine in Japanese, Maruzen pub., Tokyo), **5**(4), 51–53.

[Kid81]: Kida, S., 1981. A vortex filament moving without change of form. *J. Fluid Mech.* **112**, 397–409.

[KM87]: Kambe, T. and Minota T., 1987. Acoustic waves emitted by a vortex ring passing near a circular cylinder, *J. Sound and Vibration* **119**, 509–528.

[KMy81]: Kambe, T. and Mya Oo, U., 1981. Scattering of sound by a vortex ring. *J. Phys. Soc. Jpn.* **50**, 3507–3516.

[KNH92]: Kambe, T., Nakamura, F. and Hattori, Y., 1992. Kinematical instability and line-stretching in relation to the geodesics of fluid motion. *Topological Aspects of the Dynamics of Fluids and Plasmas* (eds. H.K. Moffatt *et al.*, Kluwer Academic Publishers), 493–504.

[Kob77]: Kobayashi, S., 1977. *Differential Geometry of Curves and Surfaces*, Shokabo, Tokyo [in Japanese].

[KT71]: Kambe T. and Takao, T., 1971. Motion of distorted vortex rings. *J. Phys. Soc. Jpn.* **31**, 591–599.

[Lamb32]: Lamb H., 1932. *Hydrodynamics*, Cambridge University Press.

[Lie1874]: Lie, S., 1874. Begründung einer Invarianten-Theorie der Berührungs-Transformationen. *Math. Ann.* VIII, 215–288.

[Lie1880]: Lie, S., 1880. Zur Theorie der Flächen konstanter Krümmung. *Arch. Math. og Naturvidenskab*, V, 282–306.

[Lin63]: Lin, C.C., 1963. Hydrodynamics of helium II. *Proc. Int. Sch. Phys. XXI*, NY: Academic, 93–146.

[Lov27]: Love, A.E.H., 1927. *A Treatise on the Mathematical Theory of Elasticity*, Cambridge Univ. Press.

[LL75]: Landau, L.D. and Lifshitz, E.M., 1975. *The Classical Theory of Fields* (4th ed), Pergamon Press.

[LL76]: Landau, L.D. and Lifshitz, E.M., 1976. *Mechanics* (3rd ed), Pergamon Press.

[LL80]: Landau, L.D. and Lifshitz, E.M., 1980. *Statistical Physics Part 1* (3rd ed), Pergamon Press.

[LL87]: Landau, L.D. and Lifshitz, E.M. *Fluid Mechanics* (2nd ed), Pergamon Press, 1987.

[LP91]: Langer, J. and Perline, R., 1991. Poisson geometry of the filament equation. *J. Nonlinear Sci.* **1**, 71–93.

[LR76]: Lund, F. and Regge, T., 1976. Unified approach to strings and vortices with soliton solutions, *Phys. Rev. D* **14**, 1524–1535.

[Luk79]: Lukatskii, A.M., 1979. Curvature of groups of diffeomorphisms preserving the measures of the 2-sphere. *Funkt. Analiz. i Ego Prilozh.* **13**, 174–177.

[Lun78]: Lund, F., 1978. Classically sovable field theory model. *Ann. of Phys.* **115**, 251–268.

[Mai1856]: Mainardi, G., 1856. Sulla teoria generale delle superficie. Ist. Lombardo, Giornale, n.s. **9**, 385–398.

[Mil63]: Milnor, J., 1963. *Morse Theory.* Princeton University Press.

[Mis93]: Misiołek, G., 1993. Stability of flows of ideal fluids and the geometry of the group of diffeomorphisms. *Indiana Univ. Math. J.* **42**, 215–235.

[Mis97]: Misiołek, G., 1997. Conjugate points in the Bott–Virasoro group and the KdV equation. *Proc. of AMS* **125**, 935–940.

[Mis98]: Misiołek, G., 1998. A shallow water equation as a geodesic flow on the Bott–Virasoro group. *J. Geometry and Phys.* **24**, 203–208.

[Moff78]: Moffat, H.K., 1978. *Magnetic Field Generation in Electrically Conducting Fluids*, Cambridge University Press.

[MR94]: Marsden, J.E. and Ratiu, T.S., 1994. *Introduction to Mechanics and Symmetries.* Springer.

[NS83]: Nash, C. and Sen, S., 1983. *Topology and Geometry for Physics.* Academic Press.

[NHK92]: Nakamura, F., Hattori, Y. and Kambe, T., 1992. Geodesics and curvature of a group of diffeomorphisms and motion of an ideal fluid. *J. Physics A* **25**, L45–L50.

[OK87]: Ovsienko, V.Yu., and Khesin, B.A., 1987. Korteweg–de Vries superequation as an Euler equation. *Funct. Anal. Appl.* **21**, 329–331.

[Ons49]: Onsager, L., 1949. Statistical hydrodynamics. *Nuovo Cimento, Supplement* **6**, 279–287.

[Ott93]: Ott, E., 1993. *Chaos in Dynamical Systems.* Cambridge University Press, Cambridge.

[Poh76]: Pohlmeyer, K., 1976. Integrable Hamiltonian systems and interactions through quadrutic constraints. *Comm. Math. Phys.* **46**, 207–221.

[PT89]: Peshkin, M. and Tonomura, A., 1989. The Aharonov–Bohm effect, *Lec. Note in Phys.* 340, Springer.

[Ptt93]: Pettini, M., 1993. Geometrical hints for a nonperturbative approach to Hamiltonian dynamics. *Phys. Rev. E* **47**, 828–850.

[Qui83]: Quigg, C., 1983. *Gauge Theories of the Strong, Weak and Electromagnetic Interactions*, Massachusetts: The Benjamin/Cummings Pub. Comp., Inc.

[Ric91]: Ricca, R.L., 1991. Rediscovery of Da Rios equations. *Nature* **352**, 561–562.

[Run59]: Rund, H., 1959. *The Differential Geometry of Finsler Spaces*, Springer-Verlag.

[Saf92]: Saffman, P.G., 1992. *Vortex Dynamics.* Cambridge University Press.

[Sal88]: Salmon, R., 1988. Hamiltonian fluid mechanics. *Ann. Rev. Fluid Mech.* **20**, 225–256.

[Sas79]: Sasaki, R., 1979. Soliton equations and pseudospherical surfaces. *Nuclear Phys. B* **154**, 343–357; 1979. Geometrization of soliton equations. *Phys. Lett.* **71**A, 390–392.

[Sch80]: Schutz, B., 1980. *Geometrical Methods of Mathematical Physics*, Cambridge University Press.

[Ser59]: Serrin, J., 1959. Mathematical principles of classical fluid mechanics. *Encyclopedia of Physics* (Ed: Flügge, CoEd: Truesdell). Springer-Verlag. 125–263.

[SOK96]: Suzuki, K., Ono, T. and Kambe, T., 1996. Riemannian geometrical analysis of the motion of a vortex filament: A system of $C^\infty(S^1, SO(3))$'. *Phys. Rev. Lett.* **77**, 1679–1682.

[Sop76]: Soper, D.E., 1976. *Classical Field Theory.* John Wiley & Sons.

[SS77]: Schutz, B.F. and Sorkin, R., 1977. Variational aspects of relativistic field theories, with application to perfect fluids. *Ann. Phys.* **107**, 1–43.

[SWK98]: Suzuki, K., Watanabe, Y. and Kambe, T., 1998. Geometrical analysis of free rotation of a rigid body. *J. Physics A: Math. Gen.* **31**, 6073–6080.

[Sym82]: Sym, A., 1982. Soliton surfaces. *Lett. Nuovo Cimento* **33**, 394–400; 1983. Soliton surfaces, II. ibid. **36**, 307–312.

[UL97]: Umeki, M. and Lund, F., 1997. Spirals and dislocations in wave-vortex systems. *Fluid Dyn. Res.* **21**, 201–210.

[Uti56]: Utiyama, R., 1956. Invariant theoretical interpretation of interaction. *Phys. Rev.* **101**(5), 1597–1607.

[vK76]: van Kampen, N.G., 1976. *Physics Rep.* **24**, 71.

[Viz01]: Vizman, C., 2001. Geodesics on extensions of Lie groups and stability: the superconductivity equation. *Phys. Lett.* A **284**, 23–30.

[Whi37]: Whittaker, E.T., 1937. *A Treatise on the Analytical Dynamics of Particles and Rigid Bodies* (4th ed.), Cambridge University Press.

[Wyb74]: Wybourne, B.G., 1974. *Classical Groups for Physicists* (New York, London); Barut, A.O. and Raczka, R., 1977. *Theory of Group Representations and Applications*, Warsaw.

[Zei92]: Zeitlin, V., 1992. On the structure of phase-space, Hamiltonian variables and statistical approach to the description of two-dimensional hydrodynamics and magnetohydrodynamics. *J. Physics* A **25**, L171–L175.

Index